D1195920

FUNDAMENTALS OF LINEAR ALGEBRA

# *Fundamentals of Linear Algebra*

KATSUMI NOMIZU

*Professor of Mathematics*
*Brown University*

CHELSEA PUBLISHING COMPANY
NEW YORK, N.Y.

SECOND EDITION



CIP

Library of Congress Cataloging in Publication Data

Nomizu, Katsumi, 1924-
Fundamentals of linear algebra.

Includes index.
1. Algebras, Linear. I. Title.
QA184.N65 512'.5 77-7468
ISBN 0-8284-0276-0

Printed on 'long-life' acid-free paper

Printed in the United States of America

# *Preface to the Second Edition*

For this edition we have not only corrected typographical misprints and a few errors in the original edition but also added answers to many problems. For theoretical problems the answers provide hints or indications of the proofs. Problems with answers at the end of the book are marked by **»**. Note that ☆ indicates that the problem is of the second or third kind as explained in the Preface to the First Edition.

I have been most encouraged by the favorable reactions of the readers who found the book useful as a reference long after they finished the course for which it was a textbook. It is my hope that the book will continue to serve those students of mathematics and science for whom a more than rudimentary background in linear algebra is an indispensable part of their training.

I thank Mr. Martin Magid for his suggestions and editorial help in writing the answers to problems.

KATSUMI NOMIZU

# Preface to the First Edition

The importance of linear algebra in the undergraduate mathematics curriculum is now so well recognized that it is hardly necessary to make any comment on why linear algebra should be taught. For the practical problem of how to teach it, one has to find an appropriate answer—depending on the purpose of a given course and the level of the students in it.

The present book is primarily intended for use in a two-semester course on linear algebra combined with the elements of modern algebra and analytic geometry of $n$ dimensions. It is designed, however, to be usable for a one-semester course on linear algebra alone or a one-semester course on linear algebra with analytic geometry. The presentation can be adjusted to an elementary level following a calculus course or to a more advanced level with a rigorous algebraic approach. Some specific recommendations are given in the Suggestions for Class Use.

The following is a brief description of the main content. After an introductory chapter (Chapter 1) explaining the motivations of the subject from various points of view, we develop the basic concepts and results on vector spaces, linear mappings, matrices, systems of linear equations, and bilinear functions in Chapters 2, 3, and 4. In Chapter 5 we introduce some basic algebraic concepts (fields, polynomials and their factorizations, rings, extensions of fields, modules) in order to make it possible to develop linear algebra in a general algebraic setting. In Chapter 6 the theory of determinants is treated for matrices over a commutative ring with identity by way of alternating $n$-linear functions. In Chapter 7 we discuss minimal polynomials and characteristic polynomials (including the Cayley-Hamilton theorem) and their applications (in particular, the Jordan forms); Schur's lemma and complex structures are also treated. In Chapter 8 we deal with inner product spaces and prove the spectral decomposition theorems for normal transformations, in particular, hermitian, unitary, symmetric, and orthogonal transformations. As explained in the Suggestions for Class Use, we indicate various proofs for these theorems in the exercises.

Chapters 9 and 10 provide a linear algebra approach to analytic geometry of $n$ dimensions, which is the most efficient way of introducing rigorously geometric concepts in affine and euclidean spaces. The introductory material in Sections 1.3 and 1.4 serves as a preview for the full geometric development of linear algebra in these last two chapters. The knowledge of $n$-dimensional analytic geometry is basic for the study of topology, algebraic geometry, and differential geometry; nevertheless, the author has often noticed remarkable lack of that basic knowledge among many students of mathematics.

We have given many examples in order to illustrate important points of discussion and computational techniques or to introduce the standard models for the concepts at hand. Exercises at the end of each section are of the following three kinds:

1. To test the understanding of basic concepts and techniques given in the text
2. To offer more challenging problems of genuine interest based on the text material
3. To provide supplementary results and alternative proofs in order to amplify the understanding of the text material

Problems of the second and third kinds are starred.

After learning the material in this book, a reader will certainly be ready to proceed to a more advanced study of linear algebra in its most prolific sense; for example, through the theory of modules to homological algebra, through the theory of matrix groups to Lie groups and Lie algebras, through the theory of exterior algebras to differential and integral calculus on differentiable manifolds, through the theory of tensor algebras, projective and other geometries to differential geometry, through the theory of Banach and Hilbert spaces to functional analysis, and so on. We should have liked to include at least an elementary introduction to some of these subjects, but they had to be left out entirely. It is hoped that the present book will give the reader a balanced background in linear algebra before he specializes in various directions.

In concluding the preface I should like to acknowledge the invaluable help I have received from Mr. Carl Pomerance, a student at Brown University, who has critically read the manuscript and suggested numerous improvements in the presentation. My thanks go also to Mrs. Marina Smyth for her expert help in proofreading.

KATSUMI NOMIZU

# *Suggestions for Class Use*

*1.* *For a short course* the following material may be used: Sections 1.1, 1.2 (or 1.3); Chapter 2; Chapter 3; Chapter 6; Section 7.1; Sections 8.1, 8.2, followed by one or two theorems selected out of Theorems 8.21, 8.22, 8.25, 8.30 (and their corollaries). For this selection observe the following:

*a.* Throughout the whole treatment, treat vector spaces and matrices over the real number field $\mathcal{R}$ or the complex number field $\mathcal{C}$. Thus the notation $\mathcal{F}$ will always stand for $\mathcal{R}$ or $\mathcal{C}$. In Chapter 6, a commutative ring with identity is to be replaced by $\mathcal{R}$ or $\mathcal{C}$. In Section 7.1, the characteristic polynomial has to be defined less formally.

*b.* Instead of (*a*), one may insert Section 5.1 between Sections 2.2 and 2.3 so that one can treat vector spaces and matrices over an arbitrary field $\mathcal{F}$.

*c.* For the proofs of Theorems 8.21, 8.22, 8.25, 8.30 follow the suggestions in suggestion 5.

*2.* *For a short course with emphasis on geometry* the following material may be used: Sections 1.3, 1.4; Chapter 2; Sections 3.1 to 3.3; Sections 6.1 to 6.4; Sections 8.1, 8.2 followed by Theorem 8.30; Chapter 9; Chapter 10.

*a.* One may treat only vector spaces and matrices over $\mathcal{R}$.

*b.* For the proof of Theorem 8.30, follow the suggestions in 5.

*3.* *A more satisfactory treatment of linear algebra with the elements of modern algebra* can be given by Chapters 1 to 8.

*a.* Sections 1.3, 1.4 may be omitted, although it is always recommended to illustrate various concepts on vector spaces and linear transformations by using geometric interpretations.

*b.* For less emphasis on algebra, one may introduce only the material in Chapter 5 that is absolutely necessary for the development of linear algebra as the need arises.

*4.* There are various ways of arriving at the *spectral theorems for normal transformations* (*matrices*), in particular hermitian and unitary transformations (matrices) in the complex case and symmetric and orthogonal transformations (matrices) in the real case. They are developed along the following lines in the main text.

*a.* Theorem 8.19 (complex normal) and Theorem 8.20 (real normal) are based on Theorem 7.8 (on the minimal polynomial). Theorem 8.21 (hermitian) and Theorem 8.22 (unitary) follow immediately by using the results in Section 8.4.

*b.* Theorem 8.25 (symmetric) follows from Theorem 8.20 as soon as Theorem 8.24 (that a symmetric transformation has the real characteristic roots) is proved, and this is proved in two ways.

*c.* Theorem 8.30 (orthogonal) is proved first for the two-dimensional case and then by using the argument on the minimal polynomial (Theorem 7.5).

5. *For alternative proofs of the spectral theorems* we suggest the following:

*a.* Theorem 8.21 (hermitian) can be proved as in Exercise 8.4, number 15 (thus before introducing normal transformations and without reference to the minimal polynomial).

*b.* Theorem 8.22 (unitary) can be proved as in Exercise 8.4, number 16, in the same way as (*a*).

*c.* Theorem 8.19 (complex normal) can be proved as in Exercise 8.5, number 8, and Theorem 8.20 (real normal) as in Exercise 8.5, number 9.

*d.* Theorem 8.25 (symmetric) can be proved as in Exercise 8.6, number 18, once the existence of an eigenvalue (a real characteristic root) is established (as in Exercise 8.6, number 16 or 17, without reference to hermitian transformations).

*e.* Theorem 8.30 (orthogonal) can be proved as in Exercise 8.7, number 8 (by using Proposition 8.16) or as in Exercise 8.7, number 9 (by using Theorem 8.25).

*f.* There are other variations; see Exercises 8.5, numbers 7 and 15, Exercise 8.6, number 19, and Exercise 8.7, numbers 12 and 13.

6. Finally, a word about the *terminology used in the text.* We assume familiarity with the notation concerning sets, mappings, and equivalence relations. We also assume that a reader is acquainted with the principle of mathematical induction. Since these ideas are now introduced at an early stage in many courses in calculus, we shall give only a very concise explanation of the terminology in the Appendix.

# Contents

FUNDAMENTALS OF LINEAR ALGEBRA

# 1 Introduction

The purpose of this chapter is to motivate the study of linear algebra. We shall show how the concepts of vector spaces, linear mappings, and matrices arise naturally from various types of problems in mathematics. Discussions are rather informal and are not to be considered part of a systematic development (which starts in Chap. 2); Definition 1.3, on matrix multiplication, and Definition 1.5, on identity matrices, will be referred to later.

## 1.1 SYSTEMS OF LINEAR EQUATIONS

To start with an easy example, we recall how we can solve a system of two linear equations in two unknowns:

$$\begin{aligned} ax + by &= u, \\ cx + dy &= v, \end{aligned} \tag{1.1}$$

where $a, b, c, d$ and $u, v$ are given (real) numbers and $x, y$ are unknowns. We assume that $a, b$ are not both 0 and $c, d$ are not both 0. Multiplying the first equation by $d$ and subtracting from it $b$ times the second equation, we obtain

$$(ad - bc)x = du - bv.$$

Similarly, by eliminating $x$, we obtain

$$(ad - bc)y = av - cu.$$

If $ad - bc \neq 0$, then we find the solution

$$x = \frac{du - bv}{ad - bc}, \qquad y = \frac{av - cu}{ad - bc}.$$

In the case where $ad - bc = 0$, we proceed as follows. If $c \neq 0$, let $k = a/c$, so that $a = ck$. Substituting this in $ad = bc$, we have $cdk = cb$. Since $c \neq 0$, we get $b = dk$. If $c = 0$, then $d \neq 0$ and we let $k = b/d$ and still obtain $a = ck$. If $k = 0$, we have $a = b = 0$, contrary to the assumption. Thus we see that there is $k \neq 0$ such that

$$a = ck \qquad \text{and} \qquad b = dk.$$

Multiplying the second equation of (1.1) by $k$, we have

$$ckx + dky = kv,$$

that is,

$$ax + by = kv.$$

If $u = kv$, then this is the same as the first equation of (1.1). In this case the system (1.1) has infinitely many solutions $(x,y)$, which are determined from the first equation by giving arbitrary values to $x$ (or $y$). On the other hand, if $u \neq kv$, then $(x,y)$ satisfying the second equation does not satisfy the first equation; that is, the system (1.1) has no solution at all.

Summing up, we have:

1. If there is no number $k \neq 0$ such that $a = ck$, $b = dk$, then (1.1) has a unique solution.

2. If there is a number $k \neq 0$ such that $a = ck$, $b = dk$, and $u = vk$, then (1.1) has infinitely many solutions.

3. If there is a number $k \neq 0$ such that $a = ck$, $b = dk$ but $u \neq vk$, then (1.1) has no solution at all.

A geometric interpretation is the following. Each of the equations in (1.1) represents a straight line on the plane with the usual coordinate system. In case 1, two straight lines are distinct and meet at one point. In case 2, two lines coincide. In case 3, two lines are distinct but parallel to each other.

We shall now consider a more general system of linear equations —a system of $m$ linear equations in $n$ unknowns $x_1, \ldots, x_n$—

$$\begin{array}{l} a_{11}x_1 + a_{12}x_2 + \cdots + a_{1n}x_n = u_1, \\ a_{21}x_1 + a_{22}x_2 + \cdots + a_{2n}x_n = u_2, \\ \cdots\cdots\cdots\cdots\cdots\cdots, \\ a_{m1}x_1 + a_{m2}x_2 + \cdots + a_{mn}x_n = u_m, \end{array} \tag{1.2}$$

where the essential data of the system are the $mn$ coefficients $a_{ij}$, $1 \leq i \leq m$, $1 \leq j \leq n$, and the numbers $u_i$, $1 \leq i \leq m$. We may express them by writing

$$\begin{bmatrix} a_{11} & a_{12} & \cdots & a_{1n} \\ a_{21} & a_{22} & \cdots & a_{2n} \\ \cdots & \cdots & \cdots & \cdots \\ a_{m1} & a_{m2} & \cdots & a_{mn} \end{bmatrix}, \quad \begin{bmatrix} u_1 \\ u_2 \\ \cdot \\ \cdot \\ \cdot \\ u_m \end{bmatrix}$$

On the other hand, the set of unknowns $x_i$, $1 \leq i \leq n$, can be written as

$$\begin{bmatrix} x_1 \\ x_2 \\ \cdot \\ \cdot \\ \cdot \\ x_n \end{bmatrix}.$$

**Definition 1.1**

A display of $mn$ numbers $a_{ij}$ in the form above is called an $m \times n$ *matrix;* we denote this matrix, for example, by

$$\mathbf{A} = \begin{bmatrix} a_{11} & a_{12} & \cdots & a_{1n} \\ a_{21} & a_{22} & \cdots & a_{2n} \\ \cdots & \cdots & \cdots & \cdots \\ a_{m1} & a_{m2} & \cdots & a_{mn} \end{bmatrix}$$

or, more briefly, by

$$\mathbf{A} = [a_{ij}], \qquad 1 \leq i \leq m, 1 \leq j \leq n.$$

Similarly, we set

$$\mathbf{u} = \begin{bmatrix} u_1 \\ u_2 \\ \cdot \\ \cdot \\ \cdot \\ u_m \end{bmatrix}, \qquad \mathbf{x} = \begin{bmatrix} x_1 \\ x_2 \\ \cdot \\ \cdot \\ \cdot \\ x_n \end{bmatrix}.$$

Although $\mathbf{u}$ is an $m \times 1$ matrix and $\mathbf{x}$ an $n \times 1$ matrix, it is more common to say that $\mathbf{u}$ is an $m$-dimensional *vector* and $\mathbf{x}$ an $n$-dimensional vector.

**Definition 1.2**

Two $m \times n$ matrices

$$\mathbf{A} = [a_{ij}] \qquad \text{and} \qquad \mathbf{B} = [b_{ij}]$$

are said to be *equal* (written $\mathbf{A} = \mathbf{B}$) if $a_{ij} = b_{ij}$ for every pair $(i,j)$, where $1 \leq i \leq m$ and $1 \leq j \leq n$.

Similarly, two $m$-dimensional vectors

$$\mathbf{y} = [y_i] \qquad \text{and} \qquad \mathbf{z} = [z_i]$$

are equal ($\mathbf{y} = \mathbf{z}$) if $y_i = z_i$ for every $i$, $1 \leq i \leq m$.

System (1.2) is thus described by an $m \times n$ matrix $\mathbf{A}$ and an $m$-dimensional vector $\mathbf{u}$; the set of unknowns is expressed by an $n$-dimensional vector $\mathbf{x}$. We shall even write system (1.2) in the form

$$\mathbf{Ax} = \mathbf{u}. \tag{1.3}$$

This matrix notation for system (1.2) will be justified when $\mathbf{Ax}$ on the left-hand side acquires the meaning of a "product" of the matrix $\mathbf{A}$ and the vector $\mathbf{x}$. For this purpose we shall define matrix multiplication in the following way.

**Definition 1.3**

Given an $m \times n$ matrix $\mathbf{A} = [a_{ij}]$, $1 \le i \le m$, $1 \le j \le n$, and an $n \times p$ matrix $\mathbf{B} = [b_{jk}]$, $1 \le j \le n$, $1 \le k \le p$, we form an $m \times p$ matrix $\mathbf{C} = [c_{ik}]$, $1 \le i \le m$, $1 \le k \le p$, where

$$(1.4) \qquad c_{ik} = \sum_{j=1}^{n} a_{ij} b_{jk}, \qquad 1 \le i \le m,\ 1 \le k \le p.$$

The matrix $\mathbf{C}$ is called the *product* of $\mathbf{A}$ and $\mathbf{B}$ and is denoted by $\mathbf{AB}$.

In the special case where $\mathbf{B}$ is an $n \times 1$ matrix, namely, an $n$-dimensional vector $[b_j]$, $1 \le j \le n$, we have an $m \times 1$ matrix, namely, an $m$-dimensional vector, as the product $\mathbf{AB} = [c_i]$, $1 \le i \le m$, where

$$(1.4') \qquad c_i = \sum_{j=1}^{n} a_{ij} b_j, \qquad 1 \le i \le m.$$

*Example 1.1*

$$\begin{bmatrix} 2 & 1 \\ 1 & 3 \end{bmatrix} \begin{bmatrix} -1 & 4 \\ 0 & -2 \end{bmatrix} = \begin{bmatrix} 2(-1) + 1(0) & 2(4) + 1(-2) \\ 1(-1) + 3(0) & 1(4) + 3(-2) \end{bmatrix} = \begin{bmatrix} -2 & 6 \\ -1 & -2 \end{bmatrix}.$$

$$\begin{bmatrix} 2 & -1 & 1 \\ 3 & -2 & 4 \end{bmatrix} \begin{bmatrix} 5 \\ -1 \\ 4 \end{bmatrix} = \begin{bmatrix} 2(5) + (-1)(-1) + 1(4) \\ 3(5) + (-2)(-1) + 4(4) \end{bmatrix} = \begin{bmatrix} 15 \\ 33 \end{bmatrix}.$$

Now (1.3) acquires the following meaning: The product $\mathbf{Ax}$ is equal to the vector $\mathbf{u}$; in fact,

$$(1.3') \qquad \sum_{j=1}^{n} a_{ij} x_j = u_i, \qquad 1 \le i \le m,$$

which is exactly the same as system (1.2).

Let us first consider a *homogeneous system:*

$$(1.5) \qquad \sum_{j=1}^{n} a_{ij} x_j = 0, \qquad 1 \le i \le m,$$

which is a special case of (1.3′) where all $u_i$'s are 0. Writing $\mathbf{0}$ for the $m$-dimensional vector (called the *$m$-dimensional zero vector*)

$$\begin{bmatrix} 0 \\ \cdot \\ \cdot \\ \cdot \\ 0 \end{bmatrix},$$

we have the matrix notation for (1.5) in the form

$$(1.5') \qquad \mathbf{Ax} = \mathbf{0}.$$

This system has an obvious solution $x_1 = x_2 = \cdots = x_n = 0$, namely,

$$\mathbf{x} = \begin{bmatrix} 0 \\ \cdot \\ \cdot \\ \cdot \\ 0 \end{bmatrix}, \qquad n\text{-dimensional zero vector.}$$

This solution is called the trivial solution of (1.5′); it may be the only solution, or there may be other solutions. At any rate, consider the set $S$ of all solutions [namely, $n$-dimensional vectors satisfying (1.5′)].

If $\mathbf{x} = [x_i]$ and $\mathbf{x}' = [x_i']$ are in $S$, then the vectors

$$\begin{bmatrix} x_1 + x_1' \\ x_2 + x_2' \\ \cdot \\ \cdot \\ \cdot \\ x_n + x_n' \end{bmatrix} \quad \text{and} \quad \begin{bmatrix} cx_1 \\ cx_2 \\ \cdot \\ \cdot \\ \cdot \\ cx_n \end{bmatrix}, \qquad c \text{ arbitrary},$$

which we denote by $\mathbf{x} + \mathbf{x}'$ and $c\mathbf{x}$, respectively, are also solutions. Thus the set of solutions $S$ has the property that $\mathbf{x}, \mathbf{x}' \in S$ implies $\mathbf{x} + \mathbf{x}' \in S$ and $c\mathbf{x} \in S$ for any number $c$.

**Definition 1.4**

Let $\mathbf{x} = [x_i]$ and $\mathbf{x}' = [x_i']$ be two $n$-dimensional vectors. The *sum* $\mathbf{x} + \mathbf{x}'$ is the $n$-dimensional vector

$$\begin{bmatrix} x_1 + x_1' \\ \cdot \\ \cdot \\ \cdot \\ x_n + x_n' \end{bmatrix},$$

and the *scalar multiple* $c\mathbf{x}$, where $c$ is an arbitrary number, is the $n$-dimensional vector

$$\begin{bmatrix} cx_1 \\ \cdot \\ \cdot \\ \cdot \\ cx_n \end{bmatrix}.$$

In connection with the sums and scalar multiples of vectors, let us observe that matrix multiplication has the following two properties:

$$\mathbf{A}(\mathbf{x} + \mathbf{x}') = \mathbf{A}\mathbf{x} + \mathbf{A}\mathbf{x}', \tag{1.6}$$

$$\mathbf{A}(c\mathbf{x}) = c(\mathbf{A}\mathbf{x}), \tag{1.7}$$

where $\mathbf{A}$ is an $m \times n$ matrix and $\mathbf{x}, \mathbf{x}'$ are $n$-dimensional vectors. In fact, the $i$th entry of the vector $\mathbf{A}(\mathbf{x} + \mathbf{x}')$ is equal to

$$\sum_{j=1}^{n} a_{ij}(x_j + x_j') = \sum_{j=1}^{n} a_{ij}x_j + \sum_{j=1}^{n} a_{ij}x_j',$$

which is equal to the $i$th entry of $\mathbf{Ax} + \mathbf{Ax}'$. Property (1.7) can be verified in a similar way.

The assertion on $S$ which we made above can be considered as a consequence of (1.6) and (1.7); in fact, if $\mathbf{Ax} = \mathbf{0}$ and $\mathbf{Ax}' = \mathbf{0}$, then

$$\mathbf{A}(\mathbf{x} + \mathbf{x}') = \mathbf{Ax} + \mathbf{Ax}' = \mathbf{0} + \mathbf{0} = \mathbf{0}$$

and

$$\mathbf{A}(c\mathbf{x}) = c\mathbf{Ax} = c\mathbf{0} = \mathbf{0},$$

showing that $\mathbf{x} + \mathbf{x}'$ and $c\mathbf{x}$ are solutions.

Let us now consider the system (1.3), called an *inhomogeneous system* if $\mathbf{u} \neq \mathbf{0}$; we call (1.5) the homogeneous system associated with (1.3). If $\mathbf{x}$ and $\mathbf{x}'$ are solutions of (1.3), that is,

$$\mathbf{Ax} = \mathbf{u} \qquad \text{and} \qquad \mathbf{Ax}' = \mathbf{u},$$

then, denoting $\mathbf{x}' + (-1)\mathbf{x}$ by $\mathbf{x}' - \mathbf{x}$, we have

$$\begin{aligned} \mathbf{A}(\mathbf{x}' - \mathbf{x}) &= \mathbf{A}(\mathbf{x}' + (-1)\mathbf{x}) = \mathbf{Ax}' + (-1)\mathbf{Ax}' \\ &= \mathbf{Ax}' - \mathbf{Ax} = \mathbf{u} - \mathbf{u} = \mathbf{0}, \end{aligned}$$

by virtue of (1.6) and (1.7). This shows that $\mathbf{x}' - \mathbf{x} = \mathbf{y}$ is a solution of (1.5). Conversely, suppose that $\mathbf{x}$ is a solution of (1.3) and $\mathbf{y}$ a solution of (1.5). Then $\mathbf{x}' = \mathbf{x} + \mathbf{y}$ is a solution of (1.5), because

$$\mathbf{Ax}' = \mathbf{A}(\mathbf{x} + \mathbf{y}) = \mathbf{Ax} + \mathbf{Ay} = \mathbf{u} + \mathbf{0} = \mathbf{u}.$$

We have seen that an arbitrary (or general) solution of system (1.3) is obtained from any particular solution by adding an arbitrary (or general) solution of the associated homogeneous system (1.5).

In order to find a solution of (1.3), one will, of course, try to reduce the system to a system of a simpler form which has the same solutions. One employs a number of elimination steps, the simplest form of which we recalled for system (1.1). One of the processes consists in multiplying one equation by a certain number and adding it to another equation. For example, we replace the $j$th equation of (1.3) by

$$(ca_{i1} + a_{j1})x_1 + \cdots + (ca_{in} + a_{jn})x_n = cu_i + u_j.$$

For the corresponding matrix $\mathbf{A}$, this process will change the $j$th row

$$[a_{j1} \quad a_{j2} \quad \cdots \quad a_{jn}]$$

into

$$[ca_{i1} + a_{j1} \quad ca_{i2} + a_{j2} \quad \cdots \quad ca_{in} + a_{jn}]$$

with the accompanying change of $\mathbf{u}$ into

$$\begin{bmatrix} u_1 \\ \cdot \\ \cdot \\ \cdot \\ u_{j-1} \\ cu_i + u_j \\ u_{j+1} \\ \cdot \\ \cdot \\ \cdot \\ u_m \end{bmatrix}.$$

It is obvious that this sort of operation does not change the solutions. The same thing is true of multiplying one equation by a nonzero number, which is the other kind of process one uses for solving a system of linear equations.

This indicates that the practical method of solving a system of linear equations can be described neatly as a sequence of certain operations performed on the matrix $\mathbf{A}$ and the vector $\mathbf{u}$.

**EXERCISE 1.1**

» **1.** Compute the following matrix products:

(*a*) $\begin{bmatrix} 2 & -1 \\ 4 & 5 \end{bmatrix}\begin{bmatrix} 1 & 0 \\ 3 & -1 \end{bmatrix}$; (*b*) $\begin{bmatrix} 3 & -1 \\ 3 & 5 \end{bmatrix}\begin{bmatrix} 1 \\ -1 \end{bmatrix}$;

(*c*) $\begin{bmatrix} 3 & 1 & 2 \\ 1 & 3 & -1 \\ 4 & 5 & 0 \end{bmatrix}\begin{bmatrix} 1 & -1 & 0 \\ 3 & 2 & 5 \\ 1 & 3 & 4 \end{bmatrix}$; (*d*) $\begin{bmatrix} 1 & -1 & 4 \\ 2 & 3 & 3 \\ 1 & 0 & 1 \end{bmatrix}\begin{bmatrix} 3 & 1 \\ 2 & -1 \\ 1 & 5 \end{bmatrix}$;

(*e*) $[1 \quad -1 \quad 1]\begin{bmatrix} 2 \\ 1 \\ 3 \end{bmatrix}$.

» **2.** Are the following products defined?

(*a*) $\begin{bmatrix} 2 & 1 \\ 1 & -3 \end{bmatrix}\begin{bmatrix} 1 \\ 0 \\ 2 \end{bmatrix}$; (*b*) $\begin{bmatrix} 2 \\ 1 \\ 3 \end{bmatrix}[1 \quad -1 \quad 1]$.

**3.** Solve the following systems of linear equations:

»(*a*)
$$\begin{aligned} x + y + z &= 1, \\ 3x - 2y + 2z &= 3, \\ 2x - y - z &= -1. \end{aligned}$$

(*b*)
$$\begin{aligned} x + y + z &= 1, \\ 3x - 2y + 2z &= 3, \\ 5y + z &= 4. \end{aligned}$$

## 1.2 DIFFERENTIAL EQUATIONS

Let us consider a differential equation

$$\frac{d^2x}{dt^2} + x = 0, \tag{1.8}$$

where $x = x(t)$ is a function of $t$ which we wish to find. It is known that a general solution is of the form

$$c_1 \cos t + c_2 \sin t, \tag{1.9}$$

where $c_1, c_2$ are arbitrary constants. In order to verify this, we see that both functions $\cos t$ and $\sin t$ satisfy (1.8), and then we may proceed as follows. We consider a mapping (or *operator*) which associates to a function $x(t)$ (which is twice differentiable, to be precise) the function

$$\frac{d^2x}{dt^2} + x(t),$$

which we denote by $Dx$. The operator $D$ then has the property that

$$D(c_1x_1 + c_2x_2) = c_1Dx_1 + c_2Dx_2,$$

where the sum of two functions $x_1 + x_2$ and the scalar multiple $cx$ are defined by

$$(x_1 + x_2)(t) = x_1(t) + x_2(t),$$
$$(cx)(t) = cx(t).$$

Thus it follows that $Dx_1 = 0$ and $Dx_2 = 0$ imply

$$D(c_1x_1 + c_2x_2) = c_1Dx_1 + c_2Dx_2 = 0;$$

in particular, any function of the form (1.9) is a solution of (1.8).

In order to prove that every solution of (1.8) is of the form (1.9), we use the result that (1.8) admits a unique solution for any initial conditions $x(0)$ and $(dx/dt)(0)$. Let $x = x(t)$ be an arbitrary solution and let $x(0) = c_1$ and $(dx/dt)(0) = c_2$. Consider the function

$$y(t) = c_1 \cos t + c_2 \sin t,$$

which is a solution. Moreover, $y(0) = c_1$ and $(dy/dt)(0) = c_2$. By the uniqueness, we must have $y(t) = x(t)$, as we wanted to show.

Summing up, we see that the set of all solutions $S$ of the differential equation (1.8) has the property that if $x, y \in S$, then $x + y$ and $cx \in S$ for every number $c$. There are two particular elements, namely, $\cos t$ and $\sin t$, in $S$, so that every element of $S$ can be written in the form (1.9) and hence is determined uniquely by the pair $(c_1,c_2)$. In this sense it is possible to consider $S$ as the set of all two-dimensional vectors $\begin{bmatrix} c_1 \\ c_2 \end{bmatrix}$.

The differential equation (1.8) is closely related to a system of differential equations:

$$\begin{aligned} \frac{dx}{dt} - y &= 0, \\ x + \frac{dy}{dt} &= 0. \end{aligned} \tag{1.10}$$

In fact, if we set $y = dx/dt$ [which is the first equation of (1.10)], then (1.8) may be written as the second equation of (1.10). In dealing with the system (1.10) let us write the pair of functions $x$ and $y$ in the form of a vector

$$\begin{bmatrix} x \\ y \end{bmatrix}.$$

If we agree that $\frac{d}{dt}\begin{bmatrix} x \\ y \end{bmatrix}$ means $\begin{bmatrix} dx/dt \\ dy/dt \end{bmatrix}$, then (1.10) takes the form

$$\frac{d}{dt}\begin{bmatrix} x \\ y \end{bmatrix} = \begin{bmatrix} 0 & 1 \\ -1 & 0 \end{bmatrix}\begin{bmatrix} x \\ y \end{bmatrix}, \tag{1.10'}$$

where on the right-hand side we multiply the matrix $\begin{bmatrix} 0 & 1 \\ -1 & 0 \end{bmatrix}$ with the "vector" $\begin{bmatrix} x \\ y \end{bmatrix}$ formally in the same way as before. We may consider (1.10′) as a differential equation for the vector-valued function $\begin{bmatrix} x \\ y \end{bmatrix}$, namely, a function whose value for each $t$ is the vector $\begin{bmatrix} x(t) \\ y(t) \end{bmatrix}$.

In order to generalize the discussions above, let $W_n$ denote the set of all functions (on the real line) which are differentiable at least $n$ times. If we define the sum $x + y$ and the scalar multiple $cx$ for functions as before, it follows that $W_n$ has the property that $x, y \in W_n$ implies $x + y \in W_n$ and $cx \in W_n$ for any number $c$.

We consider

$$D = \sum_{i=0}^{n} a^i(t)\frac{d^i}{dt^i}, \tag{1.11}$$

where $a^i(t)$ are certain functions for $0 \leq i \leq n$, $a^n(t) \equiv 1$, as a mapping of $W_n$ into the set of all functions $W_0$ which associates to $x \in W_n$ the function $Dx$:

$$(Dx)(t) = \frac{d^n x}{dt^n} + a^{n-1}(t)\frac{d^{n-1}x}{dt^{n-1}} + \cdots + a^1(t)\frac{dx}{dt} + a^0(t)x(t).$$

As before, $D$ has the property that

$$D(x_1 + x_2) = Dx_1 + Dx_2 \qquad \text{and} \qquad D(cx) = cDx.$$

Let $S$ be the set of all solutions of the differential equation $Dx = 0$, namely, the set of all functions in $W_n$ which satisfy $Dx = 0$. It follows that if $x_1, x_2 \in S$, then $x_1 + x_2 \in S$ and $cx_1 \in S$ for every number $c$.

The equation $Dx = 0$ can be described as a system of differential equations of the first order,

$$
\begin{aligned}
\frac{dx_0}{dt} &= x_1, \\
\frac{dx_1}{dt} &= x_2, \\
&\cdots\cdots, \\
\frac{dx_{n-2}}{dt} &= x_{n-1}, \\
\frac{dx_{n-1}}{dt} &= -\sum_{i=0}^{n-1} a^i x_i,
\end{aligned}
\tag{1.12}
$$

by setting $x_0 = x$, $x_1 = dx/dt, \ldots, x_{n-1} = dx^{n-1}/dt^{n-1}$. By using the matrix notation, (1.12) can be put in the form

$$
\frac{d}{dt}\begin{bmatrix} x_0 \\ x_1 \\ \cdot \\ \cdot \\ \cdot \\ x_{n-2} \\ x_{n-1} \end{bmatrix} = \begin{bmatrix} 0 & 1 & 0 & \cdots & 0 \\ 0 & 0 & 1 & \cdots & 0 \\ \cdots & \cdots & \cdots & \cdots & \cdots \\ 0 & 0 & 0 & \cdots & 1 \\ -a^0 & -a^1 & -a^2 & \cdots & -a^{n-1} \end{bmatrix} \begin{bmatrix} x_0 \\ x_1 \\ \cdot \\ \cdot \\ \cdot \\ x_{n-2} \\ x_{n-1} \end{bmatrix}. \tag{1.12'}
$$

**EXERCISE 1.2**

»› **1.** If both $x_1 = x_1(t)$ and $x_2 = x_2(t)$ are solutions of the differential equation

$$
\frac{d^2x}{dt^2} + x = 1, \tag{*}
$$

prove that $x_1 - x_2$ is a solution of (1.8). Using this fact, find all solutions of (*).

**2.** The functions $e^t$ and $e^{-t}$ are solutions of

$$
\frac{d^2x}{dt^2} - x = 0.
$$

Assuming the uniqueness of solution for any initial condition $x(0)$ and $(dx/dt)(0)$, prove that the set of all solutions of this equation consists of $c_1e^t + c_2e^{-t}$, where $c_1$ and $c_2$ are arbitrary numbers. Also, setting $y = dx/dt$, write down the matrix form for a system of differential equations for $x$ and $y$.

## 1.3 VECTORS ON THE PLANE

In this section we shall assume our intuitive knowledge of euclidean plane geometry and introduce the notion of vectors in a geometric way. Let $A$ and $B$ be two distinct points on the plane and consider the segment $AB$

with a sense of direction, either from $A$ to $B$ or from $B$ to $A$. The *directed segment* with the sense of direction from $A$ to $B$ is denoted by $\overrightarrow{AB}$; we call $A$ and $B$ the *initial point* and the *end point* of $\overrightarrow{AB}$, respectively. The directed segment with the opposite sense of direction is denoted by $\overrightarrow{BA}$. Thus $\overrightarrow{AB}$ and $\overrightarrow{BA}$ are different from each other (Fig. 1).

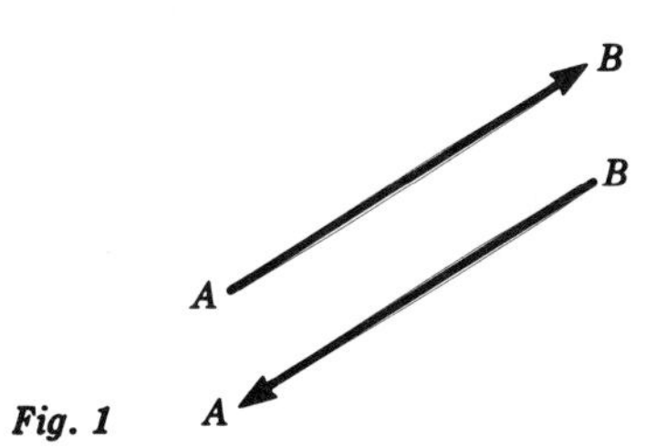


***Fig. 1***

We shall say that two directed segments $\overrightarrow{AB}$ and $\overrightarrow{CD}$ define the same *vector* if the following conditions are satisfied:

1. The lines $AB$ and $CD$ are parallel.
2. The length of $\overrightarrow{AB}$ is equal to the length of $\overrightarrow{CD}$.
3. The sense of direction of $\overrightarrow{AB}$ is the same as that of $\overrightarrow{CD}$.

Indeed, it is convenient to say that $\overrightarrow{AB}$ and $\overrightarrow{CD}$ are *parallel* when these three conditions are satisfied. Thus $\overrightarrow{AB}$ and $\overrightarrow{CD}$ define the same vector if and only if they are parallel,† and this is the case if the points $A$, $B$, $D$, and $C$ in that order form the vertices of a parallelogram (Fig. 2). We may also consider this relationship in the following way. $\overrightarrow{AB}$ and $\overrightarrow{CD}$ are parallel if

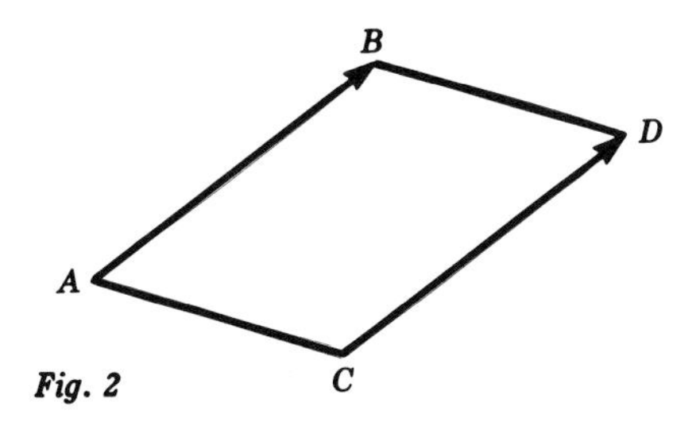


***Fig. 2***

† To be more rigorous, parallelism is an equivalence relation in the set of all directed segments and a vector is an equivalence class of a certain directed segment (see Appendix).

we can move $\overrightarrow{AB}$ to $\overrightarrow{CD}$ while preserving the distance and the sense of direction in such a way that, during the motion, the segment remains parallel to the original segment (Fig. 3).

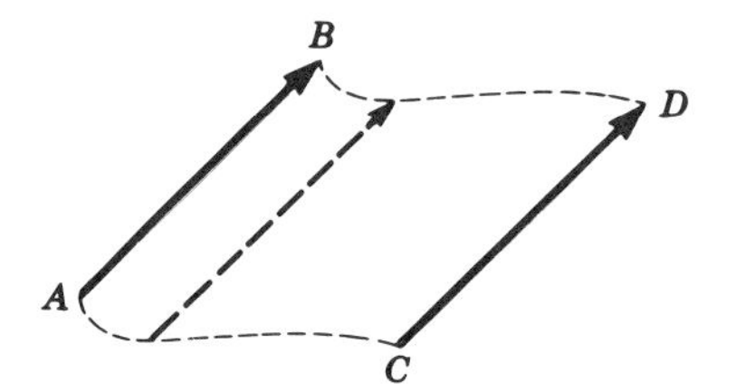


*Fig. 3*

We shall denote vectors by letters $\alpha$, $\beta$, etc. We say that the directed segment $\overrightarrow{AB}$ represents a vector and that the vector $\alpha$ is represented by $\overrightarrow{AB}$. Since any directed segment $\overrightarrow{AB}$ determines a vector, it is convenient to denote this vector itself by $\overrightarrow{AB}$. When we speak of a vector $\alpha$, it can be represented by a certain directed segment $\overrightarrow{AB}$ but equally well by any other directed segment $\overrightarrow{CD}$ which is parallel to $\overrightarrow{AB}$.

We shall need a special object called the *zero vector*. We may regard the zero vector as represented by a special kind of directed segment $\overrightarrow{AA}$, where $A$ is an arbitrary point. The zero vector is denoted by $0$; thus we may write $\overrightarrow{AA} = 0$ for any point $A$.

A basic relationship between points and vectors can be described as follows:

1. *Given two points $A$, $B$, there is a vector $\alpha = \overrightarrow{AB}$.*
2. *Given a vector $\alpha$ and a point $A$, there is one and only one point $B$ such that $\alpha = \overrightarrow{AB}$.*

The set of all vectors on the plane (including the zero vector) will be denoted by $V$. In this set $V$, we shall define two algebraic operations, namely, addition and scalar multiplication.

*Addition:* Given two vectors $\alpha$ and $\beta$, we define a third vector, called the *sum* of $\alpha$ and $\beta$ and denoted by $\alpha + \beta$, as follows. Take an arbitrary point $A$ and let $B$ be the unique point such that $\overrightarrow{AB} = \alpha$ [cf. (2)]. Let $C$ be the unique point such that $\overrightarrow{BC} = \beta$. Then $\alpha + \beta$ is the vector $\overrightarrow{AC}$. This definition seems to depend on the choice of $A$, but the vector $\overrightarrow{AC}$ is in fact independent of the choice of $A$. If we start with another point $A'$

instead of $A$ and choose $B'$ and $C'$ accordingly (that is, $\alpha = \overrightarrow{A'B'}$ and $\beta = \overrightarrow{B'C'}$), then, as we see in Fig. 4, we have $\overrightarrow{AC} = \overrightarrow{A'C'}$.

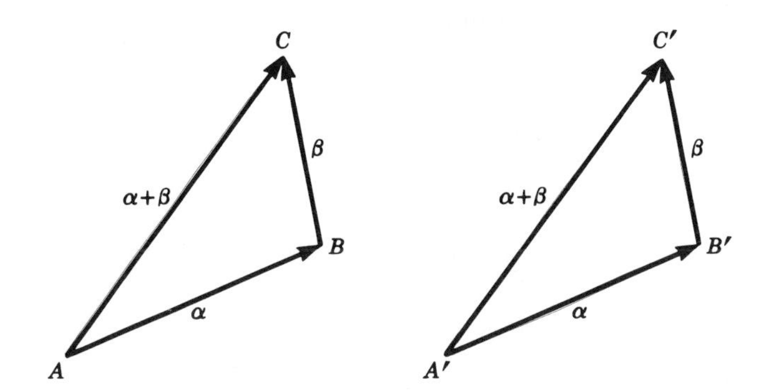

*Fig. 4*

Addition of vectors has the following properties:

a. $\alpha + \beta = \beta + \alpha$ *for all* $\alpha, \beta \in V$.
b. $(\alpha + \beta) + \gamma = \alpha + (\beta + \gamma)$ *for all* $\alpha, \beta, \gamma \in V$.
c. $\alpha + 0 = \alpha$ *for all* $\alpha \in V$.
d. *For any vector* $\alpha$, *there is a unique vector* $-\alpha$ *such that* $\alpha + (-\alpha) = 0$.

Properties $a$ and $b$ may be seen from Fig. 5 and Fig. 6, respectively. Property $c$ is obvious if we take $\alpha = \overrightarrow{AB}$ and $0 = \overrightarrow{BB}$. To verify ($d$), let $\alpha = \overrightarrow{AB}$ and consider the vector $-\alpha = \overrightarrow{BA}$ (Fig. 7). Then $\overrightarrow{AB} + \overrightarrow{BA} = \overrightarrow{AA} = 0$ implies $\alpha + (-\alpha) = 0$. It is obvious that $-\alpha$ depends only on $\alpha$ and is uniquely determined as a vector satisfying the equation $\alpha + (-\alpha) = 0$.

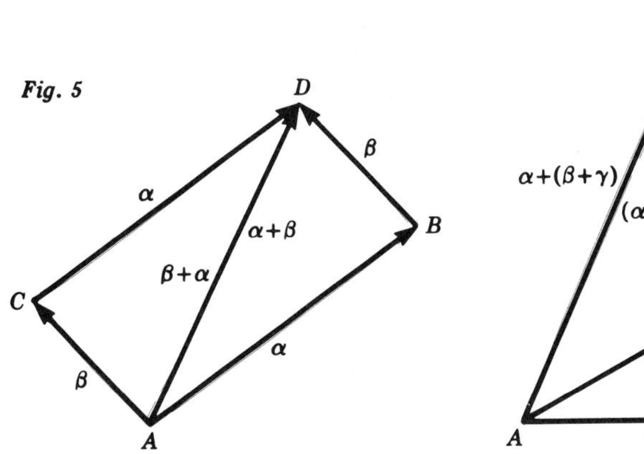

*Fig. 5*

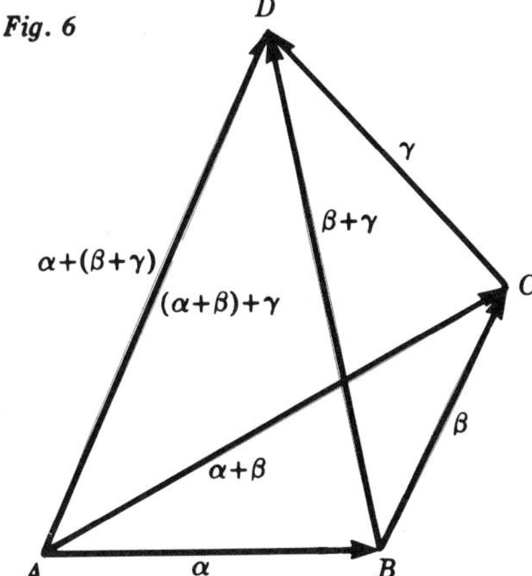

*Fig. 6*

*Scalar multiplication:* Let $\alpha$ be a vector and $c$ an arbitrary real number. (Real numbers are also called *scalars* here.) We define a vector denoted by $c\alpha$ and called the *scalar multiple* of $\alpha$ by $c$ in the following way. If $\alpha = 0$, then $c\alpha = 0$ for any scalar $c$. If $\alpha \neq 0$, let $\alpha = \overrightarrow{AB}$. If $c > 0$, then let $C$ be the unique point on the line $AB$ such that the directed segment $\overrightarrow{AC}$ has the same sense of direction as $\overrightarrow{AB}$ and such that the length of $\overrightarrow{AC}$ is $c$ times the length of $\overrightarrow{AB}$. We define $c\alpha$ to be the vector $\overrightarrow{AC}$ (cf. Fig. 8). This vector is determined independently of the choice of the point $A$.

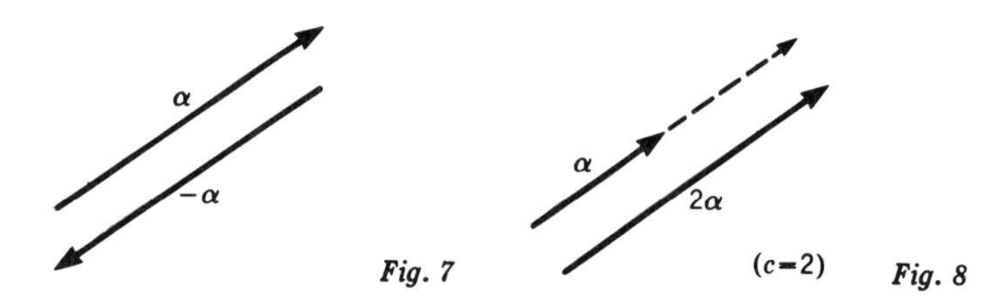

*Fig. 7* *Fig. 8*

If $c = 0$, then we define $c\alpha$ to be the zero vector. If $c < 0$, then we define $c\alpha$ to be $-((-c)\alpha)$, where $(-c)\alpha$ has already been defined, since $-c > 0$, and the first minus sign is that used in (*d*) of the properties of addition.

We may easily verify the following properties of scalar multiplication:

*e.* $(ab)\alpha = a(b\alpha)$ *for any* $\alpha \in V$ *and any scalars* $a, b$.

*f.* $1\alpha = \alpha$ *for any vector* $\alpha$.

*g.* $a(\alpha + \beta) = a\alpha + a\beta$ *for any scalar* $a$ *and* $\alpha, \beta \in V$.

*h.* $(a + b)\alpha = a\alpha + b\alpha$ *for any* $\alpha \in V$ *and scalars* $a, b$.

For example, (*g*) is illustrated by Fig. 9.

In this way the set of vectors $V$ can be made into an algebraic system with addition and scalar multiplication which satisfy conditions *a* to *h*.

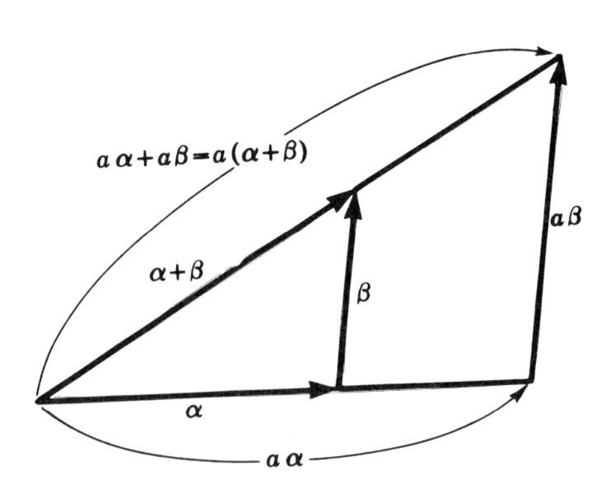

*Fig. 9*

These algebraic operations, together with the basic relationships 1 and 2 between vectors and points, are indeed sufficient to deduce all the geometric properties of the plane that do not depend on the notions of length and angle; these properties are called *affine properties* of the plane.

***Example 1.2.*** If $\triangle ABC$ is a triangle, then

$$\overrightarrow{AB} + \overrightarrow{BC} + \overrightarrow{CA} = 0.$$

***Example 1.3.*** For any points $A$, $B$ ($A \neq B$), a point $P$ is on the straight line determined by $A$ and $B$ if and only if $\overrightarrow{AP} = c\overrightarrow{AB}$ for some scalar $c$. In particular, the point $P$ such that $\overrightarrow{AP} = t\overrightarrow{AB}$, where $0 < t < 1$, divides the segment $AB$ in the ratio $t:(1 - t)$. If $t = \frac{1}{2}$, the corresponding point is the midpoint of $\overrightarrow{AB}$.

***Example 1.4.*** Let $A$ and $B$ be two distinct points and let $O$ be an arbitrary point. Then a point $P$ is on the straight line through $A$ and $B$ if and only if

$$\overrightarrow{OP} = (1 - t)\overrightarrow{OA} + t\overrightarrow{OB},$$

where $t$ is a scalar. In particular, $P$ is on the segment $AB$ if and only if $0 \leq t \leq 1$.

***Example 1.5.*** Let $\triangle ABC$ be a triangle and let $L$, $M$, $N$ be the midpoints of the sides $BC$, $CA$, $AB$, respectively. Then the medians $AL$, $BM$, and $CN$ meet at one point, called the *center of gravity* (or barycenter) of the triangle. We may prove this fact as follows. Let $G$ be the intersection of $BM$ and $CN$. Let $p:q$ be the ratio of $BG$ to $GM$. Since the line $MN$ is parallel to the line $BC$, the ratio of $CG$ to $GN$ is also $p:q$. Let $\beta = AC$ and $\gamma = AB$. Then $AM = \beta/2$ and $AN = \gamma/2$. By Example 1.4 we have [taking $t = p/(p + q)$]

$$\overrightarrow{AG} = \frac{q}{p+q}\gamma + \frac{p}{p+q}\frac{\beta}{2}$$

and also

$$\overrightarrow{AG} = \frac{q}{p+q}\beta + \frac{p}{p+q}\frac{\gamma}{2}.$$

We obtain from these two equations

$$(2q - p)(\beta - \gamma) = 0.$$

If $2q - p \neq 0$, it would follow that $\beta = \gamma$, which is impossible. Thus $2q - p = 0$; that is, $p = 2q$. In other words, $CN$ divides $BM$ in the ratio 2:1. Similarly, we may prove that $AL$ divides $BM$ in the ratio 2:1. This means that the three medians meet at one point, namely, $G$.

We shall say that two vectors $\alpha_1$ and $\alpha_2$ are *linearly dependent* if one is a scalar multiple of the other. Otherwise, we say that $\alpha_1$ and $\alpha_2$ are *linearly independent*. Geometrically speaking, let

$$\alpha_1 = \overrightarrow{AB_1} \qquad \text{and} \qquad \alpha_2 = \overrightarrow{AB_2}.$$

Then $\alpha_1$ and $\alpha_2$ are linearly dependent if and only if $A$, $B_1$, and $B_2$ are collinear (that is, they lie on one straight line). Any ordered pair of vectors $\{\alpha_1,\alpha_2\}$ which are linearly independent is called a *basis* of $V$.

Given a basis $\{\alpha_1,\alpha_2\}$ and an arbitrary point $O$, we may set up a coordinate system of the plane, called an *affine coordinate system*, as follows. Let $E_1$ and $E_2$ be points such that $\alpha_1 = \overrightarrow{OE_1}$ and $\alpha_2 = \overrightarrow{OE_2}$. Given any point $P$, we draw a straight line through $P$ parallel to the line $OE_2$ and call $P_1$ its intersection with the line $OE_1$. Similarly, let $P_2$ be a point on the line $OE_2$ such that the line $PP_2$ is parallel to the line $OE_1$. Then we have (Fig. 10)

$$\overrightarrow{OP} = \overrightarrow{OP_1} + \overrightarrow{OP_2}, \qquad \overrightarrow{OP_1} = c_1\overrightarrow{OE_1}, \qquad \text{and} \qquad \overrightarrow{OP_2} = c_2\overrightarrow{OE_2}$$

for some scalars $c_1$ and $c_2$. We thus obtain

$$\overrightarrow{OP} = c_1\overrightarrow{OE_1} + c_2\overrightarrow{OE_2}. \tag{1.13}$$

It is obvious that the pair of scalars $(c_1,c_2)$ is determined uniquely by $P$ and that, conversely, any pair of scalars $(c_1,c_2)$ determines a unique point $P$ for which (1.13) is satisfied. We call $(c_1,c_2)$ the *coordinates* of $P$. It is convenient to write $P = (c_1,c_2)$. Two points $A = (a_1,a_2)$ and $B = (b_1,b_2)$ coincide if and only if $a_1 = b_1$ and $a_2 = b_2$.

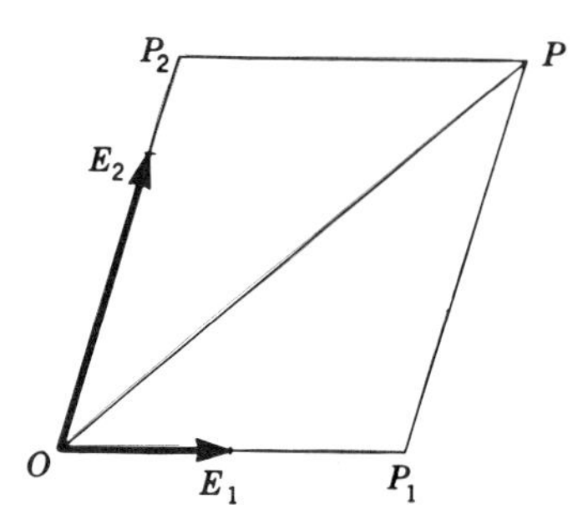


*Fig. 10*

Starting from a basis $\{\alpha_1,\alpha_2\}$ and an arbitrary point $O$, we have thus obtained a rule which associates with each point $P$ its coordinates $(c_1,c_2)$. This rule is called the *affine coordinate system determined by* $\{\alpha_1,\alpha_2\}$ *and* $O$. The point $O$ is called the *origin*, and the points $E_1$ and $E_2$ are called the *unit points*. We have $O = (0,0)$, $E_1 = (1,0)$, and $E_2 = (0,1)$.

Let $\{\alpha_1,\alpha_2\}$ be a basis of $V$. If $\alpha$ is an arbitrary vector, then $\alpha$ can be represented in the form

$$\alpha = c_1\alpha_1 + c_2\alpha_2, \tag{1.14}$$

where $c_1$, $c_2$ are certain scalars which are uniquely determined by $\alpha$. Equation (1.14) is nothing but (1.13) if we let $\alpha_1 = \overrightarrow{OE_1}$, $\alpha_2 = \overrightarrow{OE_2}$, and $\alpha = \overrightarrow{OP}$.

***Example 1.6.*** With respect to an affine coordinate system with origin $O$ and corresponding to a basis $\{\alpha_1,\alpha_2\}$, let $A = (a_1,a_2)$ and $B = (b_1,b_2)$. Then

$$\overrightarrow{AB} = (b_1 - a_1)\alpha_1 + (b_2 - a_2)\alpha_2. \tag{1.15}$$

This follows from

$$\overrightarrow{AB} = \overrightarrow{OB} - \overrightarrow{OA} = (b_1\alpha_1 + b_2\alpha_2) - (a_1\alpha_1 + a_2\alpha_2)$$
$$= (b_1 - a_1)\alpha_1 + (b_2 - a_2)\alpha_2.$$

***Example 1.7.*** With respect to an affine coordinate system, a straight line $l$ can be expressed by a linear equation

$$c_1x_1 + c_2x_2 = d, \tag{1.16}$$

where $c_1$ and $c_2$ are not both 0 [that is, $l$ is the set of all points whose coordinates $(x_1,x_2)$ satisfy (1.16)]. To prove this, let $A = (a_1,a_2)$ and $B = (b_1,b_2)$ be two distinct points on $l$. A point $P = (x_1,x_2)$ is on $l$ if and only if $\overrightarrow{AP} = c\overrightarrow{AB}$ for some scalar $c$. By (1.15) this can be expressed by

$$x_1 - a_1 = c(b_1 - a_1) \quad \text{and} \quad x_2 - a_2 = c(b_2 - a_2).$$

Assuming $b_1 \neq a_1$, we have

$$x_2 - a_2 = \frac{(x_1 - a_1)(b_2 - a_2)}{b_1 - a_1},$$

which is a linear equation of the form (1.16). If $b_1 = a_1$, then we get $x_1 = a_1$, which is a special case of (1.16).

***Example 1.8.*** In order to illustrate the usefulness of an affine coordinate system, we give another proof of the result in Example 1.5. Given a triangle $\triangle ABC$, we take the affine coordinate system with origin $B$ and the basis $\{\overrightarrow{BC},\overrightarrow{BA}\}$. Thus $B = (0,0)$, $C = (1,0)$, and $A = (0,1)$. We then have $L = (\frac{1}{2},0)$, $M = (\frac{1}{2},\frac{1}{2})$, and $N = (0,\frac{1}{2})$. The equations for the medians $AL$, $BM$, and $CN$ are, respectively,

$$2x + y = 1, \qquad x - y = 0, \qquad \text{and} \qquad x + 2y = 1.$$

We find that these three equations have a common solution $x = \frac{1}{3}$, $y = \frac{1}{3}$. This means that the three medians meet at the point $(\frac{1}{3},\frac{1}{3})$. We also see that $\overrightarrow{BG} = \frac{2}{3}\overrightarrow{BM}$, $\overrightarrow{AG} = \frac{2}{3}\overrightarrow{AL}$, and $\overrightarrow{CG} = \frac{2}{3}\overrightarrow{CN}$.

We shall now treat the *metric* properties of the plane, namely, properties concerning or involving the notions of length and angle.

For a vector $\alpha$, the *length* of $\alpha$, denoted by $\|\alpha\|$, is defined to be the length of a directed segment $\overrightarrow{AB}$ which represents $\alpha$. (If $\overrightarrow{AB} = \overrightarrow{CD}$, then, of course, $\overrightarrow{AB}$ and $\overrightarrow{CD}$ have the same length.) We have

$$\|c\alpha\| = |c|\,\|\alpha\|, \tag{1.17}$$

$$\|\alpha + \beta\| \leq \|\alpha\| + \|\beta\|, \qquad \textit{triangle inequality}. \tag{1.18}$$

In fact, these follow from the definitions of $c\alpha$ and $\alpha + \beta$ (Fig. 4).

Given two vectors $\alpha$ and $\beta$, we define the *inner product* $(\alpha,\beta)$ as follows. If $\alpha = 0$ or $\beta = 0$, we set $(\alpha,\beta) = 0$. If $\alpha \neq 0$, $\beta \neq 0$, then choose any point $O$ and let $\alpha = \overrightarrow{OA}$ and $\beta = \overrightarrow{OB}$. The two half-lines $OA$ and $OB$

determine two angles; let $\theta$ be the smaller angle, so that $0 \leq \theta \leq \pi$ (Fig. 11). Then we set

$$(1.19) \qquad (\alpha,\beta) = \|\alpha\| \, \|\beta\| \cos \theta.$$

Note that this definition is independent of the choice of $O$. For $\alpha = \beta$ (and $\theta = 0$), we have

$$(1.20) \qquad (\alpha,\alpha) = \|\alpha\|^2.$$

Two nonzero vectors $\alpha$ and $\beta$ are said to be *orthogonal* (or *perpendicular* or *normal*) if the angle $\theta$ equals $\pi/2$, and this condition is equivalent to $(\alpha,\beta) = 0$. A vector of length 1 is called a *unit vector*.

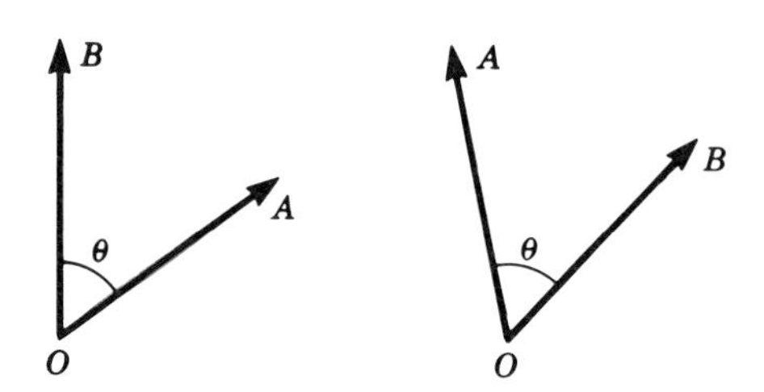


***Fig. 11***

***Example 1.9.*** Let $\alpha = \overrightarrow{OA}$ be a unit vector. For any nonzero vector $\beta = \overrightarrow{OB}$, let $C$ be the projection of $B$ on the line $OA$ (that is, the foot of the perpendicular line to $OA$ drawn from $B$; see Fig. 12). If $\overrightarrow{OC} = c\,\overrightarrow{OA}$ and if $\theta$ is the angle between $OA$ and $OB$, then

$$\cos \theta = \frac{c}{\|\beta\|}$$

and

$$(\beta,\alpha) = \|\beta\| \, \|\alpha\| \cos \theta = c \, \|\alpha\| = c,$$

which implies $c = (\beta,\alpha)$. Thus

$$(1.21) \qquad \overrightarrow{OC} = (\beta,\alpha)\overrightarrow{OA}.$$

The inner product has the following properties:

$$(1.22) \qquad (\beta,\alpha) = (\alpha,\beta).$$

$$(1.23) \qquad (c\alpha,\beta) = c(\alpha,\beta) = (\alpha,c\beta).$$

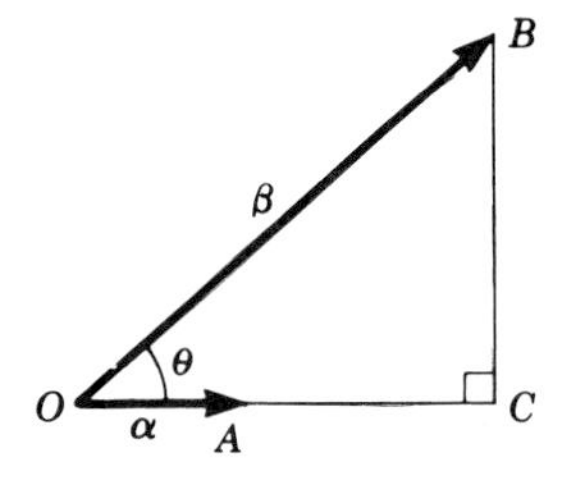


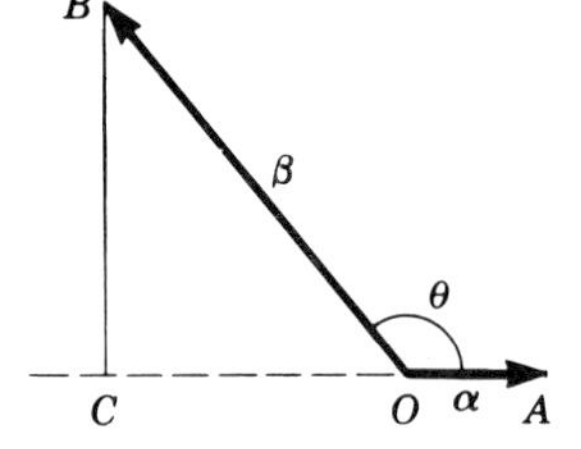


***Fig. 12***

$$(1.24)\qquad \begin{aligned} (\alpha_1 + \alpha_2, \beta) &= (\alpha_1,\beta) + (\alpha_2,\beta), \\ (\alpha, \beta_1 + \beta_2) &= (\alpha,\beta_1) + (\alpha,\beta_2). \end{aligned}$$

Property (1.22) is obvious from the definition, and (1.23) is also obvious if $c > 0$. For $c < 0$, we note that the angle between $c\alpha$ and $\beta$ is $\pi - \theta$, where $\theta$ is the angle between $\alpha$ and $\beta$. Thus $(c\alpha,\beta) = \|c\alpha\|\ \|\beta\| \cos(\pi - \theta) = -|c|\ \|\alpha\|\ \|\beta\| \cos\theta = c(\alpha,\beta)$. Similarly, $(\alpha,c\beta) = c(\alpha,\beta)$.

It is sufficient to prove (1.24) in the case where $\beta$ is a unit vector. For a nonzero vector $\beta$, let $\gamma = \beta/\|\beta\|$. If $(\alpha_1 + \alpha_2, \gamma) = (\alpha_1,\gamma) + (\alpha_2,\gamma)$, then we may multiply both sides by $\|\beta\|$ and find $(\alpha_1 + \alpha_2, \beta) = (\alpha_1,\beta) + (\alpha_2,\beta)$ by using (1.23). Assume $\beta$ is a unit vector. Then the identity follows from the geometric interpretation of the inner product in Example 1.9 (see Fig. 13).

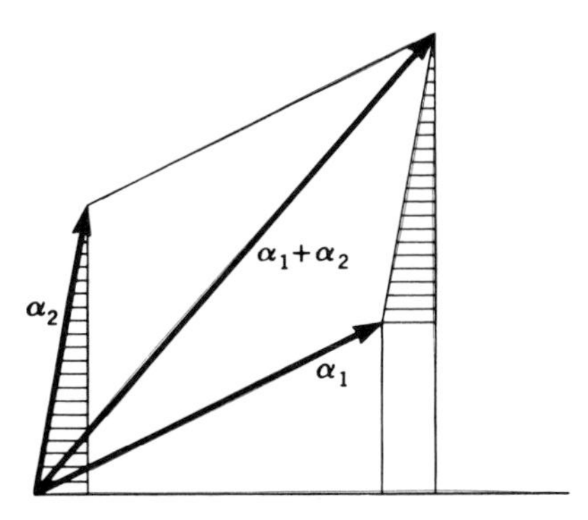


***Fig. 13***

All the metric properties of the plane can be derived from the properties of the inner product. Let $\alpha_1$ and $\alpha_2$ be two unit vectors which are perpendicular [that is, $(\alpha_1,\alpha_1) = (\alpha_2,\alpha_2) = 1$ and $(\alpha_1,\alpha_2) = 0$]. Taking any point $O$ as the origin, we construct an affine coordinate system. This coordinate system is a *rectangular coordinate system*, namely, an ordinary coordinate system based on two perpendicular lines with equal unit length on each line (coordinate axis) (see Fig. 14).

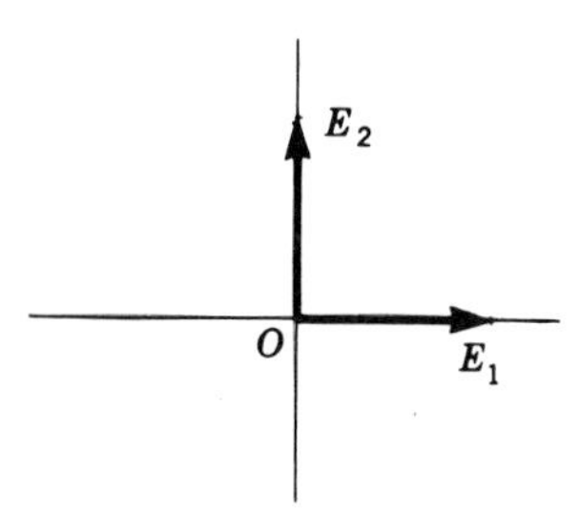


***Fig. 14***

Let $A = (a_1,a_2)$ and $B = (b_1,b_2)$ with respect to a rectangular coordinate system. Then we have

$$\overrightarrow{AB} = (b_1 - a_1)\alpha_1 + (b_2 - a_2)\alpha_2$$

and

$$\begin{aligned}(\overrightarrow{AB},\overrightarrow{AB}) &= ((b_1 - a_1)\alpha_1 + (b_2 - a_2)\alpha_2, (b_1 - a_1)\alpha_1 + (b_2 - a_2)\alpha_2) \\ &= (b_1 - a_1)^2(\alpha_1,\alpha_1) + (b_2 - a_2)^2(\alpha_2,\alpha_2) \\ &= (b_1 - a_1)^2 + (b_2 - a_2)^2,\end{aligned}$$

by using $(\alpha_1,\alpha_2) = 0$ and $(\alpha_1,\alpha_1) = (\alpha_2,\alpha_2) = 1$. Thus the distance $d(A,B)$ between $A$ and $B$ is given by

(1.25) $$d(A,B) = \sqrt{(b_1 - a_1)^2 + (b_2 - a_2)^2}.$$

***Example 1.10.*** Fix a rectangular coordinate system with origin $O$. For a line $l$ which does not pass through $O$, let $H = (h_1,h_2)$ be the foot of the perpendicular drawn from $O$ to $l$ (Fig. 15). A point $P = (x_1,x_2)$ is on the line $l$ if and only if $(\overrightarrow{OH},\overrightarrow{HP}) = 0$. This condition may be expressed by

$$(x_1 - h_1)h_1 + (x_2 - h_2)h_2 = 0,$$

that is,

$$\frac{h_1}{\sqrt{h_1^2 + h_2^2}}x_1 + \frac{h_2}{\sqrt{h_1^2 + h_2^2}}x_2 = \sqrt{h_1^2 + h_2^2}.$$

We rewrite this in the form

(1.26) $$a_1x_1 + a_2x_2 = d, \qquad \text{where } d > 0 \text{ and } a_1^2 + a_2^2 = 1.$$

This equation is called the *normal form* of the line $l$. Note that $d$ is the distance from $O$ to the line $l$. The vector $a_1\alpha_1 + a_2\alpha_2$ is a unit vector perpendicular to $l$, and is called a *unit normal vector* of $l$. Equation (1.26) is still valid if $l$ goes through $O$, in which case $d = 0$.

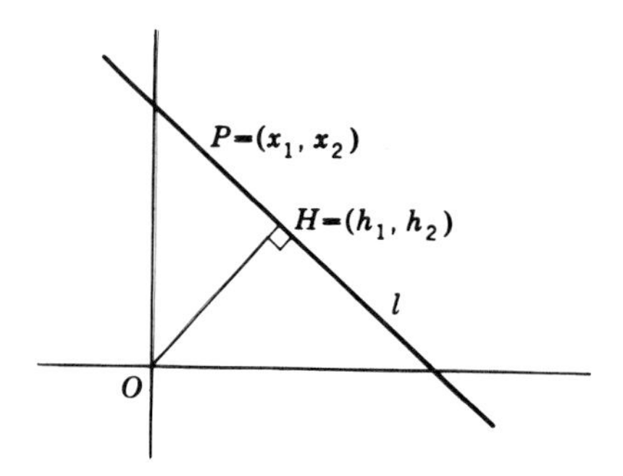


***Fig. 15***

**EXERCISE 1.3**

**1.** Verify $(\overrightarrow{AB} + \overrightarrow{BC}) + \overrightarrow{CD} = \overrightarrow{AD} = \overrightarrow{AB} + (\overrightarrow{BC} + \overrightarrow{CD})$. Draw a geometric illustration for these identities.

**2.** If $\triangle ABC$ is a triangle, show that $\overrightarrow{AB} = \overrightarrow{AC} - \overrightarrow{BC}$.

**3.** For any points $P_i$, $1 \leq i \leq n$, show that

$$\overrightarrow{P_1P_2} + \overrightarrow{P_2P_3} + \cdots + \overrightarrow{P_{n-1}P_n} + \overrightarrow{P_nP_1} = 0.$$

» **4.** Let $G$ be the center of gravity of a triangle $\triangle ABC$. Prove that $\overrightarrow{GA} + \overrightarrow{GB} + \overrightarrow{GC} = 0$.

**5.** For a vector $\alpha$ and for a scalar $c$, prove that

$$-(c\alpha) = (-c)\alpha = c(-\alpha).$$

**6.** In an affine coordinate system, show that the equation

$$\frac{x_1}{a_1} + \frac{x_2}{a_2} = 1$$

represents a line which goes through $(a_1,0)$ and $(0,a_2)$.

» **7.** For a triangle $\triangle ABC$, where $A = (a_1,a_2)$, $B = (b_1,b_2)$, $C = (c_1,c_2)$, find the coordinates of the center of gravity.

**8.** Suppose $(\alpha,\beta) \neq 0$ and set $\gamma = \alpha - \{(\alpha,\alpha)/(\alpha,\beta)\}\beta$. Show that $(\alpha,\gamma) = 0$. Prove also that $\gamma = 0$ if and only if $\beta = c\alpha$ for some scalar $c$.

**9.** For two vectors $\alpha$, $\beta$, prove that

$$\left\|\frac{\alpha+\beta}{2}\right\|^2 + \left\|\frac{\alpha-\beta}{2}\right\|^2 = \frac{\|\alpha\|^2 + \|\beta\|^2}{2}.$$

**10.** Given $n$ distinct points $A_1, \ldots, A_n$, show that there is one and only one point $G$ such that $\sum_{i=1}^{n} d(P,A_i)^2$ is minimum at $P = G$. For this point show that $\sum_{i=1}^{n} \overrightarrow{GA_i} = 0$.

» **11.** Let $A$ and $B$ be two given points. For an arbitrary point $P$, let $Q$ be the point symmetric to $P$ with respect to $A$ (that is, $A$ is the midpoint of $PQ$) and let $R$ be the point symmetric to $Q$ with respect to $B$. Show that $\overrightarrow{PR}$ is a constant vector.

**12.** Let $\triangle ABC$ be a triangle. By using vectors, prove that the perpendiculars $AA'$, $BB'$, and $CC'$ drawn from $A$, $B$, and $C$ to the opposite sides meet at one point. [*Hint:* Let $\beta = \overrightarrow{CA}$, $\gamma = \overrightarrow{AB}$. Let $H$ be the intersection of $BB'$ and $CC'$ and set $\overrightarrow{BH} = \xi$. Write down $(\overrightarrow{BH},\overrightarrow{CA}) = (\overrightarrow{CH},\overrightarrow{AB}) = 0$ and derive $(\overrightarrow{AH},\overrightarrow{BC}) = 0$.]

**13.** Let $O$ be the intersection of the perpendicular bisectors of the three sides of a triangle $\triangle ABC$. Show that

$$\overrightarrow{OA} + \overrightarrow{OB} + \overrightarrow{OC} = 3\,\overrightarrow{OG},$$

where $G$ is the center of gravity of $\triangle ABC$. [*Hint:* Let $\alpha = \overrightarrow{OA}$, $\beta = \overrightarrow{OB}$, $\gamma = \overrightarrow{OC}$. Express $\overrightarrow{OL}$ ($L$: midpoint of $BC$) and then $\overrightarrow{OG}$ in terms of $\alpha$, $\beta$, $\gamma$.]

**14.** Let $\alpha = \overrightarrow{AB}$ and $\beta = \overrightarrow{AD}$. Show that the area of the parallelogram $ABCD$ (with vertices in this order) is equal to $\sqrt{(\alpha,\alpha)(\beta,\beta) - (\alpha,\beta)^2}$.

» **15.** Find the normal form of each of the straight lines:

(*a*) $3x_1 + 4x_2 = 10$;

(*b*) $3x_1 + 4x_2 + 10 = 0$.

» **16.** Find the distance from a point $(y_1,y_2)$ to the line given by the normal form (1.26).

**17.** Suppose that for an affine coordinate system the distance $d(A,B)$ between any two points $A = (a_1,a_2)$ and $B = (b_1,b_2)$ is given by the formula (1.25). Prove that the coordinate system is a rectangular coordinate system.

## 1.4 CHANGE OF COORDINATES

An affine coordinate system $\{x_1,x_2\}$ is constructed from a point $O$ (which will be the origin) and a basis $\{\alpha_1,\alpha_2\}$ of $V$. For each point $P$ we denote the coordinates of $P$ by $(x_1(P),x_2(P))$. We shall study how two affine coordinate systems are related to each other.

Let $\{x_1,x_2\}$ be an affine coordinate system constructed from $O$ and $\{\alpha_1,\alpha_2\}$. Let $\{x_1',x_2'\}$ be an affine coordinate system constructed from $O'$ and $\{\alpha_1',\alpha_2'\}$. We first consider two special cases.

Case 1. Assume $\alpha_1 = \alpha_1'$ and $\alpha_2 = \alpha_2'$. Let $O'$ have coordinates $(a_1,a_2)$ with respect to $\{x_1,x_2\}$. This means

$$\overrightarrow{OO'} = a_1\alpha_1 + a_2\alpha_2.$$

For any point $P$ we have

$$\overrightarrow{OP} = x_1(P)\alpha_1 + x_2(P)\alpha_2$$

and

$$\overrightarrow{O'P} = x_1'(P)\alpha_1 + x_2'(P)\alpha_2.$$

Since $\overrightarrow{OP} = \overrightarrow{OO'} + \overrightarrow{O'P}$, we obtain

$$x_1(P)\alpha_1 + x_2(P)\alpha_2 = (a_1 + x_1'(P))\alpha_1 + (a_2 + x_2'(P))\alpha_2,$$

that is, $$x_1(P) = x_1'(P) + a_1, \qquad x_2(P) = x_2'(P) + a_2.$$

This being the case for any point $P$, we may simply write

$$\text{(1.27)} \qquad \begin{aligned} x_1 &= x_1' + a_1, \\ x_2 &= x_2' + a_2, \end{aligned} \qquad \textit{translation of coordinate axes,}$$

or

$$\text{(1.27')} \qquad \begin{aligned} x_1' &= x_1 - a_1, \\ x_2' &= x_2 - a_2. \end{aligned}$$

The relationship is illustrated in Fig. 16.

Case 2. Assume $O = O'$. Since $\{\alpha_1,\alpha_2\}$ is a basis of $V$, every vector can be written uniquely in the form (1.14). Let

$$\text{(1.28)} \qquad \begin{aligned} \alpha_1' &= a_{11}\alpha_1 + a_{21}\alpha_2, \\ \alpha_2' &= a_{12}\alpha_1 + a_{22}\alpha_2. \end{aligned}$$

For any point $P$ we have (since $O = O'$)

$$\begin{aligned} \overrightarrow{OP} &= x_1'(P)\alpha_1' + x_2'(P)\alpha_2' \\ &= x_1'(P)(a_{11}\alpha_1 + a_{21}\alpha_2) + x_2'(P)(a_{12}\alpha_1 + a_{22}\alpha_2) \\ &= (a_{11}x_1'(P) + a_{12}x_2'(P))\alpha_1 + (a_{21}x_1'(P) + a_{22}x_2'(P))\alpha_2. \end{aligned}$$

Comparing this with

$$\overrightarrow{OP} = x_1(P)\alpha_1 + x_2(P)\alpha_2,$$

we obtain

$$x_1(P) = a_{11}x_1'(P) + a_{12}x_2'(P),$$
$$x_2(P) = a_{21}x_1'(P) + a_{22}x_2'(P).$$

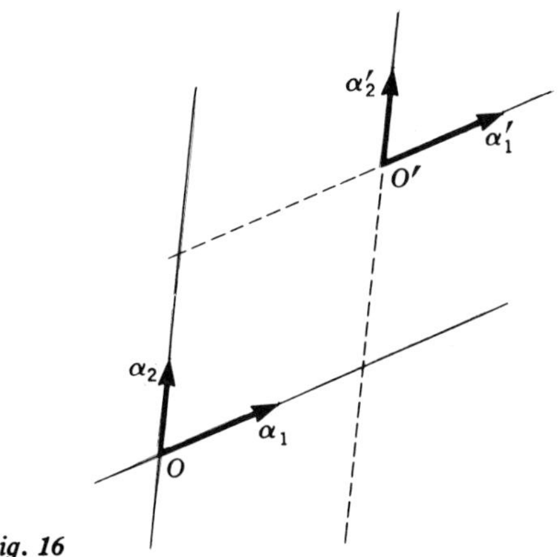


*Fig. 16*

Dropping $P$, we may write

(1.29)
$$x_1 = a_{11}x_1' + a_{12}x_2',$$
$$x_2 = a_{21}x_1' + a_{22}x_2'.$$

See Fig. 17.

Similarly, starting from

(1.28′)
$$\alpha_1 = b_{11}\alpha_1' + b_{21}\alpha_2',$$
$$\alpha_2 = b_{12}\alpha_1' + b_{22}\alpha_2',$$

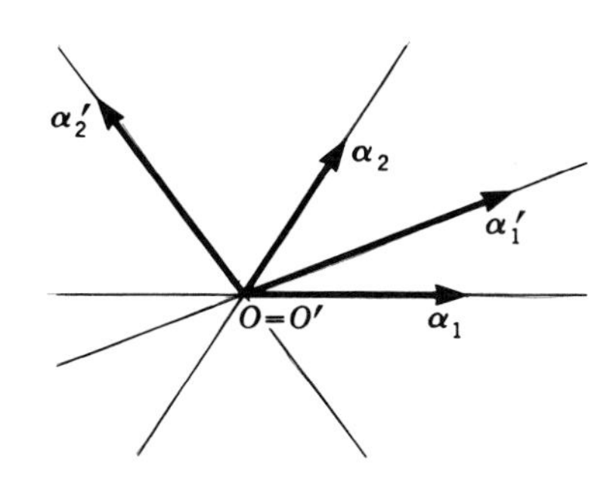


*Fig. 17*

we obtain

$$(1.29') \qquad \begin{aligned} x_1' &= b_{11}x_1 + b_{12}x_2, \\ x_2' &= b_{21}x_1 + b_{22}x_2. \end{aligned}$$

By using the matrix notation in Sec. 1.1, (1.29) and (1.29′) can be expressed, respectively, by

$$(1.30) \qquad \begin{bmatrix} x_1 \\ x_2 \end{bmatrix} = \begin{bmatrix} a_{11} & a_{12} \\ a_{21} & a_{22} \end{bmatrix} \begin{bmatrix} x_1' \\ x_2' \end{bmatrix},$$

$$(1.30') \qquad \begin{bmatrix} x_1' \\ x_2' \end{bmatrix} = \begin{bmatrix} b_{11} & b_{12} \\ b_{21} & b_{22} \end{bmatrix} \begin{bmatrix} x_1 \\ x_2 \end{bmatrix}.$$

The matrices

$$\mathbf{A} = \begin{bmatrix} a_{11} & a_{12} \\ a_{21} & a_{22} \end{bmatrix} \quad \text{and} \quad \mathbf{B} = \begin{bmatrix} b_{11} & b_{12} \\ b_{21} & b_{22} \end{bmatrix}$$

are related by

$$(1.31) \qquad \mathbf{AB} = \mathbf{BA} = \begin{bmatrix} 1 & 0 \\ 0 & 1 \end{bmatrix}.$$

In fact, substituting (1.29′) in (1.29), we have

$$\begin{aligned} x_1 &= a_{11}(b_{11}x_1 + b_{12}x_2) + a_{12}(b_{21}x_1 + b_{22}x_2) \\ &= (a_{11}b_{11} + a_{12}b_{21})x_1 + (a_{11}b_{12} + a_{12}b_{22})x_2 \end{aligned}$$

and, similarly,

$$x_2 = (a_{21}b_{11} + a_{22}b_{21})x_1 + (a_{21}b_{12} + a_{22}b_{22})x_2.$$

Putting $x_1 = 1$, $x_2 = 0$, we get

$$a_{11}b_{11} + a_{12}b_{21} = 1, \qquad a_{21}b_{11} + a_{22}b_{21} = 0.$$

Putting $x_1 = 0$, $x_2 = 1$, we get

$$a_{11}b_{12} + a_{12}b_{22} = 0, \qquad a_{21}b_{11} + a_{22}b_{22} = 1.$$

These four conditions are equivalent to the matrix equation

$$\mathbf{AB} = \begin{bmatrix} 1 & 0 \\ 0 & 1 \end{bmatrix}.$$

Similarly, we get

$$\mathbf{BA} = \begin{bmatrix} 1 & 0 \\ 0 & 1 \end{bmatrix}.$$

Now the general case can be treated by combining the special cases 1 and 2. One passes from the coordinate system $\{x_1',x_2'\}$ to the coordinate system $\{y_1,y_2\}$ constructed from $O'$ and $\{\alpha_1,\alpha_2\}$ and then from $\{y_1,y_2\}$ to the system $\{x_1,x_2\}$. (See Fig. 18.) The first passage is related as in case 2, so that

$$y_1 = a_{11}x_1' + a_{12}x_2',$$
$$y_2 = a_{21}x_1' + a_{22}x_2'.$$

The second passage is related as in case 1, so that

$$x_1 = y_1 + a_1,$$
$$x_2 = y_2 + a_2.$$

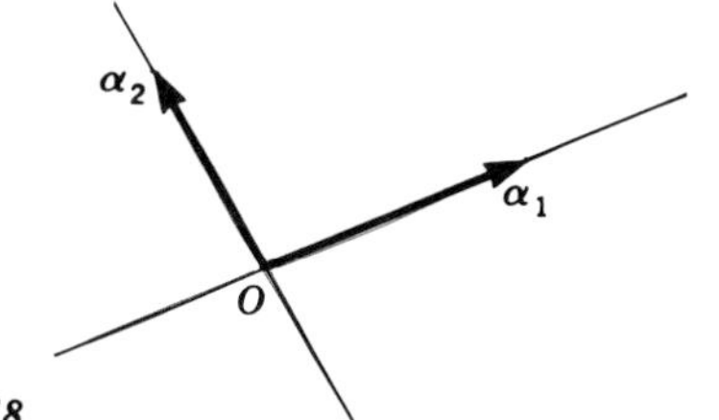


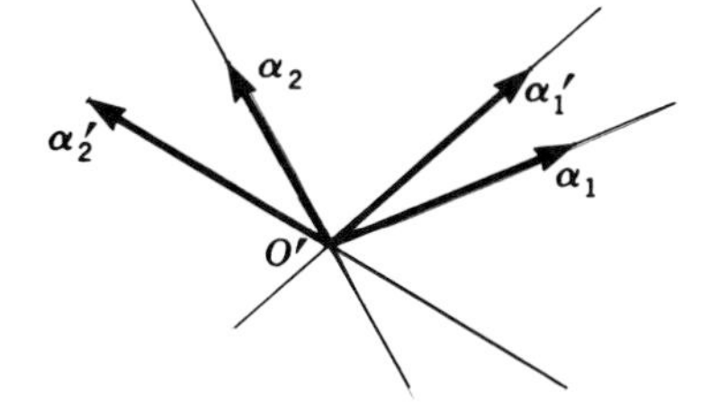


*Fig. 18*

Therefore we obtain

$$\begin{aligned} x_1 &= a_{11}x_1' + a_{12}x_2' + a_1, \\ x_2 &= a_{21}x_1' + a_{22}x_2' + a_2 \end{aligned} \tag{1.32}$$

or

$$\begin{aligned} x_1' &= b_{11}x_1 + b_{12}x_2 + b_1, \\ x_2' &= b_{21}x_1 + b_{22}x_2 + b_2. \end{aligned} \tag{1.32'}$$

(1.32) can be expressed in matrix notation by

$$\begin{bmatrix} x_1 \\ x_2 \\ 1 \end{bmatrix} = \begin{bmatrix} a_{11} & a_{12} & a_1 \\ a_{21} & a_{22} & a_2 \\ 0 & 0 & 1 \end{bmatrix} \begin{bmatrix} x_1' \\ x_2' \\ 1 \end{bmatrix}. \tag{1.33}$$

Similarly, (1.32′) can be expressed by

$$\begin{bmatrix} x_1' \\ x_2' \\ 1 \end{bmatrix} = \begin{bmatrix} b_{11} & b_{12} & b_1 \\ b_{21} & b_{22} & b_2 \\ 0 & 0 & 1 \end{bmatrix} \begin{bmatrix} x_1 \\ x_2 \\ 1 \end{bmatrix}. \tag{1.33'}$$

If we let

$$\tilde{\mathbf{A}} = \begin{bmatrix} a_{11} & a_{12} & a_1 \\ a_{21} & a_{22} & a_2 \\ 0 & 0 & 1 \end{bmatrix}, \qquad \tilde{\mathbf{B}} = \begin{bmatrix} b_{11} & b_{12} & b_1 \\ b_{21} & b_{22} & b_2 \\ 0 & 0 & 1 \end{bmatrix},$$

then we can verify

$$\tilde{\mathbf{A}}\tilde{\mathbf{B}} = \tilde{\mathbf{B}}\tilde{\mathbf{A}} = \begin{bmatrix} 1 & 0 & 0 \\ 0 & 1 & 0 \\ 0 & 0 & 1 \end{bmatrix}. \tag{1.34}$$

On the other hand, writing

$$\tilde{\mathbf{A}} = \begin{bmatrix} \mathbf{A} & \alpha \\ \mathbf{0} & 1 \end{bmatrix}, \qquad \text{where } \alpha = \begin{bmatrix} a_1 \\ a_2 \end{bmatrix},$$

$$\tilde{\mathbf{B}} = \begin{bmatrix} \mathbf{B} & \beta \\ \mathbf{0} & 1 \end{bmatrix}, \qquad \text{where } \beta = \begin{bmatrix} b_1 \\ b_2 \end{bmatrix},$$

we see that

$$\tilde{\mathbf{A}}\tilde{\mathbf{B}} = \begin{bmatrix} \mathbf{AB} & \mathbf{A}\alpha + \beta \\ \mathbf{0} & 1 \end{bmatrix}$$

so that

$$\mathbf{A}\alpha + \beta = \mathbf{0},$$

that is,

$$\beta = -\mathbf{A}\alpha \qquad \text{or} \qquad \alpha = -\mathbf{B}\beta. \tag{1.35}$$

Thus two affine coordinate systems $\{x_1,x_2\}$ and $\{x_1',x_2'\}$ are related by (1.33) and (1.33′), where the matrices **A**, **B** and vectors $\alpha$, $\beta$ are related by (1.31) and (1.35).

We shall make the following definitions.

**Definition 1.5**

For any positive integer $n$, the $n \times n$ matrix of the form

$$\mathbf{I}_n = \begin{bmatrix} 1 & 0 & \cdots & 0 \\ 0 & 1 & \cdots & 0 \\ \cdots & \cdots & \cdots & \cdots \\ 0 & 0 & \cdots & 1 \end{bmatrix}$$

(that is, the entries on the diagonal are 1 and all the other entries are 0) is called the *identity matrix* (or *unit matrix*) of *degree n*.

**Definition 1.6**

For an $n \times n$ matrix **A**, an $n \times n$ matrix **B** such that $\mathbf{AB} = \mathbf{BA} = \mathbf{I}_n$ is called an *inverse* of **A**.

The identity matrix $\mathbf{I}_n$ has a remarkable property that

$$\mathbf{AI}_n = \mathbf{I}_n\mathbf{A} = \mathbf{A} \qquad \text{for any } n \times n \text{ matrix } \mathbf{A}. \tag{1.36}$$

If we denote the $(i,j)$ entry of $\mathbf{I}_n$ by $\delta_{ij}$, then we have

$$\delta_{ij} = \begin{cases} 1 & \text{for } i = j, \\ 0 & \text{for } i \neq j. \end{cases}$$

($\delta_{ij}$ is called the *Kronecker symbol.*) Condition (1.36) can be easily verified by going back to Definition 1.3 of the matrix product.

Let us also observe

(1.37) $\quad (\mathbf{AB})\mathbf{C} = \mathbf{A}(\mathbf{BC}) \qquad$ for any $n \times n$ matrices **A**, **B**, and **C**.

In fact, if $\mathbf{A} = [a_{ij}]$, $\mathbf{B} = [b_{ij}]$, and $\mathbf{C} = [c_{ij}]$ and if we denote by $(\mathbf{AB})_{ij}$, $((\mathbf{AB})\mathbf{C})_{ij}$, etc., the $(i,j)$ entry of **AB**, (**AB**)**C**, etc., then we have

$$((\mathbf{AB})\mathbf{C})_{ij} = \sum_{k=1}^{n} (\mathbf{AB})_{ik}c_{kj} = \sum_{m,k=1}^{n} a_{im}b_{mk}c_{kj}$$

and

$$(\mathbf{A}(\mathbf{BC}))_{ij} = \sum_{m=1}^{n} a_{im}(\mathbf{BC})_{mj} = \sum_{m,k=1}^{n} a_{im}b_{mk}c_{kj},$$

so that (1.37) is valid.

By using (1.37), we may show that when an $n \times n$ matrix **A** has an inverse **B**, that is, a matrix **B** such that $\mathbf{AB} = \mathbf{BA} = \mathbf{I}_n$, such a matrix **B** is uniquely determined. In fact, if **C** is another $n \times n$ matrix such that $\mathbf{AC} = \mathbf{CA} = \mathbf{I}_n$, then

$$(\mathbf{CA})\mathbf{B} = \mathbf{C}(\mathbf{AB})$$

by (1.37). The left-hand side is equal to $\mathbf{I}_n\mathbf{B} = \mathbf{B}$, while the right-hand side is equal to $\mathbf{CI}_n = \mathbf{C}$, thus proving $\mathbf{B} = \mathbf{C}$.

The inverse of **A** is denoted by $\mathbf{A}^{-1}$.

***Example 1.11.*** For

$$\mathbf{A} = \begin{bmatrix} 1 & 0 & a_1 \\ 0 & 1 & a_2 \\ 0 & 0 & 1 \end{bmatrix} \qquad \text{we have} \qquad \mathbf{A}^{-1} = \begin{bmatrix} 1 & 0 & -a_1 \\ 0 & 1 & -a_2 \\ 0 & 0 & 1 \end{bmatrix}.$$

This fact conforms with (1.27) and (1.27′).

***Example 1.12.*** In order to see whether

$$\mathbf{A} = \begin{bmatrix} a_{11} & a_{12} \\ a_{21} & a_{22} \end{bmatrix}$$

has an inverse, let $\mathbf{X} = \begin{bmatrix} x_{11} & x_{12} \\ x_{21} & x_{22} \end{bmatrix}$ and write down $\mathbf{AX} = \mathbf{I}_2$. Then we obtain a system of four linear equations in four unknowns $x_{ij}$, $1 \leq i, j \leq 2$. If this system has no solution, then **A** has no inverse. For example, $\begin{bmatrix} 4 & 6 \\ 2 & 3 \end{bmatrix}$ has no inverse.

We shall now consider how two rectangular coordinate systems $\{x_1,x_2\}$ and $\{x'_1,x'_2\}$ are related to each other. Let $O$, $\alpha_1$, $\alpha_2$ and $O'$, $\alpha'_1$, $\alpha'_2$ be as before, except that now we have $(\alpha_i,\alpha_j) = (\alpha'_i,\alpha'_j) = \delta_{ij}$ (using Kronecker's symbol).

In case 1 (that is, $\alpha_1 = \alpha'_1$ and $\alpha_2 = \alpha'_2$), we have (1.27) and (1.27′) just as before.

In case 2 (that is, $O = O'$) we proceed as follows. Let us consider $(\alpha_1,\alpha_2)$ as illustrated in Fig. 19. To any vector $\beta$ let us associate the angle $\varphi$ $(0 \leq \varphi < 2\pi)$ measured from $\alpha_1$ to $\beta$ in the counterclockwise direction (that is, in the direction from $\alpha_1$ to $\alpha_2$). (This notion of angle is different from that in Sec. 1.3 and is based on the notion of orientation of the plane given by $\{\alpha_1,\alpha_2\}$. We shall not elaborate on this, however.)

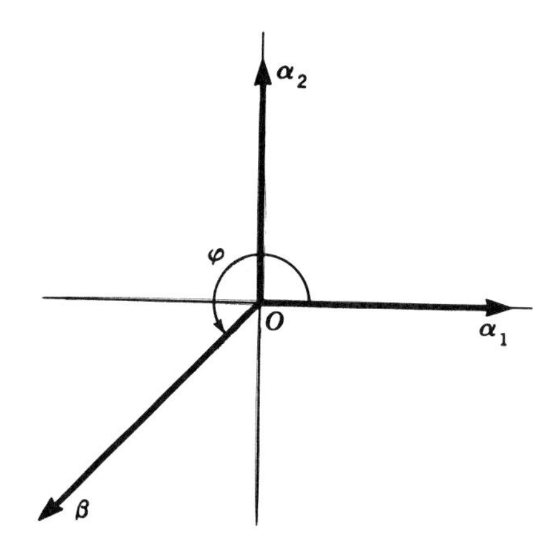


*Fig. 19*

Then we have

$$\beta = c\,((\cos\varphi)\alpha_1 + (\sin\varphi)\alpha_2), \qquad \text{where } c = \|\beta\|.$$

In particular, let $\theta$ be the angle from $\alpha_1$ to $\alpha_1'$ in this sense. Then, in (1.28), we have

$$a_{11} = \cos\theta, \qquad a_{21} = \sin\theta.$$

Let $\omega$ be the angle from $\alpha_1$ to $\alpha_2'$ in the sense above. There are two cases:

*a.* $\omega = \theta + \pi/2$ [if $\theta + \pi/2 > 2\pi$, $\omega = (\theta + \pi/2) - 2\pi$] as in Fig. 20.

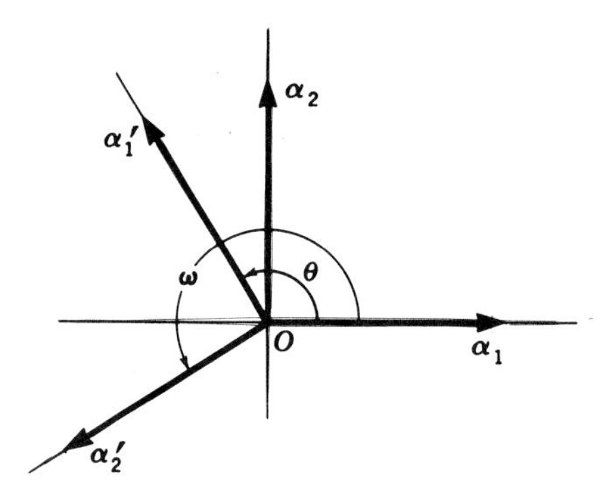


*Fig. 20*

*b*. $\omega = \theta + 3\pi/2$ [if $\theta + 3\pi/2 > 2\pi$, $\omega = (\theta + 3\pi/2) - 2\pi$] as in Fig. 21.

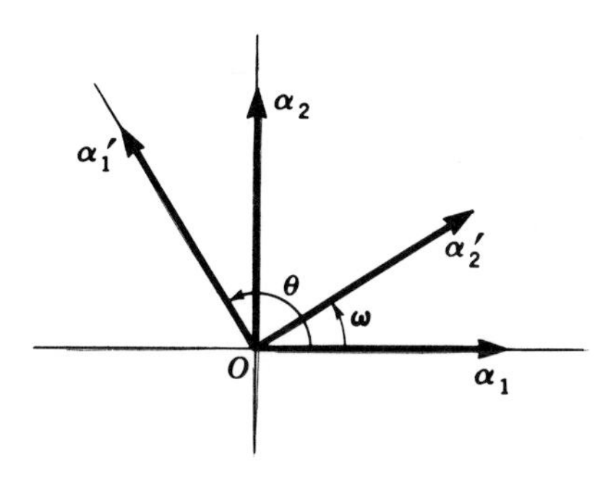


***Fig. 21***

In case *a* we have in (1.28)

$$a_{12} = \cos\left(\theta + \frac{\pi}{2}\right) = -\sin\theta,$$

$$a_{22} = \sin\left(\theta + \frac{\pi}{2}\right) = \cos\theta.$$

In case *b* we have

$$a_{12} = \cos\left(\theta + \frac{3\pi}{2}\right) = \sin\theta,$$

$$a_{22} = \sin\left(\theta + \frac{3\pi}{2}\right) = -\cos\theta.$$

Thus the matrix **A** in (1.30) is equal to

$$\mathbf{R}(\theta) = \begin{bmatrix} \cos\theta & -\sin\theta \\ \sin\theta & \cos\theta \end{bmatrix} \quad \text{in case } a, \tag{1.38}$$

$$\mathbf{S}(\theta) = \begin{bmatrix} \cos\theta & \sin\theta \\ \sin\theta & -\cos\theta \end{bmatrix} \quad \text{in case } b. \tag{1.39}$$

In particular, for $\theta = 0$, we have

$$\mathbf{S}(0) = \begin{bmatrix} 1 & 0 \\ 0 & -1 \end{bmatrix}.$$

We note

$$\mathbf{S}(\theta) = \mathbf{R}(\theta)\mathbf{S}(0). \tag{1.40}$$

The matrix $\mathbf{S}(0)$ represents the change of axes illustrated in Fig. 22. Thus (1.30) takes the form

$$\begin{bmatrix} x_1 \\ x_2 \end{bmatrix} = \begin{bmatrix} \cos\theta & -\sin\theta \\ \sin\theta & \cos\theta \end{bmatrix} \begin{bmatrix} x_1' \\ x_2' \end{bmatrix}, \tag{1.41}$$

which is the rotation of the coordinate axes by angle $\theta$, or

$$(1.42)\qquad \begin{bmatrix} x_1 \\ x_2 \end{bmatrix} = \begin{bmatrix} \cos\theta & -\sin\theta \\ \sin\theta & \cos\theta \end{bmatrix} \begin{bmatrix} 1 & 0 \\ 0 & -1 \end{bmatrix} \begin{bmatrix} x_1 \\ x_2 \end{bmatrix},$$

which is the change of direction of the second axis followed by the rotation (1.41).

Finally, a general change of rectangular coordinate systems is given by (1.33), where the matrix $\mathbf{A} = \begin{bmatrix} a_{11} & a_{12} \\ a_{21} & a_{22} \end{bmatrix}$ is either $\mathbf{R}(\theta)$ or $\mathbf{S}(\theta)$.

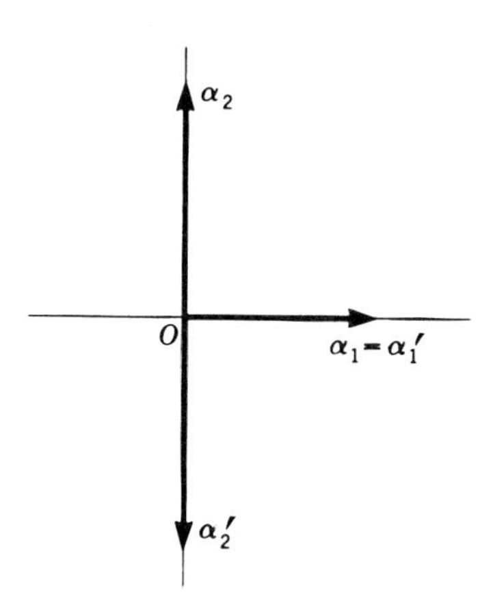


*Fig. 22*

***Example 1.13.*** For any angles $\theta$ and $\omega$, we have

$$\mathbf{R}(\theta)\mathbf{R}(\omega) = \mathbf{R}(\theta + \omega).$$

This is obvious in view of the geometric interpretation of $\mathbf{R}(\theta)$ as the rotation by angle $\theta$ of the coordinate axes. The identity is equivalent to the following formulas in trigonometry:

$$\cos(\theta + \omega) = \cos\theta\cos\omega - \sin\theta\sin\omega,$$

$$\sin(\theta + \omega) = \sin\theta\cos\omega + \cos\theta\sin\omega.$$

***Example 1.14.*** Fixing a rectangular coordinate system $\{x_1,x_2\}$, the equation

$$(1.43)\qquad \begin{bmatrix} y_1 \\ y_2 \end{bmatrix} = \mathbf{R}(\theta) \begin{bmatrix} x_1 \\ x_2 \end{bmatrix}$$

may be interpreted as a point transformation $f$ of the plane which maps a point $(x_1,x_2)$ into $(y_1,y_2)$ given by (1.43). The distance between the points remains unchanged by this transformation, namely, $d(f(A),f(B)) = d(A,B)$. ($f$ is called the *rotation* of the plane by angle $\theta$.) This property can be verified as follows. If $A = (a_1,a_2)$, $B = (b_1,b_2)$, then (1.25) implies

$$d(A,B)^2 = (b_1 - a_1)^2 + (b_2 - a_2)^2.$$

By the same formula, we have

$$\begin{aligned} d(f(A),f(B))^2 &= ((b_1\cos\theta - b_2\sin\theta) - (a_1\cos\theta - a_2\sin\theta))^2 \\ &\qquad + ((b_1\sin\theta + b_2\cos\theta) - (a_1\sin\theta + a_2\cos\theta))^2 \\ &= ((b_1 - a_1)\cos\theta - (b_2 - a_2)\sin\theta)^2 \\ &\qquad + ((b_1 - a_1)\sin\theta + (b_2 - a_2)\cos\theta)^2 \\ &= (b_1 - a_1)^2 + (b_2 - a_2)^2, \end{aligned}$$

so that $d(f(A),f(B)) = d(A,B)$.

**EXERCISE 1.4**

**1.** Find the inverse of

»(*a*) $\begin{bmatrix} 2 & 0 \\ 0 & 1 \end{bmatrix}$; (*b*) $\begin{bmatrix} 1 & 2 \\ 0 & 1 \end{bmatrix}$; »(*c*) $\begin{bmatrix} -1 & 0 \\ 0 & 1 \end{bmatrix}$.

**2.** Does $\begin{bmatrix} 3 & -2 \\ -6 & 4 \end{bmatrix}$ have an inverse?

» **3.** Show that A $= \begin{bmatrix} a & b \\ c & d \end{bmatrix}$ has an inverse if and only if $ad - bc \neq 0$.

**4.** Show that the inverse of $\mathbf{R}(\theta)$ is $\mathbf{R}(2\pi - \theta)$.

» **5.** Write down $\mathbf{R}(\theta)$ for $\theta = \pi/3, \pi/2, \pi$.

**6.** With respect to a rectangular coordinate system, consider a point transformation $f$ which maps $(x_1,x_2)$ upon $(x_1 + a_1, x_2 + a_2)$, where $a_1$, $a_2$ are given numbers. Show that the distance between the points remains unchanged by $f$. ($f$ is called a *translation.*)

» **7.** The equation $x_1^2/9 + x_2^2/4 = 1$ represents an ellipse with respect to a rectangular coordinate system. Let $O = (0,0)$, $A_1 = (3,0)$, $A_2 = (0,2)$. What is the equation of the given ellipse with respect to the affine coordinate system based on $O$, $\overrightarrow{OA_1}$, $\overrightarrow{OA_2}$?

» **8.** Let $\{x_1,x_2\}$ be a rectangular coordinate system and consider the figure $S$ consisting of all points whose coordinates satisfy the equation $x_1^2 - 2x_1 + x_2^2 - 4x_2 = 20$. Taking a new rectangular coordinate system $x_1'$, $x_2'$ such that

$$x_1' = x_1 - 1, \qquad x_2' = x_2 - 2,$$

express $S$ by a new equation. What is $S$ geometrically?

# 2 Vector Spaces

Assuming familiarity with the real and complex number systems $\mathcal{R}$ and $\mathcal{C}$, we first define vector spaces $\mathcal{R}^n$ and $\mathcal{C}^n$, which will be the models of $n$-dimensional vector spaces over $\mathcal{R}$ and $\mathcal{C}$, respectively. We then proceed to an axiomatic definition of vector space and introduce related basic concepts.

With the notion of a field, which will be introduced in Chap. 3, discussion of vector spaces over an arbitrary field will present no further difficulty. Those who wish to study vector spaces over an arbitrary field from the beginning should insert Sec. 5.1 perhaps between Secs. 2.2 and 2.3. In Definition 2.1, as elsewhere, $\mathcal{F}$ can be an arbitrary field.

## 2.1 $\mathcal{R}^n$ AND $\mathcal{C}^n$

Let $\mathcal{R}$ and $\mathcal{C}$ denote the real and complex number systems, respectively. Let $n$ be an arbitrary but fixed positive integer. We shall denote by $\mathcal{R}^n$ the set of all ordered $n$-tuples $(a_1, \ldots ,a_n)$ of real numbers. Each $n$-tuple is often denoted by a single Greek letter, say $\alpha$, so that we write $\alpha = (a_1, \ldots ,a_n)$ or, more briefly, $\alpha = (a_i)$. Two ordered $n$-tuples $\alpha = (a_i)$ and $\beta = (b_i)$ are regarded as equal to each other if and only if $a_i = b_i$ for every $i$, $1 \leq i \leq n$. Similarly, we shall denote by $\mathcal{C}^n$ the set of all ordered $n$-tuples $(a_1, \ldots ,a_n)$, where $a_1, \ldots , a_n$ are complex numbers, and use the same sort of notation $\alpha = (a_1, \ldots ,a_n)$ or $\alpha = (a_i)$.

From this point on, our discussions go quite parallelly for $\mathcal{R}^n$ and $\mathcal{C}^n$. We shall therefore use the letter $\mathcal{F}$ to denote either $\mathcal{R}$ or $\mathcal{C}$ as we choose, and denote by $\mathcal{F}^n$ the set of all $n$-tuples $(a_1, \ldots ,a_n)$ of elements of $\mathcal{F}$. Instead of speaking of real or complex numbers each time, it is convenient to speak of *scalars* when we refer to elements of $\mathcal{F}$. On the other hand, each element $\alpha = (a_i)$ of $\mathcal{F}^n$ is called a *vector*; the scalar $a_i$ is called the $i$th *component* of the vector $\alpha$. Thus each vector $\alpha$ has $n$ components and is determined by these $n$ components.

We now define two algebraic operations, addition and scalar multiplication, in the set $\mathcal{F}^n$.

*Addition:* For $\alpha = (a_i)$ and $\beta = (b_i)$ in $\mathcal{F}^n$, we denote by $\alpha + \beta$ the vector $(a_i + b_i)$, that is, the vector whose $i$th component is $a_i + b_i$ for each $i$, $1 \leq i \leq n$. $\alpha + \beta$ is called the *sum* of $\alpha$ and $\beta$.

*Scalar multiplication:* For $\alpha = (a_i)$ in $\mathcal{F}^n$ and for $c$ in $\mathcal{F}$, we denote by $c\alpha$ the vector $(ca_i)$, that is, the vector whose $i$th component is $ca_i$ for each $i$, $1 \leq i \leq n$. $c\alpha$ is called a *scalar multiple* of $\alpha$.

In this way we have defined addition, which associates to any pair of vectors $\alpha$ and $\beta$ a third vector $\alpha + \beta$ [or, what amounts to the same thing, a mapping $(\alpha,\beta) \in \mathcal{F}^n \times \mathcal{F}^n \rightarrow \alpha + \beta \in \mathcal{F}^n$], and scalar multiplication, which associates to a scalar $c$ and a vector $\alpha$ a second vector $c\alpha$ [or, what amounts to the same thing, a mapping $(c,\alpha) \in \mathcal{F} \times \mathcal{F}^n \rightarrow c\alpha \in \mathcal{F}^n$].

We shall denote by 0 the $n$-tuple $(0, \ldots, 0)$ and call it the *zero vector.* It is now quite straightforward to prove the following

***Proposition 2.1.*** *Addition and scalar multiplication in $\mathcal{F}^n$ have the following properties:*

**1.** $\alpha + \beta = \beta + \alpha$ *(commutativity).*
**2.** $(\alpha + \beta) + \gamma = \alpha + (\beta + \gamma)$ *(associativity).*
**3.** $\alpha + 0 = \alpha$ *for any vector* $\alpha$.
**4.** *For any* $\alpha \in \mathcal{F}^n$, *there is a vector denoted by* $-\alpha$ *with* $\alpha + (-\alpha) = 0$; *in fact, if* $\alpha = (a_i)$, *then* $-\alpha = (-a_i)$.
**5.** $c(\alpha + \beta) = c\alpha + c\beta$.
**6.** $(c + d)\alpha = c\alpha + d\alpha$.
**7.** $(cd)\alpha = c(d\alpha)$.
**8.** $1\alpha = \alpha$.

*In all the identities,* $\alpha, \beta, \gamma$ *are arbitrary vectors and* $c, d$ *are arbitrary scalars.*

**Definition 2.1**

The set $\mathcal{F}^n$ with addition and scalar multiplication as defined above is called the *n-dimensional standard vector space over* $\mathcal{F}$. Of course, according as $\mathcal{F} = \mathcal{R}$ or $\mathcal{C}$, we shall denote the corresponding vector space by $\mathcal{F}^n$ (*n-dimensional standard real vector space*) or $\mathcal{C}^n$ (*n-dimensional standard complex vector space*).

In the vector space $\mathcal{F}^n$, the sum of vectors can be defined for any number of vectors. For $k$ vectors $\alpha_1, \ldots, \alpha_k$, the sum $\alpha_1 + \cdots + \alpha_k$ is the vector whose $i$th component is equal to the sum of the $i$th components of $\alpha_1, \ldots, \alpha_k$. We can easily verify that

$$\alpha_1 + \cdots + \alpha_k = (\alpha_1 + \cdots + \alpha_{k-1}) + \alpha_k.$$

It is also clear that in taking the sum of any number of vectors the order of the vectors does not matter; for example,

$$\alpha_1 + \alpha_2 + \alpha_3 = \alpha_2 + \alpha_1 + \alpha_3 = \alpha_3 + \alpha_2 + \alpha_1.$$

By a *linear combination* of $\alpha_1, \ldots, \alpha_k$ we mean a vector of the form $c_1\alpha_1 + \cdots + c_k\alpha_k$, where $c_1, \ldots, c_k$ are $k$ arbitrary scalars, called the *coefficients* of the linear combination.

We shall often use the summation notation. For any number of vectors $\alpha_1, \ldots, \alpha_k$, the sum $\alpha_1 + \cdots + \alpha_k$ is denoted by $\sum_{i=1}^{k} \alpha_i$ and a linear combination $c_1\alpha_1 + \cdots + c_k\alpha_k$ by $\sum_{i=1}^{k} c_i\alpha_i$.

In $\mathcal{F}^n$ we define $n$ special vectors $\epsilon_1, \ldots, \epsilon_n$ as follows. For each $i$, $\epsilon_i$ is the vector whose components are all 0 except for the $i$th component, which is equal to 1. The set of vectors $\{\epsilon_1, \ldots, \epsilon_n\}$ is called the *standard basis* of $\mathcal{F}^n$ for the following reason

***Proposition 2.2.*** *Every vector $\alpha$ in $\mathcal{F}^n$ can be expressed as a linear combination of $\epsilon_1, \ldots, \epsilon_n$ in a unique manner (that is, the set of coefficients is uniquely determined by $\alpha$).*

*Proof.* Given any vector $\alpha = (a_1, a_2, \ldots, a_n)$ in $\mathcal{F}^n$, we have

$$\alpha = \sum_{i=1}^{n} a_i\epsilon_i,$$

where the coefficient for $\epsilon_i$ is nothing but the $i$th component of $\alpha$. The uniqueness of this representation follows from the definition of equality for vectors in $\mathcal{F}^n$.

**EXERCISE 2.1**

›› **1.** In $\mathcal{R}^3$, calculate the following linear combinations: $2(1,0,4) + (-4)(2,1,5)$; $3(2,-1,-2) + 2(-3,3/2,3)$.

**2.** In $\mathcal{C}^2$, calculate $i(2, -1 + 3i) + (3 - i)(1 + i, 4)$.

›› **3.** In $\mathcal{R}^3$, write $(1,0,0)$ as a linear combination of $(0,1,1)$, $(1,0,1)$, and $(1,1,0)$.

**4.** In $\mathcal{F}^n$, prove that for any vectors $\alpha$ and $\beta$ there is one and only one vector $\xi$ such that $\alpha + \xi = \beta$. (This vector is denoted by $\beta - \alpha$.) Show that $\beta - \alpha = \beta + (-\alpha)$.

## 2.2 VECTOR SPACES

Before proceeding to an axiomatic definition of vector space we shall define the concept of additive group.

**Definition 2.2**

By an *additive group* we mean a set $V$ together with a rule, called *addition*, which associates to any pair of elements $a$ and $b$ in $V$ a unique element denoted by $a + b$ in $V$ so as to satisfy the following conditions:

**1.** $a + b = b + a$.

**2.** $(a + b) + c = a + (b + c)$.
**3.** There exists a certain element, denoted by 0, such that $a + 0 = a$ for every $a \in V$.
**4.** For any $a \in V$ there exists an element $a' \in V$ such that $a + a' = 0$.

***Example 2.1.*** The set of all integers $\mathbb{Z}$ is an additive group where addition is the usual one. The set of all rational numbers $\mathbb{Q}$ is also an additive group with respect to the usual addition. Similarly, $\mathcal{R}$ and $\mathcal{C}$ are additive groups.

***Example 2.2.*** The vector space $\mathfrak{F}^n$ is an additive group (when we just consider vector addition and forget about scalar multiplication).

A vector space, which we shall soon define in an abstract way, is an additive group with a further property, as illustrated by Example 2.2. For this reason we shall first study some of the basic properties of an arbitrary additive group.

***Proposition 2.3.*** *Let $V$ be an additive group.*
**1.** *There exists one and only one element* 0 *in $V$ such that $a + 0 = a$ for every $a \in V$.* (This unique element is called the *zero element* of $V$.)
**2.** *For any $a \in V$ there exists one and only one element $a' \in V$ such that $a + a' = 0$.* (This unique element is called the *inverse* of $a$ and is denoted by $-a$.)
**3.** *For any $a, b \in V$ there exists one and only one element $x \in V$ such that $a + x = b$.* (This element is called the *difference of $b$ and $a$* and is denoted by $b - a$.)
**4.** *If $a + x = a$ for some $a$, then $x = 0$.*

*Proof.* 1. We know that there is at least one such element 0. Suppose there is another element $0'$ with the same property. Then $0 + 0' = 0$. By condition 1 of Definition 2.2, we have $0 + 0' = 0' + 0 = 0'$, so that $0 = 0'$.

2. We know that there is at least one such element $a'$ (condition 4 of Definition 2.2). Suppose that $a''$ is another such element. Using condition 2 of Definition 2.2, we have $a' + 0 = a' + (a + a'') = (a' + a) + a'' = 0 + a''$ and hence $a' = a''$.

3. Suppose there is an element $x$ such that $a + x = b$. By adding $-a$ to both sides, we get $(-a) + (a + x) = (-a) + b$. By associativity (that is, condition 2 of Definition 2.2), we have $(-a) + (a + x) = ((-a) + a) + x = 0 + x = x$, so that $x = b + (-a)$. Conversely, the element $b + (-a)$ actually satisfies $a + x = b$, because

$$\begin{aligned} a + (b + (-a)) &= a + ((-a) + b) = (a + (-a)) + b \\ &= 0 + b = b. \end{aligned}$$

4. By (3) of this proposition, there is only one element $x$ such that $a + x = a$, and, in fact, $x = a + (-a) = 0$.

***Remark.*** For any $a \in V$, it is clear that $-(-a) = a$.

In an additive group $V$ we shall define the sum of any number of elements $a_1, \ldots, a_k$ inductively† by

$$(2.1) \qquad a_1 + \cdots + a_k = (a_1 + \cdots + a_{k-1}) + a_k.$$

We denote this sum by $\sum_{i=1}^{k} a_i$.

***Proposition 2.4.*** *Let $a_1, \ldots, a_{k+m}$ be arbitrary elements in an additive group $V$. Then*

$$(2.2) \qquad \left(\sum_{i=1}^{k} a_i\right) + \left(\sum_{i=k+1}^{k+m} a_i\right) = \sum_{i=1}^{k+m} a_i.$$

*Proof.* We use induction† on $m$. For $m = 1$, formula (2.2) is nothing but (2.1.) Assuming that (2.2) is valid for some $m$, we shall prove (2.2) for $m + 1$. We have

$$\begin{aligned} \left(\sum_{i=1}^{k} a_i\right) + \left(\sum_{i=k+1}^{k+m+1} a_i\right) &= \left(\sum_{i=1}^{k} a_i\right) + \left(\left(\sum_{i=k+1}^{k+m} a_i\right) + a_{k+m+1}\right) \\ &= \left(\left(\sum_{i=1}^{k} a_i\right) + \left(\sum_{i=k+1}^{k+m} a_i\right)\right) + a_{k+m+1} \\ &= \left(\sum_{i=1}^{k+m} a_i\right) + a_{k+m+1} = \sum_{i=1}^{k+m+1} a_i. \end{aligned}$$

We also observe that the sum $\sum_{i=1}^{k} a_i$ is independent of the order of the elements $a_1, \ldots, a_k$, as the following examples illustrate:

$$\begin{aligned} a_2 + a_1 + a_4 + a_3 &= (a_2 + a_1) + (a_4 + a_3) \\ &= (a_1 + a_2) + (a_3 + a_4) = a_1 + a_2 + a_3 + a_4; \\ a_4 + a_1 + a_3 + a_2 &= a_4 + (a_1 + a_3 + a_2) \\ &= (a_1 + a_3 + a_2) + a_4 \\ &= a_1 + a_3 + a_2 + a_4. \end{aligned}$$

Now we shall define the notion of a vector space. Let $\mathcal{F}$ be $\mathcal{R}$ or $\mathcal{C}$ as in Sec. 2.1 ($\mathcal{F}$ can be an arbitrary field, which will be defined in Sec. 5.1).

**Definition 2.3**

A *vector space over* $\mathcal{F}$ is an additive group $V$ for which there is, furthermore, a mapping $(c,\alpha) \in \mathcal{F} \times V \to c\alpha \in V$, called *scalar multiplication,* which has the following properties:

**1.** $c(\alpha + \beta) = c\alpha + c\beta$.

† For "induction" and "inductive definition" see Appendix.

**2.** $(c + d)\alpha = c\alpha + d\alpha$.
**3.** $(cd)\alpha = c(d\alpha)$.
**4.** $1\alpha = \alpha$.

Here $\alpha, \beta$ are arbitrary elements in $V$ and $c, d$ are arbitrary elements in $\mathcal{F}$.

As in the case of $\mathcal{F}^n$, elements of $\mathcal{F}$ are called *scalars* and elements of $V$ *vectors*. When $\mathcal{F} = \mathcal{R}$, $V$ is called a *real vector space*; and when $\mathcal{F} = \mathcal{C}$, $V$ is called a *complex vector space*.

All the properties of an additive group are, of course, valid for a vector space with respect to addition. The zero element 0 of $V$, called the *zero vector*, is a vector such that $\alpha + 0 = \alpha$ for every $\alpha \in V$. We are using the symbol 0 for this zero vector as well as for the scalar zero, but this should not cause any confusion. We now prove some more basic rules concerning a vector space; for example, we have $0\alpha = 0$ for any $\alpha \in V$. In this equation, 0 on the left-hand side is the scalar 0 and 0 on the right-hand side is the zero vector.

***Proposition 2.5.*** *Let $V$ be a vector space over $\mathcal{F}$.*
**1.** $0\alpha = 0$ *for every* $\alpha \in V$.
**2.** $c\,0 = 0$ *for every* $c \in \mathcal{F}$.
**3.** *If* $c\alpha = 0$, *then* $c = 0$ *or* $\alpha = 0$.
**4.** *For any* $c \in \mathcal{F}$ *and* $\alpha \in V$, $(-c)\alpha = -(c\alpha)$.

*Proof.* 1. $0\alpha = (0 + 0)\alpha = 0\alpha + 0\alpha$ by condition 2 of Definition 2.3. By (4) of Proposition 2.3 we have $0\alpha = 0$.

2. $c\,0 = c\,(0 + 0) = c\,0 + c\,0$ by condition 1 of Definition 2.3. Again by (4) of Proposition 2.3 we have $c\,0 = 0$.

3. Suppose $c\alpha = 0$ and $c \neq 0$. Multiplying $c\alpha = 0$ by $c^{-1}$, we have

$$0 = c^{-1}(c\alpha) = (c^{-1}c)\alpha = 1\alpha = \alpha$$

by conditions 3 and 4 of Definition 2.3.

4. $c\alpha + (-c)\alpha = ((c + (-c))\alpha = 0\alpha = 0$ by condition 2 of Definition 2.3 and (1) of this proposition. Recalling (2) of Proposition 2.3, we have $(-c)\alpha = -(c\alpha)$.

**Definition 2.4**

Given a finite number of vectors $\alpha_1, \ldots, \alpha_k$, a vector of the form $\sum_{i=1}^{k} c_i\alpha_i$ is called a *linear combination* of $\alpha_1, \ldots, \alpha_k$ with *coefficients* $c_1, \ldots, c_k \in \mathcal{F}$.

We now give some examples of vector spaces.

***Example 2.3.*** $\mathcal{F}^n$ is a vector space over $\mathcal{F}$.

***Example 2.4.*** Let $\mathfrak{F} = \mathfrak{R}$ or $\mathfrak{C}$ (or, more generally, an arbitrary field). Denote by $\mathfrak{F}^\infty$ the set of all sequences $\alpha = (a_0, a_1, a_2, \ldots, a_n, \ldots)$, where $a_i$'s are elements of $\mathfrak{F}$ and are 0 except for a finite number of them. For $\alpha = (a_i)$ and $\beta = (b_i)$, we define

$$\alpha + \beta = (a_0 + b_0, a_1 + b_1, a_2 + b_2, \ldots, a_n + b_n, \ldots)$$

and

$$c\alpha = (ca_0, ca_1, ca_2, \ldots, ca_n, \ldots),$$

where $c \in \mathfrak{F}$. Then these sequences $\alpha + \beta$ and $c\alpha$ are again elements of $\mathfrak{F}^\infty$. It is easy to verify that $\mathfrak{F}^\infty$ is a vector space over $\mathfrak{F}$ with respect to these operations. The zero vector is $(0,0,0, \ldots, 0, \ldots)$.

***Example 2.5.*** For each $\alpha = (a_i)$ in $\mathfrak{F}^\infty$, we consider the expression of a usual polynomial

$$f_\alpha(x) = a_0 + a_1x + a_2x^2 + \cdots + a_nx^n,$$

where $a_n$ is the last nonzero term of the sequence $\alpha = (a_i)$. It is obvious that, conversely, any polynomial expression with coefficients in $\mathfrak{F}$ corresponds to an element of $\mathfrak{F}^\infty$ in this fashion. One can also easily verify that $\alpha + \beta$ and $c\alpha$ correspond to the sum $f_\alpha(x) + f_\beta(x)$ of two polynomials and the scalar multiple $cf_\alpha(x)$, respectively. This means that the set of all polynomials with coefficients in $\mathfrak{F}$ is a vector space over $\mathfrak{F}$, which is essentially the same vector space as $\mathfrak{F}^\infty$.

***Example 2.6.*** Let $A$ be an arbitrary set. Consider the set $\mathfrak{F}^A$ of all functions on $A$ with values in $\mathfrak{F}$. For two such functions $f$ and $g$, we define

$$(f + g)(x) = f(x) + g(x), \qquad (cf)(x) = cf(x)$$

where $x \in A$ and $c \in \mathfrak{F}$. With respect to these operations $\mathfrak{F}^A$ is a vector space over $\mathfrak{F}$. The zero vector is the function which is identically 0 on $A$.

***Example 2.7.*** By specifying the set $A$ and also by restricting the nature of functions in Example 2.6, we can obtain many examples of a vector space.

(*a*) Let $A = \mathfrak{R}$ and $\mathfrak{F} = \mathfrak{R}$. Then we have a real vector space consisting of all real-valued functions on the real line. Let $A$ be the unit interval $[0,1]$ and $\mathfrak{F} = \mathfrak{R}$. Then we have a real vector space consisting of all real-valued functions on the unit interval.

(*b*) Let $A = \mathfrak{R}$ (or $[0,1]$) and $\mathfrak{F} = \mathfrak{R}$ and consider only those functions which are continuous. Since the functions $f(x) + g(x)$ and $cf(x)$ are continuous together with $f(x)$ and $g(x)$, we obtain a real vector space consisting of all real-valued continuous functions on $\mathfrak{R}$ (or $[0,1]$).

(*c*) Let $A$ be the set of natural numbers and consider the set of all functions on $A$ with values in $\mathfrak{F}$ each of which has the following property: the values $f(n)$ are zero except for a finite number of natural numbers $n$. This set becomes a vector space over $\mathfrak{F}$. With any such function $f$, we associate $(a_i) \in \mathfrak{F}^\infty$, where $a_i = f(i)$. It is obvious that our vector space is essentially the same as the vector space $\mathfrak{F}^\infty$ in Example 2.4.

***Example 2.8.*** Another special case of Example 2.6 is the following. Let $A$ be the set of all pairs $(i,j)$ where $i$ and $j$ are integers such that $1 \leq i \leq m$, $1 \leq j \leq n$, and $m$ and $n$ are two fixed positive integers. A function $f$ on $A$ with values in $\mathfrak{F}$ can be expressed by writing down all the values $a_{ij} = f(i,j)$ for the pairs $(i,j) \in A$ in the form

$$\mathbf{A} = \begin{bmatrix} a_{11} & \cdots & a_{1n} \\ a_{21} & \cdots & a_{2n} \\ \cdots & \cdots & \cdots \\ a_{m1} & \cdots & a_{mn} \end{bmatrix}, \qquad \text{or} \qquad \mathbf{A} = [a_{ij}]. \tag{2.3}$$

A display $\mathbf{A}$ of $mn$ elements of $\mathfrak{F}$ in this kind of form is called an $m \times n$ *matrix over* $\mathfrak{F}$. For each pair $(i,j)$, $1 \leq i \leq m$, $1 \leq j \leq n$, the element $a_{ij}$ is called the $(i,j)$ *component* of the matrix **A**. The set of all $m \times n$ matrices over $\mathfrak{F}$ will be denoted by $\mathfrak{F}^m{}_n$. The addition and scalar multiplication in the vector space $\mathfrak{F}^A$ will correspond to the following operations in the set $\mathfrak{F}^m{}_n$:

For $\mathbf{A} = [a_{ij}]$ and $\mathbf{B} = [b_{ij}]$, $\mathbf{A} + \mathbf{B}$ is the matrix $[c_{ij}]$, where $c_{ij} = a_{ij} + b_{ij}$ for each pair $(i,j)$.

For $\mathbf{A} = [a_{ij}]$ and $c \in \mathfrak{F}$, $c\mathbf{A}$ is the matrix whose $(i,j)$ component is equal to $ca_{ij}$.

The zero element of the vector space is the $m \times n$ *zero matrix* $\mathbf{O}_{m,n}$, namely, the $m \times n$ matrix whose components are all 0. In the special case where $m = 1$, a $1 \times n$ matrix is simply a vector $(a_1, \ldots, a_n)$ in $\mathfrak{F}^n$ in our previous notation. On the other hand, when $n = 1$, an $m \times 1$ matrix is of the form

$$\begin{bmatrix} a_1 \\ \cdot \\ \cdot \\ \cdot \\ a_m \end{bmatrix}. \tag{2.4}$$

We may observe that $\mathfrak{F}^1{}_n$ and $\mathfrak{F}^n{}_1$ are different expressions of the same vector space $\mathfrak{F}^n$; that is, a vector $(a_1, \ldots, a_n)$ in $\mathfrak{F}^n$ can be expressed either by a vector $[a_1, \ldots, a_n] \in \mathfrak{F}^1{}_n$ or by a vector of the form (2.4) in $\mathfrak{F}^n{}_1$.

For an $m \times n$ matrix (2.3), the vectors in $\mathfrak{F}^1{}_n$,

$$[a_{11}, \ldots, a_{1n}], \quad \ldots, \quad [a_{m1}, \ldots, a_{mn}]$$

are called the *row vectors*. Similarly, the vectors in $\mathfrak{F}^m{}_1$,

$$\begin{bmatrix} a_{11} \\ \cdot \\ \cdot \\ \cdot \\ a_{m1} \end{bmatrix}, \quad \ldots, \quad \begin{bmatrix} a_{1n} \\ \cdot \\ \cdot \\ \cdot \\ a_{mn} \end{bmatrix}$$

are called the *column vectors*.

## EXERCISE 2.2

**1.** In an additive group, prove the following identities:

$$-(x - y) = y - x; \qquad -\left(\sum_{i=1}^{k} x_i\right) = \sum_{i=1}^{k} (-x_i).$$

☆ **2.** In an additive group $V$ we make the following definition. For any $x \in V$ and for any positive integer $n$, define $nx$ inductively by

$$1x = x \qquad \text{and} \qquad nx = (n-1)x + x.$$

For a negative integer $n$, define

$$nx = -((-n)x),$$

where $-n$ is a positive integer and hence $(-n)x$ has already been defined. Denoting by $\mathbb{Z}$ the set of all integers, show that the mapping $(n,x) \in \mathbb{Z} \times V \to nx \in V$ has the following properties:

$$n(x + y) = nx + ny; \qquad (m + n)x = mx + nx; \qquad (mn)x = m(nx); \qquad 1x = x.$$

(Note that these are precisely the same formal properties assumed for scalar multiplication in Definition 2.3. We do not say, however, that $V$ is a vector space over $\mathbb{Z}$; $\mathbb{Z}$ is not a field.)

**3.** In a vector space $V$ over $\mathfrak{F}$, prove the following identities:

$$a(b\alpha + c\beta) = (ab)\alpha + (ac)\beta; \qquad a(\alpha - \beta) = a\alpha - a\beta,$$

where $a, b, c \in \mathfrak{F}$ and $\alpha, \beta \in V$.

»  **4.** In the set $\mathfrak{R}^3$, keep the usual addition and define

$$c(a_1,a_2,a_3) = (ca_1,ca_2,2ca_3)$$

for $c \in \mathfrak{R}$. Do we obtain a vector space? Explain.

**5.** In the vector space $\mathfrak{F}^m{}_n$, let $\mathbf{E}_{i,j}$ be the matrix whose components are all 0 except for the $(i,j)$ component, which is equal to 1. Prove that every matrix in $\mathfrak{F}^m{}_n$ is a linear combination of $mn$ elements $\mathbf{E}_{i,j}$, $1 \leq i \leq m$, $1 \leq j \leq n$.

**6.** Let $V$ be a complex vector space. By keeping the same addition and defining scalar multiplication $c\alpha$, $c \in \mathfrak{R}$, as the restriction of scalar multiplication $c\alpha$, $c \in \mathfrak{C}$, we obtain a vector space over $\mathfrak{R}$. Verify this statement.

»  **7.** In the complex vector space $\mathfrak{C}^n$, let $\{\epsilon_1, \ldots, \epsilon_n\}$ be the standard basis. Let $\epsilon_{n+1} = i\epsilon_1$, $\epsilon_{n+2} = i\epsilon_2$, $\ldots$, $\epsilon_{2n} = i\epsilon_n$, where $i = \sqrt{-1}$. Show that every vector in $\mathfrak{C}^n$ can be expressed as a linear combination of $\epsilon_1, \ldots, \epsilon_{2n}$ with real numbers as coefficients and that such an expression is unique. (For the special case $n = 1$, this simply means that every complex number can be expressed uniquely as $a + bi$, where $a$ and $b$ are real numbers.)

☆ **8.** To obtain another example of an additive group, let $m$ be an arbitrary positive integer and denote by $\mathbb{Z}_m$ the set of all residues modulo $m$, namely, $\mathbb{Z}_m = \{0, 1, 2, \ldots, m - 1\}$. For any two elements $p$ and $q$ in $\mathbb{Z}_m$, define $p + q$ to be the residue of $p + q$ modulo $m$ so that $p + q \in \mathbb{Z}_m$. Show that $\mathbb{Z}_m$ is an additive group with respect to this operation.

## 2.3 SUBSPACES

Let $V$ be a vector space over $\mathfrak{F}$.

### Definition 2.5

A nonempty subset $W$ of $V$ is called a *subspace* (or *linear subspace*) of $V$ if it has the following properties:

**1.** For $\alpha, \beta \in W$, $\alpha + \beta \in W$.

**2.** For $\alpha \in W$ and $c \in \mathfrak{F}$, $c\alpha \in W$.

A subspace $W$ contains the zero vector 0 of $V$. In fact, let $\alpha \in W$. Then $0 = 0\alpha \in W$ by property 2. We also have $-\alpha = (-1)\alpha \in W$. Thus a subspace $W$ is itself a vector space over $\mathfrak{F}$ with respect to addition and scalar multiplication of $V$ restricted to $W$.

The subset consisting of 0 alone is clearly a subspace of $V$. This subspace, denoted by (0) or 0, and $V$ itself are called *trivial subspaces* of $V$. Any other subspace is said to be *proper*.

***Example 2.9.*** Let $1 \leq m \leq n$. Let $W$ be the set of all vectors in $\mathfrak{F}^n$ whose last $n - m$ components are 0. $W$ is then a subspace of $\mathfrak{F}^n$.

***Example 2.10.*** In the vector space $V$ of all real-valued functions on the unit interval $[0,1]$, let $W_1$ be the set of all $f \in V$ such that $f(0) = 0$. $W_1$ is a (proper) subspace of $V$. The set $W_2$ of all continuous functions $f \in V$ is also a (proper) subspace of $V$.

***Proposition 2.6.*** *Let $S$ be an arbitrary nonempty subset of a vector space $V$ over $\mathfrak{F}$. The set of all linear combinations $\sum c_i\alpha_i$, where $\alpha_i$'s are a finite number of elements taken from $S$ and $c_i$'s are arbitrary scalars, is a subspace of $V$.*

*Proof.* This is obvious from

$$\left(\sum_{i=1}^{k} c_i\alpha_i\right) + \left(\sum_{i=k+1}^{k+m} c_i\alpha_i\right) = \sum_{i=1}^{k+m} c_i\alpha_i$$

and

$$c\left(\sum_{i=1}^{k} c_i\alpha_i\right) = \sum_{i=1}^{k} (cc_i)\alpha_i.$$

***Corollary.*** *Let $\alpha_1, \ldots, \alpha_k \in V$. A linear combination of $\beta_1, \ldots, \beta_m$, where each $\beta_i$ is a linear combination of $\alpha_1, \ldots, \alpha_k$, is a linear combination of $\alpha_1, \ldots, \alpha_k$.*

*Proof.* This is a special case where $S = \{\alpha_1, \ldots, \alpha_k\}$.

**Definition 2.6**

For a nonempty subset $S$ of $V$, the subspace consisting of all linear combinations of any finite number of elements in $S$ is called the *subspace spanned by* $S$, or the *span of* $S$, and is denoted by $\mathrm{Sp}(S)$. When $S = \{\alpha_1, \ldots, \alpha_k\}$, we shall denote $\mathrm{Sp}(S)$ by $\mathrm{Sp}\{\alpha_1, \ldots, \alpha_k\}$.

***Proposition 2.7.*** *Let $V$ be a vector space over $\mathfrak{F}$.*

**1.** *If $W_\lambda$, $\lambda \in \Lambda$, is any indexed family† of subspaces of $V$, then the intersection $\bigcap_\lambda W_\lambda$ is a subspace of $V$.*

**2.** *Let $S$ be a nonempty subset of $V$. Then $\mathrm{Sp}(S)$ is the smallest subspace of $V$ containing $S$; that is, if $W$ is a subspace containing $S$, then $W$ contains $\mathrm{Sp}(S)$.*

*Proof.* 1. Let $\alpha, \beta \in \bigcap_\lambda W_\lambda$. For every $\lambda$, we have $\alpha, \beta \in W_\lambda$. Since $W_\lambda$ is a subspace, we have $\alpha + \beta \in W_\lambda$ and $c\alpha \in W_\lambda$ for any scalar $c$. Thus $\alpha + \beta \in \bigcap_\lambda W_\lambda$ and $c\alpha \in \bigcap_\lambda W_\lambda$, proving that $\bigcap_\lambda W_\lambda$ is a subspace.

2. Let $W$ be a subspace containing $S$. For any elements $\alpha_1, \ldots, \alpha_k \in S$, we have $\alpha_1, \ldots, \alpha_k \in W$ and hence any linear combination of $\alpha_1, \ldots, \alpha_k$ is in $W$. Since $\mathrm{Sp}(S)$ is the set of all linear combination of elements of $S$, we see that $\mathrm{Sp}(S) \subseteq W$.

† See Appendix.

**Corollary.** *For a finite number of subspaces $W_1, \ldots, W_k$ of $V$, the intersection $\bigcap_{i=1}^{k} W_i$ is a subspace. The set*

$$W = \left\{\sum_{i=1}^{k} \alpha_i;\ \alpha_i \in W_i,\ 1 \leq i \leq k\right\}$$

*is a subspace of $V$* (called the *sum* of $W_1, \ldots, W_k$ and denoted by $W_1 + \cdots + W_k$).

**EXERCISE 2.3**

» 1. In $\mathcal{R}^n$, determine which of the following subsets are subspaces:
   (*a*) $\{\alpha = (x_i);\ x_1 + \cdots + x_n = 0\}$.
   (*b*) $\{\alpha = (x_i);\ x_1 + \cdots + x_n = 1\}$.
   (*c*) $\{\alpha = (x_i);\ x_1 > 0\}$.
   (*d*) $\{\alpha = (x_i);\ x_1 = 0\}$.
   (*e*) $\{\alpha = (x_i)$; all $x_i$'s are rational numbers$\}$.

2. Let $W$ be the set of all $2 \times 2$ matrices of the form $\begin{bmatrix} x & z \\ -z & y \end{bmatrix}$, where $x$, $y$, and $z$ are arbitrary elements of $\mathfrak{F}$. Prove that $W$ is a subspace of $\mathfrak{F}^2_2$.

» 3. When $W_1$ and $W_2$ are subspaces of a vector space $V$, is the union $W_1 \cup W_2$ (set of all vectors which belong to $W_1$ or $W_2$) a subspace?

» 4. The set $\mathcal{R}^n$ can be considered as a subset of the complex vector space $\mathcal{C}^n$. Is it a subspace of $\mathcal{C}^n$? What is $\mathrm{Sp}(\mathcal{R}^n)$ in $\mathcal{C}^n$?

5. Let $W$ be a subspace of a vector space $V$ over $\mathfrak{F}$. For $\alpha, \beta \in W$, we shall write $\alpha \equiv \beta \bmod W$ if $\beta - \alpha \in W$. Prove that this relation, called *congruence* mod $W$, has the following properties:
   (*a*) $\alpha \equiv \alpha \bmod W$ for every $\alpha \in V$.
   (*b*) $\alpha \equiv \beta \bmod W$ implies $\beta \equiv \alpha \bmod W$.
   (*c*) $\alpha \equiv \beta \bmod W$ and $\beta \equiv \gamma \bmod W$ imply $\alpha \equiv \gamma \bmod W$.
   (*d*) $\alpha_1 \equiv \beta_1 \bmod W$ and $\alpha_2 \equiv \beta_2 \bmod W$ imply $\alpha_1 + \alpha_2 \equiv \beta_1 + \beta_2 \bmod W$.
   (*e*) $\alpha \equiv \beta \bmod W$ implies $c\alpha \equiv c\beta \bmod W$, where $c \in \mathfrak{F}$.

6. Let $\alpha \approx \beta$ be a relation among elements in a vector space $V$ which has the properties *a* to *e* in number 5. Prove that

$$W = \{\alpha;\ \alpha \approx 0\}$$

is a subspace and that the given relation $\alpha \approx \beta$ is equivalent to $\alpha \equiv \beta \bmod W$.

7. Let $V$ be a vector space and let $\alpha_1, \ldots, \alpha_{k+1}, \beta$ be elements in $V$. If $\beta \in \mathrm{Sp}\{\alpha_1, \ldots, \alpha_{k+1}\}$ and if $\beta \notin \mathrm{Sp}\{\alpha_1, \ldots, \alpha_k\}$, prove that $\alpha_{k+1} \in \mathrm{Sp}\{\alpha_1, \ldots, \alpha_k, \beta\}$.

## 2.4 LINEAR INDEPENDENCE; BASIS

Let $V$ be a vector space over $\mathfrak{F}$.

### Definition 2.7

A set of elements $\{\alpha_1, \ldots, \alpha_k\}$ in $V$ is said to be *linearly independent* if $\sum_{i=1}^{k} c_i\alpha_i = 0$ implies that $c_1 = \cdots = c_k = 0$. Otherwise, the set is said to be *linearly dependent*.

In other words, if $\alpha_1, \ldots, \alpha_k$ are linearly dependent, then there exist $c_1, \ldots, c_k$ in $\mathcal{F}$ which are not all 0 such that $\sum_{i=1}^{k} c_i\alpha_i = 0$.

***Example 2.11.*** In $\mathcal{R}^3$, $\epsilon_1$, $\epsilon_2$, $\epsilon_3$ of the standard basis are linearly independent, because $\sum_{i=1}^{3} c_i\epsilon_i = (c_1,c_2,c_3)$ is 0 if and only if $c_1 = c_2 = c_3 = 0$. On the other hand, the vectors

$$\alpha_1 = (1,2,-3), \qquad \alpha_2 = (2,-1,1), \qquad \alpha_3 = (-1,8,-11)$$

are linearly dependent, since $3\alpha_1 + (-2)\alpha_2 + (-1)\alpha_3 = 0$.

For one vector $\alpha$, $\alpha$ is linearly independent if and only if $\alpha \neq 0$. In fact, if $\alpha \neq 0$ and $c\alpha = 0$, then $c = 0$. On the other hand, if $\alpha = 0$, then $1\alpha = 0$, so that $\alpha$ is linearly dependent. It is also clear that if $\alpha_1, \ldots, \alpha_k$ are linearly independent, then $\alpha_1, \ldots, \alpha_m$, where $1 \leq m < k$, are linearly independent.

**Definition 2.8**

More generally, let $S$ be an arbitrary subset of $V$. We say that $S$ is *linearly independent* if any finite set of elements in $S$ are linearly independent. (For a finite set $S$, this coincides with the definition already given.) Otherwise, $S$ is said to be *linearly dependent.*

The following proposition is easy to prove, but useful.

***Proposition 2.8.*** *Let $S$ and $S'$ be subsets of a vector space $V$ such that $S \subseteq S'$.*

**1.** *If $S'$ is linearly independent, so is $S$.*

**2.** *If $S$ is linearly dependent, so is $S'$.*

**Definition 2.9**

A subset $S$ of $V$ is called a *basis* of $V$ if it has the following two properties:

**1.** $S$ is linearly independent.

**2.** $\mathrm{Sp}(S) = V$, in other words, every $\alpha \in V$ is a linear combination of a finite number of vectors in $S$.

***Example 2.12.*** The standard basis $\{\epsilon_1, \ldots, \epsilon_n\}$ of the vector space $\mathcal{F}^n$ is a basis in the sense of Definition 2.9.

***Example 2.13.*** In the vector space $\mathcal{F}^\infty$, let $\epsilon_n$ be the element $(0, \ldots, 0,1,0, \ldots)$, where 1 is the $n$th component and all other components are 0. Then $\{\epsilon_n\}$, $n = 1, 2, \ldots$, is a basis. In terms of the vector space of all polynomials with coefficients in $\mathcal{F}_i$, the set of elements $\{1,x,x^2, \ldots, x^n, \ldots\}$ is a basis.

***Remark.*** Here and in the subsequent discussions, we shall exclude the trivial vector space consisting of the zero vector only. When we define the notion of dimension after Theorem 2.11, we shall consider the dimension of the trivial vector space to be 0.

***Proposition 2.9.*** *A set of elements* $\{\alpha_1, \ldots, \alpha_n\}$ *of* $V$ *forms a basis if and only if every element of* $V$ *can be expressed uniquely as a linear combination of* $\alpha_1, \ldots, \alpha_n$.

*Proof.* Assume that $\alpha_1, \ldots, \alpha_n$ form a basis. We know that every element can be expressed as a linear combination of $\alpha_1, \ldots, \alpha_n$. Suppose an element $\alpha \in V$ is written in two ways:

$$\alpha = \sum_{i=1}^{n} a_i\alpha_i = \sum_{i=1}^{n} b_i\alpha_i.$$

Then we have

$$\sum_{i=1}^{n} (b_i - a_i)\alpha_i = 0.$$

Since $\alpha_1, \ldots, \alpha_n$ are linearly independent, we must have $b_i - a_i = 0$, that is, $b_i = a_i$ for every $i$.

Conversely, assume that $\{\alpha_1, \ldots, \alpha_n\}$ has the property that every element of $V$ can be uniquely expressed as a linear combination of $\alpha_1, \ldots, \alpha_n$. It is clear that $\mathrm{Sp}\{\alpha_1, \ldots, \alpha_n\} = V$. In order to prove linear independence, suppose

$$\sum_{i=1}^{n} c_i\alpha_i = 0.$$

Since the zero vector 0 can be written as

$$0 = \sum_{i=1}^{n} 0\alpha_i,$$

the uniqueness assumption implies that all $c_i$'s are 0. Hence $\alpha_1, \ldots, \alpha_n$ are linearly independent. Thus $\{\alpha_1, \ldots, \alpha_n\}$ is a basis.

A vector space $V$ is said to be *finite-dimensional* (or *of finite dimension*) if there is a basis consisting of a finite number of elements. For example, $\mathcal{F}^n$ is finite-dimensional since it has the standard basis $\{\epsilon_1, \ldots, \epsilon_n\}$, whereas the vector space $\mathcal{F}^\infty$ is not finite-dimensional.

***Remark.*** Let $V$ be a finite-dimensional vector space. A basis $\{\alpha_1, \ldots, \alpha_n\}$ is called an *ordered basis* when we consider the order of the elements which make up the basis. Thus two ordered bases $\{\alpha_1, \ldots, \alpha_n\}$ and $\{\beta_1, \ldots, \beta_n\}$ are equal if $\alpha_1 = \beta_1, \ldots, \alpha_n = \beta_n$. However, we shall simply speak of a basis unless we wish to emphasize the order.

We now proceed to prove that if $V$ is finite-dimensional, then every basis of $V$ is indeed finite and has the same number of elements.

We first prove:

***Lemma.*** *Let* $\beta$ *be a linear combination of* $\alpha_1, \ldots, \alpha_n$: $\beta = \sum_{i=1}^{n} c_i\alpha_i$ *and* $c_1 \neq 0$. *Then we have*

$$\mathrm{Sp}\,\{\alpha_1, \alpha_2, \ldots, \alpha_n\} = \mathrm{Sp}\,\{\beta, \alpha_2, \ldots, \alpha_n\}.$$

*Proof.* By the corollary to Proposition 2.6 we have

$$\text{Sp}\ \{\beta,\alpha_2,\ .\ .\ .\ ,\alpha_n\} \subseteq \text{Sp}\ \{\alpha_1,\alpha_2,\ .\ .\ .\ ,\alpha_n\}.$$

On the other hand, from $\beta = \sum_{i=1}^{n} c_i\alpha_i$, $c_1 \neq 0$, we have

$$\alpha_1 = \frac{1}{c_1}\beta - \sum_{i=2}^{n} \frac{c_i}{c_1}\alpha_i.$$

Thus any linear combination of $\alpha_1, \alpha_2, \ldots, \alpha_n$ is a linear combination of $\beta, \alpha_2, \ldots, \alpha_n$ if we replace $\alpha_1$ by the expression above. Thus

$$\text{Sp}\ \{\alpha_1,\alpha_2,\ .\ .\ .\ ,\alpha_n\} \subseteq \text{Sp}\ \{\beta,\alpha_2,\ .\ .\ .\ ,\alpha_n\}.$$

***Theorem 2.10.*** *Let V be a vector space over $\mathfrak{F}$. Assume that*
**1.** $\beta_1, \ldots, \beta_m \in \text{Sp}\{\alpha_1, \ldots, \alpha_n\}$.
**2.** $\beta_1, \ldots, \beta_m$ *are linearly independent.*
*Then* $m \leq n$.

*Proof.* Assuming that $m > n$, we shall derive a contradiction. The element $\beta_1$ is a linear combination of $\alpha_1, \ldots, \alpha_n$: $\beta_1 = \sum_{i=1}^{n} c_i\alpha_i$. Since $\beta_1 \neq 0$, at least one of the coefficients $c_i$ is not 0. By rearranging the indices, we may assume that $c_1 \neq 0$. By the lemma above, we have

$$\text{Sp}\ \{\beta_1,\alpha_2,\ .\ .\ .\ ,\alpha_n\} = \text{Sp}\ \{\alpha_1,\alpha_2,\ .\ .\ .\ ,\alpha_n\}.$$

Since $\beta_2$ belongs to this subspace, it is a linear combination of $\beta_1, \alpha_2, \ldots, \alpha_n$: $\beta_2 = a\beta_1 + \sum_{i=2}^{n} c_i\alpha_i$. Not all $c_i$'s are 0, because otherwise we would have $\beta_2 = a\beta_1$ contrary to the second assumption. Again by rearranging the indices, we may assume that $c_2 \neq 0$. Again by the lemma we have

$$\text{Sp}\ \{\beta_1,\beta_2,\alpha_3,\ .\ .\ .\ ,\alpha_n\} = \text{Sp}\ \{\alpha_1,\alpha_2,\ .\ .\ .\ ,\alpha_n\}.$$

Now assume that we have proved for some $k < m$

$$\text{Sp}\ \{\beta_1,\ .\ .\ .\ ,\beta_k,\alpha_{k+1},\ .\ .\ .\ ,\alpha_n\} = \text{Sp}\ \{\alpha_1,\ .\ .\ .\ ,\alpha_k,\alpha_{k+1},\ .\ .\ .\ ,\alpha_n\}.$$

The vector $\beta_{k+1}$ belonging to this subspace is a linear combination of $\beta_1, \ldots, \beta_k, \alpha_{k+1}, \ldots, \alpha_n$ but not a linear combination of $\beta_1, \ldots, \beta_k$, since $\beta_1, \ldots, \beta_{k+1}$ are linearly independent. Thus, by rearranging the indices, we may assume that

$$\beta_{k+1} = \sum_{i=1}^{k} a_i\beta_i + \sum_{i=k+1}^{n} c_i\alpha_i, \qquad \text{where } c_{k+1} \neq 0.$$

By the lemma it follows that

$$\begin{aligned}\text{Sp}\ \{\beta_1,\beta_2,\ .\ .\ .\ ,\beta_{k+1},\alpha_{k+2},\ .\ .\ .\ ,\alpha_n\} &= \text{Sp}\ \{\beta_1,\beta_2,\ .\ .\ .\ ,\beta_k,\alpha_{k+1},\ .\ .\ .\ ,\alpha_n\}\\ &= \text{Sp}\ \{\alpha_1,\alpha_2,\ .\ .\ .\ ,\alpha_n\}.\end{aligned}$$

Since we are assuming $m > n$, we may finally arrive at the set $\{\beta_1,\beta_2, \ldots ,\beta_n\}$ which spans $\mathrm{Sp}\{\alpha_1, \ldots ,\alpha_n\}$. But this is a contradiction, since then $\beta_{n+1}$ would be a linear combination of $\beta_1, \ldots, \beta_n$.

We now prove the following important theorem.

***Theorem 2.11.*** *Let $V$ be a vector space over $\mathcal{F}$ of finite dimension and let $\{\alpha_1, \ldots ,\alpha_n\}$ be a basis of $V$. Then every basis is finite and has exactly $n$ elements.*

*Proof.* Let $S$ be any basis of $V$. Since $S$ cannot contain more than $n$ elements which are linearly independent by Theorem 2.10, it follows that $S$ must be finite. Let $S = \{\beta_1, \ldots ,\beta_m\}$. Again by Theorem 2.10, we have $m \leq n$. Now interchanging the roles of $\{\alpha_i\}$ and $\{\beta_i\}$, we have $n \leq m$ by the same theorem. Hence $m = n$.

Thus for a finite-dimensional vector space $V$ the number of elements in any basis is the same. This number is called the *dimension* of $V$. If $V$ has a basis consisting of $n$ vectors, then every basis of $V$ has exactly $n$ vectors; $V$ is said to be *$n$-dimensional* and we write $\dim V = n$. [For the trivial vector space $V = (0)$, we consider $\dim V = 0$, as we remarked earlier.] For example, $\dim \mathcal{F}^n = n$ (cf. Example 2.12).

***Theorem 2.12.*** *Let $V$ be a vector space over $\mathcal{F}$. If*

$$V = \mathrm{Sp}\,\{\alpha_1, \ldots ,\alpha_m\},$$

*then we may choose a subset of $\{\alpha_1, \ldots ,\alpha_m\}$ which is a basis. In particular, $V$ is finite-dimensional and* $\dim V \leq m$.

*Proof.* We may assume that none of $\alpha_i$'s is 0. If $\alpha_2$ is a linear combination of $\alpha_1$ (that is, a scalar multiple of $\alpha_1$), then we omit $\alpha_2$. Otherwise, consider $\{\alpha_1,\alpha_2\}$. If $\alpha_3$ is a linear combination of $\alpha_1$ and $\alpha_2$, then we omit $\alpha_3$. Otherwise, we consider $\{\alpha_1,\alpha_2,\alpha_3\}$. Proceeding in this way, we obtain a subset $\{\alpha_{i_1}, \ldots ,\alpha_{i_n}\}$, where $1 = i_1 < i_2 < \cdots < i_n \leq m$, such that each $\alpha_{i_k}$ is not a linear combination of the preceding elements and such that $\mathrm{Sp}\,\{\alpha_{i_1}, \ldots ,\alpha_{i_n}\} = V$. We have only to show that $\alpha_{i_1}, \ldots, \alpha_{i_n}$ are linearly independent; this will follow from the following lemma.

***Lemma.*** *If $\alpha_1, \ldots, \alpha_k$ are linearly dependent, then there is an $s$, $1 < s \leq k$, such that $\alpha_s$ is a linear combination of $\alpha_1, \ldots, \alpha_{s-1}$.*

*Proof.* Let $\sum_{i=1}^{k} c_i\alpha_i = 0$, where not all $c_i$'s are 0. Let $c_s$ be the last nonzero coefficient among $c_1, \ldots, c_k$, so that $\sum_{i=1}^{s} c_i\alpha_i = 0$ with $c_s \neq 0$. Then we have

$$\alpha_s = \sum_{i=1}^{s-1} \frac{-c_i}{c_s}\,\alpha_i.$$

***Corollary.*** *Any finite set of elements* $\alpha_1, \ldots, \alpha_m$ *in a vector space* $V$ *has a maximal subset of linearly independent vectors. The number of vectors in such a subset is uniquely determined.*

*Proof.* Apply Theorem 2.12 to the subspace $\text{Sp}\{\alpha_1, \ldots, \alpha_m\}$. A subset of $\alpha_1, \ldots, \alpha_m$ which forms a basis of this subspace is, of course, linearly independent. Now let this subset be $\{\alpha_{i_1}, \ldots, \alpha_{i_n}\}$. Then $n = \dim \text{Sp}\{\alpha_1, \ldots, \alpha_m\}$. If $\beta$ is any element of the subspace, then it is a linear combination of $\alpha_{i_1}, \ldots, \alpha_{i_n}$, and hence $\alpha_{i_1}, \ldots, \alpha_{i_n}, \beta$ are not linearly independent. This shows that $\{\alpha_{i_1}, \ldots, \alpha_{i_n}\}$ is a maximal linearly independent subset of the given set $\{\alpha_1, \ldots, \alpha_m\}$.

Theorem 2.12 says that we can extract a basis from any finite set of vectors spanning a vector space. On the other hand, given a finite set of linearly independent vectors in a finite-dimensional vector space, we can enlarge it to a basis. More specifically, we have:

***Theorem 2.13.*** *Let* $V$ *be an n-dimensional vector space over* $\mathfrak{F}$. *Given any linearly independent set* $\alpha_1, \ldots, \alpha_m$ *in* $V$, *we may find* $\alpha_{m+1}, \ldots, \alpha_n$ *so that* $\{\alpha_1, \ldots, \alpha_m, \alpha_{m+1}, \ldots, \alpha_n\}$ *is a basis of* $V$.

*Proof.* If every vector is a linear combination of $\alpha_1, \ldots, \alpha_m$, then these elements form a basis of $V$. Otherwise, let $\alpha_{m+1}$ be a vector which is not in $\text{Sp}\{\alpha_1, \ldots, \alpha_m\}$. Then $\alpha_1, \ldots, \alpha_m, \alpha_{m+1}$ are linearly independent by the lemma for Theorem 2.12; in fact, otherwise, since $\alpha_1, \ldots, \alpha_m$ are linearly independent, $\alpha_{m+1}$ would be a linear combination of $\alpha_1, \ldots, \alpha_m$. If $\text{Sp}\{\alpha_1, \ldots, \alpha_{m+1}\} = V$, then $\alpha_1, \ldots, \alpha_{m+1}$ form a basis. Otherwise, we pick an element $\alpha_{m+2} \notin \text{Sp}\{\alpha_1, \ldots, \alpha_{m+1}\}$, so that $\alpha_1, \ldots, \alpha_{m+2}$ are linearly independent. Since $V$ does not contain more than $n$ linearly independent vectors by Theorem 2.10, it follows that we arrive at a set $\{\alpha_1, \ldots, \alpha_m, \alpha_{m+1}, \ldots, \alpha_n\}$ which forms a basis.

**EXERCISE 2.4**

» **1.** In a vector space $V$ over $\mathfrak{F}$, let $\alpha, \beta, \gamma$ be linearly independent. Are the following sets of vectors linearly independent?

$$\{\alpha + \beta - \gamma, \beta + \gamma, 2\alpha\}; \qquad \{\alpha + 2\beta + 3\gamma, 2\alpha, -\alpha + 2\beta + 3\gamma\}.$$

**2.** In a vector space $V$ over $\mathfrak{F}$, assume that $\alpha_1, \ldots, \alpha_n$ are linearly independent but that $\beta, \alpha_1, \ldots, \alpha_n$ are linearly dependent for each $\beta \in V$. Prove that $\alpha_1, \ldots, \alpha_n$ form a basis of $V$.

» **3.** Consider $\mathcal{C}^n$ as a vector space over $\mathcal{R}$ by restricting scalar multiplication only to real numbers. What is the dimension of this vector space? Exhibit one of the bases. (See Exercise 2.2, number 7.)

» **4.** In $\mathfrak{F}^n{}_n$, let $W$ be the set of all matrices $\mathbf{A} = [a_{ij}]$ such that $a_{ij} = 0$ for $i < j$. Show that $W$ is a subspace and find dim $W$.

**5.** In a vector space $V$ over $\mathfrak{F}$, assume that $\text{Sp}\{\alpha_1, \ldots, \alpha_n\} = V$ but that $\text{Sp}\{\alpha_1, \ldots, \hat{\alpha}_i, \ldots, \alpha_n\} \neq V$ for any $i$. (Recall that the symbol ^ means that the element marked is omitted.) Show that $\alpha_1, \ldots, \alpha_n$ form a basis of $V$.

»☆ **6.** Let $V$ be an $n$-dimensional vector space over $\mathfrak{F}$. Show that the following three conditions on a set of $n$ vectors in $V$ are mutually equivalent:

(*a*) They form a basis of $V$; (*b*) they are linearly independent; (*c*) they span $V$.

☆ **7.** Let $\alpha_1, \ldots, \alpha_n$ be $n$ elements in a vector space $V$ over $\mathfrak{F}$. If they are linearly dependent and if any $n - 1$ vectors among them are linearly independent, then there exist $c_1, \ldots, c_n \in \mathfrak{F}$, none of which is 0, such that $\sum_{i=1}^{n} c_i\alpha_i = 0$. If $d_1, \ldots, d_n \in \mathfrak{F}$ satisfy the same requirements, then there is a $k \neq 0$ in $\mathfrak{F}$ such that $d_i = kc_i$, $1 \leq i \leq n$.

## 2.5 DIMENSIONS OF SUBSPACES

We shall prove a few basic theorems concerning the dimensions of subspaces.

***Theorem 2.14.*** *Let $W$ be a subspace of an $n$-dimensional vector space $V$ over $\mathfrak{F}$. Then*

**1.** *There exists a basis $\{\alpha_1, \ldots, \alpha_m, \alpha_{m+1}, \ldots, \alpha_n\}$ of $V$ such that $\{\alpha_1, \ldots, \alpha_m\}$ is a basis of $W$; in particular, $W$ is finite-dimensional and* $\dim W \leq n$.

**2.** $\dim W = \dim V$ *if and only if* $W = V$.

*Proof.* 1. If $W = (0)$, the assertion is trivial. If $W \neq (0)$, let $\alpha_1$ be a nonzero vector in $W$. If $W = \text{Sp}\{\alpha_1\}$, then $\alpha_1$ is a basis of $W$ and it can be enlarged to a basis of $V$ by virtue of Theorem 2.13. If $W \neq \text{Sp}\{\alpha_1\}$, we choose an $\alpha_2 \in W$, which is not in $\text{Sp}\{\alpha_1\}$. Then $\alpha_1, \alpha_2$ are linearly independent. If $W = \text{Sp}\{\alpha_1,\alpha_2\}$, then $\{\alpha_1,\alpha_2\}$ is a basis of $W$ and it can be enlarged to a basis of $V$, again by virtue of Theorem 2.13. Otherwise, we continue the same process by picking up an $\alpha_3 \in W$ which is not in $\text{Sp}\{\alpha_1,\alpha_2\}$. Since $\dim V = n$, there are not more than $n$ linearly independent vectors in $V$ (a fortiori, in $W$). This means that there exist $\alpha_1, \ldots, \alpha_m$ in $W$ which form a basis of $W$, where $1 \leq m \leq n$. Enlarging this to a basis $\{\alpha_1, \ldots, \alpha_m, \alpha_{m+1}, \ldots, \alpha_n\}$ of $V$, we have (1).

2. Choose a basis $\{\alpha_1, \ldots, \alpha_m, \alpha_{m+1}, \ldots, \alpha_n\}$ as in (1), where $m = \dim W$ and $n = \dim V$. If $\dim W = \dim V$, then $\{\alpha_1, \ldots, \alpha_n\}$ is actually a basis of $W$ and hence $W = V$. The converse is obvious.

***Theorem 2.15.*** *Let $W_1$ and $W_2$ be finite-dimensional subspaces of a vector space $V$ over $\mathfrak{F}$. Then both $W_1 + W_2$ and $W_1 \cap W_2$ are finite-dimensional and*

(2.5) $\dim (W_1 + W_2) = \dim W_1 + \dim W_2 - \dim (W_1 \cap W_2)$.

*Proof.* We first note that $W_1 \cap W_2$ is a subspace of $W_1$ and $W_2$ and that $W_1$ and $W_2$ are subspaces of $W_1 + W_2$. Let $\dim W_1 = r$ and $\dim W_2 = s$. Since $W_1 \cap W_2$ is a subspace of $W_1$, it is finite-dimensional by Theorem 2.14. Let $t = \dim (W_1 \cap W_2)$, so that $t \leq r$ and $t \leq s$. Let

$\{\alpha_1, \ldots, \alpha_t\}$ be a basis of $W_1 \cap W_2$. By virtue of Theorem 2.13 we can take $\beta_1, \ldots, \beta_{r-t}$ and $\gamma_1, \ldots, \gamma_{s-t}$ such that $\{\alpha_1, \ldots, \alpha_t, \beta_1, \ldots, \beta_{r-t}\}$ and $\{\alpha_1, \ldots, \alpha_t, \gamma_1, \ldots, \gamma_{s-t}\}$ form a basis of $W_1$ and a basis of $W_2$, respectively. All these elements are in $W_1 + W_2$. We shall show that $S = \{\alpha_1, \ldots, \alpha_t, \beta_1, \ldots, \beta_{r-t}, \gamma_1, \ldots, \gamma_{s-t}\}$ is a basis of $W_1 + W_2$; it will then follow that $W_1 + W_2$ is finite-dimensional and

$$\begin{aligned}\dim (W_1 + W_2) &= t + (r - t) + (s - t) = r + s - t \\ &= \dim W_1 + \dim W_2 - \dim (W_1 \cap W_2).\end{aligned}$$

First, we show that every element $\alpha \in W_1 + W_2$ is a linear combination of elements in $S$. Let $\alpha = \beta + \gamma$, where $\beta \in W_1$ and $\gamma \in W_2$. Since $\beta$ is a linear combination of $\alpha_1, \ldots, \alpha_t, \beta_1, \ldots, \beta_{r-t}$ and since $\gamma$ is a linear combination of $\alpha_1, \ldots, \alpha_t, \gamma_1, \ldots, \gamma_{s-t}$, it follows that $\alpha$ is a linear combination of $\alpha_1, \ldots, \alpha_t, \beta_1, \ldots, \beta_{r-t}, \gamma_1, \ldots, \gamma_{s-t}$. Second, we show that these elements are linearly independent. Assume that

$$\left(\sum_{i=1}^{t} x_i\alpha_i\right) + \left(\sum_{j=1}^{r-t} y_j\beta_j\right) + \left(\sum_{k=1}^{s-t} z_k\gamma_k\right) = 0.$$

Denoting by $\alpha$, $\beta$, and $\gamma$ the three terms on the left-hand side, we have $\alpha + \beta + \gamma = 0$. Then $\alpha + \gamma = -\beta$ is in $W_1 \cap W_2$ because $\beta \in W_1$ and $\alpha + \gamma \in W_2$. Since $\alpha \in W_1 \cap W_2$, $\gamma = (\alpha + \gamma) - \alpha$ is in $W_1 \cap W_2$. But this implies that $\gamma = 0$, because $\alpha_1, \ldots, \alpha_t, \gamma_1, \ldots, \gamma_{s-t}$ are linearly independent and $\{\alpha_1, \ldots, \alpha_t\}$ is a basis of $W_1 \cap W_2$. Thus we have $\gamma = 0$ and $\beta = -\alpha$. Since $\alpha_1, \ldots, \alpha_t, \beta_1, \ldots, \beta_{r-t}$ are linearly independent, we have $\alpha = \beta = 0$. Thus all the coefficients $x_i$, $y_j$, and $z_k$ are 0. This proves Theorem 2.15.

**Definition 2.10**

Let $V$ be a vector space. We say that $V$ is the *direct sum* of two subspaces $W_1$ and $W_2$ if $V = W_1 + W_2$ and $W_1 \cap W_2 = (0)$. In this case, we write $V = W_1 \oplus W_2$.

***Proposition 2.16.*** *Let $W_1$ and $W_2$ be subspaces of a vector space $V$.*
**1.** *$V = W_1 \oplus W_2$ if and only if every $\alpha \in V$ can be written uniquely as $\alpha = \alpha_1 + \alpha_2$ with $\alpha_1 \in W_1$ and $\alpha_2 \in W_2$.*
**2.** *If $V = W_1 \oplus W_2$ and if $W_1, W_2$ are finite-dimensional, then $V$ is finite-dimensional and* $\dim V = \dim W_1 + \dim W_2$.

*Proof.* 1. Assume that $V = W_1 \oplus W_2$, that is, $V = W_1 + W_2$ and $W_1 \cap W_2 = (0)$. Every $\alpha \in V$ can be written as $\alpha = \alpha_1 + \alpha_2$ with $\alpha_1 \in W_1$ and $\alpha_2 \in W_2$. Suppose $\alpha = \beta_1 + \beta_2$ with $\beta_1 \in W_1$ and $\beta_2 \in W_2$ at the same time. Then $\alpha_1 - \beta_1 = \beta_2 - \alpha_2$ belongs to $W_1$ and $W_2$. Since $W_1 \cap W_2 = (0)$, we have $\alpha_1 - \beta_1 = \beta_2 - \alpha_2 = 0$, that is, $\alpha_1 = \beta_1$ and $\alpha_2 = \beta_2$.

Conversely, suppose that every $\alpha \in V$ can be written in the form $\alpha = \alpha_1 + \alpha_2$ in a unique way. Then, of course, $V = W_1 + W_2$. To show that $W_1 \cap W_2 = (0)$, assume that $\alpha \in W_1 \cap W_2$. Then $\alpha = \alpha + 0$, where $\alpha \in W_1$ and $0 \in W_2$, and $\alpha = 0 + \alpha$, where $0 \in W_1$ and $\alpha \in W_2$. By the uniqueness, we must have $\alpha = 0$. Thus $W_1 \cap W_2 = (0)$ and $V = W_1 \oplus W_2$.

2. This follows from Theorem 2.15.

The notion of direct sum can be generalized as follows. We say that $V$ is the *direct sum* of subspaces $W_1, \ldots, W_k$ and write $V = W_1 \oplus W_2 \oplus \cdots \oplus W_k$ if the following conditions are satisfied:

1. $V = W_1 + \cdots + W_k$.
2. For each $j$, $1 \leq j \leq k - 1$, we have

$$(W_1 + \cdots + W_j) \cap W_{j+1} = (0).$$

We have:

***Proposition 2.17.*** *Let $W_1, \ldots, W_k$ be subspaces of a vector space $V$.*
**1.** *$V = W_1 \oplus \cdots \oplus W_k$ if and only if every $\alpha \in V$ can be expressed uniquely as $\alpha = \sum_{i=1}^{k} \alpha_i$ with $\alpha_i \in W_i$.*
**2.** *If $V = W_1 \oplus \cdots \oplus W_k$ and if each $W_i$ is finite-dimensional, then $V$ is finite-dimensional and*

$$\dim V = \sum_{i=1}^{k} \dim W_i.$$

*Proof.* We shall prove only the following assertion; the rest is similar to the proof of Proposition 2.16. If $V = W_1 \oplus \cdots \oplus W_k$, then the expression $\alpha = \sum_{i=1}^{k} \alpha_i$ with $\alpha_i \in W_i$ is unique. Suppose we have

$$\alpha = \sum_{i=1}^{k} \alpha_i = \sum_{i=1}^{k} \beta_i, \qquad \text{where } \alpha_i, \beta_i \in W_i.$$

Then we have

$$\beta_k - \alpha_k = \sum_{i=1}^{k-1} (\alpha_i - \beta_i),$$

where the left-hand side belongs to $W_k$ and the right-hand side belongs to $W_1 + \cdots + W_{k-1}$. By condition 2 for $j = k - 1$ in the definition of direct sum, we have

$$\beta_k - \alpha_k = \sum_{i=1}^{k-1} (\alpha_i - \beta_i) = 0.$$

In particular, $\alpha_k = \beta_k$. Now writing

$$\beta_{k-1} - \alpha_{k-1} = \sum_{i=1}^{k-2} (\alpha_i - \beta_i)$$

and using condition 2 for $j = k - 2$, we see that

$$\beta_{k-1} - \alpha_{k-1} = \sum_{i=1}^{k-2} (\alpha_i - \beta_i) = 0,$$

so that $\alpha_{k-1} = \beta_{k-1}$. Continuing this process, we have $\alpha_i = \beta_i$ for every $i$, $1 \leq i \leq k$.

***Example 2.14.*** Let $\{\alpha_1, \ldots, \alpha_n\}$ be a basis of a vector space $V$. Now if $W_1 = \mathrm{Sp}\,\{\alpha_1, \ldots, \alpha_m\}$ and $W_2 = \mathrm{Sp}\,\{\alpha_{m+1}, \ldots, \alpha_n\}$, then $V = W_1 \oplus W_2$. We have also

$$V = \mathrm{Sp}\,\{\alpha_1\} \oplus \mathrm{Sp}\,\{\alpha_2\} \oplus \cdots \oplus \mathrm{Sp}\,\{\alpha_n\}.$$

**EXERCISE 2.5**

›› **1.** Let $W$ be a subspace of an $n$-dimensional vector space $V$. Show that there is a subspace $U$ such that $V = U \oplus W$. Is such a subspace $U$ unique?

**2.** Let $W_1$ and $W_2$ be subspaces of an $n$-dimensional vector space $V$ such that $W_1 \cap W_2 = (0)$. Show that:

(*a*) There is a subspace $U$ such that $V = W_1 \oplus U$ and $U \supseteq W_2$.

(*b*) There is a subspace $W_3$ such that $V = W_1 \oplus W_2 \oplus W_3$.

**3.** Let $W_1$ and $W_2$ be subspaces of an $n$-dimensional vector space $V$ such that $V = W_1 + W_2$. Prove that $V = W_1 \oplus W_2$ if and only if $\dim V = \dim W_1 + \dim W_2$.

›› **4.** In the vector space $\mathfrak{F}^n{}_n$, let $W_1$ be the subspace consisting of all matrices $[a_{ij}]$ such that $a_{ij} = 0$ for $i > j$. Let $W_2$ be the subspace consisting of all matrices $[a_{ij}]$ such that $a_{ij} = 0$ for $i < j$.

(*a*) Describe the subspace $W_1 \cap W_2$.

(*b*) Find $\dim W_1$, $\dim W_2$, and $\dim (W_1 \cap W_2)$.

›› **5.** In $\mathcal{R}^3$ with standard basis $\{\epsilon_1,\epsilon_2,\epsilon_3\}$, let $W_1 = \mathrm{Sp}\,\{\epsilon_1,\epsilon_2\}$, $W_2 = \mathrm{Sp}\,\{\epsilon_3\}$, and $W_3 = \mathrm{Sp}\,\{\epsilon_2 + \epsilon_3\}$. Show that $\mathcal{R}^3 = W_1 + W_2 + W_3$, $W_1 \cap W_2 = W_2 \cap W_3 = W_1 \cap W_3 = (0)$. Is $\mathcal{R}^3 = W_1 \oplus W_2 \oplus W_3$?

**6.** If $V = W_1 \oplus W_2 \oplus W_3$, then does one have $W_2 \cap (W_1 + W_3) = (0)$?

# 3 Linear Mappings and Matrices

We introduce the notion of linear mapping and study basic properties of linear mappings. We interpret matrices as linear mappings of the standard vector space $\mathfrak{F}^m$ into $\mathfrak{F}^n$; on the other hand, linear mappings can be represented by matrices with respect to any bases we choose.

A change of bases can also be represented by a matrix. The effect of such a change on the representation of linear mappings is studied; this will lead to some interesting results on matrices, namely, Theorems 3.14, 3.15 (on row-reduced echelon forms), and 3.17 (on ranks).

We then discuss elementary operations, which are useful in many ways—in particular, in finding a practical method of solving a system of linear equations. Systems of linear equations are studied in close connection with linear mappings.

## 3.1 LINEAR MAPPINGS

Let $V$ and $W$ be vector spaces over $\mathfrak{F}$.

**Definition 3.1**

A mapping $A$ of $V$ into $W$ is called a *linear mapping* if it has the following properties:

**1.** $A(\alpha + \beta) = A(\alpha) + A(\beta)$ for all $\alpha, \beta \in V$.

**2.** $A(c\alpha) = cA(\alpha)$ for all $\alpha \in V$ and $c \in \mathfrak{F}$.

In particular, when $V = W$, we often use the term *linear transformation of* $V$ or *linear endomorphism of* $V$.

It is easy to observe that conditions 1 and 2 can be combined in one condition as follows: A mapping $A$ of $V$ into $W$ is linear if

$$A(a\alpha + b\beta) = aA(\alpha) + bA(\beta) \qquad \text{for all } \alpha, \beta \in V \text{ and } a, b \in \mathfrak{F}.$$

***Example 3.1.*** The mapping $A(\alpha) = 0$ for every $\alpha \in V$, where 0 is the zero vector of $W$, is a linear mapping, called the *zero mapping*. We shall denote it by 0.

***Example 3.2.*** For any vector space $V$, the mapping $I$ defined by $I(\alpha) = \alpha$ for every $\alpha \in V$ is linear. It is called the *identity transformation* of $V$. For any scalar $c$, the mapping $A$ defined by $A(\alpha) = c\alpha$ for every $\alpha \in V$ is also linear.

***Example 3.3.*** The mapping $A_1$ of $\mathfrak{F}^n$ into $\mathfrak{F}$ defined by $A_1(\alpha) = a_1$ for every $\alpha = (a_i) \in \mathfrak{F}^n$ is linear. More generally, the mapping $A_i$ defined by $A_i(\alpha) = a_i$ is linear for each $i$, $1 \leq i \leq n$.

When $A$ is a linear mapping of $V$ into $W$, we shall often express this fact by simply writing $A : V \to W$. We prove:

***Proposition 3.1.*** *Let $A: V \to W$ be a linear mapping.*

**1.** $A\left(\sum_{i=1}^{k} c_i\alpha_i\right) = \sum_{i=1}^{k} c_iA(\alpha_i)$.

**2.** $A(0) = 0$, *where* 0 *on the left-hand side is the zero vector of $V$ and that on the right-hand side is the zero vector of $W$.*

**3.** *For any subspace $V_1$ of $V$, $A(V_1) = \{A(\alpha); \alpha \in V_1\}$ is a subspace of $W$.*

**4.** *For any subspace $W_1$ of $W$ the set $A^{-1}(W_1) = \{\alpha \in V; A(\alpha) \in W_1\}$ is a subspace of $V$.*

**5.** *$A$ is one-to-one (that is, $A(\alpha) = A(\beta)$ implies $\alpha = \beta$) if and only if $A(\alpha) = 0$ implies $\alpha = 0$.*

**6.** *If $A$ is one-to-one and if $\alpha_1, \ldots, \alpha_k$ are linearly independent in $V$, then $A(\alpha_1), \ldots, A(\alpha_k)$ are linearly independent in $W$.*

*Proof.* Parts 1, 3, and 4 are easy and omitted. As for part 2, we have $A(0) = A(0 + 0) = A(0) + A(0)$. Thus $A(0) = 0$. To prove (5), assume that $A$ is one-to-one and $A(\alpha) = 0$. Since $A(0) = 0$ as in (2), we must have $\alpha = 0$. Conversely, assume that $A(\alpha) = 0$ implies $\alpha = 0$ and that we have $A(\alpha) = A(\beta)$. Then $A(\alpha - \beta) = A(\alpha) - A(\beta) = 0$ and hence $\alpha - \beta = 0$, that is, $\alpha = \beta$. We finally prove (6). Assume that $\sum_{i=1}^{k} c_iA(\alpha_i) = 0$. Then

$$A\left(\sum_{i=1}^{k} c_i\alpha_i\right) = \sum_{i=1}^{k} c_iA(\alpha_i) = 0.$$

Since $A$ is one-to-one, we have $\sum_{i=1}^{k} c_i\alpha_i = 0$ and $c_i = 0$ for every $i$, because $\alpha_1, \ldots, \alpha_k$ are linearly independent.

**Definition 3.2**

For $A: V \to W$, $A(\alpha)$ is called the *image* of $\alpha \in V$ by $A$. The subset $A(S) = \{A(\alpha); \alpha \in S\}$ of $W$ is called the *image* of the subset $S$ of $V$. The subspace $A^{-1}(0) = \{\alpha \in V; A(\alpha) = 0\}$ of $V$ is called the *null space* (or *kernel*) of $A$. The subspace $A(V) = \{A(\alpha); \alpha \in V\}$ of $W$ is called the *range* of $A$.

**Definition 3.3**

If the null space $A^{-1}(0)$ is finite-dimensional, its dimension is called the *nullity* of $A$. If the range $A(V)$ is finite-dimensional, its dimension is called the *rank* of $A$. If $A(V) = W$, we say that $A$ is *surjective* (or $A$ is *onto* $W$). If $A$ is one-to-one, we also say that $A$ is *injective* (or *nonsingular*).

***Example 3.4.*** Let $\mathbf{A} = [a_{ij}]$ be an arbitrary $m \times n$ matrix over $\mathfrak{F}$. We define a mapping of the vector space $\mathfrak{F}^n$ into the vector space $\mathfrak{F}^m$ by

$$\mathbf{x} = \begin{bmatrix} x_1 \\ \cdot \\ \cdot \\ \cdot \\ x_n \end{bmatrix} \to \mathbf{y} = \begin{bmatrix} y_1 \\ \cdot \\ \cdot \\ \cdot \\ y_m \end{bmatrix} = \begin{bmatrix} a_{11} & \cdots & a_{1n} \\ \cdots & \cdots & \cdots \\ a_{m1} & \cdots & a_{mn} \end{bmatrix} \begin{bmatrix} x_1 \\ \cdot \\ \cdot \\ \cdot \\ x_n \end{bmatrix}. \tag{3.1}$$

We can easily verify that this mapping is linear; in fact, it depends on the following two properties of matrix multiplication [cf. Definition 1.3 and (1.6) and (1.7)]

$$\mathbf{A}(\mathbf{x} + \mathbf{y}) = \mathbf{A}\mathbf{x} + \mathbf{A}\mathbf{y}; \qquad \mathbf{A}(c\mathbf{x}) = c\mathbf{A}\mathbf{x}.$$

We shall denote this linear mapping also by $\mathbf{A}$. (Thus we may always consider an $m \times n$ matrix $\mathbf{A}$ over $\mathfrak{F}$ as a linear mapping $\mathbf{A}\colon \mathfrak{F}^m \to \mathfrak{F}^n$ in this manner.) Writing (3.1) in the form

$$y_i = \sum_{j=1}^{n} a_{ij}x_j, \qquad 1 \leq i \leq m, \tag{3.2}$$

we have the following interpretation. Given $\mathbf{y} = (y_i)$ in $\mathfrak{F}^m$, we wish to find all $\mathbf{x} = (x_i)$ in $\mathfrak{F}^n$ which satisfy (3.2). This problem of solving the system of linear equations (3.2) amounts to finding the set $\mathbf{A}^{-1}(\mathbf{y}) \subset \mathfrak{F}^n$ for the linear mapping $\mathbf{A}$. If (3.2) has a solution, it means that $\mathbf{y}$ is in the range of the mapping $\mathbf{A}$.

In particular, suppose $\mathbf{y} = 0$ and consider the system of *homogeneous* linear equations

$$\sum_{j=1}^{n} a_{ij}x_j = 0, \qquad 1 \leq i \leq m. \tag{3.3}$$

The set of all solutions of (3.3) is nothing but the null space of the mapping $\mathbf{A}$.

We prove:

***Proposition 3.2.*** *Let $\{\alpha_1, \ldots, \alpha_n\}$ be a basis of a vector space $V$ over $\mathfrak{F}$. Let $\beta_1, \ldots, \beta_n$ be any $n$ vectors in a vector space $W$ over $\mathfrak{F}$. Then there exists one and only one linear mapping $A$ of $V$ into $W$ such that $A(\alpha_i) = \beta_i$ for every $i$, $1 \leq i \leq n$.*

*Proof.* Assume that there is a linear mapping $A\colon V \to W$ such that $A(\alpha_i) = \beta_i$. For any $\alpha \in V$, we have $\alpha = \sum_{i=1}^{n} c_i\alpha_i$ uniquely, and hence

$$A(\alpha) = A\left(\sum_{i=1}^{n} c_i\alpha_i\right) = \sum_{i=1}^{n} c_iA(\alpha_i) = \sum_{i=1}^{n} c_i\beta_i.$$

This shows that the image of any $\alpha \in V$ is completely determined by the images $\beta_i$ of $\alpha_i$, proving the uniqueness of the mapping $A$. Now the formula

$$A\left(\sum_{i=1}^{n} c_i\alpha_i\right) = \sum_{i=1}^{n} c_i\beta_i$$

actually defines a linear mapping $A: V \to W$, as can be easily verified. We have $A(\alpha_i) = \beta_i$ for each $i$.

***Theorem 3.3.*** *Let $A: V \to W$ be a linear mapping of a finite-dimensional vector space $V$ into a vector space $W$. Then*

$$\text{rank } A + \text{nullity } A = \dim V. \tag{3.4}$$

*Proof.* Let $n = \dim V$ and $s = \text{nullity } A$. We may choose a basis $\{\alpha_1, \ldots, \alpha_s, \alpha_{s+1}, \ldots, \alpha_n\}$ of $V$ such that $\{\alpha_1, \ldots, \alpha_s\}$ is a basis of the null space $N$ of $A$ [(1) of Theorem 2.14]. We shall prove that $A(\alpha_{s+1}), \ldots, A(\alpha_n)$ form a basis of the range $A(V)$; then it will follow that $\text{rank } A = \dim A(V) = n - s$, so that (3.4) is proved. For any $\alpha \in V$, we have $\alpha = \sum_{i=1}^{n} c_i\alpha_i$ and hence

$$A(\alpha) = \sum_{i=1}^{n} c_iA(\alpha_i) = \sum_{i=s+1}^{n} c_iA(\alpha_i),$$

so that $A(V) = \text{Sp}\,\{A(\alpha_{s+1}), \ldots, A(\alpha_n)\}$. To prove that

$$A(\alpha_{s+1}), \ldots, A(\alpha_n)$$

are linearly independent, assume that $\sum_{i=s+1}^{n} c_iA(\alpha_i) = 0$. Then

$$A\left(\sum_{i=s+1}^{n} c_i\alpha_i\right) = 0,$$

so that $\sum_{i=s+1}^{n} c_i\alpha_i$ is in $N$. Since $\{\alpha_1, \ldots, \alpha_s\}$ is a basis of $N$, there exist $d_1, \ldots, d_s$ such that $\sum_{i=s+1}^{n} c_i\alpha_i = \sum_{i=1}^{s} d_i\alpha_i$. Since $\{\alpha_1, \ldots, \alpha_n\}$ is a basis of $V$, we must have $c_i = 0$ (and $d_i = 0$) for all $i$.

**Definition 3.4**

A linear mapping $A: V \to W$ is called a *linear isomorphism of $V$ onto $W$* if it is injective and surjective at the same time. If such an $A$ exists, we say that $V$ is *isomorphic* to $W$.

For a linear isomorphism $A: V \to W$ we can define a linear mapping of $W$ into $V$ (which is also a linear isomorphism) in the following way. For any

$\beta \in W$, there is at least one $\alpha \in V$ such that $A(\alpha) = \beta$, since $A$ is surjective. Such an $\alpha$ is unique, since $A$ is injective. Define $A^{-1}(\beta) = \alpha$. It is easily verified that the mapping $A^{-1}$ so defined is a linear mapping, indeed, a linear isomorphism of $W$ onto $V$. It is called the *inverse* of $A$. From its definition we have

$$A(A^{-1}(\beta)) = \beta \qquad \text{for every } \beta \in W$$

and

$$A^{-1}(A(\alpha)) = \alpha \qquad \text{for every } \alpha \in V.$$

***Example 3.5.*** Suppose that $A: V \to W$ and $B: W \to V$ are linear mappings such that $A(B(\beta)) = \beta$ for every $\beta \in W$ and $B(A(\alpha)) = \alpha$ for every $\alpha \in V$. Then $A$ is a linear isomorphism of $V$ onto $W$ and its inverse is $B$.

***Example 3.6.*** Let both $V$ and $W$ be $n$-dimensional vector spaces. For any basis $\{\alpha_i\}$ of $V$ and $\{\beta_i\}$ of $W$, the linear mapping $A: V \to W$ such that $A(\alpha_i) = \beta_i$, $1 \leq i \leq n$, is a linear isomorphism. (For the existence of $A$, see Proposition 3.2.)

**Theorem 3.4.** *Let $V$ and $W$ be vector spaces over $\mathcal{F}$ of the same dimension $n$. The following conditions for a linear mapping $A: V \to W$ are equivalent:*

1. *$A$ is a linear isomorphism of $V$ onto $W$.*
2. *$A$ is injective.*
3. *$A(\alpha) = 0$ for $\alpha \in V$ implies that $\alpha = 0$.*
4. nullity $A = 0$.
5. rank $A = n$.
6. *$A$ is surjective.*

*Proof.* The implications $(1) \to (2)$, $(2) \to (3)$, and $(3) \to (4)$ are obvious. The implication $(4) \to (5)$ follows from formula (3.4), in which $\dim V = n$. The implication $(5) \to (6)$ follows from (2) of Theorem 2.14. Finally, (6) implies (5) and (4) by virtue of formula (3.4), and hence (1).

**Corollary.** *Let $A: V \to W$ be a linear mapping, where* $\dim V = \dim W$. *If there exists a linear mapping $B: W \to V$ such that $A(B(\beta)) = \beta$ for every $\beta \in W$ or $B(A(\alpha)) = \alpha$ for every $\alpha \in V$, then $A$ is a linear isomorphism and $A^{-1}$ is equal to $B$.*

*Proof.* If $A(B(\beta)) = \beta$ for every $\beta \in W$, then $A$ is surjective. If $B(A(\alpha)) = \alpha$ for every $\alpha \in V$, then $A$ is injective. In either case, $A$ is a linear isomorphism and, obviously, $B = A^{-1}$.

Let $A: V \to W$ and $B: W \to U$ be linear mappings, where $V$, $W$, and $U$ are vector spaces over $\mathcal{F}$. The composed mapping $BA$ defined by $BA(\alpha) = B(A(\alpha))$ for every $\alpha \in V$ is a linear mapping $V \to U$. We call $BA$ the *composite* or *product* of $A$ and $B$ (or, more precisely, product $B$ times $A$, meaning that $B$ follows $A$). If $A$ and $B$ are linear transformations of the same vector space $V$, then both $AB$ and $BA$ are defined. They do not coincide in general, as the following example shows.

***Example 3.7.*** Let $V$ be a vector space with a basis $\{\alpha_1,\alpha_2\}$. Let $A$ and $B$ be linear transformations of $V$ such that

$$A(\alpha_1) = \alpha_2, \qquad B(\alpha_1) = -\alpha_1,$$
and
$$A(\alpha_2) = \alpha_1, \qquad B(\alpha_2) = \alpha_2.$$

Then

$$BA(\alpha_1) = \alpha_2, \qquad AB(\alpha_1) = -\alpha_2,$$
and
$$BA(\alpha_2) = -\alpha_1, \qquad AB(\alpha_2) = \alpha_1,$$

so that $AB \neq BA$.

***Proposition 3.5.*** *Let $A\colon V \to W$ and $B\colon W \to U$ be linear mappings, where $V$, $W$, and $U$ are finite-dimensional. Then*

$$\text{rank } BA \leq \min\{\text{rank } A, \text{rank } B\}.$$

*Proof.* Since $A(V) \subset W$, we have $(BA)(V) = B(A(V)) \subset B(W)$ and $\dim B(A(V)) \leq \dim B(W)$, that is, $\text{rank } BA \leq \text{rank } B$. From Theorem 3.3 applied to the linear mapping $B$ restricted to $A(V)$ [that is, the linear mapping $B\colon A(V) \to W$], it follows that $\dim B(A(V)) \leq \dim A(V)$, which means $\text{rank } BA \leq \text{rank } A$. Hence, $\text{rank } BA \leq \min\{\text{rank } A, \text{rank } B\}$.

Given two vector spaces $V$ and $W$ over $\mathcal{F}$, we shall denote by $\text{Hom}(V,W)$ the set of all linear mappings of $V$ into $W$. In the set $\text{Hom}(V,W)$ we shall define two operations, addition and scalar multiplication, in the following way.

For $A, B \in \text{Hom}(V,W)$, $A + B$ is a mapping defined by

$$(A + B)(\alpha) = A(\alpha) + B(\alpha) \qquad \text{for every } \alpha \in V.$$

For $A \in \text{Hom}(V,W)$ and $c \in \mathcal{F}$, $cA$ is a mapping defined by

$$(cA)(\alpha) = c(A(\alpha)) \qquad \text{for every } \alpha \in V.$$

It is easy to verify that both $A + B$ and $cA$ are linear mappings. For example, we have

$$\begin{aligned}(A + B)(a\alpha + b\beta) &= A(a\alpha + b\beta) + B(a\alpha + b\beta)\\ &= aA(\alpha) + bA(\beta) + aB(\alpha) + bB(\beta)\\ &= aA(\alpha) + aB(\alpha) + bA(\beta) + bB(\beta)\\ &= a(A + B)(\alpha) + b(A + B)(\beta).\end{aligned}$$

It is also straightforward to show that $\text{Hom}(V,W)$ is a vector space over $\mathcal{F}$ with respect to these two operations.

In particular, for a vector space $V$ over $\mathcal{F}$, $\text{Hom}(V,V)$ is a vector space over $\mathcal{F}$. We shall often denote this by $\mathfrak{gl}(V)$. $\mathfrak{gl}(V)$ has a natural rule of multiplication, namely, that of taking the product $AB$ of any two elements $A, B \in \mathfrak{gl}(V)$.

***Proposition 3.6.*** *The set $\mathfrak{gl}(V)$ of all linear transformations of a vector space $V$ over $\mathcal{F}$ is a vector space over $\mathcal{F}$ in which multiplication $(A,B) \to AB$ is so defined as to satisfy the following conditions:*

**1.** $A(B + C) = AB + AC$; $(B + C)A = BA + CA$.

**2.** $(AB)C = A(BC)$.

**3.** $(cA)B = A(cB) = c(AB)$, $c \in \mathcal{F}$.

*Furthermore,*

**4.** $AI = IA = A$, *where $I$ is the identity transformation.*

**5.** *If $A \in \mathfrak{gl}(V)$ is injective or surjective, then it is a linear isomorphism and there is $A^{-1} \in \mathfrak{gl}(V)$ such that $AA^{-1} = A^{-1}A = I$.*

**6.** *If $A$ and $B$ have inverses, then $AB$ has an inverse and $(AB)^{-1} = B^{-1}A^{-1}$.*

*Proof.* The verification of (1), (2), and (3) is straightforward. Condition 4 is obvious, and (5) follows from the corollary to Theorem 3.4. As for (6), we have

$$(B^{-1}A^{-1})(AB) = B^{-1}A^{-1}AB = B^{-1}IB = B^{-1}B = I,$$

and similarly

$$(AB)(BA)^{-1} = I,$$

showing that $AB$ has an inverse and that $(AB)^{-1} = B^{-1}A^{-1}$.

We say that a set such as $\mathfrak{gl}(V)$ having the properties of a vector space together with a multiplication satisfying (1), (2), and (3) is an *algebra over* $\mathcal{F}$.

***Remark 1.*** Condition 2 can be proved more generally in the following form: *If $A: V \to W$, $B: W \to U$, and $C: U \to X$ are linear mappings, then*

$$C(BA) = (CB)A.$$

This is the *associative law* for compositions. Condition 2 is called the *associative law* for multiplication in $\mathfrak{gl}(V)$, and condition 1 is called the *distributive law*.

***Remark 2.*** *For $A \in \mathfrak{gl}(V)$, all the following conditions are equivalent:*

*1. $A$ is injective (or nonsingular).*

*2. $A$ is surjective.*

*3. $A$ is invertible.*

*4. $A$ is a linear isomorphism.*

This is a special case of Theorem 3.4.

## EXERCISE 3.1

**1.** Let $A$ be a mapping of a vector space $V$ into a vector space $W$. Prove that if $A(\alpha + c\beta) = A(\alpha) + cA(\beta)$ for all $\alpha, \beta \in V$ and $c \in \mathfrak{F}$, then $A$ is linear.

**2.** Let $(c_1, \ldots, c_n)$ be an arbitrary set of $n$ elements in $\mathfrak{F}$. Prove that the mapping $A: \mathfrak{F}^n \to \mathfrak{F}$ defined by

$$A(\alpha) = \sum_{i=1}^{n} c_i a_i \qquad \text{for } \alpha = (a_1, \ldots, a_n) \in \mathfrak{F}^n$$

is linear.

**3.** Prove that in the vector space $\mathrm{Hom}(\mathfrak{F}^n, \mathfrak{F})$ the $n$ linear mappings $A_i$, $1 \leq i \leq n$, defined in Example 3.3 form a basis.

›› **4.** Let $V$ be a finite-dimensional vector space and let $\alpha_1, \ldots, \alpha_k$ be linearly independent vectors in $V$. For any $k$ vectors $\beta_1, \ldots, \beta_k$ in a vector space $W$, show that there is a linear mapping $A: V \to W$ such that $A(\alpha_i) = \beta_i$ for each $i$.

›› **5.** Let $A: V \to W$ be a linear mapping. Prove that if $V_1$ is a finite-dimensional subspace of $V$, then $A(V_1)$ is finite-dimensional and $\dim A(V_1) \leq \dim V_1$.

››☆ **6.** If both the null space and the range of a linear mapping $A: V \to W$ are finite-dimensional, then $V$ is finite-dimensional [and formula (3.4) holds].

**7.** Let $A: V \to W$ be a linear mapping. If $V$ or $W$ is finite-dimensional, then $A(V)$ is finite-dimensional [and hence rank $A = \dim A(V)$ can be defined].

›› **8.** For two linear mappings $A, B: V \to W$, prove that

$$\mathrm{rank}(A + B) \leq \mathrm{rank}\, A + \mathrm{rank}\, B.$$

(We assume both or one of $V$ and $W$ to be finite-dimensional.)

››☆ **9.** Prove Proposition 3.5 assuming only that $W$ is finite-dimensional.

›› **10.** Let $\mathbf{A}$ be an $m \times n$ matrix and $\mathbf{B}$ an $n \times m$ matrix, both over $\mathfrak{F}$. Prove that if $\mathbf{AB} = \mathbf{I}_m$ and $\mathbf{BA} = \mathbf{I}_n$, then $m = n$. (Recall that $\mathbf{I}_n$ denotes the identity matrix of degree $n$; cf. Definition 1.5.)

**11.** Let $V$ be the direct sum of two subspaces $U$ and $W$: $V = U \oplus W$. Define $P_U$ and $P_W$ by

$$P_U(\alpha) = \beta \qquad \text{and} \qquad P_W(\alpha) = \gamma$$

for $\alpha \in V$, where $\alpha = \beta + \gamma$ with $\beta \in U$ and $\gamma \in W$. Show that $P_U$ and $P_W$ are linear transformations of $V$. Find the ranges and the null spaces of $P_U$ and $P_W$. Also prove that $P_U P_W = P_W P_U = 0$, $(P_U)^2 = P_U$, and $(P_W)^2 = P_W$. ($P_U$ is called the *projection* of $V$ onto $U$ along $W$.)

››☆ **12.** Let $A$ be a linear transformation of an $n$-dimensional vector space $V$ such that $A^2 = A$. Setting

$$U = \{\alpha \in V; A(\alpha) = \alpha\} \qquad \text{and} \qquad W = \{\alpha \in V; A(\alpha) = 0\},$$

prove that $V = U \oplus W$ and that $A$ is the projection of $V$ onto $U$ along $W$.

››☆ **13.** Let $\mathfrak{F} = \mathfrak{R}$ or $\mathfrak{C}$ (or more generally, any field of characteristic $\neq 2$; cf. Sec. 5.1) and let $V$ be an $n$-dimensional vector space over $\mathfrak{F}$. For any linear transformation $A$ of $V$ such that $A^2 = I$ prove that $V = U \oplus W$, where

$$U = \{\alpha \in V; A(\alpha) = \alpha\} \qquad \text{and} \qquad W = \{\alpha \in V; A(\alpha) = -\alpha\}.$$

Also prove that $\mathrm{rank}(A - I) + \mathrm{rank}(A + I) = n$.

**14.** If there is an injective linear mapping of an $n$-dimensional vector space $V$ into an $m$-dimensional vector space $W$, then $\dim V \leq \dim W$. What about the case where there is a surjective linear mapping $V \to W$?

**15.** Let $V$ and $W$ be vector spaces with the same dimension. Prove that if there exist linear mappings $A: V \to W$ and $B: W \to V$ such that $BA(\alpha) = \alpha$ for every $\alpha \in V$, then $A$ and $B$ are linear isomorphisms.

**16.** Does the conclusion of number 15 hold without assuming $\dim V = \dim W$?

››☆ **17.** For two linear transformations $A$ and $B$ of an $n$-dimensional vector space $V$ such that $AB = 0$, prove that rank $A$ + rank $B \leq n$.

☆ **18.** For three linear transformations $A$, $B$, and $C$ of an $n$-dimensional vector space $V$ such that $ABC = 0$, prove that rank $A$ + rank $B$ + rank $C \leq 2n$.

☆ **19.** For two linear transformations $A$ and $B$ of an $n$-dimensional vector space $V$ such that $A + B = I$ prove that rank $A$ + rank $B \geq n$. Show that if the equality holds, then $A^2 = A$, $B^2 = B$, and $AB = BA = 0$.

»☆**20.** Let $A$ be a linear transformation of an $n$-dimensional vector space $V$ such that $A^{m-1}(\alpha) \neq 0$ but $A^m(\alpha) = 0$ for some positive integer $m$ and for some vector $\alpha$. Prove that $\{\alpha, A(\alpha), A^2(\alpha), \ldots, A^{m-1}(\alpha)\}$ are linearly independent.

☆**21.** Let $A$ be a linear transformation of an $n$-dimensional real vector space $V$ such that $A^2 = -I$.

»(*a*) Show that there is a basis of the form $\{\alpha_1, \ldots, \alpha_m, A(\alpha_1), \ldots, A(\alpha_m)\}$ in $V$. (It follows that $n = 2m$ and hence $n$ is even.)

(*b*) By defining $(a + bi)\alpha = a\alpha + bA(\alpha)$ for any complex number $a + bi$, where $a, b \in \mathfrak{R}$, and by keeping the original addition, $V$ becomes a vector space over $\mathfrak{C}$.

## 3.2 MATRICES

In Example 3.4 we have seen that any $m \times n$ matrix $\mathbf{A} = [a_{ij}]$ over $\mathfrak{F}$ gives rise to a linear mapping $\mathfrak{F}^n \to \mathfrak{F}^m$. We shall now show that, conversely, any linear mapping $A: \mathfrak{F}^n \to \mathfrak{F}^m$ can be expressed by a matrix.

Let $\{\epsilon_1, \ldots, \epsilon_n\}$ be the standard basis of $\mathfrak{F}^n$ and $\{\epsilon'_1, \ldots, \epsilon'_m\}$ that of $\mathfrak{F}^m$. We may express each $A(\epsilon_j)$, $1 \leq j \leq n$, as a linear combination of $\epsilon'_1, \ldots, \epsilon'_m$ in a unique way:

$$A(\epsilon_j) = \sum_{i=1}^{m} a_{ij}\epsilon'_i. \tag{3.5}$$

We form an $m \times n$ matrix

$$\mathbf{A} = \begin{bmatrix} a_{11} & \cdots & a_{1n} \\ a_{21} & \cdots & a_{2n} \\ \cdots & \cdots & \cdots \\ a_{m1} & \cdots & a_{mn} \end{bmatrix} \tag{3.6}$$

We observe that the $j$th column vector

$$\alpha_j = \begin{bmatrix} a_{1j} \\ \cdot \\ \cdot \\ \cdot \\ a_{mj} \end{bmatrix} \tag{3.7}$$

is indeed the image $A(\epsilon_j)$, $1 \leq j \leq n$. For any $\xi \in \mathfrak{F}^n$, write it as a linear combination of $\epsilon_i$'s *or* as a column vector

$$\xi = \sum_{j=1}^{n} x_j\epsilon_j = \begin{bmatrix} x_1 \\ \cdot \\ \cdot \\ \cdot \\ x_n \end{bmatrix} \tag{3.8}$$

We then have

$$A(\xi) = A\left(\sum_{j=1}^{n} x_j\epsilon_j\right) = \sum_{j=1}^{n} x_j A(\epsilon_j)$$
$$= \sum_{j=1}^{n} x_j \left(\sum_{i=1}^{m} a_{ij}\epsilon_i'\right) = \sum_{i=1}^{m}\left(\sum_{j=1}^{n} a_{ij}x_j\right)\epsilon_i',$$

or, in the form of a column vector,

$$(3.9) \qquad \eta = A(\xi) = \begin{bmatrix} y_1 \\ \cdot \\ \cdot \\ \cdot \\ y_m \end{bmatrix}, \qquad \text{where } y_i = \sum_{j=1}^{n} a_{ij}x_j.$$

This means that $\eta = A(\xi)$ is obtained by multiplication of the matrix $\mathbf{A}$ in (3.6) with the vector $\xi$ in (3.8).

Thus, as long as we express vectors in $\mathfrak{F}^n$ and $\mathfrak{F}^m$ by column vectors (by using the standard bases), there is a one-to-one correspondence between the set of $m \times n$ matrices $\mathfrak{F}^m{}_n$ and the set of linear mappings $\mathrm{Hom}(\mathfrak{F}^n,\mathfrak{F}^m)$.

Before we proceed to the discussion of matrix representation of linear mappings for arbitrary vector spaces, we shall give the following interpretation of the notion of basis.

***Proposition 3.7.*** *Let $V$ be an $n$-dimensional vector space over $\mathfrak{F}$. There is a one-to-one correspondence between the set of bases in $V$ and the set of all linear isomorphisms of $\mathfrak{F}^n$ onto $V$ in the following way: For any basis $\{\alpha_1, \ldots, \alpha_n\}$ there is a unique linear isomorphism $\Phi$: $\mathfrak{F}^n \to V$ such that $\Phi(\epsilon_i) = \alpha_i$ for $1 \leq i \leq n$; conversely, any linear isomorphism $\Phi$: $\mathfrak{F}^n \to V$ gives rise to a basis, namely, $\{\Phi(\epsilon_i), \ldots, \Phi(\epsilon_n)\}$.*

*Proof.* For any basis $\{\alpha_1, \ldots, \alpha_n\}$ there is a linear isomorphism $\Phi$: $\mathfrak{F}^n \to V$ such that $\Phi(\epsilon_i) = \alpha_i$, $1 \leq i \leq n$, as in Example 3.6. Such $\Phi$ is uniquely determined. Conversely, given a linear isomorphism $\Phi$: $\mathfrak{F}^n \to V$, we show that $\Phi(\epsilon_1), \ldots, \Phi(\epsilon_n)$ form a basis. By (6) of Proposition 3.1 they are linearly independent. For any $\alpha \in V$, consider $\Phi^{-1}(\alpha) \in \mathfrak{F}^n$, where $\Phi^{-1}$ is, of course, the inverse of $\Phi$. If we write $\Phi^{-1}(\alpha) = \sum_{i=1}^{n} a_i\epsilon_i$, then

$$\alpha = \Phi(\Phi^{-1}(\alpha)) = \Phi\left(\sum_{i=1}^{n} a_i\epsilon_i\right) = \sum_{i=1}^{n} a_i\Phi(\epsilon_i),$$

showing that $V = \mathrm{Sp}\,\{\Phi(\epsilon_1), \ldots, \Phi(\epsilon_n)\}$. This proves our assertion.

Now let us consider linear mappings $V \to W$, where $\dim V = n$ and $\dim W = m$. Choose an arbitrary basis $\{\alpha_1, \ldots, \alpha_n\}$ in $V$, which we denote by the same letter, $\Phi$, as the linear isomorphism $\Phi: \mathfrak{F}^n \to V$ corresponding to the basis in Proposition 3.7. Similarly, choose an arbitrary basis $\{\beta_1, \ldots, \beta_m\}$ in $W$ corresponding to a linear isomorphism $\Psi: \mathfrak{F}^m \to W$ and denoted by the same letter. Given any linear mapping $A: V \to W$, we consider the composite $\Psi^{-1}A\Phi: \mathfrak{F}^n \to \mathfrak{F}^m$, which can be illustrated by the diagram

$$\begin{array}{ccc} \mathfrak{F}^n & \xrightarrow{\Psi^{-1}A\Phi} & \mathfrak{F}^m \\ \Phi \downarrow & & \uparrow \Psi^{-1} \\ V & \xrightarrow{A} & W \end{array} . \tag{3.10}$$

As we already know, the linear mapping $\Psi^{-1}A\Phi: \mathfrak{F}^n \to \mathfrak{F}^m$ can be expressed by a certain $m \times n$ matrix $\mathbf{A} = [a_{ij}]$ in the form

$$\begin{bmatrix} x_1 \\ \cdot \\ \cdot \\ \cdot \\ x_n \end{bmatrix} \in \mathfrak{F}^n \to \begin{bmatrix} y_1 \\ \cdot \\ \cdot \\ \cdot \\ y_m \end{bmatrix} = \mathbf{A} \begin{bmatrix} x_1 \\ \cdot \\ \cdot \\ \cdot \\ x_n \end{bmatrix} \in \mathfrak{F}^m,$$

where the components $a_{ij}$ of the matrix $\mathbf{A}$ are determined by

$$A(\alpha_j) = \sum_{i=1}^{m} a_{ij}\beta_i, \qquad 1 \leq j \leq n. \tag{3.11}$$

Indeed, we have

$$\begin{aligned} A(\alpha_j) &= A\Phi(\epsilon_j) = \Psi(\Psi^{-1}A\Phi(\epsilon_j)) \\ &= \Psi\left(\sum_{i=1}^{m} a_{ij}\epsilon'_i\right) = \sum_{i=1}^{m} a_{ij}\Psi(\epsilon'_i) = \sum_{i=1}^{m} a_{ij}\beta_i, \end{aligned}$$

where $\{\epsilon_j\}$ and $\{\epsilon'_i\}$ are the standard bases in $\mathfrak{F}^n$ and $\mathfrak{F}^m$ as before.

Conversely, given any $m \times n$ matrix $\mathbf{A} = [a_{ij}]$, the composite $\Psi A \Phi^{-1}$ is a linear mapping $V \to W$. Thus, whenever a basis $\Phi$ in $V$ and a basis $\Psi$ in $W$ are fixed, there is a one-to-one correspondence between the set $\mathfrak{F}^m{}_n$ and the set $\mathrm{Hom}(V,W)$. For any $A \in \mathrm{Hom}(V,W)$ we shall denote the corresponding matrix by $\mathbf{A} = A_{\Phi,\Psi}$, referring to the bases we are using. It is called the *matrix representation* of $A$ with respect to $\Phi$, $\Psi$.

Summing up, we shall state:

***Proposition 3.8.*** *Given a basis $\Phi$ in $V$ and given a basis $\Psi$ in $W$, there is a one-to-one correspondence between* $\mathrm{Hom}(V,W)$ *and* $\mathfrak{F}^m{}_n$*: $A \in \mathrm{Hom}(V,W) \to A_{\Phi,\Psi} \in \mathfrak{F}^m{}_n$, where the components $a_{ij}$ of $A$ are given by* (3.11). *Moreover, the mapping $A \to A_{\Phi,\Psi}$ is a linear isomorphism of the vector space* $\mathrm{Hom}(V,W)$ *onto the vector space* $\mathfrak{F}^m{}_n$.

*Proof.* As for the second assertion we have

$$(A + B)(\alpha_j) = A(\alpha_j) + B(\alpha_j) = \sum_{i=1}^{m} a_{ij}\beta_i + \sum_{i=1}^{m} b_{ij}\beta_i$$
$$= \sum_{i=1}^{m} (a_{ij} + b_{ij})\beta_i,$$

and similarly

$$(cA)(\alpha_j) = c\left(\sum_{i=1}^{m} a_{ij}\beta_i\right) = \sum_{i=1}^{m} (ca_{ij})\beta_i,$$

which show that $A + B$ and $cA$ correspond to $A_{\Phi,\Psi} + B_{\Phi,\Psi}$ and $cA_{\Phi,\Psi}$, respectively.

The composition of linear mappings is related to matrix multiplication defined in Definition 1.3.

***Proposition 3.9.*** *Let $V$, $W$, and $U$ be vector spaces with bases $\Phi$, $\Psi$, and $\Theta$, respectively. For any*

$$A\colon V \to W \qquad \text{and} \qquad B\colon W \to U,$$

*we have*

$$(BA)_{\Phi,\Theta} = B_{\Psi,\Theta}A_{\Phi,\Psi};$$

*in other words, the composite of two linear mappings corresponds to the product of the corresponding matrices (for proper use of bases).*

*Proof.* Let $\Phi = \{\alpha_j\}$, $\Psi = \{\beta_i\}$, $\Theta = \{\gamma_k\}$. Let $A_{\Phi,\Psi} = [a_{ij}]$ and $B_{\Psi,\Theta} = [b_{ki}]$, where $1 \leq i \leq m$, $1 \leq j \leq n$, and $1 \leq k \leq p$ (so that $n = \dim V$, $m = \dim W$, and $p = \dim U$). Then

$$(BA)(\alpha_j) = B(A(\alpha_j)) = B\left(\sum_{i=1}^{m} a_{ij}\beta_i\right)$$
$$= \sum_{i=1}^{m} a_{ij}B(\beta_i) = \sum_{i=1}^{m} a_{ij}\left(\sum_{k=1}^{p} b_{ki}\gamma_k\right)$$
$$= \sum_{k=1}^{p}\left(\sum_{i=1}^{m} b_{ki}a_{ij}\right)\gamma_k,$$

which shows that the components $c_{kj}$ of $(BA)_{\Phi,\Theta}$ are equal to $\sum_{i=1}^{m} b_{ki}a_{ij}$. Recalling Definition 1.3, we have

$$(BA)_{\Phi,\Theta} = B_{\Psi,\Theta}A_{\Phi,\Psi}.$$

As a special case of the situations above, let $V$ be a vector space with a basis $\Phi = \{\alpha_1, \ldots, \alpha_n\}$. With any $A \in \mathfrak{gl}(V)$ we associate the matrix $A_{\Phi,\Phi} = [a_{ij}] \in \mathcal{F}^n{}_n$, where $a_{ij}$'s are determined by

$$A(\alpha_j) = \sum_{i=1}^{n} a_{ij}\alpha_i, \qquad 1 \leq j \leq n. \tag{3.12}$$

We shall call $A_{\Phi,\Phi}$ the *matrix representation of A with respect to the basis* $\Phi$ and denote it by $A_\Phi$ for simplicity. As a corollary of Propositions 3.8 and 3.9, we have:

***Corollary.*** *Let $\Phi$ be a basis of a vector space V. The mapping $A \in \mathfrak{gl}(V) \to A_\Phi \in \mathcal{F}^n_n$ is a linear isomorphism of the vector space $\mathfrak{gl}(V)$ onto the vector space $\mathcal{F}^n_n$ which preserves multiplication.*

As in the case of $\mathfrak{gl}(V)$, we shall speak of the *algebra* $\mathcal{F}^n_n$. The identity transformation $I$ of $V$ corresponds to the identity matrix $\mathbf{I}_n \in \mathcal{F}^n_n$. A matrix $\mathbf{A} \in \mathcal{F}^n_n$ is said to be *nonsingular* (or *invertible*) if it has an inverse, that is, a matrix $\mathbf{B} \in \mathcal{F}^n_n$ such that $\mathbf{AB} = \mathbf{BA} = \mathbf{I}_n$. For the correspondence in the corollary above, it corresponds to a nonsingular transformation of $V$. The inverse $\mathbf{A}^{-1}$ of a nonsingular matrix is uniquely determined.

Suppose, for a given $A \in \mathfrak{gl}(V)$, there is a basis $\Phi = \{\alpha_1, \ldots, \alpha_n\}$ such that $A(\alpha_j) = c_j\alpha_j$, $1 \leq j \leq n$, where $c_1, \ldots, c_n \in \mathcal{F}$. Then

$$A_\Phi = \begin{bmatrix} c_1 & & & \\ & \cdot & & \\ & & \cdot & \\ & & & \cdot \\ & & & & c_n \end{bmatrix} \qquad \text{(all entries are 0 except those on the diagonal).}$$

A matrix of this form is called a *diagonal matrix*.

Finally, we shall interpret the rank of a linear mapping $A: V \to W$ in terms of the corresponding matrix $A_{\Phi,\Psi}$. First, consider the linear mapping $\mathcal{F}^n \to \mathcal{F}^m$ defined by the matrix $\mathbf{A}$. Since $\mathbf{A}(\epsilon_j)$ equals the $j$th column vector of $\mathbf{A}$ and since $\mathbf{A}(\epsilon_1), \ldots, \mathbf{A}(\epsilon_n)$ span the range $\mathbf{A}(\mathcal{F}^n)$, we see that the range of $\mathbf{A}$ is equal to the subspace of $\mathcal{F}^m$ spanned by $n$ column vectors of $\mathbf{A}$ (each of which is a vector in $\mathcal{F}^m$, of course). The rank of the mapping $\mathbf{A}$ is thus equal to the maximum number of linearly independent column vectors of the matrix $\mathbf{A}$.

**Definition 3.5**

For any $m \times n$ matrix $\mathbf{A}$ over $\mathcal{F}$ the *column space* [*row space*] of $\mathbf{A}$ is the subspace of $\mathcal{F}^m$ [$\mathcal{F}^n$] spanned by $n$ column vectors [$m$ row vectors] of $\mathbf{A}$. The *column rank* (*row rank*) is the dimension of the column space [row space], namely, the [maximum] number of linearly independent column vectors [row vectors].

Thus, the rank of the linear mapping $\mathbf{A}: \mathcal{F}^n \to \mathcal{F}^m$ is equal to the column rank of the matrix $\mathbf{A}$.

***Remark.*** We shall prove later (Theorem 3.17) that the row rank and the column rank coincide for each matrix.

Now going back to an arbitrary linear mapping $A: V \to W$, $A_{\Phi,\Psi}$ is the matrix which defines $\Psi^{-1}A\Phi: \mathfrak{F}^n \to \mathfrak{F}^m$. Since $\Phi$ and $\Psi$ are linear isomorphisms, we have

$$\dim A(V) = \dim A\Phi(\mathfrak{F}^n) = \dim \Psi^{-1}A\Phi(\mathfrak{F}^n),$$

which is equal to the column rank of the matrix $A_{\Phi,\Psi}$ by what we have already seen. Hence we have:

***Proposition 3.10.*** *In the correspondence $A \to A_{\Phi,\Psi}$ in Proposition* 3.8, *the rank of $A$ equals the column rank of $A_{\Phi,\Psi}$.*

**EXERCISE 3.2**

›› **1.** Let $\mathbf{A}: \mathbb{R}^3 \to \mathbb{R}^2$ be defined by the matrix

$$\mathbf{A} = \begin{bmatrix} 2 & 0 & 4 \\ -1 & 1 & -2 \end{bmatrix}.$$

(*a*) Find the images of $\begin{bmatrix} 1 \\ 0 \\ 1 \end{bmatrix}, \begin{bmatrix} 2 \\ 1 \\ -3 \end{bmatrix}$.

(*b*) Find the rank of **A**.

(*c*) Find the nullity of **A**.

**2.** Let $\mathbf{A}: \mathbb{R}^4 \to \mathbb{R}^3$ be defined by $\mathbf{A} = \begin{bmatrix} 1 & 0 & 3 & -1 \\ 2 & 1 & 0 & -4 \\ 3 & 0 & 2 & -2 \end{bmatrix}$.

(*a*) Find the images of $\begin{bmatrix} 1 \\ 0 \\ -1 \\ 2 \end{bmatrix}, \begin{bmatrix} 2 \\ 1 \\ -2 \\ 0 \end{bmatrix}$.

(*b*) Find the rank of **A**.

(*c*) Find the nullity of **A**.

**3.** Compute the following matrix products or powers:

››(*a*) $\begin{bmatrix} a_{11} & a_{12} & \cdots & a_{1n} \\ 0 & a_{22} & \cdots & a_{2n} \\ \cdots & \cdots & \cdots & \cdots \\ 0 & 0 & \cdots & a_{nn} \end{bmatrix} \begin{bmatrix} b_{11} & b_{12} & \cdots & b_{1n} \\ 0 & b_{22} & \cdots & b_{2n} \\ \cdots & \cdots & \cdots & \cdots \\ 0 & 0 & \cdots & b_{nn} \end{bmatrix}$;

(*b*) $\begin{bmatrix} 0 & a & b \\ 0 & 0 & c \\ 0 & 0 & 0 \end{bmatrix} \begin{bmatrix} 0 & 0 & 0 \\ a' & 0 & 0 \\ b' & c' & 0 \end{bmatrix}$; ››(*c*) $\begin{bmatrix} 0 & a & b \\ 0 & 0 & c \\ 0 & 0 & 0 \end{bmatrix}^3$; (*d*) $\begin{bmatrix} 0 & 1 \\ -1 & 0 \end{bmatrix}^2$;

››(*e*) $\begin{bmatrix} 0 & 0 & 1 \\ 0 & 1 & 0 \\ 1 & 0 & 0 \end{bmatrix}^n$; (*f*) $\begin{bmatrix} 0 & 1 & 0 & \cdots & 0 \\ 0 & 0 & 1 & \cdots & 0 \\ & & 0 & \cdot & \\ & & & \ddots & \\ & & & & \ddots \\ & & & \cdot & 1 \\ & & & & 0 \end{bmatrix}^n$ ($n \times n$ matrix).

**4.** Show that the linear transformation $cI: V \to V$, where $\dim V = n$, can be expressed by the matrix $c\mathbf{I}_n$ with respect to an arbitrary basis of $V$.

**5.** Show that the column rank of a matrix $\mathbf{A} = [a_{ij}] \in \mathfrak{F}^n{}_n$, where $a_{ij} = b_i c_j$, $1 \leq i, j \leq n$, for some $(b_1, \ldots, b_n)$ and $(c_1, \ldots, c_n)$ in $\mathfrak{F}^n$, is at most 1.

»☆ **6.** Let $A$ be a linear transformation of an $n$-dimensional vector space $V$ over $\mathfrak{F}$. If rank $A \leq 1$, then the matrix $A_\Phi = [a_{ij}]$ for any basis $\Phi$ of $V$ is of the form $a_{ij} = b_i c_j$ as in number 5.

**7.** Let $\mathbf{A}$ be an element in $\mathfrak{F}^n{}_n$. Show that both the mappings $\mathbf{X} \in \mathfrak{F}^n{}_n \to \mathbf{AX} \in \mathfrak{F}^n{}_n$ and $\mathbf{X} \in \mathfrak{F}^n{}_n \to \mathbf{XA} \in \mathfrak{F}^n{}_n$ are linear.

**8.** Let $\mathbf{A} = \begin{bmatrix} 2 & 1 \\ 0 & -1 \end{bmatrix}$. Find the matrix representations of the linear mappings $\mathbf{X} \to \mathbf{AX}$ and $\mathbf{X} \to \mathbf{XA}$ of the vector space $\mathfrak{R}^2{}_2$ with respect to the following basis:

$$\begin{bmatrix} 1 & 0 \\ 0 & 0 \end{bmatrix}, \quad \begin{bmatrix} 0 & 1 \\ 0 & 0 \end{bmatrix}, \quad \begin{bmatrix} 0 & 0 \\ 1 & 0 \end{bmatrix}, \quad \begin{bmatrix} 0 & 0 \\ 0 & 1 \end{bmatrix}.$$

☆ **9.** Let $\mathbf{E}_{i,j}$, $1 \leq i, j \leq n$, be the $n^2$ matrices defined in Exercise 2.2, number 5. (They form a basis of $\mathfrak{F}^n{}_n$, called the *standard basis.*) Prove that

$$\mathbf{E}_{i,j}\mathbf{E}_{p,q} = \delta_{jp}\mathbf{E}_{i,q},$$

where $\delta_{jp}$ is the Kronecker delta.

**10.** Prove that $\mathbf{A} \in \mathfrak{F}^n{}_n$ is nonsingular if there exists $\mathbf{B} \in \mathfrak{F}^n{}_n$ such that $\mathbf{AB} = \mathbf{I}_n$ (or $\mathbf{BA} = \mathbf{I}_n$).

☆**11.** Let $\mathbf{A} = [a_{ik}]$ be an $m \times n$ matrix and $\mathbf{B} = [b_{kj}]$ be an $n \times p$ matrix.

(*a*) Show that the $i$th row vector of $\mathbf{C} = \mathbf{AB}$ is equal to $\sum_{k=1}^{n} a_{ik}\beta_k$, where $\beta_1, \ldots, \beta_n$ are the row vectors of $\mathbf{B}$.

(*b*) Show that the $j$th column vector of $\mathbf{C} = \mathbf{AB}$ is equal to $\sum_{k=1}^{n} b_{kj}\alpha_k$, where $\alpha_1, \ldots, \alpha_n$ are the column vectors of $\mathbf{A}$.

(*c*) Show that the $i$th row vector of $\mathbf{C} = \mathbf{AB}$ is equal to $\alpha_i'\mathbf{B}$, where $\alpha_i'$ is the $i$th row vector of $\mathbf{A}$.

(*d*) Show that the $j$th column vector of $\mathbf{C} = \mathbf{AB}$ is equal to $\mathbf{A}\beta_j'$, where $\beta_j'$ is the $j$th column vector of $\mathbf{B}$.

**12.** For

$$\mathbf{A} = \begin{bmatrix} \mathbf{A}_{11} & \mathbf{A}_{12} \\ \mathbf{A}_{21} & \mathbf{A}_{22} \end{bmatrix} \quad \text{and} \quad \mathbf{B} = \begin{bmatrix} \mathbf{B}_{11} & \mathbf{B}_{12} \\ \mathbf{B}_{21} & \mathbf{B}_{22} \end{bmatrix},$$

where $\mathbf{A}_{11}$ is $m_1 \times n_1$, $\mathbf{A}_{12}$ is $m_1 \times n_2$, $\mathbf{A}_{21}$ is $m_2 \times n_1$, $\mathbf{A}_{22}$ is $m_2 \times n_2$, $\mathbf{B}_{11}$ is $n_1 \times p_1$, $\mathbf{B}_{12}$ is $n_1 \times p_2$, $\mathbf{B}_{21}$ is $n_2 \times p_1$, and $\mathbf{B}_{22}$ is $n_2 \times p_2$, show that

$$\mathbf{AB} = \begin{bmatrix} \mathbf{A}_{11}\mathbf{B}_{11} + \mathbf{A}_{12}\mathbf{B}_{21} & \mathbf{A}_{11}\mathbf{B}_{12} + \mathbf{A}_{12}\mathbf{B}_{22} \\ \mathbf{A}_{21}\mathbf{B}_{11} + \mathbf{A}_{22}\mathbf{B}_{21} & \mathbf{A}_{21}\mathbf{B}_{12} + \mathbf{A}_{22}\mathbf{B}_{22} \end{bmatrix}.$$

☆**13.** If an $n \times n$ matrix $\mathbf{A}$ over $\mathfrak{F}$ commutes with every $n \times n$ diagonal matrix $\mathbf{D}$ (that is, $\mathbf{AD} = \mathbf{DA}$), show that $\mathbf{A}$ is a diagonal matrix.

☆**14.** In $\mathfrak{F}^n{}_n$, we define the *bracket operation* $[\mathbf{A},\mathbf{B}]$ by

$$[\mathbf{A},\mathbf{B}] = \mathbf{AB} - \mathbf{BA}, \quad \text{where} \quad \mathbf{A}, \mathbf{B} \in \mathfrak{F}^n{}_n.$$

Prove:

(*a*) $[\mathbf{A},\mathbf{B}]$ is linear in $\mathbf{A}$ when $\mathbf{B}$ is fixed and is linear in $\mathbf{B}$ when $\mathbf{A}$ is fixed.

(*b*) $[\mathbf{A},\mathbf{A}] = 0$ for every $\mathbf{A}$, and $[\mathbf{B},\mathbf{A}] = -[\mathbf{A},\mathbf{B}]$ for all $\mathbf{A},\mathbf{B}$.

(*c*) $[[\mathbf{A},\mathbf{B}],\mathbf{C}] + [[\mathbf{B},\mathbf{C}],\mathbf{A}] + [[\mathbf{C},\mathbf{A}],\mathbf{B}] = 0$ for all $\mathbf{A},\mathbf{B}$, and $\mathbf{C}$.

(Thus $\mathfrak{F}^n{}_n$ is what is called a *Lie algebra.*)

☆**15.** For any $\mathbf{A} \in \mathfrak{F}^n{}_n$ let $f$ be the mapping of $\mathfrak{F}^n{}_n$ into itself defined by $f(\mathbf{X}) = [\mathbf{A},\mathbf{X}]$ for all $\mathbf{X} \in \mathfrak{F}^n{}_n$. Prove that $f$ is a linear transformation satisfying

$$f(\mathbf{XY}) = f(\mathbf{X})\mathbf{Y} + \mathbf{X}f(\mathbf{Y}) \qquad \text{for all } \mathbf{X}, \mathbf{Y} \in \mathfrak{F}^n{}_n.$$

☆**16.** Let $f$ be any linear transformation of the vector space $\mathfrak{F}^n{}_n$ which satisfies

$$f(\mathbf{XY}) = f(\mathbf{X})\mathbf{Y} + \mathbf{X}f(\mathbf{Y}) \qquad \text{for all } \mathbf{X}, \mathbf{Y}.$$

Prove that there is an $\mathbf{A} \in \mathfrak{F}^n{}_n$ such that $f(\mathbf{X}) = [\mathbf{A},\mathbf{X}]$.

››☆**17.** For any $\mathbf{A} \in \mathfrak{F}^n{}_n$, let $f_{\mathbf{A}}$ be the linear transformation of the vector space $\mathfrak{F}^n{}_n$ defined by $f_{\mathbf{A}}(\mathbf{X}) = \mathbf{AX}$. Prove that rank of $f_{\mathbf{A}} = n \cdot \text{rank } \mathbf{A}$, where rank $\mathbf{A}$ is the column rank of $\mathbf{A}$.

## 3.3 CHANGE OF BASES

Given an $n$-dimensional vector space $V$, there are many choices of bases in $V$. We shall study how one basis is related to another and how the components of a vector and the representing matrix for a linear mapping will change when we make a change of bases.

Let $\{\alpha_1, \ldots, \alpha_n\}$ and $\{\alpha'_1, \ldots, \alpha'_n\}$ be two bases of $V$ and let also $\Phi: \mathcal{F}^n \to V$ and $\Phi': \mathcal{F}^n \to V$ be the corresponding linear isomorphisms. The composite $P = \Phi'^{-1}\Phi$, illustrated by

$$\begin{array}{ccc} \mathcal{F}^n & \xrightarrow{\Phi'^{-1}\Phi} & \mathcal{F}^n \\ {\scriptstyle\Phi}\searrow & & \nearrow{\scriptstyle\Phi'^{-1}} \\ & V & \end{array}, \tag{3.13}$$

is a linear isomorphism of $\mathcal{F}^n$ and can be represented by an $n \times n$ matrix $\mathbf{P}$. For any $\alpha \in V$, let

$$\alpha = \sum_{i=1}^{n} x_i\alpha_i = \sum_{i=1}^{n} x'_i\alpha'_i, \tag{3.14}$$

so that

$$\Phi^{-1}(\alpha) = \begin{bmatrix} x_1 \\ \cdot \\ \cdot \\ \cdot \\ x_n \end{bmatrix} \quad \text{and} \quad \Phi'^{-1}(\alpha) = \begin{bmatrix} x'_1 \\ \cdot \\ \cdot \\ \cdot \\ x'_n \end{bmatrix}.$$

Then

$$\begin{bmatrix} x'_1 \\ \cdot \\ \cdot \\ \cdot \\ x'_n \end{bmatrix} = \mathbf{P} \begin{bmatrix} x_1 \\ \cdot \\ \cdot \\ \cdot \\ x_n \end{bmatrix}. \tag{3.15}$$

In order to determine the matrix $\mathbf{P}$, let us write each $\alpha_j$ as a linear combination of $\alpha'_1, \ldots, \alpha'_n$:

$$\alpha_j = \sum_{i=1}^{n} p_{ij}\alpha'_i, \qquad 1 \leq j \leq n. \tag{3.16}$$

Then we have

$$\begin{aligned} P(\epsilon_j) &= \Phi'^{-1}\Phi(\epsilon_j) = \Phi'^{-1}(\alpha_j) \\ &= \Phi'^{-1}\left(\sum_{i=1}^{n} p_{ij}\alpha'_i\right) = \sum_{i=1}^{n} p_{ij}\Phi'^{-1}(\alpha'_i) \\ &= \sum_{i=1}^{n} p_{ij}\epsilon_i, \end{aligned}$$

which shows that the matrix $\mathbf{P}$ is $[p_{ij}]$.

Summing up, we have:

***Proposition 3.11.*** *If two bases* $\{\alpha_1, \ldots, \alpha_n\}$ *and* $\{\alpha'_1, \ldots, \alpha'_n\}$ *are related by* (3.16), *the components* $\{x_i\}$ *and* $\{x'_i\}$ *of any vector* $\alpha$ *with respect to* $\{\alpha_i\}$ *and* $\{\alpha'_i\}$ *are related by* (3.15).

By reversing the roles of $\{\alpha_i\}$ and $\{\alpha'_i\}$, we can write

$$\alpha'_j = \sum_{i=1}^{n} p'_{ij}\alpha_i, \qquad 1 \leq j \leq n. \tag{3.16$'$}$$

With the matrix $\mathbf{P}' = [p'_{ij}]$, we have

$$\begin{bmatrix} x_1 \\ \cdot \\ \cdot \\ \cdot \\ x_n \end{bmatrix} = \mathbf{P}' \begin{bmatrix} x'_1 \\ \cdot \\ \cdot \\ \cdot \\ x'_n \end{bmatrix}. \tag{3.15$'$}$$

From (3.16) and (3.16′) we have

$$\begin{aligned} \alpha_j &= \sum_{i=1}^{n} p_{ij}\left(\sum_{k=1}^{n} p'_{ki}\alpha_k\right) \\ &= \sum_{k=1}^{n}\left(\sum_{i=1}^{n} p'_{ki}p_{ij}\right)\alpha_k. \end{aligned}$$

Therefore, we have

$$\sum_{i=1}^{n} p'_{ki}p_{ij} = \delta_{kj}, \qquad \text{that is,} \qquad \mathbf{P}'\mathbf{P} = \mathbf{I}_n. \tag{3.17}$$

Similarly, we have

$$\sum_{i=1}^{n} p_{ki}p'_{ij} = \delta_{kj}, \qquad \text{that is,} \qquad \mathbf{P}\mathbf{P}' = \mathbf{I}_n. \tag{3.17$'$}$$

In other words, the matrices $\mathbf{P}$ and $\mathbf{P}'$ are inverse to each other: $\mathbf{P}' = \mathbf{P}^{-1}$.

Another way of looking at the change of bases is the following. Given two bases $\Phi = \{\alpha_1, \ldots, \alpha_n\}$ and $\Phi' = \{\alpha_1', \ldots, \alpha_n'\}$, we know that there is a unique isomorphism $A$ of $V$ such that $A(\alpha_i) = \alpha_i'$ for $1 \leq i \leq n$. We have $A\Phi = \Phi'$ as illustrated by

(3.18)
$$\begin{array}{ccc} & \mathcal{F}^n & \\ {\scriptstyle\Phi}\swarrow & & \searrow{\scriptstyle\Phi'} \\ V & \xrightarrow{A} & V \end{array} .$$

We leave it to the reader to verify

(3.19)
$$A_\Phi = \mathbf{P}' \quad \text{and} \quad A_{\Phi'} = \mathbf{P}'.$$

Now we shall consider matrix representations of a linear mapping $A\colon V \to W$. Let $\Phi = \{\alpha_1, \ldots, \alpha_n\}$ and $\Phi' = \{\alpha_1', \ldots, \alpha_n'\}$ be two bases of $V$ related by a matrix $\mathbf{P}$ as before, that is,

(3.16)
$$\alpha_j = \sum_{i=1}^{n} p_{ij}\alpha_i'.$$

Similarly, let $\Psi = \{\beta_1, \ldots, \beta_m\}$ and $\Psi' = \{\beta_1', \ldots, \beta_m'\}$ be two bases of $W$ related by a matrix $\mathbf{Q} = [q_{ij}]$, that is,

(3.20)
$$\beta_j = \sum_{i=1}^{n} q_{ij}\beta_i'.$$

We prove:

***Proposition 3.12.*** *Let $\Phi$ and $\Phi'$ be two bases in $V$ related by* (3.16) *and let $\Psi$ and $\Psi'$ be two bases of $W$ related by* (3.20). *Then for any linear mapping $A\colon V \to W$, we have*

(3.21)
$$A_{\Phi',\Psi'} = \mathbf{Q}A_{\Phi,\Psi}\mathbf{P}^{-1}.$$

*Proof.* We have by definition

$$A_{\Phi,\Psi} = \Psi^{-1}A\Phi \quad \text{and} \quad A_{\Phi',\Psi'} = \Psi'^{-1}A\Phi'.$$

On the other hand, we have from the diagram (3.13)

$$\mathbf{P} = \Phi'^{-1}\Phi \quad \text{and} \quad \mathbf{Q} = \Psi'^{-1}\Psi$$

or

$$\Phi' = \Phi\mathbf{P}^{-1} \quad \text{and} \quad \Psi'^{-1} = \mathbf{Q}\Psi^{-1}.$$

Therefore, we get

$$A_{\Phi',\Psi'} = \mathbf{Q}\Psi^{-1}A\Phi\mathbf{P}^{-1} = \mathbf{Q}A_{\Phi,\Psi}\mathbf{P}^{-1}.$$

***Corollary.*** *Assume that two bases $\Phi$ and $\Phi'$ in a vector space $V$ are related by a matrix $\mathbf{P}$ as in* (3.16). *For any linear transformation $A$ of $V$, we have*

(3.22)
$$A_{\Phi'} = \mathbf{P}A_\Phi\mathbf{P}^{-1}.$$

***Example 3.8.*** In $\mathcal{R}^2$, let $\alpha_1' = (1,2)$ and $\alpha_2' = (-1,1)$. The relationship between the standard basis $\{\epsilon_1,\epsilon_2\}$ and a new basis $\{\alpha_1',\alpha_2'\}$ can be given by

$$\alpha_1' = \epsilon_1 + 2\epsilon_2, \qquad \alpha_2' = -\epsilon_1 + \epsilon_2 \tag{3.23}$$

and

$$\epsilon_1 = \tfrac{1}{3}\alpha_1' - \tfrac{2}{3}\alpha_2', \qquad \epsilon_2 = \tfrac{1}{3}\alpha_1' + \tfrac{1}{3}\alpha_2'. \tag{3.23'}$$

The matrix $\mathbf{P} = [p_{ij}]$ in the formula (3.16), where we now consider $\alpha_1 = \epsilon_1$, $\alpha_2 = \epsilon_2$, is

$$\mathbf{P} = \begin{bmatrix} \frac{1}{3} & \frac{1}{3} \\ -\frac{2}{3} & \frac{1}{3} \end{bmatrix}.$$

The matrix $\mathbf{P}'$ in the formula (3.16′) is

$$\mathbf{P}' = \begin{bmatrix} 1 & -1 \\ 2 & 1 \end{bmatrix}.$$

(The reader can directly verify $\mathbf{PP}' = \mathbf{P}'\mathbf{P} = \mathbf{I}_2$.)

Now consider the linear mapping $A\colon \mathcal{R}^2 \to \mathcal{R}^2$ defined by the matrix $\begin{bmatrix} 3 & 2 \\ 2 & -4 \end{bmatrix}$ with respect to $\{\epsilon_1,\epsilon_2\}$. The matrix representing $A$ with respect to $\{\alpha_1',\alpha_2'\}$ is

$$\begin{bmatrix} \frac{1}{3} & \frac{1}{3} \\ -\frac{2}{3} & \frac{1}{3} \end{bmatrix}\begin{bmatrix} 3 & 2 \\ 2 & -4 \end{bmatrix}\begin{bmatrix} 1 & -1 \\ 2 & 1 \end{bmatrix} = \begin{bmatrix} \frac{1}{3} & -\frac{7}{3} \\ -\frac{20}{3} & -\frac{4}{3} \end{bmatrix}$$

by the corollary to Proposition 3.12.

***Remark 1.*** We shall later discuss a practical method of finding the inverse $\mathbf{P}^{-1}$ of a given matrix $\mathbf{P}$ (when it exists).

***Remark 2.*** It is convenient to memorize the following rule to find the matrix $A_{\Phi,\Psi}$ representing $A\colon V \to W$ with respect to the bases $\Phi = \{\alpha_i\}$ and $\Psi = \{\beta_j\}$. Write down as the $j$th *column* vector the components of $A(\alpha_j)$ with respect to $\{\beta_1, \ldots, \beta_m\}$. Likewise, the matrix $\mathbf{P}$ in the formula (3.16) can be obtained by writing down as the $j$th *column* vector the components of $\alpha_j$ with respect to $\{\alpha_1', \ldots, \alpha_n'\}$.

We shall now illustrate how we may make use of the change of bases in order to prove some important results on matrices.

***Proposition 3.13.*** *Let $A\colon V \to W$ be a linear mapping, where* $\dim V = n$ *and* $\dim W = m$. *Then there exists a basis $\Phi = \{\alpha_1, \ldots, \alpha_n\}$ in $V$ and a basis $\Psi = \{\beta_1, \ldots, \beta_m\}$ in $W$ for which $A_{\Phi,\Psi}$ is of the following form with $r =$* rank $A$

$$\begin{bmatrix} \mathbf{I}_r & \mathbf{O}_{r,n-r} \\ \mathbf{O}_{m-r,r} & \mathbf{O}_{m-r,n-r} \end{bmatrix}, \tag{3.24}$$

*where $I_r$ is the unit matrix of degree $r$ and $\mathbf{O}_{p,q}$ is the $p \times q$ zero matrix.*

*Proof.* Let $\Phi = \{\alpha_1, \ldots, \alpha_r, \alpha_{r+1}, \ldots, \alpha_n\}$ be a basis of $V$ such that $\alpha_{r+1}, \ldots, \alpha_n$ form a basis of the null space $N$ of $A$. From the proof of Theorem 3.3 we know that $A(\alpha_1), \ldots, A(\alpha_r)$ form a basis of the range $A(V)$. Let $\beta_i = A(\alpha_i)$ for $1 \le i \le r$ and choose $\beta_{r+1}, \ldots, \beta_m$ in $W$ so that $\Psi = \{\beta_1, \ldots, \beta_r, \beta_{r+1}, \ldots, \beta_m\}$ is a basis of $W$ (by virtue of Theorem 2.13). Since $A(\alpha_i) = \beta_i$ for $1 \le i \le r$ and $A(\alpha_j) = 0$ for $r + 1 \le j \le n$, we have $A_{\Phi,\Psi}$ in the desired form (3.24).

Now we obtain:

***Theorem 3.14.*** *For any $m \times n$ matrix* $\mathbf{A}$ *over* $\mathfrak{F}$, *there exists a nonsingular $m \times m$ matrix* $\mathbf{Q}$ *and a nonsingular $n \times n$ matrix* $\mathbf{P}$, *both over* $\mathfrak{F}$, *such that* $\mathbf{QAP}^{-1}$ *is of the form* (3.24), *where $r$ equals the column rank of* $\mathbf{A}$.

*Proof.* Consider $\mathbf{A}$ as a linear mapping $A: \mathfrak{F}^n \to \mathfrak{F}^m$. The rank of $A$ is equal to the column rank of $\mathbf{A}$. Now by Proposition 3.13 there exist a basis $\Phi$ in $\mathfrak{F}^n$ and a basis $\Psi$ in $\mathfrak{F}^m$ such that $A_{\Phi,\Psi}$ is of the form (3.24). By Proposition 3.12 we have

$$A_{\Phi,\Psi} = \mathbf{QAP}^{-1},$$

where $\mathbf{Q}$ is a certain $m \times m$ nonsingular matrix and $\mathbf{P}$ is a certain $n \times n$ nonsingular matrix.

We shall now consider what we can do to a given matrix $\mathbf{A}$ by multiplying it on the left by a nonsingular matrix $\mathbf{Q}$. In terms of the linear mapping $A: \mathfrak{F}^n \to \mathfrak{F}^m$ in the preceding proof, this means that we keep the standard basis in $\mathfrak{F}^n$ and make a change of basis in $\mathfrak{F}^m$ only. We shall obtain:

***Theorem 3.15.*** *For any $m \times n$ matrix* $\mathbf{A}$ *over* $\mathfrak{F}$, *there exists a nonsingular $m \times m$ matrix* $\mathbf{Q}$ *over* $\mathfrak{F}$ *such that* $\mathbf{QA}$ *is of the following form: there exist an integer $r$, $0 \le r \le m$, and a set of integers $\{k_1, \ldots, k_r\}$, $1 \le k_1 < \cdots < k_r \le n$, such that:*

**1.** *For each $i$, $1 \le i \le r$, the first nonzero component of the $i$th row is* 1 *and it is on the $k_i$th column.*

**2.** *For each $i$, $1 \le i \le r$, all the components of the $k_i$th column are* 0 *except for the $(i,k_i)$ component (which is* 1 *as already stated).*

**3.** *The last $m - r$ rows are all* 0.

Thus, this matrix (namely, $\mathbf{QA}$) looks like

$$\begin{array}{c} \\ r\left\{\begin{array}{c} \\ \\ \\ \\ \end{array}\right. \\ m-r\left\{\begin{array}{c} \\ \\ \\ \end{array}\right. \end{array}
\begin{array}{c}
\begin{array}{ccccccccccccc} & & & k_1 & & & & k_2 & & \cdots & & k_r & \end{array} \\
\begin{bmatrix}
0 & \cdots & 0 & 1 & * & \cdots & 0 & * & \cdots & 0 & * & \cdots \\
0 & \cdots & 0 & 0 & 0 & \cdots & 1 & * & \cdots & 0 & * & \cdots \\
\cdot & \cdot & \cdot & \cdot & \cdot & \cdot & \cdot & \cdot & \cdot & \cdot & \cdot & \cdot \\
0 & \cdots & 0 & 0 & 0 & \cdots & 0 & 0 & \cdots & 1 & * & \cdots \\
0 & \cdots & 0 & 0 & 0 & \cdots & 0 & 0 & \cdots & 0 & 0 & \cdots \\
\cdot & \cdot & \cdot & \cdot & \cdot & \cdot & \cdot & \cdot & \cdot & \cdot & \cdot & \cdot \\
0 & \cdots & 0 & 0 & 0 & \cdots & 0 & 0 & \cdots & 0 & 0 & \cdots
\end{bmatrix}
\end{array},$$

where the components * are arbitrary. A matrix satisfying the above conditions is called a *row-reduced echelon matrix.*

*Proof.* Consider any linear mapping $A: V \to W$, where $n = \dim V$ and $m = \dim W$. Let $\Phi = \{\alpha_1, \ldots, \alpha_n\}$ be any given basis in $V$. We recall that the rank of $A$ is equal to the maximum number of linearly independent vectors among $A(\alpha_i)$, $1 \leq i \leq n$. Let $A(\alpha_{k_1})$ be the first nonzero vector among $A(\alpha_i)$, $1 \leq i \leq n$. Let $A(\alpha_{k_2})$ be the first vector among $A(\alpha_j)$, $k_1 + 1 \leq j \leq n$, which is not linearly dependent on $A(\alpha_{k_1})$. In general, let $A(\alpha_{k_i})$ be the first vector among $A(\alpha_j)$, $k_{i-1} + 1 \leq j \leq n$, which is not linearly dependent on $A(\alpha_{k_1}), \ldots, A(\alpha_{k_{i-1}})$. In this way [and we might note that this procedure is the same as that in the proof of Theorem 2.12, which is now applied to the range $A(V) = \mathrm{Sp}\,\{A(\alpha_1), \ldots, A(\alpha_n)\}$], we obtain $A(\alpha_{k_1}), \ldots, A(\alpha_{k_r})$, which form a basis of $A(V)$, where $1 \leq k_1 < \cdots < k_r \leq n$ and $r = \text{rank } A$, such that $A(\alpha_j)$ is linearly dependent on $A(\alpha_{k_1}), \ldots, A(\alpha_{k_i})$, where $k_i \leq j$. Now let

$$\Psi = \{\beta_1, \ldots, \beta_r, \beta_{r+1}, \ldots, \beta_m\}$$

be a basis of $W$ such that $\beta_i = A(\alpha_{k_i})$ for $1 \leq i \leq r$.

We shall show that the matrix $A_{\Phi,\Psi}$ is a row-reduced echelon matrix. Let us recall that the $j$th column vector of $A_{\Phi,\Psi}$ is obtained by writing down vertically the coefficients we have in representing $A(\alpha_j)$ as a linear combination of $\beta_1, \ldots, \beta_m$.

For each $j$, $1 \leq j \leq n$, $A(\alpha_j)$ is a linear combination of $\beta_1, \ldots, \beta_r$, since $\beta_1, \ldots, \beta_r$ form a basis of $A(V)$. Thus the last $m - r$ components of the $j$th column vector of $A_{\Phi,\Psi}$ are 0. This means that the last $m - r$ row vectors of $A_{\Phi,\Psi}$ are all equal to 0, that is, condition 3 is satisfied.

For each $i$, $1 \leq i \leq r$, the image $A(\alpha_{k_i})$ is nothing but $\beta_i$. Thus the $k_i$th column vector of $A_{\Phi,\Psi}$ has 1 as the $i$th component and all other components are 0, that is, $A_{\Phi,\Psi}$ satisfies condition 2.

Finally, for each $i$, $1 \leq i \leq r$, let $j$ be any index such that $j < k_i$. Then $A(\alpha_j)$ is a linear combination of $A(\alpha_{k_1}), \ldots, A(\alpha_{k_p})$, where $p$ is the largest integer such that $k_p \leq j$ (and hence $p \leq i - 1$). This means that the $i$th component of the $j$th column vector of $A_{\Phi,\Psi}$ is 0. Since $j$ is arbitrary as long as $j < k_i$, we see that all the components of the $i$th row vector of $A_{\Phi,\Psi}$ before the $k_i$th component are 0, proving that $A_{\Phi,\Psi}$ satisfies condition 1. Thus $A_{\Phi,\Psi}$ is a row-reduced echelon matrix.

We now prove Theorem 3.15. Given an $m \times n$ matrix $\mathbf{A}$ over $\mathfrak{F}$, consider the linear mapping $A\colon \mathfrak{F}^n \to \mathfrak{F}^m$ defined by $\mathbf{A}$ with respect to the standard basis $\Phi$ in $\mathfrak{F}^n$ and the standard basis $\Psi'$ of $\mathfrak{F}^m$. We have just shown that there is a basis $\Psi$ of $\mathfrak{F}^m$ such that $A_{\Phi,\Psi}$ is a row-reduced echelon matrix. By Proposition 3.12 there is a nonsingular $m \times m$ matrix $\mathbf{Q}$ such that $A_{\Phi,\Psi} = \mathbf{QA}$. This completes the proof of Theorem 3.15.

In order to derive an application of Theorem 3.15, we prove:

***Proposition 3.16.*** *Let* $\mathbf{Q}$ *be a nonsingular* $m \times m$ *matrix and* $\mathbf{A}$ *an* $m \times n$ *matrix, both over* $\mathfrak{F}$. *Then the row space of* $\mathbf{QA}$ *is equal to the row space of* $\mathbf{A}$. *In particular,* $\mathbf{QA}$ *and* $\mathbf{A}$ *have the same row rank.*

*Proof.* If $\mathbf{Q} = [q_{ij}]$ and $\mathbf{A} = [a_{jk}]$, then we have

$$(\mathbf{QA})_{ik} = \sum_{j=1}^{m} q_{ij}a_{jk}.$$

This means that the $m$ row vectors $\beta_1, \ldots, \beta_m$ of $\mathbf{QA}$, where $\beta_i = ((\mathbf{QA})_{i1}, \ldots, (\mathbf{QA})_{in})$, are linear combinations of the $m$ row vectors $\alpha_1, \ldots, \alpha_m$ of $\mathbf{A}$; that is,

$$\beta_i = \sum_{j=1}^{m} q_{ij}\alpha_j.$$

Thus the row space of $\mathbf{QA}$ is contained in the row space of $\mathbf{A}$. Since $\mathbf{Q}$ is nonsingular, we may write $\mathbf{A} = \mathbf{Q}^{-1}(\mathbf{QA})$. By applying the same argument, we see that the row space of $\mathbf{A}$ is contained in the row space of $\mathbf{QA}$. Hence $\mathbf{QA}$ and $\mathbf{A}$ have the same row space.

We shall now prove:

***Theorem 3.17.*** *For any $m \times n$ matrix $\mathbf{A}$ over $\mathcal{F}$, the row rank and the column rank are equal to each other.*

*Proof.* We consider $\mathbf{A}$ as a linear mapping $A: \mathcal{F}^n \to \mathcal{F}^m$ (with respect to the standard bases in $\mathcal{F}^n$ and $\mathcal{F}^m$). By choosing a different basis in $\mathcal{F}^n$ and keeping the standard basis in $\mathcal{F}^m$, the linear mapping $A$ can be represented by a row-reduced echelon matrix $\mathbf{QA}$ as in Theorem 3.15. We know that the column rank of $\mathbf{QA}$ equals the rank of the mapping $A$, which is equal to the column rank of $\mathbf{A}$. On the other hand, the row rank of $\mathbf{A}$ is equal to the row rank of $\mathbf{QA}$ by Proposition 3.16. Now for the row-reduced echelon matrix $\mathbf{QA}$, it is obvious that its row rank and column rank coincide. Hence the column rank and the row rank of $\mathbf{A}$ coincide.

In view of Theorem 3.17, we shall henceforth speak of the *rank* of a matrix.

A practical method of obtaining the row-reduced echelon form of a given matrix will be given in the next section.

### EXERCISE 3.3

›› **1.** In $\mathcal{R}^2$, let $\alpha_1' = (3,1)$ and $\alpha_2' = (1,-1)$.

(*a*) Find the matrix $\mathbf{P} = [p_{ij}]$ in (3.16) which expresses the standard basis with respect to $\{\alpha_1', \alpha_2'\}$.

(*b*) Consider the linear transformation $A$ of $\mathcal{R}^2$ defined by the matrix $\begin{bmatrix} 1 & -1 \\ 2 & 0 \end{bmatrix}$ with respect to the standard basis and represent it with respect to $\{\alpha_1', \alpha_2'\}$.

›› **2.** Enumerate all possible $2 \times 3$ row-reduced echelon matrices over $\mathcal{R}$.

**3.** Enumerate all possible $3 \times 2$ row-reduced echelon matrices over $\mathcal{R}$.

»☆ **4.** If an $n \times n$ matrix $\mathbf{A}$ over $\mathfrak{F}$ satisfies $\mathbf{A}^2 = \mathbf{A}$, show that there is a nonsingular $n \times n$ matrix $\mathbf{P}$ such that $\mathbf{P}^{-1}\mathbf{AP}$ equals

$$\begin{bmatrix} \mathbf{I}_r & \mathbf{O}_{r,n-r} \\ \mathbf{O}_{n-r,r} & \mathbf{O}_{n-r,n-r} \end{bmatrix},$$

where $r = \text{rank } \mathbf{A}$ (cf. Exercise 3.1, number 12).

»☆ **5.** If an $n \times n$ matrix $\mathbf{A}$ over $\mathfrak{F}$ satisfies $\mathbf{A}^2 = \mathbf{I}_n$, show that there is a nonsingular $n \times n$ matrix $\mathbf{P}$ such that $\mathbf{P}^{-1}\mathbf{AP}$ equals

$$\begin{bmatrix} \mathbf{I}_r & \mathbf{O}_{r,n-r} \\ \mathbf{O}_{n-r,r} & -\mathbf{I}_{n-r} \end{bmatrix},$$

where $r = \text{rank } \mathbf{A}$. Here $\mathfrak{F}$ is $\mathfrak{R}$, $\mathfrak{C}$, or, more generally, an arbitrary field of characteristic $\neq 2$ (cf. Sec. 5.1).

»☆ **6.** Let $\mathbf{A}$ be an $n \times n$ real matrix such that $\mathbf{A}^2 = -\mathbf{I}_n$. Prove that $n$ is even. Prove that there exists an $n \times n$ nonsingular matrix $\mathbf{P}$ such that $\mathbf{PAP}^{-1}$ is equal to

$$\begin{bmatrix} \mathbf{O}_m & -\mathbf{I}_m \\ \mathbf{I}_m & \mathbf{O}_m \end{bmatrix},$$

where $m = n/2$. (See Exercise 3.1, number 21.)

» **7.** Let $\mathbf{P}$ be a nonsingular $n \times n$ matrix and $\mathbf{A}$ an $m \times n$ matrix, both over $\mathfrak{F}$. Then the column space of $\mathbf{AP}$ is equal to the column space of $\mathbf{A}$. In particular, $\mathbf{AP}$ and $\mathbf{A}$ have the same column rank. Prove this analogue of Proposition 3.16.

☆ **8.** Let $f$ be a nonzero linear transformation of the vector space $\mathfrak{F}^n{}_n$ such that $f(\mathbf{XY}) = f(\mathbf{X})f(\mathbf{Y})$. Prove that there is a nonsingular matrix $\mathbf{P} \in \mathfrak{F}^n{}_n$ such that

$$f(\mathbf{X}) = \mathbf{PXP}^{-1} \qquad \text{for all } \mathbf{X}.$$

☆ **9.** Let $V$ be an $n$-dimensional real (or complex) vector space. A mapping $\alpha: t \to \alpha(t)$ of the unit interval $I = [0,1]$ $[I_0 = (0,1)]$ into $V$ is said to be *continuous* [*differentiable*] if the following condition is satisfied: For a basis $\Phi = \{\alpha_1, \ldots, \alpha_n\}$ of $V$ let $\alpha(t) = \sum_{i=1}^{n} f_i(t)\alpha_i$; then each function $f_i(t)$, $1 \leq i \leq n$, is continuous [differentiable]. Prove that this condition is independent of the choice of $\Phi$. If $\alpha(t)$ is differentiable, we define the derivative $d\alpha/dt$ to be $\sum_{i=1}^{n} (df_i/dt)\alpha_i$. Prove that this is independent of the choice of $\Phi$.

☆**10.** (Continuation of 9) For two mappings $\alpha$ and $\beta$ of $I$ $[I_0]$ into $V$ define the mapping $\alpha + \beta$ by $(\alpha + \beta)(t) = \alpha(t) + \beta(t)$ for each $t$. For a real-valued [or complex-valued] function $a(t)$, define the mapping $a\alpha$ by $(a\alpha)(t) = a(t)\alpha(t)$ for each $t$. Prove:

(*a*) If $\alpha$ and $\beta$ are continuous [differentiable], then $\alpha + \beta$ is continuous [differentiable and $d(\alpha + \beta)/dt = d\alpha/dt + d\beta/dt$].

(*b*) If $\alpha$ and $a$ are continuous [differentiable], then $a\alpha$ is continuous [differentiable and $d(a\alpha)/dt = (da/dt)\alpha + a(d\alpha/dt)$.]

## 3.4 ELEMENTARY OPERATIONS

Let $V$ be a vector space over $\mathcal{F}$.

### Definition 3.6

An *elementary operation* on a system of $m$ vectors $\{\alpha_1, \ldots, \alpha_m\}$ in $V$ is an operation of one of the following three kinds:

1. To replace one of the vectors $\alpha_i$ by $c\alpha_i$, where $c \neq 0$ is in $\mathcal{F}$.

**2.** To replace one of the vectors $\alpha_i$ by $\alpha_i + c\alpha_j$, where $j \neq i$ and $c \in \mathcal{F}$.
**3.** To interchange $\alpha_i$ and $\alpha_j$, where $i \neq j$.
A system $\{\alpha_1, \ldots, \alpha_m\}$ is said to be *equivalent* to a system $\{\beta_1, \ldots, \beta_m\}$ if one can get the latter from the former by a finite number of successive applications of elementary operations.

Each elementary operation, denoted by $e$, admits the inverse operation, which is again an elementary operation. In fact, if $e$ is of type 1, the inverse $e^{-1}$ is multiplication of the $i$th vector by $c^{-1}$. If $e$ is of type 2, then $e^{-1}$ is to replace the $i$th vector, namely, $\alpha_i + c\alpha_j$, by this $i$th vector plus $-c$ times the $j$th vector, namely, $\alpha_i + c\alpha_j + (-c)\alpha_j = \alpha_i$. If $e$ is of type 3, then $e^{-1}$ is the same operation as $e$.

It follows that if $\{\alpha_1, \ldots, \alpha_m\}$ is equivalent to $\{\beta_1, \ldots, \beta_m\}$, then $\{\beta_1, \ldots, \beta_m\}$ is equivalent to $\{\alpha_1, \ldots, \alpha_m\}$. Indeed, if we express this relation by $\{\alpha_1, \ldots, \alpha_m\} \approx \{\beta_1, \ldots, \beta_m\}$, then the relation $\approx$ is an equivalence relation.†

We shall prove:

***Proposition 3.18.*** *Two systems $\{\alpha_1, \ldots, \alpha_m\}$ and $\{\beta_1, \ldots, \beta_m\}$ are equivalent if and only if*

$$\mathrm{Sp}\{\alpha_1, \ldots, \alpha_m\} = \mathrm{Sp}\{\beta_1, \ldots, \beta_m\}.$$

*Proof.* It is obvious that a single elementary operation does not change the subspace spanned by the system of vectors. Hence, if $\{\alpha_1, \ldots, \alpha_m\} \approx \{\beta_1, \ldots, \beta_m\}$, then $\mathrm{Sp}\{\alpha_1, \ldots, \alpha_m\} = \mathrm{Sp}\{\beta_1, \ldots, \beta_m\}$. To prove the converse, assume that

$$W = \mathrm{Sp}\,\{\alpha_1, \ldots, \alpha_m\} = \mathrm{Sp}\,\{\beta_1, \ldots, \beta_m\}$$

and that $\dim W = r$. By rearranging the indices (which means that we perform operations of type 3 a number of times), we may assume that $\alpha_1, \ldots, \alpha_r$ form a basis of $W$ (cf. Theorem 2.12). For each $j$, $r + 1 \leq j \leq m$, $\alpha_j$ is a linear combination of $\alpha_1, \ldots, \alpha_r$: $\alpha_j = \sum_{i=1}^{r} c_i\alpha_i$. Replacing $\alpha_j$ by $\alpha_j - c_1\alpha_1$ and then by $\alpha_j - c_1\alpha_1 - c_2\alpha_2$, and so on, we have $\{\alpha_1, \ldots, \alpha_r, \alpha_{r+1}, \ldots, \alpha_m\} \approx \{\alpha_1, \ldots, \alpha_r, 0, \ldots, 0\}$. Similarly, we may assume that $\{\beta_1, \ldots, \beta_r\}$ is a basis of $W$ and that

$$\{\beta_1, \ldots, \beta_r, \beta_{r+1}, \ldots, \beta_m\} \approx \{\beta_1, \ldots, \beta_r, 0, \ldots, 0\}.$$

Since $\beta_1 \neq 0$ is a linear combination of $\alpha_1, \ldots, \alpha_r$, we may assume that the coefficient $c_1$ of

$$\beta_1 = c_1\alpha_1 + c_2\alpha_2 + \cdots + c_r\alpha_r$$

† See Appendix.

is not 0 (this can be achieved by interchanging $\alpha_1$ and $\alpha_i$ in case $c_1 = 0$ and $c_i \neq 0$). Then we have

$$\{\alpha_1, \ldots, \alpha_r, 0, \ldots, 0\} \approx \{\beta_1, \alpha_2, \ldots, \alpha_r, 0, \ldots, 0\}$$

by first replacing $\alpha_1$ by $c_1\alpha_1$ ($c_1 \neq 0$ so that this is an elementary operation of type 1) and then adding $c_2\alpha_2, \ldots, c_r\alpha_r$ successively to $c_1\alpha_1$ (this can be done by performing an elementary operation of type 2 a number of times). Next consider $\beta_2$, which is a linear combination of $\alpha_1, \ldots, \alpha_r$ and hence of $\beta_1, \alpha_2, \ldots, \alpha_r$ (note that $\alpha_1$ is a linear combination of $\beta_1, \alpha_2, \ldots, \alpha_r$). Since $\beta_2$ is not a linear combination of $\beta_1$ alone, we may assume that

$$\beta_2 = d_1\beta_1 + d_2\alpha_2 + \cdots + d_r\alpha_r$$

with $d_2 \neq 0$. By the same kind of elementary operations as before, we obtain

$$\{\beta_1, \alpha_2, \alpha_3, \ldots, \alpha_r, 0, \ldots, 0\} \approx \{\beta_1, \beta_2, \alpha_3, \ldots, \alpha_r, 0, \ldots, 0\}.$$

Continuing this process, we obtain

$$\{\alpha_1, \alpha_2, \ldots, \alpha_r, 0, \ldots, 0\} \approx \{\beta_1, \beta_2, \ldots, \beta_r, 0, \ldots, 0\}.$$

Since $\approx$ is an equivalence relation, we have

$$\{\alpha_1, \ldots, \alpha_m\} \approx \{\beta_1, \ldots, \beta_m\}.$$

We shall now apply the considerations above to the system formed by $m$ row vectors of any $m \times n$ matrix $\mathbf{A}$.

**Definition 3.7**

An *elementary row operation* on an $m \times n$ matrix $\mathbf{A}$ is an elementary operation on the system of $m$ row vectors (vectors in $\mathfrak{F}^n$). An $m \times n$ matrix $\mathbf{A}$ with $m$ row vectors $\alpha_1, \ldots, \alpha_m$ is said to be *row-equivalent* to an $m \times n$ matrix $\mathbf{B}$ with $m$ row vectors $\beta_1, \ldots, \beta_m$ if $\{\alpha_1, \ldots, \alpha_m\} \approx \{\beta_1, \ldots, \beta_m\}$. We write $\mathbf{A} \approx \mathbf{B}$. This is an equivalence relation.

The notions of elementary column operation and column equivalence can be defined in a similar manner. For the time being, however, we shall consider only row operations and row equivalence (cf. Exercise 3.4, number 6).

From Proposition 3.18, we have immediately:

***Theorem 3.19.*** *Two $m \times n$ matrices $\mathbf{A}$ and $\mathbf{B}$ over $\mathfrak{F}$ are row-equivalent if and only if they have the same row space.*

We shall now prove:

***Theorem 3.20.*** *Any $m \times n$ matrix* $\mathbf{A}$ *over* $\mathcal{F}$ *is row-equivalent to a row-reduced echelon matrix* $\mathbf{R}$. *The matrix* $\mathbf{R}$ *is determined uniquely by* $\mathbf{A}$.

*Proof.* By Theorem 3.15 there exists a nonsingular matrix $\mathbf{Q}$ such that $\mathbf{R} = \mathbf{QA}$ is a row-reduced echelon matrix. By Proposition 3.16, $\mathbf{R}$ and $\mathbf{A}$ have the same row space. By Theorem 3.19 this means that $\mathbf{A}$ and $\mathbf{R}$ are row-equivalent. In order to prove the uniqueness of a row-reduced echelon matrix $\mathbf{R}$ to which $\mathbf{A}$ is row-equivalent, it is sufficient to prove that if $\mathbf{R}$ and $\mathbf{R}'$ are $m \times n$ row-reduced echelon matrices which are row-equivalent, then $\mathbf{R} = \mathbf{R}'$. First of all, $\mathbf{R}$ and $\mathbf{R}'$ have the same row space so that they have the same rank, say, $r$. We may assume $r \neq 0$. Let $k_1 < k_2 < \cdots < k_r$ be the set of integers which determine the position of the first nonzero entry (namely, 1) of each of the first $r$ row vectors $\rho_1, \ldots, \rho_r$. Let $k_1' < k_2' < \cdots < k_r'$ be the corresponding set of integers for the matrix $\mathbf{R}'$. We shall show that $k_i = k_i'$ for $1 \leq i \leq r$. First, assume that $k_1 < k_1'$. The first row vector $\rho_1$ of $\mathbf{R}$, whose $k_1$th component is 1, cannot be a linear combination of the row vectors $\rho_1', \ldots, \rho_r'$ of $\mathbf{R}'$, whose $k_1$th components are all 0. This is a contradiction, since $\rho_1$ is in the row space of $\mathbf{R}'$. Thus $k_1 \geq k_1'$. By interchanging the roles of $\mathbf{R}$ and $\mathbf{R}'$, we see that $k_1' \geq k_1$, which implies $k_1 = k_1'$. Next, assume that $k_2 < k_2'$. The second row vector $\rho_2$ of $\mathbf{R}$ is a linear combination of $\rho_1', \ldots, \rho_r'$, and, in this linear combination, the first row vector $\rho_1'$ of $\mathbf{R}'$ must appear with a nonzero coefficient, because the other row vectors of $\mathbf{R}'$ have 0 as their $k_2$th component. But then the $k_1$th component of $\rho_2$ will not be 0, which is a contradiction. This proves $k_2 \geq k_2'$, and hence $k_2 = k_2'$ by the same argument as before. Continuing in this manner, we obtain $k_3 = k_3', \ldots, k_r = k_r'$.

We shall now show that $\rho_i = \rho_i'$ for $1 \leq i \leq r$ so that $\mathbf{R} = \mathbf{R}'$. The vector $\rho_1$ is a linear combination of $\rho_1', \rho_2', \ldots, \rho_r'$. Looking at the $k_2$th, $\ldots$, $k_r$th components of this linear combination, we see that the coefficients for $\rho_2', \ldots, \rho_r'$ are 0. Thus $\rho_1 = c\rho_1'$ with some scalar $c$. Comparing the $k_1$th components of both sides, we see that $c = 1$, that is, $\rho_1 = \rho_1'$. This argument can be applied to each row vector $\rho_i$ of $\mathbf{R}$ and hence $\mathbf{R} = \mathbf{R}'$, concluding the proof of Theorem 3.20.

We now illustrate how a given matrix $\mathbf{A}$ can be row-reduced to its echelon form $\mathbf{R}$. In practice a number of elementary row operations can be performed simultaneously to save time and space.

***Example 3.9***

$$\begin{bmatrix} 3 & 0 & 3 & 6 \\ 1 & 5 & -4 & 7 \\ 4 & 3 & 1 & 11 \end{bmatrix} \xrightarrow{(1)} \begin{bmatrix} 1 & 0 & 1 & 2 \\ 1 & 5 & -4 & 7 \\ 4 & 3 & 1 & 11 \end{bmatrix} \xrightarrow{(2)} \begin{bmatrix} 1 & 0 & 1 & 2 \\ 0 & 5 & -5 & 5 \\ 0 & 3 & -3 & 3 \end{bmatrix}$$

$$\xrightarrow{(3)} \begin{bmatrix} 1 & 0 & 1 & 2 \\ 0 & 1 & -1 & 1 \\ 0 & 1 & -1 & 1 \end{bmatrix} \xrightarrow{(4)} \begin{bmatrix} 1 & 0 & 1 & 2 \\ 0 & 1 & -1 & 1 \\ 0 & 0 & 0 & 0 \end{bmatrix},$$

where we have

1. Divided the first row by 3.
2. Subtracted the first row from the second and 4 times the first row from the third.
3. Divided the second row by 5 and the third by 3.
4. Subtracted the second row from the third.

Of course, there are other ways of row-reducing the given matrix; for example, we could have interchanged the first and second rows in the beginning.

***Example 3.10.*** The elementary operation of type 3 can be obtained as a succession of operations of types 1 and 2, as is shown by

$$\begin{bmatrix}\alpha_i\\ \alpha_j\end{bmatrix} \xrightarrow{(1)} \begin{bmatrix}\alpha_i+\alpha_j\\ \alpha_j\end{bmatrix} \xrightarrow{(2)} \begin{bmatrix}\alpha_i+\alpha_j\\ \alpha_j-(\alpha_i+\alpha_j)\end{bmatrix} = \begin{bmatrix}\alpha_i+\alpha_j\\ -\alpha_i\end{bmatrix}$$

$$\xrightarrow{(3)} \begin{bmatrix}\alpha_i+\alpha_j+(-\alpha_i)\\ -\alpha_i\end{bmatrix} = \begin{bmatrix}\alpha_j\\ -\alpha_i\end{bmatrix} \xrightarrow{(4)} \begin{bmatrix}\alpha_j\\ \alpha_i\end{bmatrix},$$

where the other row vectors remain as they are in the beginning; (1), (2), and (3) are of type 2, and (4) is of type 1.

One application of row reduction is that we can find the dimension of $\text{Sp}\{\alpha_1, \ldots, \alpha_m\}$, where $\alpha_1, \ldots, \alpha_m$ are given vectors in $\mathcal{F}^n$. Write down an $m \times n$ matrix $\mathbf{A}$ with $m$ row vectors $\alpha_1, \ldots, \alpha_m$. Row-reduce $\mathbf{A}$ to its echelon form $\mathbf{R}$ and let $\rho_1, \ldots, \rho_r$ be the set of nonzero row vectors of $\mathbf{R}$. Then $\dim \text{Sp}\{\alpha_1, \ldots, \alpha_m\} = r$ and, indeed, $\rho_1, \ldots, \rho_r$ form a basis of $\text{Sp}\{\alpha_1, \ldots, \alpha_m\}$.

We can also find the rank of a given $m \times n$ matrix $\mathbf{A}$ by row-reducing it to its echelon form.

***Example 3.11.*** For the matrix given in Example 3.9, the rank is equal to 2. The vectors $(1,0,1,2)$ and $(0,1,-1,1)$ form a basis of the row space.

***Example 3.12.*** In $\mathcal{R}^3$, let

$$\alpha_1 = (1,0,-1), \qquad \alpha_2 = (1,2,1), \qquad \alpha_3 = (0,-3,2).$$

By row-reducing the matrix $\mathbf{A} = \begin{bmatrix}1 & 0 & -1\\ 1 & 2 & 1\\ 0 & -3 & 2\end{bmatrix}$, we find the unit matrix $\mathbf{I}_3$. This means that $\alpha_1, \alpha_2, \alpha_3$ are linearly independent; hence they form a basis of $\mathcal{R}^3$.

Another related application of elementary row operations is to the practical method of deciding whether a given matrix of degree $n$ is nonsingular and of finding its inverse when it is nonsingular. We begin with:

***Proposition 3.21.*** *Let* $\mathbf{A}$ *be an* $m \times m$ *matrix and* $\mathbf{B}$ *an* $m \times n$ *matrix. For any elementary row operation* $e$ *we have* $e(\mathbf{AB}) = e(\mathbf{A})\mathbf{B}$. *In particular, for* $\mathbf{A} = \mathbf{I}_m$ *we have* $e(\mathbf{B}) = e(\mathbf{I}_m)\mathbf{B}$.

*Proof.* Let $\mathbf{A} = [a_{ij}]$ and let $\beta_1, \ldots, \beta_m$ be the row vectors of $\mathbf{B}$. Then the row vectors of $\mathbf{AB}$ are $\sum_{j=1}^{n} a_{ij}\beta_j$, $1 \leq i \leq m$. Our assertion can easily be verified for each type of elementary row operation.

**Definition 3.8**

An *elementary matrix* of degree $n$ is a matrix of the form $e(\mathbf{I}_n)$, where $e$ is an arbitrary elementary row operation.

***Example 3.13.*** We enumerate all elementary matrices of degree 2:

$$\begin{bmatrix} c & 0 \\ 0 & 1 \end{bmatrix}, c \neq 0; \quad \begin{bmatrix} 1 & 0 \\ 0 & c \end{bmatrix}, c \neq 0; \quad \begin{bmatrix} 1 & c \\ 0 & 1 \end{bmatrix}; \quad \begin{bmatrix} 1 & 0 \\ c & 1 \end{bmatrix}; \quad \begin{bmatrix} 0 & 1 \\ 1 & 0 \end{bmatrix}.$$

From $e(\mathbf{B}) = e(\mathbf{I}_m)\mathbf{B}$ in Proposition 3.21, we see that an elementary row operation $e$ on any $m \times n$ matrix $\mathbf{B}$ is the same as the left multiplication by the corresponding elementary matrix $e(\mathbf{I}_m)$. From the same formula we have

$$e^{-1}(e(\mathbf{I}_m)) = e^{-1}(\mathbf{I}_m)e(\mathbf{I}_m).$$

Since $e^{-1}(e(\mathbf{I}_m)) = \mathbf{I}_m$, we see that $e^{-1}(\mathbf{I}_m)e(\mathbf{I}_m) = \mathbf{I}_m$, so that $e(\mathbf{I}_m)$ is nonsingular and its inverse is equal to $e^{-1}(\mathbf{I}_m)$. Summing up, we have:

***Proposition 3.22***

1. *An elementary matrix* $e(\mathbf{I}_m)$ *is nonsingular and its inverse is* $e^{-1}(\mathbf{I}_m)$, *where* $e^{-1}$ *is the inverse of the elementary row operation* $e$.
2. *For any* $m \times n$ *matrix* $\mathbf{A}$ *there exist elementary matrices* $\mathbf{P}_1, \ldots, \mathbf{P}_s$ *of degree* $m$ *such that* $\mathbf{R} = \mathbf{P}_s \cdots \mathbf{P}_1\mathbf{A}$ *is the row-reduced echelon form of* $\mathbf{A}$.

We now prove:

***Theorem 3.23.*** *For an* $n \times n$ *matrix* $\mathbf{A}$ *over* $\mathcal{F}$, *the following conditions are equivalent:*

1. *The rank of* $\mathbf{A}$ *is* $n$.
2. $\mathbf{A}$ *is nonsingular.*
3. $\mathbf{A}$ *is row-equivalent to the identity matrix* $\mathbf{I}_n$.
4. $\mathbf{A}$ *is a product of elementary matrices.*

*When condition* 3 *is satisfied, let* $e_s \cdots e_1(\mathbf{A}) = \mathbf{I}_n$ *with elementary row operations* $e_1, \ldots, e_s$. *Then*

$$\mathbf{A}^{-1} = e_s \cdots e_1(\mathbf{I}_n).$$

*Proof.* The equivalence of (1) and (2) is a special case of the equivalence of (1) and (5) in Theorem 3.4 applied to the linear mapping $\mathbf{A}: \mathcal{F}^n \to \mathcal{F}^n$. Assume (1). Then the row-reduced echelon form $\mathbf{R}$ also has rank $n$ and hence $\mathbf{R} = \mathbf{I}_n$. Thus (1) implies (3). If (3) holds, then this means that there exist, by (2) of Proposition 3.22, elementary matrices $\mathbf{P}_1, \ldots, \mathbf{P}_s$ such that $\mathbf{P}_s \cdots \mathbf{P}_1\mathbf{A} = \mathbf{I}_n$. Since $\mathbf{P}_i$'s are nonsingular by (1) of Proposition 3.22, we have $\mathbf{A} = \mathbf{P}_1^{-1} \cdots \mathbf{P}_s^{-1}$, where $\mathbf{P}_i^{-1}$'s are also elementary matrices. Thus (3) implies (4). Finally, if (4) holds, then $\mathbf{A}$ is also nonsingular by (5) of Proposition 3.6. We have thus proved the equivalence of (1), (2), (3), and (4).

Now assume that (3) is satisfied and that $e_s \cdots e_1(\mathbf{A}) = \mathbf{I}_n$. Then, as we already know, $\mathbf{P}_s \cdots \mathbf{P}_1\mathbf{A} = \mathbf{I}_n$, where $\mathbf{P}_i = e_i(\mathbf{I}_n)$ for each $i$. This means that $\mathbf{A}^{-1} = \mathbf{P}_s \cdots \mathbf{P}_1$. By Proposition 3.21 we have $\mathbf{P}_s \cdots \mathbf{P}_1 = e_s \cdots e_1(\mathbf{I}_n)$.

We shall illustrate practical applications of Theorem 3.23.

***Example 3.14.*** Let $\mathbf{A} = \begin{bmatrix} 1 & 2 & 1 \\ 3 & 7 & 4 \\ 2 & -1 & 3 \end{bmatrix}$. We row-reduce $\mathbf{A}$ as follows:

$$\begin{bmatrix} 1 & 2 & 1 \\ 3 & 7 & 4 \\ 2 & -1 & 3 \end{bmatrix} \xrightarrow[e_2]{e_1} \begin{bmatrix} 1 & 2 & 1 \\ 0 & 1 & 1 \\ 0 & -5 & 1 \end{bmatrix} \xrightarrow[e_4]{e_3} \begin{bmatrix} 1 & 0 & -1 \\ 0 & 1 & 1 \\ 0 & 0 & 6 \end{bmatrix}$$

$$\xrightarrow{e_5} \begin{bmatrix} 1 & 0 & -1 \\ 0 & 1 & 1 \\ 0 & 0 & 1 \end{bmatrix} \xrightarrow[e_7]{e_6} \begin{bmatrix} 1 & 0 & 0 \\ 0 & 1 & 0 \\ 0 & 0 & 1 \end{bmatrix},$$

where we have

$e_1$: row 2 $-$ (3 $\times$ row 1),
$e_2$: row 3 $-$ (2 $\times$ row 1),
$e_3$: row 1 $-$ (2 $\times$ row 2),
$e_4$: row 3 $+$ (5 $\times$ row 2),
$e_5$: $\frac{1}{6} \times$ row 3,
$e_6$: row 1 $+$ row 3,
$e_7$: row 2 $-$ row 3.

We see that $\mathbf{R} = \mathbf{I}_3$ and hence $\mathbf{A}$ is nonsingular. To obtain $\mathbf{A}^{-1}$, we perform the same sequence of elementary row operations to $\mathbf{I}_3$ as follows.

$$\begin{bmatrix} 1 & 0 & 0 \\ 0 & 1 & 0 \\ 0 & 0 & 1 \end{bmatrix} \xrightarrow[e_2]{e_1} \begin{bmatrix} 1 & 0 & 0 \\ -3 & 1 & 0 \\ -2 & 0 & 1 \end{bmatrix} \xrightarrow[e_4]{e_3} \begin{bmatrix} 7 & -2 & 0 \\ -3 & 1 & 0 \\ -17 & 5 & 1 \end{bmatrix}$$

$$\xrightarrow{e_5} \begin{bmatrix} 7 & -2 & 0 \\ -3 & 1 & 0 \\ -\frac{17}{5} & \frac{5}{6} & \frac{1}{6} \end{bmatrix} \xrightarrow[e_7]{e_6} \begin{bmatrix} \frac{25}{6} & -\frac{7}{6} & \frac{1}{6} \\ -\frac{1}{6} & \frac{1}{6} & -\frac{1}{6} \\ -\frac{17}{6} & \frac{5}{6} & \frac{1}{6} \end{bmatrix} = \mathbf{A}^{-1}.$$

If $\mathbf{P}_i = e_i(\mathbf{I}_3)$, $1 \le i \le 7$, then $\mathbf{P}_7 \cdots \mathbf{P}_1\mathbf{A} = \mathbf{I}_3$ so that $\mathbf{A} = \mathbf{P}_1^{-1} \cdots \mathbf{P}_7^{-1}$. Each $\mathbf{P}_i^{-1}$ is equal to $e_i^{-1}(\mathbf{I}_3)$ so that

$$\mathbf{P}_1^{-1} = \begin{bmatrix} 1 & 0 & 0 \\ 3 & 1 & 0 \\ 0 & 0 & 1 \end{bmatrix}, \qquad \mathbf{P}_2^{-1} = \begin{bmatrix} 1 & 0 & 0 \\ 0 & 1 & 0 \\ 2 & 0 & 1 \end{bmatrix},$$

$$\mathbf{P}_3^{-1} = \begin{bmatrix} 1 & 2 & 0 \\ 0 & 1 & 0 \\ 0 & 0 & 1 \end{bmatrix}, \qquad \mathbf{P}_4^{-1} = \begin{bmatrix} 1 & 0 & 0 \\ 0 & 1 & 0 \\ 0 & -5 & 1 \end{bmatrix},$$

$$\mathbf{P}_5^{-1} = \begin{bmatrix} 1 & 0 & 0 \\ 0 & 1 & 0 \\ 0 & 0 & 6 \end{bmatrix}, \qquad \mathbf{P}_6^{-1} = \begin{bmatrix} 1 & 0 & -1 \\ 0 & 1 & 0 \\ 0 & 0 & 1 \end{bmatrix},$$

$$\mathbf{P}_7^{-1} = \begin{bmatrix} 1 & 0 & 0 \\ 0 & 1 & 1 \\ 0 & 0 & 1 \end{bmatrix}.$$

The matrix $\mathbf{A}$ is the product of these matrices in this order.

**EXERCISE 3.4**

» **1.** Find the rank of the following matrices:

$$\begin{bmatrix} 2 & 3 \\ -1 & 1 \\ 5 & 15 \end{bmatrix}; \quad \begin{bmatrix} 1 & 2 & -1 & 11 \\ 3 & -1 & 4 & 0 \\ -3 & 8 & -11 & 3 \end{bmatrix}; \quad \begin{bmatrix} 2 & -3 & 5 \\ 1 & 2 & -1 \\ 3 & 1 & 6 \end{bmatrix}.$$

**2.** (*a*) In $\mathcal{R}^4$, find the dimension of the subspace $W$ spanned by

$$(1,2,0,-1),\ (2,1,-1,0),\ (9,3,7,10),\ (3,0,2,4).$$

(*b*) Find a basis of $W$.

(*c*) Does $(2,7,-3,-7)$ belong to $W$?

(*d*) Does $(12,1,4,13)$ belong to $W$?

**3.** Decide whether the following matrices are nonsingular and find the inverses if they exist:

$$\begin{bmatrix} 2 & -1 & 4 \\ 1 & 0 & 3 \\ -3 & 1 & 0 \end{bmatrix}; \quad \begin{bmatrix} 4 & -2 & 1 \\ 11 & 2 & -1 \\ 1 & 2 & -1 \end{bmatrix}; \quad \begin{bmatrix} 1 & 1 & 1 \\ 0 & 1 & 1 \\ 0 & 0 & -1 \end{bmatrix}.$$

» **4.** Write down each of the following matrices as a product of elementary matrices:

$$\begin{bmatrix} 2 & 1 \\ 1 & 4 \end{bmatrix}; \quad \begin{bmatrix} 3 & -1 \\ 1 & 5 \end{bmatrix}; \quad \begin{bmatrix} 1 & 1 & 1 \\ 0 & 1 & 1 \\ 0 & 0 & 1 \end{bmatrix}.$$

**5.** Express $\begin{bmatrix} 0 & 1 \\ 1 & 0 \end{bmatrix}$ as a product of the following matrices:

$$\begin{bmatrix} 1 & 1 \\ 0 & 1 \end{bmatrix}, \quad \begin{bmatrix} 1 & 0 \\ -1 & 1 \end{bmatrix}, \quad \begin{bmatrix} 1 & 0 \\ 0 & -1 \end{bmatrix} \qquad \text{(see Example 3.13).}$$

☆ **6.** (*a*) Give the definitions of elementary column operation and column equivalence for matrices.

(*b*) State the analogue of Theorem 3.19 for column equivalence.

(*c*) Define a column-reduced echelon matrix.

(*d*) State and prove the analogue of Theorem 3.20 for column equivalence.

(*e*) Define elementary matrices in terms of elementary column operations and state the analogue of Theorem 3.23.

☆ **7.** (Continuation of 6) We shall say that an $m \times n$ matrix $\mathbf{A}$ is *equivalent* to an $m \times n$ matrix $\mathbf{B}$ if it is possible to obtain $\mathbf{B}$ by a finite number of elementary row or column operations on $\mathbf{A}$. Show that this is an equivalence relation. Prove that any $m \times n$ matrix $\mathbf{A}$ over $\mathfrak{F}$ is equivalent to the matrix

$$\begin{bmatrix} \mathbf{I}_r & \mathbf{O}_{r,n-r} \\ \mathbf{O}_{m-r,r} & \mathbf{O}_{m-r,n-r} \end{bmatrix}$$

where $r = \text{rank } \mathbf{A}$ (cf. Theorem 3.14).

## 3.5 SYSTEMS OF LINEAR EQUATIONS

We shall now apply row reduction of matrices to the problem of solving a system of linear equations. Let us first consider a homogeneous system of linear equations

$$\begin{aligned} a_{11}x_1 + \cdots + a_{1n}x_n &= 0, \\ \cdots\cdots\cdots\cdots &, \\ a_{m1}x_1 + \cdots + a_{mn}x_n &= 0, \end{aligned} \tag{3.25}$$

with coefficients $a_{ij}$ in $\mathcal{F}$. We wish to find the set of all $n$-tuples $(x_1, \ldots, x_n)$, $x_i \in \mathcal{F}$, which satisfy (3.25), namely, the set of all solutions of (3.25). As we have already seen in Example 3.4, system (3.25) can be expressed in matrix form

$$\mathbf{Ax} = \mathbf{0}, \tag{3.26}$$

wher

$$\mathbf{A} = \begin{bmatrix} a_{11} & \cdots & a_{1n} \\ \cdots & \cdots & \cdots \\ a_{m1} & \cdots & a_{mn} \end{bmatrix}, \qquad \mathbf{x} = \begin{bmatrix} x_1 \\ \cdot \\ \cdot \\ \cdot \\ x_n \end{bmatrix}, \qquad \mathbf{0} = \begin{bmatrix} 0 \\ \cdot \\ \cdot \\ \cdot \\ 0 \end{bmatrix}.$$

Considering the linear mapping $\mathbf{A}\colon \mathcal{F}^n \to \mathcal{F}^m$, $\mathbf{x} \in \mathcal{F}^n$ is a solution of (3.25) if and only if it is in the kernel of $\mathbf{A}$. The zero vector $\mathbf{0}$ in $\mathcal{F}^n$ is called a *trivial solution.*

Let $\mathbf{R} = [r_{ij}]$ be the row-reduced echelon form of the matrix $\mathbf{A}$ and consider the system of linear equations

$$\begin{aligned} r_{11}x_1 + \cdots + r_{1n}x_n &= 0, \\ \cdots\cdots\cdots\cdots &\ , \\ r_{m1}x_1 + \cdots + r_{mn}x_n &= 0, \end{aligned} \tag{3.27}$$

or, in matrix form,

$$\mathbf{Rx} = \mathbf{0}. \tag{3.28}$$

We assert that systems (3.25) and (3.27) have the same solutions (that is, a solution of one of the systems is a solution of the other). In order to prove this, it is sufficient to see that each elementary row operation on the matrix $\mathbf{A}$ will change system (3.25) into a system which has the same solutions. Multiplying the $i$th row by $c \neq 0$ simply replaces the $i$th equation of (3.25) by

$$ca_{i1}x_1 + \cdots + ca_{in}x_n = 0.$$

The operation of multiplying the $i$th row by $c$ and adding to the $j$th row replaces the $j$th equation of (3.25) by

$$(ca_{i1} + a_{j1})x_1 + \cdots + (ca_{in} + a_{jn})x_n = 0.$$

Finally, interchanging two rows of $\mathbf{A}$ means interchanging the corresponding two equations of system (3.24). It is obvious that each of these operations on system (3.25) does not change the solutions.

Let $r$ be the rank of $\mathbf{A}$ and $\mathbf{R}$, and let $k_1 < \cdots < k_r$ be the integers which determine the first nonzero entry, namely, 1, of each of the first $r$ row vectors of $\mathbf{R}$; the last row vectors of $\mathbf{R}$ are 0. Denoting by $u_1, \ldots, u_{n-r}$ the unknowns other than $x_{k_1}, x_{k_2}, \ldots, x_{k_r}$ and by $r'_{ij}$

the components of $\mathbf{R}$ on the $i$th row and the column corresponding to $u_1, \ldots, u_{n-r}$, we see that system (3.27) can be written in the form

$$
\begin{aligned}
x_{k_1} + \sum_{j=1}^{n-r} r'_{1j} u_j &= 0, \\
x_{k_2} + \sum_{j=1}^{n-r} r'_{2j} u_j &= 0, \\
\cdots\cdots\cdots &, \\
x_{k_r} + \sum_{j=1}^{n-r} r'_{rj} u_j &= 0.
\end{aligned}
\tag{3.29}
$$

Thus all the solutions of (3.25) can be obtained by assigning arbitrary values (scalars in $\mathcal{F}$) to $u_1, \ldots, u_{n-r}$ and determining $x_{k_1}, \ldots, x_{k_r}$ by (3.29) accordingly.

We make the following observations. If the number of equations $m$ in (3.25) is smaller than the number of unknowns $x_1, \ldots, x_n$ in the system, then $r \leq m < n$ and hence (3.25) has nontrivial solutions. If $m = n$, then (3.25) has a nontrivial solution if and only if rank $\mathbf{A} < n$ (that is, if and only if $\mathbf{A}$ is singular). If $\mathbf{A}$ is nonsingular, then $\mathbf{R} = \mathbf{I}_n$ and (3.25) has the unique solution $\mathbf{x} = \mathbf{0}$.

The set of all solutions of (3.25) is a subspace of $\mathcal{F}^n$, called the *solution space*. It is nothing but the kernel of the linear mapping $\mathbf{A}\colon \mathcal{F}^n \to \mathcal{F}^m$. The dimension of the solution space, which is nothing but the nullity of the mapping $\mathbf{A}$, is equal to $n - r$ (as is also clear from Theorem 3.3). One can find a basis for the solution space as follows. We know that for any $(u_1, \ldots, u_{n-r}) \in \mathcal{F}^{n-r}$ there is a solution determined by (3.29). This solution, as a vector in $\mathcal{F}^n$, has $u_1, \ldots, u_{n-r}$ as some of the components and $x_{k_1}, \ldots, x_{k_r}$, determined by (3.29), as the other components. If we denote by $\mathbf{x}^{(i)}$ the solution corresponding to the $i$th vector of the standard basis of $\mathcal{F}^{n-r}$ [namely, the solution corresponding to $(u_1, \ldots, u_{n-r}) = (0, \ldots, 1, \ldots, 0)$ with 1 as the $i$th entry], then it is clear that $\mathbf{x}^{(1)}, \mathbf{x}^{(2)}, \ldots, \mathbf{x}^{(n-r)}$ are linearly independent and hence form a basis of the solution space.

***Example 3.15.*** Consider the system

$$
\begin{aligned}
x_1 + 2x_2 + 3x_3 &= 0, \\
2x_1 + x_2 - x_3 &= 0.
\end{aligned}
$$

Since the matrix $\mathbf{A} = \begin{bmatrix} 1 & 2 & 3 \\ 2 & 1 & -1 \end{bmatrix}$ is row-reduced to $\mathbf{R} = \begin{bmatrix} 1 & 0 & -\frac{5}{3} \\ 0 & 1 & \frac{7}{3} \end{bmatrix}$, the given system has the same solutions as

$$
\begin{aligned}
x_1 - \tfrac{5}{3}x_3 &= 0, \\
x_2 + \tfrac{7}{3}x_3 &= 0,
\end{aligned}
$$

whose solutions are

$$x_1 = \frac{5c}{3}, \qquad x_2 = \frac{-7c}{3}, \qquad x_3 = c, \qquad c \text{ arbitrary},$$

or, as vectors in $\mathcal{F}^3$, $(5c/3, -7c/3, c)$. The dimension of the solution space is 1, and the solution $(\frac{5}{3}, -\frac{7}{3}, 1)$ corresponding to $c = 1$ constitutes a basis.

We shall now consider an inhomogeneous system of linear equations:

$$(3.30) \qquad \begin{aligned} a_{11}x_1 + \cdots + a_{1n}x_n &= y_1, \\ \cdots\cdots\cdots\cdots &\quad , \\ a_{m1}x_1 + \cdots + a_{mn}x_n &= y_m, \end{aligned}$$

where $a_{ij} \in \mathcal{F}$ and $(y_1, \ldots, y_m) \neq 0$ is in $\mathcal{F}^m$. If we change $y_i$'s into 0, we get the homogeneous system (3.25), which is called the *associated homogeneous system.*

In matrix notation, (3.30) can be expressed by

$$(3.31) \qquad \mathbf{Ax} = \mathbf{y},$$

where

$$\mathbf{A} = \begin{bmatrix} a_{11} & \cdots & a_{1n} \\ \cdots & \cdots & \cdots \\ a_{m1} & \cdots & a_{mn} \end{bmatrix}, \qquad \mathbf{x} = \begin{bmatrix} x_1 \\ \cdot \\ \cdot \\ \cdot \\ x_n \end{bmatrix}, \qquad \mathbf{y} = \begin{bmatrix} y_1 \\ \cdot \\ \cdot \\ \cdot \\ y_m \end{bmatrix}.$$

As we observed in Example 3.4, Eq. (3.31) has a solution $\mathbf{x}$ for a given $\mathbf{y}$ if and only if $\mathbf{y} \in \mathcal{F}^m$ lies in the range of the linear mapping $\mathbf{A}: \mathcal{F}^n \to \mathcal{F}^m$. We shall soon give a practical method of determining whether a given $\mathbf{y}$ is in the range of $\mathbf{A}$. Now assuming that it is, we can find all $\mathbf{x} \in \mathcal{F}^n$ such that $\mathbf{Ax} = \mathbf{y}$ [namely, the set of all solutions $\mathbf{x}$ of (3.30)] as follows. If $\mathbf{x}$ and $\mathbf{x}'$ are two solutions, then $\mathbf{Ax} = \mathbf{Ax}' = \mathbf{y}$ so that $\mathbf{A}(\mathbf{x}' - \mathbf{x}) = \mathbf{0}$, that is, $\mathbf{x}'' = \mathbf{x}' - \mathbf{x}$ is a solution of the associated homogeneous system (3.26). Conversely, if $\mathbf{x}$ is a solution of $\mathbf{Ax} = \mathbf{y}$ and if $\mathbf{x}''$ is a solution of $\mathbf{Ax} = \mathbf{0}$, then $\mathbf{x}' = \mathbf{x} + \mathbf{x}''$ is a solution of $\mathbf{Ax} = \mathbf{y}$, because

$$\mathbf{A}(\mathbf{x}') = \mathbf{A}(\mathbf{x} + \mathbf{x}'') = \mathbf{Ax} + \mathbf{Ax}'' = \mathbf{y} + \mathbf{0} = \mathbf{y}.$$

We have thus seen that *an arbitrary solution* (or a *general solution* as we call it) *of* (3.26) *is obtained by adding a solution of the homogeneous system* (3.25) *to a particular solution of* (3.26).

In order to find a practical method of solving (3.26), we consider the following matrix, called the *augmented matrix,*

$$\tilde{\mathbf{A}} = \begin{bmatrix} a_{11} & \cdots & a_{1n} & y_1 \\ \cdots & \cdots & \cdots & \cdots \\ a_{m1} & \cdots & a_{mn} & y_m \end{bmatrix} \qquad \text{or, more briefly,} \qquad \tilde{\mathbf{A}} = [\mathbf{A} \quad \mathbf{y}].$$

Let us perform a sequence of elementary row operations to bring **A** to its row-reduced echelon form **R**. By applying the same sequence of operations on the augmented matrix $\tilde{\mathbf{A}}$, we shall then obtain a matrix of the form

$$\tilde{\mathbf{R}} = \begin{bmatrix} r_{11} & \cdots & r_{1n} & z_1 \\ \cdot & \cdots & \cdot & \cdot \\ r_{m1} & \cdots & r_{mn} & z_m \end{bmatrix}$$

or, briefly,

$$\tilde{\mathbf{R}} = [\mathbf{R} \quad \mathbf{z}]$$

where $\mathbf{z}$ is a certain vector in $\mathcal{F}^m$. (Note that $\tilde{\mathbf{R}}$ is not in general the row-reduced echelon form of $\tilde{\mathbf{A}}$.) Each component $z_i$ of $\mathbf{z}$ is a linear expression $\sum_{j=1}^{m} c_{ij}y_j$. By the same kind of reasoning we had for (3.25) and (3.27), we see that system (3.30) has the same solutions as the new system

$$\begin{aligned} r_{11}x_1 + \cdots + r_{1n}x_n &= z_1, \\ \cdots\cdots\cdots\cdots &, \\ r_{m1}x_1 + \cdots + r_{mn}x_n &= z_m, \end{aligned} \tag{3.32}$$

or, more briefly,

$$\mathbf{R}\mathbf{x} = \mathbf{z}. \tag{3.33}$$

If $r = \text{rank } \mathbf{R} = \text{rank } \mathbf{A}$, then the coefficients $r_{ij}$ of the last $m - r$ equations in (3.32) are 0. If there is a solution $\mathbf{x}$ for (3.32), we must have

$$z_{r+1} = 0, \ \ldots, \ z_m = 0. \tag{3.34}$$

If (3.34) is satisfied, then, using the same notation as in (3.29), we see that system (3.32) can be expressed by

$$\begin{aligned} x_{k_1} + \sum_{j=1}^{n-r} r'_{1j}u_j &= z_1, \\ x_{k_2} + \sum_{j=1}^{n-r} r'_{2j}u_j &= z_2, \\ \cdots\cdots\cdots\cdots &, \\ x_{k_r} + \sum_{j=1}^{n-r} r'_{rj}u_j &= z_r. \end{aligned} \tag{3.35}$$

Here, whatever $z_1, \ldots, z_r$ may be, we may find a solution by determining $x_{k_1}, \ldots, x_{k_r}$ for any values of $u_1, \ldots, u_{n-r}$ we choose. In other words, if (3.34) is satisfied, system (3.30) has a solution and we know how to find all solutions.

Since each $z_i$ is of the form $\sum_{j=1}^{m} c_{ij}y_j$, condition (3.34) is a homogeneous system of linear equations,

$$\begin{aligned} \sum_{j=1}^{m} c_{r+1,j}y_j &= 0, \\ \cdots\cdots\cdots &, \\ \sum_{j=1}^{m} c_{mj}y_j &= 0, \end{aligned} \tag{3.36}$$

whose solutions constitute the range of the mapping **A**. It is the same as the set of all **y**'s for which the given inhomogeneous system (3.30) has a solution.

***Example 3.16.*** Consider the system over $\mathcal{R}$:

$$\begin{aligned} x_1 - 2x_2 + x_3 &= y_1, \\ 2x_1 + x_2 + x_3 &= y_2, \\ 5x_2 - x_3 &= y_3. \end{aligned}$$

When we reduce

$$\mathbf{A} = \begin{bmatrix} 1 & -2 & 1 \\ 2 & 1 & 1 \\ 0 & 5 & -1 \end{bmatrix}$$

to its row-reduced echelon form

$$\mathbf{R} = \begin{bmatrix} 1 & 0 & \frac{3}{5} \\ 0 & 1 & -\frac{1}{5} \\ 0 & 0 & 0 \end{bmatrix},$$

the augmented matrix

$$\tilde{\mathbf{A}} = \begin{bmatrix} 1 & -2 & 1 & y_1 \\ 2 & 1 & 1 & y_2 \\ 0 & 5 & -1 & y_3 \end{bmatrix}$$

is changed into

$$\tilde{\mathbf{R}} = \begin{bmatrix} 1 & 0 & \frac{3}{5} & (y_1 + 2y_2)/5 \\ 0 & 1 & -\frac{1}{5} & (y_2 - 2y_1)/5 \\ 0 & 0 & 0 & y_3 - y_2 + 2y_1 \end{bmatrix}.$$

Thus a necessary and sufficient condition for the given system to have a solution is

$$2y_1 - y_2 + y_3 = 0.$$

For $\mathbf{y} = (y_1, y_2, y_3)$ satisfying the condition above, a general solution of the system is

$$\begin{aligned} x_1 &= \frac{-3c}{5} + \frac{y_1 + 2y_2}{5}, \\ x_2 &= \frac{c}{5} + \frac{y_2 - 2y_1}{5}, \\ x_3 &= c, \end{aligned}$$

where $c$ is an arbitrary scalar. By choosing an arbitrary value of $c$, for example, $c = 0$, we obtain one particular solution:

$$\begin{aligned} x_1 &= \frac{y_1 + 2y_2}{5}, \\ x_2 &= \frac{y_2 - 2y_1}{5}, \\ x_3 &= 0 \end{aligned}$$

of the given inhomogeneous system. On the other hand,

$$x_1 = \frac{-3c}{5},$$

$$x_2 = \frac{c}{5},$$

$$x_3 = c$$

is a solution of the associated homogeneous system for any value of $c$.

**EXERCISE 3.5**

›› **1.** Find all solutions of the system over $\mathcal{R}$:

$$\begin{aligned} 2x_1 - 3x_2 + x_3 &= 0, \\ x_1 + 2x_2 - 3x_3 &= 0, \\ x_1 \qquad - x_3 &= 0. \end{aligned}$$

**2.** Find the kernel of the linear mapping $\mathcal{R}^3 \to \mathcal{R}^4$ defined by

$$\begin{bmatrix} x_1 \\ x_2 \\ x_3 \end{bmatrix} \to \begin{bmatrix} 1 & -1 & 2 \\ 2 & 1 & 3 \\ -3 & 0 & 2 \\ -2 & -4 & 5 \end{bmatrix} \begin{bmatrix} x_1 \\ x_2 \\ x_3 \end{bmatrix}.$$

Also find its rank.

›› **3.** For what values of $(y_1,y_2,y_3)$ does the system over $\mathcal{R}$,

$$\begin{aligned} 2x_1 - 3x_2 + x_3 &= y_1, \\ x_1 + 2x_2 - 3x_3 &= y_2, \\ x_1 \qquad - x_3 &= y_3, \end{aligned}$$

have a solution? For such $(y_1,y_2,y_3)$, find all solutions.

**4.** Find the range of the linear mapping $\mathcal{R}^2 \to \mathcal{R}^3$ defined by

$$\begin{bmatrix} x_1 \\ x_2 \end{bmatrix} \to \begin{bmatrix} 2 & -1 \\ 1 & 3 \\ 3 & 2 \end{bmatrix} \begin{bmatrix} x_1 \\ x_2 \end{bmatrix}.$$

Also find its kernel.

**5.** Determine the values of $c$ so that the system over $\mathcal{R}$,

$$\begin{aligned} x_1 + \qquad 2x_2 &= 1, \\ 2x_1 + (1 + c)x_2 &= -1, \\ (1 - c)x_1 + \qquad 3x_2 &= 2, \end{aligned}$$

has solutions. For such $c$, find all solutions.

››☆ **6.** Prove that the inhomogeneous system (3.30) has a solution if and only if rank $\tilde{\mathbf{A}}$ = rank $\mathbf{A}$, where $\tilde{\mathbf{A}}$ is the augmented matrix $[\mathbf{A} \quad \mathbf{y}]$.

# 4 Construction of Vector Spaces

We shall discuss several ways of constructing a vector space from a given family of vector spaces: the dual space (or conjugate space) $V^*$ of a given vector space $V$, the direct sum of two given vector spaces $V_1 \oplus V_2$, the quotient space $V/W$, where $W$ is a subspace of a vector space $V$, and many others which are obtained by considering multilinear mappings. In the last section we shall discuss bilinear functions in detail and prove the classical results about bilinear (or quadratic) forms and symmetric matrices (Theorems 4.17 to 4.19 and their corollaries).

## 4.1 DUAL SPACE

Let $V$ be a vector space over $\mathcal{F}$.

**Definition 4.1**

By a *linear function* (or *linear functional*) on $V$, we mean a linear mapping $f: V \to \mathcal{F}$, where $\mathcal{F}$ is considered as a one-dimensional vector space over $\mathcal{F}$ itself.

In other words, a linear function $f$ on $V$ is a function on $V$ with values in $\mathcal{F}$ satisfying

$$f(\alpha + \beta) = f(\alpha) + f(\beta),$$

$$f(c\alpha) = cf(\alpha),$$

for all $\alpha, \beta \in V$ and $c \in \mathcal{F}$.

We shall denote by $V^*$ the set of all linear functions on $V$. In our previous notation introduced in Sec. 3.1, we have $V^* = \mathrm{Hom}(V,\mathcal{F})$. As we already know, it is a vector space over $\mathcal{F}$, which we shall call the *dual space* (or *conjugate space*) of $V$.

**Definition 4.2**

$V^* = \mathrm{Hom}(V,\mathcal{F})$ is called the *dual space* (or *conjugate space*) of $V$.

We shall now adopt the following notation. Elements in $V^*$ are generally denoted by $\alpha^*, \beta^*, \ldots$, and the value which the linear function $\alpha^*$ takes on $\alpha \in V$ is denoted by $\langle \alpha,\alpha^* \rangle$ instead of $\alpha^*(\alpha)$. The fact that $\alpha^*$ is linear is expressed by

$$\text{(4.1)} \qquad \begin{aligned} \langle \alpha + \beta, \alpha^* \rangle &= \langle \alpha,\alpha^* \rangle + \langle \beta,\alpha^* \rangle, \\ \langle c\alpha,\alpha^* \rangle &= c\langle \alpha,\alpha^* \rangle, \end{aligned}$$

where $\alpha, \beta \in V$ and $c \in \mathcal{F}$. By definition of addition and scalar multiplication in $V^* = \mathrm{Hom}(V,\mathcal{F})$, we have

$$\text{(4.2)} \qquad \begin{aligned} \langle \alpha, \alpha^* + \beta^* \rangle &= \langle \alpha,\alpha^* \rangle + \langle \alpha,\beta^* \rangle, \\ \langle \alpha, c\alpha^* \rangle &= c\langle \alpha,\alpha^* \rangle, \end{aligned}$$

where $\alpha^*, \beta^* \in V^*$, $\alpha \in V$, and $c \in \mathcal{F}$.

We shall now prove:

***Theorem 4.1.*** *For any basis* $\{\alpha_1, \ldots ,\alpha_n\}$ *in* $V$, *there exists a basis* $\{\alpha_1^*, \ldots ,\alpha_n^*\}$ *in* $V^*$ *such that* $\langle \alpha_i,\alpha_j^* \rangle = \delta_{ij}$ *(Kronecker's delta). Such a basis is uniquely determined. In particular,* $\dim V = \dim V^*$.

*Proof.* For each $j$, $1 \leq j \leq n$, there exists a unique element $\alpha_j^* \in V^*$ such that $\langle \alpha_i,\alpha_j^* \rangle = \delta_{ij}$, $1 \leq i \leq n$, by virtue of Proposition 3.2. We show that $\alpha_1^*, \ldots, \alpha_n^*$ form a basis in $V^*$. To prove linear independence, suppose that $\sum_{j=1}^n c_j\alpha_j^* = 0$. Then we have

$$\begin{aligned} 0 = \langle \alpha_i, \sum_{j=1}^n c_j\alpha_j^* \rangle &= \sum_{j=1}^n c_j\langle \alpha_i,\alpha_j^* \rangle \\ &= \sum_{j=1}^n c_j\delta_{ij} = c_i \end{aligned}$$

for each $i$, $1 \leq i \leq n$. To show that they span $V^*$, let $\alpha^*$ be an arbitrary element of $V^*$ and let $c_j = \langle \alpha_j,\alpha^* \rangle$. The element $\sum_{j=1}^n c_j\alpha_j^*$ in $V^*$ has the value $c_i$ for $\alpha_i$ for each $i$, $1 \leq i \leq n$. Thus $\sum_{j=1}^n c_j\alpha_j^*$ and $\alpha^*$ have the same value on each $\alpha_i$ and hence coincide by virtue of Proposition 3.2. Hence we have shown that $\alpha_1^*, \ldots, \alpha_n^*$ form a basis of $V^*$.

**Definition 4.3**

The basis $\{\alpha_1^*, \ldots ,\alpha_n^*\}$ in Theorem 4.1 is called the *dual basis* corresponding to $\{\alpha_1, \ldots ,\alpha_n\}$.

***Proposition 4.2.*** *If $V$ is finite-dimensional, then for any $\alpha \neq 0$ in $V$ there exists an element $\alpha^* \in V^*$ such that $\langle \alpha,\alpha^* \rangle \neq 0$.*

*Proof.* Take a basis $\{\alpha_1, \ldots, \alpha_n\}$ of $V$ with $\alpha_1 = \alpha$ (this is possible by Theorem 2.13). Let $\alpha^*$ be the first element $\alpha_1^*$ of the corresponding dual basis. Then $\langle \alpha,\alpha^* \rangle = 1$.

***Example 4.1.*** We have so far identified the vector spaces $\mathfrak{F}^n{}_1$ and $\mathfrak{F}^1{}_n$ with the standard vector space $\mathfrak{F}^n$. Here we shall distinguish $\mathfrak{F}^n{}_1$ and $\mathfrak{F}^1{}_n$ by using the concept of dual space. Let $V = \mathfrak{F}^n{}_1$ and represent its elements by column vectors

$$\xi = \begin{bmatrix} x_1 \\ \cdot \\ \cdot \\ \cdot \\ x_n \end{bmatrix}.$$

Represent elements of $\mathfrak{F}^1{}_n$ by row vectors

$$\alpha^* = [a_1 \cdots a_n].$$

Each $\alpha^* \in \mathfrak{F}^1{}_n$ determines a linear function by

$$\langle \xi,\alpha^* \rangle = \sum_{i=1}^{n} a_i x_i = [a_1 \cdots a_n] \begin{bmatrix} x_1 \\ \cdot \\ \cdot \\ \cdot \\ x_n \end{bmatrix}. \tag{4.3}$$

Conversely, given a linear function on $V$, we may find $\alpha^* \in \mathfrak{F}^1{}_n$ so that the given linear function can be represented in the manner above. In this way, we may identify $\mathfrak{F}^1{}_n$ with the dual space $V^*$ of $V = \mathfrak{F}^n{}_1$; more precisely, $\mathfrak{F}^1{}_n$ and $V^*$ are isomorphic in the natural fashion. The dual basis in $\mathfrak{F}^1{}_n$ corresponding to the standard basis in $\mathfrak{F}^n{}_1$ is nothing but the standard basis in $\mathfrak{F}^1{}_n$.

We now prove:

***Theorem 4.3.*** *Let $V$ be an $n$-dimensional vector space $V$ over $\mathfrak{F}$. Then there is a linear isomorphism $A$ of $V$ onto the dual space $(V^*)^*$ of the dual space $V^*$ such that*

$$\langle \alpha^*,A(\alpha) \rangle = \langle \alpha,\alpha^* \rangle \tag{4.4}$$

*for all $\alpha \in V$ and $\alpha^* \in V^*$.*

*Proof.* For each $\alpha \in V$, we define a function $A(\alpha)$ on $V^*$ by formula (4.4). To show that this function is linear, let $\alpha^*, \beta^* \in V^*$ and $a, b \in \mathfrak{F}$. Then

$$\begin{aligned} \langle a\alpha^* + b\beta^*, A(\alpha) \rangle &= \langle \alpha, a\alpha^* + b\beta^* \rangle \\ &= a\langle \alpha,\alpha^* \rangle + b\langle \alpha,\beta^* \rangle = a\langle \alpha^*,A(\alpha) \rangle + b\langle \beta^*,A(\alpha) \rangle. \end{aligned}$$

Thus $A(\alpha) \in (V^*)^*$. We now show that $A$ is injective. If $A(\alpha) = 0$, then

$$\langle \alpha^*,A(\alpha) \rangle = \langle \alpha,\alpha^* \rangle = 0$$

for every $\alpha^* \in V^*$. This implies $\alpha = 0$ by virtue of the corollary to

Theorem 4.1. Since $\dim (V^*)^* = \dim V^* = \dim V$ by Theorem 4.1, we see that $A$ is a linear isomorphism (cf. Theorem 3.4).

From now on, we may identify $(V^*)^*$ with $V$ by means of the isomorphism $A$ in Theorem 4.3. The notation $\langle \alpha,\alpha^* \rangle$ will then signify that whenever $\alpha$ is fixed, it denotes the value of the linear function $\alpha$ $(= A(\alpha)) \in (V^*)^*$ on $\alpha^* \in V^*$.

**Definition 4.4**

For any subset $S$ of $V$, the subset $S' = \{\alpha^* \in V^*; \langle \alpha,\alpha^* \rangle = 0$ for every $\alpha \in S\}$ is called the *annihilator* of $S$. For any subset of $V^*$, the annihilator is similarly defined as a subset of $V$ because of the identification $(V^*)^* = V$.

We prove:

***Theorem 4.4.*** *Let $V$ be an $n$-dimensional vector space over $\mathfrak{F}$, and let $S$ be a subset of $V$.*

**1.** *The annihilator $S'$ of $S$ is a subspace of $V^*$.*

**2.** *If $S$ is a subspace of $V$, then* $\dim S' = n - \dim S$.

**3.** *The annihilator $S''$ of $S'$ is equal to the subspace of $V$ spanned by $S$. In particular, if $S$ is a subspace, then $S'' = S$.*

*Proof.* The verification of (1) is immediate. To prove (2), let $\{\alpha_1, \ldots ,\alpha_n\}$ be a basis of $V$ such that $\{\alpha_1, \ldots ,\alpha_k\}$ is a basis of $S$, where $k = \dim S$. Let $\{\alpha_1^*, \ldots ,\alpha_n^*\}$ be the dual basis corresponding to $\{\alpha_1, \ldots ,\alpha_n\}$. It is then easy to show that $\alpha_{k+1}^*, \ldots , \alpha_n^*$ form a basis of $S'$. Hence $\dim S' = n - k$. (It is also easy to show that $S'' = \mathrm{Sp}\{\alpha_1, \ldots ,\alpha_k\} = S$ in this case.) We prove (3). By applying (1) to $S'$, we know that $S''$ is a subspace of $V$ which obviously contains $S$. We shall show that if $T$ is a subspace containing $S$, then $S'' \subseteq T$ so that $S''$ is the smallest subspace containing $S$. Since $S \subseteq T$, we have clearly $S' \supseteq T'$ and hence $S'' \subseteq T''$. By the assertion in parentheses in the proof of (2), we know that $T'' = T$. Hence $S'' \subseteq T$.

***Example 4.2.*** We shall look at a system of linear equations

$$\sum_{j=1}^{n} a_{ij}x_j = 0, \qquad 1 \leq i \leq m, \tag{4.5}$$

from the point of view developed in this section. Let $V = \mathfrak{F}^n{}_1$ and $V^* = \mathfrak{F}^1{}_n$ as in Example 4.1. Let

$$\alpha_i^* = [a_{i1}\ a_{i2}\ \cdots\ a_{in}] \in \mathfrak{F}^1{}_n, \qquad 1 \leq i \leq m.$$

System (4.5) is then equivalent to

$$\langle \xi,\alpha_i^* \rangle = 0, \qquad 1 \leq i \leq m,$$

where $\xi$ is the column vector with components $x_1, \ldots, x_n$. Thus the solution space of (4.5) is equal to the annihilator of the subset $\{\alpha_1^*, \ldots, \alpha_m^*\}$ of $V^*$. If $r = \dim \mathrm{Sp}\{\alpha_1^*, \ldots, \alpha_m^*\}$, which is in fact equal to the (row) rank of the matrix $\mathbf{A} = [a_{ij}]$, then the solution space has dimension $n - r$ by (2) of Theorem 4.4.

***Example 4.3.*** Given a $k$-dimensional subspace $W$ of an $n$-dimensional vector space $V$, there exist $n - k$ elements $\alpha_1^*, \ldots, \alpha_{n-k}^*$ such that

$$W = \{\xi \in V; \langle \xi, \alpha_i^* \rangle = 0, 1 \leq i \leq n - k\},$$

that is, $W$ is the annihilator of the subset $\{\alpha_1^*, \ldots, \alpha_{n-k}^*\}$ of $V^*$. (Indeed, take $\alpha_1^*, \ldots, \alpha_{n-k}^*$ to be a basis of the annihilator $W'$ of $W$.) In particular, if $V = \mathfrak{F}^n$ and if $W$ is a subspace of $V$, then there exist, in the notation of Example 4.2, $\alpha_i^* = [a_{i1} \cdots a_{in}]$, $1 \leq i \leq n - k$, such that $W$ is the solution space of the system of linear equations

$$\sum_{j=1}^{n} a_{ij} x_j = 0, \qquad 1 \leq i \leq n - k.$$

Recall that in Sec. 3.5 we found that the range of any linear mapping $\mathfrak{F}^n \to \mathfrak{F}^m$ can be described by a certain set of linear equations.

We now define the notion of transpose of a linear mapping $A: V \to W$. Let $V^*$ and $W^*$ be the dual spaces of $V$ and $W$, respectively. For any $\beta^* \in W^*$, the function

$$\alpha \in V \to \langle A(\alpha), \beta^* \rangle \in \mathfrak{F}$$

is obviously linear and hence an element of $V^*$. We denote this element of $V^*$ by ${}^tA(\beta^*)$ so that

$$(4.6) \qquad \langle \alpha, {}^tA(\beta^*) \rangle = \langle A(\alpha), \beta^* \rangle \qquad \text{for every } \alpha \in V.$$

It is straightforward to verify that ${}^tA$ is a linear mapping of $W^*$ into $V^*$.

**Definition 4.5**

${}^tA: W^* \to V^*$ is called the *transpose* of $A: V \to W$.

We prove:

***Theorem 4.5.*** *Let $V$ and $W$ be finite-dimensional vector spaces over $\mathfrak{F}$, and let $A: V \to W$ be a linear mapping.*

**1.** *The annihilator of the range of $A$ coincides with the kernel of ${}^tA$.*

**2.** *The annihilator of the kernel of $A$ coincides with the range of ${}^tA$.*

*Proof.* 1. Let $\beta^*$ be in the annihilator of the range of $A$. Then the right-hand side of (4.6) is 0, and so is the left-hand side, for each $\alpha \in V$. This means that ${}^tA(\beta^*) = 0$, that is, $\beta^*$ is in the kernel of ${}^tA$. By reversing the argument, we see that the kernel of ${}^tA$ is contained in the annihilator of the range of $A$.

2. We first observe that, when we identify $V$ with $(V^*)^*$, the transpose of ${}^tA$ coincides with $A$. Then, by applying (1) to ${}^tA$ instead of $A$,

we see that the annihilator of the range of ${}^tA$ coincides with the kernel of $A$. Taking the annihilators of both subspaces and using (3) of Theorem 4.4, we have (2).

**Definition 4.6**

For any $m \times n$ matrix $\mathbf{A} = [a_{ij}]$, the *transpose* of $\mathbf{A}$ is the $n \times m$ matrix $\mathbf{B} = [b_{ij}]$, where $b_{ij} = a_{ji}$. It is denoted by ${}^t\mathbf{A}$.

We prove:

***Proposition 4.6.*** *Let $A: V \to W$ be a linear mapping, $\Phi = \{\alpha_1, \ldots, \alpha_n\}$ a basis of $V$, and also $\Psi = \{\beta_1, \ldots, \beta_m\}$ a basis of $W$. Let $\Phi^* = \{\alpha_1^*, \ldots, \alpha_n^*\}$ and $\Psi^* = \{\beta_1^*, \ldots, \beta_m^*\}$ be the corresponding dual bases. Then the matrix $({}^tA)_{\Psi^*,\Phi^*}$ is the transpose of the matrix $A_{\Phi,\Psi}$.*

*Proof.* Let $A_{\Phi,\Psi} = [a_{ij}]$, that is,

$$A(\alpha_i) = \sum_{j=1}^{m} a_{ji}\beta_j, \qquad 1 \leq i \leq n.$$

Then

$$\langle \alpha_i, {}^tA(\beta_j^*) \rangle = \langle A(\alpha_i), \beta_j^* \rangle = \left\langle \sum_{k=1}^{m} a_{ki}\beta_k, \beta_j^* \right\rangle$$

$$= \sum_{k=1}^{m} a_{ki}\delta_{kj} = a_{ji},$$

so that

$${}^tA(\beta_j^*) = \sum_{i=1}^{n} a_{ji}\alpha_i^*, \qquad 1 \leq j \leq m.$$

If $({}^tA)_{\Phi^*,\Psi^*} = [b_{ij}]$, then

$${}^tA(\beta_j^*) = \sum_{i=1}^{n} b_{ij}\alpha_i^*, \qquad 1 \leq j \leq m,$$

so that $b_{ij} = a_{ji}$, proving our assertion.

**EXERCISE 4.1**

» **1.** In $V = \mathcal{R}^2{}_1$, let $\alpha_1 = \begin{bmatrix} 1 \\ 1 \end{bmatrix}$ and $\alpha_2 = \begin{bmatrix} 2 \\ -1 \end{bmatrix}$. Find the dual basis in $V^* = \mathcal{R}^1{}_2$ corresponding to $\{\alpha_1, \alpha_2\}$.

**2.** In $V = \mathcal{R}^3{}_1$, let

$$\alpha_1 = \begin{bmatrix} 1 \\ 0 \\ 1 \end{bmatrix}, \qquad \alpha_2 = \begin{bmatrix} 0 \\ 1 \\ 1 \end{bmatrix}, \qquad \alpha_3 = \begin{bmatrix} 1 \\ -1 \\ 2 \end{bmatrix}.$$

Show that $\{\alpha_1, \alpha_2, \alpha_3\}$ is a basis in $V$ and find the dual basis in $V^* = \mathcal{R}^1{}_3$.

**3.** Prove that the transpose of ${}^tA$ is $A$ (as asserted in the proof of Theorem 4.5).

›› **4.** In $\mathcal{R}^4$, $\{\xi = (x_1,x_2,x_3,x_4);\ x_1 + x_2 - x_3 = 0\}$ is a subspace. Find its annihilator in the dual space.

**5.** Let $W_1$ and $W_2$ be subspaces of a vector space $V$ over $\mathfrak{F}$. Find the relationship among the annihilators $W_1'$, $W_2'$, $(W_1 + W_2)'$, and $(W_1 \cap W_2)'$.

››☆ **6.** Let $A: V \to W$ and $B: W \to U$ be linear mappings. Prove that ${}^t(BA) = {}^tA{}^tB$.

☆ **7.** For two matrices $\mathbf{A}$ and $\mathbf{B}$ such that $\mathbf{BA}$ is defined, prove that ${}^t(\mathbf{BA}) = {}^t\mathbf{A}{}^t\mathbf{B}$.

››☆ **8.** Suppose that $V$ and $W$ are finite-dimensional vector spaces over $\mathfrak{F}$, and assume that there is given a mapping $f(\alpha,\beta)$ of $V \times W$ into $\mathfrak{F}$ having the following properties:

(*a*) When $\beta$ is fixed, $f(\alpha,\beta)$ is linear in $\alpha$.
(*b*) When $\alpha$ is fixed, $f(\alpha,\beta)$ is linear in $\beta$.
(*c*) If $f(\alpha,\beta) = 0$ for all $\beta \in W$, then $\alpha = 0$.
(*d*) If $f(\alpha,\beta) = 0$ for all $\alpha \in V$, then $\beta = 0$.

Prove that there is a linear isomorphism $C$ of $W$ onto $V^*$ such that $\langle \alpha, C(\beta)\rangle = f(\alpha,\beta)$. (In other words, $W$ is essentially the dual space of $V$; in particular, dim $W$ = dim $V$.)

☆ **9.** For any $n \times n$ matrix $\mathbf{A} = [a_{ij}]$ over $\mathfrak{F}$, the *trace* of $\mathbf{A}$ is defined by

$$\text{trace } \mathbf{A} = \sum_{i=1}^{n} a_{ii}.$$

Prove that $\mathbf{A} \in \mathfrak{F}^n{}_n \to f(\mathbf{A}) = \text{trace } \mathbf{A} \in \mathfrak{F}$ is a linear function on the vector space $\mathfrak{F}^n{}_n$ such that $f(\mathbf{AB}) = f(\mathbf{BA})$ for all $\mathbf{A}$, $\mathbf{B}$.

››☆ **10.** For any $\mathbf{A} \in \mathfrak{F}^n{}_n$, show that

$$f(\mathbf{X}) = \text{trace}(\mathbf{AX}), \qquad \mathbf{X} \in \mathfrak{F}^n{}_n,$$

is a linear function on $\mathfrak{F}^n{}_n$. Conversely, for any linear function $f$ on $\mathfrak{F}^n{}_n$, there is a uniquely determined $\mathbf{A}$ in $\mathfrak{F}^n{}_n$ such that $f(\mathbf{X}) = \text{trace}(\mathbf{AX})$ for all $\mathbf{X} \in \mathfrak{F}^n{}_n$.

☆ **11.** If $f$ is a linear function on $\mathfrak{F}^n{}_n$ satisfying

$$f(\mathbf{XY}) = f(\mathbf{YX}) \qquad \text{for all } \mathbf{X}, \mathbf{Y} \in \mathfrak{F}^n{}_n,$$

then $f(\mathbf{X}) = a$ trace $\mathbf{X}$, where $a$ is a certain element in $\mathfrak{F}$.

☆ **12.** For any linear transformation $A$ of an $n$-dimensional vector space $V$ over $\mathfrak{F}$, we wish to define trace $A$. For this purpose, show that if $\Phi$ is any basis of $V$, then trace $A_\Phi$ is independent of $\Phi$. We define trace $A$ = trace $A_\Phi$.

**13.** Find the trace of the linear transformation $\mathbf{X} \in \mathfrak{F}^2{}_2 \to \mathbf{AX} \in \mathfrak{F}^2{}_2$, where

$$\mathbf{A} = \begin{bmatrix} a & b \\ c & d \end{bmatrix}.$$

**14.** Let $A$ be a linear mapping of a vector space $U$ into a vector space $V$. If rank $A = 1$, show that there exist $\alpha^* \in U^*$ and $\beta \in V$ such that $A(\xi) = \langle \xi,\alpha^*\rangle\beta$ for all $\xi \in U$.

››☆ **15.** Let $V$ be an $n$-dimensional vector space and let $\alpha \in V$ and $\alpha^* \in V^*$. Show that the trace of the linear transformation of $V$: $\xi \to \langle \xi,\alpha^*\rangle\alpha$, where $\xi \in V$, is equal to $\langle \alpha,\alpha^*\rangle$.

☆ **16.** Let $W = \{\mathbf{A} \in \mathfrak{F}^n{}_n;\ \text{trace } \mathbf{A} = 0\}$. Show that $W$ is a subspace of $\mathfrak{F}^n{}_n$ and find its dimension. Prove that $[\mathbf{A},\mathbf{B}] \in W$ for all $\mathbf{A}$, $\mathbf{B} \in \mathfrak{F}^n{}_n$. ($[\mathbf{A},\mathbf{B}] = \mathbf{AB} - \mathbf{BA}$ as was defined in Exercise 3.2, number 14.)

## 4.2 DIRECT SUM AND QUOTIENT SPACE

Let $V_1$ and $V_2$ be vector spaces over $\mathfrak{F}$. In the cartesian product $V = V_1 \times V_2$ [namely, the set of all ordered pairs $(\alpha_1,\alpha_2)$ with $\alpha_1 \in V_1$ and $\alpha_2 \in V_2$], we shall define addition and scalar multiplication (over $\mathfrak{F}$) as follows:

For $\alpha = (\alpha_1,\alpha_2)$ and $\beta = (\beta_1,\beta_2)$ in $V$,

$$\alpha + \beta = (\alpha_1 + \beta_1,\ \alpha_2 + \beta_2) \in V.$$

For $\alpha = (\alpha_1,\alpha_2)$ in $V$ and for $c \in \mathfrak{F}$,

$$c\alpha = (c\alpha_1,c\alpha_2).$$

It is easy to verify that $V$ then becomes a vector space over $\mathfrak{F}$.

**Definition 4.7**

The vector space $V$ constructed above is called the *direct sum* of $V_1$ and $V_2$ and is denoted by $V_1 \oplus V_2$.

We recall that we have already used the terminology "direct sum" and the notation $V_1 \oplus V_2$ in Sec. 2.5. The relationship between Definitions 4.7 and 2.10 is given by the following.

***Proposition 4.7***

**1.** *Let $V_1$ and $V_2$ be vector spaces over $\mathfrak{F}$ and let $W = V_1 \oplus V_2$ in the sense of Definition* 4.7. *Let*

$$W_1 = \{(\alpha_1,0);\ \alpha_1 \in V_1\} \qquad \textit{and} \qquad W_2 = \{(0,\alpha_2);\ \alpha_2 \in V_2\}.$$

*Then $W_1$ and $W_2$ are subspaces of $W$, and $W = W_1 \oplus W_2$ in the sense of Definition* 2.10.

**2.** *Let $V$ be a vector space which is the direct sum of two subspaces $V_1$ and $V_2$ in the sense of Definition* 2.10. *Then the direct sum $W = V_1 \oplus V_2$ in the sense of Definition* 4.7 *is isomorphic with $V$.*

*Proof.* 1. Clearly, $W_1$ and $W_2$ are subspaces of $W$. If $\alpha = (\alpha_1,\alpha_2)$ is an arbitrary element in $W$, then $\alpha = (\alpha_1,0) + (0,\alpha_2)$, where $(\alpha_1,0) \in W_1$ and $(0,\alpha_2) \in W_2$. If $\alpha = (\alpha_1,\alpha_2)$ is in $W_1 \cap W_2$, then we must have $\alpha_1 = 0$ and $\alpha_2 = 0$, that is, $\alpha = 0$. Thus $W = W_1 \oplus W_2$ in the sense of Definition 2.10.

2. We shall define a linear mapping $C$ of $V$ into $W$ as follows. By assumption and by (1) of Proposition 2.16, every $\alpha \in V$ is of the form $\alpha = \alpha_1 + \alpha_2$, where $\alpha_1 \in V_1$ and $\alpha_2 \in V_2$ are uniquely determined. Define

$$C(\alpha) = (\alpha_1,\alpha_2) \in W.$$

It is straightforward to verify that $C$ is a linear isomorphism of $V$ onto $W$.

It is not difficult at all to extend the notion of direct sum to the case of more than two vector spaces. Let $V_1, \ldots, V_k$ be vector spaces over $\mathfrak{F}$. In the cartesian product $V = V_1 \times \cdots \times V_k$ [namely, the set of all

$k$-tuples $(\alpha_1, \ldots, \alpha_k)$, where $\alpha_i \in V_i$ for each $i$, $1 \leq i \leq k$], we define addition and scalar multiplication by

$$(\alpha_1, \ldots, \alpha_k) + (\beta_1, \ldots, \beta_k) = (\alpha_1 + \beta_1, \ldots, \alpha_k + \beta_k),$$
$$c(\alpha_1, \ldots, \alpha_k) = (c\alpha_1, \ldots, c\alpha_k), \qquad c \in \mathfrak{F}.$$

$V$ is then a vector space over $\mathfrak{F}$, called the *direct sum* $V_1 \oplus \cdots \oplus V_k$ of $V_1, \ldots, V_k$. We leave it to the reader to formulate and prove an extension of Proposition 4.7 to the case of $V_1 \oplus \cdots \oplus V_k$ so as to relate the definition of direct sum here to that used in Proposition 2.17.

***Example 4.4.*** If we regard $\mathfrak{F}$ as a one-dimensional vector space over $\mathfrak{F}$, then $\mathfrak{F} \oplus \cdots \oplus \mathfrak{F}$ ($n$ times) is identical with $\mathfrak{F}^n$.

***Example 4.5.*** Let $V_1$, $V_2$, $W_1$, and $W_2$ be vector spaces over $\mathfrak{F}$. For any linear mappings $A_1\colon V_1 \to W_1$ and $A_2\colon V_2 \to W_2$, define a mapping $A$ of $V_1 \oplus V_2$ into $W_1 \oplus W_2$ by

$$A(\alpha_1,\alpha_2) = (A_1(\alpha_1), A_2(\alpha_2)) \qquad \text{for } (\alpha_1,\alpha_2) \in V_1 \oplus V_2.$$

It is easy to verify that $A\colon V_1 \oplus V_2 \to W_1 \oplus W_2$ is a linear mapping. $A$ is called the *direct sum* of $A_1$ and $A_2$ and is denoted by $A_1 \oplus A_2$.

***Example 4.6.*** Let $V_1$ and $V_2$ be vector spaces over $\mathfrak{F}$ and let $V_1^*$ and $V_2^*$ be their dual spaces, respectively. Then the dual space $(V_1 \oplus V_2)^*$ of $V_1 \oplus V_2$ is isomorphic with $V_1^* \oplus V_2^*$. More precisely, there is a linear isomorphism $C\colon V_1^* \oplus V_2^*$ onto $(V_1 \oplus V_2)^*$ such that

$$\langle(\alpha_1,\alpha_2), C(\alpha_1^*,\alpha_2^*)\rangle = \langle\alpha_1,\alpha_1^*\rangle + \langle\alpha_2,\alpha_2^*\rangle$$

for all $(\alpha_1,\alpha_2) \in V_1 \oplus V_2$ and $(\alpha_1^*,\alpha_2^*) \in V_1^* \oplus V_2^*$.

We shall now introduce the notion of quotient space. Let $V$ be a vector space over $\mathfrak{F}$ and $W$ a subspace of $V$. We shall say that $\alpha \in V$ is *congruent to* $\beta \in V$ *modulo* $W$ if $\beta - \alpha \in W$. In this case we shall write $\alpha \equiv \beta \pmod{W}$.

***Proposition 4.8.*** $\alpha \equiv \beta \pmod{W}$ *is an equivalence relation,*† *that is, it satisfies the following three conditions:*

**1.** $\alpha \equiv \alpha \pmod{W}$ *for every* $\alpha \in V$.

**2.** *If* $\alpha \equiv \beta \pmod{W}$, *then* $\beta \equiv \alpha \pmod{W}$.

**3.** *If* $\alpha \equiv \beta \pmod{W}$ *and* $\beta \equiv \gamma \pmod{W}$, *then* $\alpha \equiv \gamma \pmod{W}$.

*Moreover, this equivalence relation satisfies:*

**4.** *If* $\alpha_1 \equiv \beta_1 \pmod{W}$ *and* $\alpha_2 \equiv \beta_2 \pmod{W}$, *then* $\alpha_1 + \alpha_2 \equiv \beta_1 + \beta_2 \pmod{W}$.

**5.** *If* $\alpha \equiv \beta \pmod{W}$, *then* $c\alpha \equiv c\beta \pmod{W}$ *for every* $c \in \mathfrak{F}$.

*Proof.* (This was given as Exercise 2.3, number 5.) The verification is straightforward.

† See Appendix.

Let us denote by $V' = V/W$ the set of all equivalence classes of $V$ with respect to the equivalence relation above. Each element $\alpha' \in V'$ is the equivalence class of a certain element $\alpha \in V$. For each $\alpha \in V$, denote by $\alpha' = \pi(\alpha)$ the equivalence class (hence an element of $V'$) containing $\alpha$. Then $\pi$ is a mapping of $V$ onto $V'$, and $\pi(\alpha) = \pi(\beta)$ if and only if $\alpha \equiv \beta$ (mod $W$). We now wish to make $V' = V/W$ into a vector space over $\mathfrak{F}$ in such a way that $\pi$ is linear.

Let $\alpha', \beta' \in V'$. Choose $\alpha, \beta \in V$ such that $\pi(\alpha) = \alpha'$ and $\pi(\beta) = \beta'$. Then the element $\pi(\alpha + \beta)$ is independent of the choice of $\alpha$ and $\beta$ and depends only on $\alpha'$ and $\beta'$. In fact, let $\pi(\alpha_1) = \alpha'$ and $\pi(\beta_1) = \beta'$. By condition 4 of Proposition 4.8, we have $\pi(\alpha + \beta) = \pi(\alpha_1 + \beta_1)$. Thus it is possible to define $\alpha' + \beta'$ to be $\pi(\alpha + \beta)$ without any ambiguity. It is also clear that with this definition of addition in $V'$ we have $\pi(\alpha + \beta) = \pi(\alpha) + \pi(\beta)$ for all $\alpha, \beta \in V$.

Similarly, we can define, by virtue of condition 5 of Proposition 4.8, scalar multiplication $c\alpha'$ for $\alpha' \in V'$ and $c \in \mathfrak{F}$ in such a way that $\pi(c\alpha) = c\pi(\alpha)$ for all $\alpha \in V$ and $c \in \mathfrak{F}$.

It is not difficult to see that $V'$ becomes a vector space over $\mathfrak{F}$ in this manner. For example, $(\alpha' + \beta') + \gamma' = \alpha' + (\beta' + \gamma')$ can be verified as follows. Choose $\alpha$, $\beta$, and $\gamma$ in $V$ so that $\pi(\alpha) = \alpha'$, $\pi(\beta) = \beta'$, and $\pi(\gamma) = \gamma'$. Then $\alpha' + \beta' = \pi(\alpha + \beta)$, and hence $(\alpha' + \beta') + \gamma' = \pi((\alpha + \beta) + \gamma)$. On the other hand, $\beta' + \gamma' = \pi(\beta + \gamma)$ and hence $\alpha' + (\beta' + \gamma') = \pi(\alpha + (\beta + \gamma))$. Now since $(\alpha + \beta) + \gamma = \alpha + (\beta + \gamma)$ in $V$, we have $(\alpha' + \beta') + \gamma' = \alpha' + (\beta' + \gamma')$ in $V'$. The other conditions for $V'$ to be a vector space follow from the corresponding conditions which are valid in $V$. Finally, it is obvious that the mapping $\pi$ is linear.

We have thus proved:

***Proposition 4.9.*** *Let $V$ be a vector space over $\mathfrak{F}$ and let $W$ be a subspace of $V$. Then the set of equivalence classes $V' = V/W$ with respect to congruence* mod $W$ *forms a vector space such that the mapping $\pi$ of $V$ onto $V'$ is linear.*

**Definition 4.8**

The vector space $V' = V/W$ is called the *quotient space of $V$ over $W$*. The linear mapping $\pi\colon V \to V'$ is called the *natural projection.*

We remark that the kernel of $\pi$ is precisely the subspace $W$. We now prove:

***Proposition 4.10.*** *Let $W$ be a subspace of a vector space $V$ over $\mathfrak{F}$.*

**1.** *If $V$ is finite-dimensional, so is the quotient space $V/W$ and* $\dim (V/W) = \dim V - \dim W$.

**2.** *If $V/W$ is finite-dimensional, there exists a subspace $U$ of $V$ such that $V = U \oplus W$ and $\pi: V \to V/W$ induces a linear isomorphism of $U$ onto $V/W$ (that is, $\pi$ restricted to $U$ is a linear isomorphism of $U$ onto $V/W$).*

**3.** *If both $W$ and $V/W$ are finite-dimensional, so is $V$.*

*Proof.* 1. Since $\pi$ is surjective, it follows from Theorem 3.3 (rather, from its proof) that $V/W$ is finite-dimensional. Also, $\dim (V/W) + \dim W = \dim V$, because rank $A = \dim (V/W)$ and nullity $A = \dim W$.

2. Let $\alpha_1', \ldots, \alpha_r'$ be a basis of $V/W$ and let $\alpha_1, \ldots, \alpha_r$ be elements of $V$ such that $\pi(\alpha_i) = \alpha_i'$, $1 \leq i \leq r$. Let $U = \mathrm{Sp}\,\{\alpha_1, \ldots, \alpha_r\}$. First we show that $U \cap W = (0)$. If $\sum_{i=1}^{r} c_i\alpha_i \in W$, then we have

$$0 = \pi\left(\sum_{i=1}^{r} c_i\alpha_i\right) = \sum_{i=1}^{r} c_i\pi(\alpha_i) = \sum_{i=1}^{r} c_i\alpha_i',$$

which implies that $c_i = 0$, $1 \leq i \leq r$. Thus $U \cap W = (0)$. Let $\alpha$ be an arbitrary element of $V$. We may write

$$\pi(\alpha) = \sum_{i=1}^{r} c_i\alpha_i'$$

with suitable coefficients $c_i$, $1 \leq i \leq r$. Then

$$\pi\left(\alpha - \sum_{i=1}^{r} c_i\alpha_i\right) = 0,$$

which implies that $\alpha - \sum_{i=1}^{r} c_i\alpha_i \in W$. Hence $\alpha$ is the sum of this element in $W$ and the element $\sum_{i=1}^{r} c_i\alpha_i$ in $U$. This proves that $V = U \oplus W$. Clearly, $\pi(V) = \pi(U) = V/W$; that is, $\pi: U \to V/W$ is surjective. If $\alpha \in U$ satisfies $\pi(\alpha) = 0$, then $\alpha \in W \cap U$ and hence $\alpha = 0$. This shows that $\pi: U \to V/W$ is a linear isomorphism.

3. This follows from (2).

***Proposition 4.11.*** *Let $A: V \to V_1$ be a linear mapping, and let $W$ be a subspace of $V$ and $W_1$ a subspace of $V_1$ containing $A(W)$. Then there exists a unique linear mapping $A': V/W \to V_1/W_1$ such that the diagram*

$$\begin{array}{ccc} V & \xrightarrow{A} & V_1 \\ \pi \downarrow & & \downarrow \pi_1 \\ V/W & \xrightarrow{A'} & V_1/W_1 \end{array}$$

*is commutative (that is, $A'\pi = \pi_1 A$), where $\pi$ and $\pi_1$ are the natural projections.*

*Proof.* For any $\alpha' \in V/W$, choose $\alpha \in V$ such that $\pi(\alpha) = \alpha'$. The element $\pi_1(A(\alpha))$ is independent of the choice of $\alpha$, for if $\pi(\beta) = \alpha'$, then $\beta - \alpha \in W$ and hence $A(\beta) - A(\alpha) = A(\beta - \alpha) \in A(W) \subset W_1$ so that

$\pi_1(A(\alpha)) = \pi_1(A(\beta))$. We may thus define a mapping $A'$ of $V/W$ into $V_1/W_1$ such that $A'\pi(\alpha) = \pi_1 A(\alpha)$ for every $\alpha \in V$. It is easy to verify that $A'$ is a linear mapping. Finally, suppose that $A'': V/W \to V_1/W_1$ is a linear mapping such that $A''\pi = \pi_1 A$. Then for any $\alpha' \in V/W$, choose $\alpha \in V$ such that $\pi(\alpha) = \alpha'$. We have

$$A''(\alpha') = A''\pi(\alpha) = \pi_1 A(\alpha) = A'\pi(\alpha) = A'(\alpha'),$$

which shows that $A'$ and $A''$ coincide.

**Corollary.** *Let $A$ be a linear transformation of a vector space $V$ and let $W$ be a subspace of $V$ invariant by $A$ [that is, $A(W) \subseteq W$]. Then there is a unique linear transformation $A'$ of $V/W$ such that $A'\pi = \pi A$, where $\pi$ is the natural projection: $V \to V/W$.*

In the corollary above, let $\{\alpha_1, \ldots, \alpha_m, \alpha_{m+1}, \ldots, \alpha_n\}$ be a basis of $V$ such that $\{\alpha_1, \ldots, \alpha_m\}$ is a basis of $W$. Let $\mathbf{A} = [a_{ij}]$ be the matrix representing $A$ with respect to $\{\alpha_1, \ldots, \alpha_n\}$. Since $A(W) \subseteq W$, $\mathbf{A}$ is of the form

$$\mathbf{A} = \begin{bmatrix} \mathbf{A}_1 & \mathbf{B}_1 \\ \mathbf{O} & \mathbf{A}_2 \end{bmatrix},$$

where the $m \times m$ matrix $\mathbf{A}_1$ represents the restriction $A_W$ of $A$ to the subspace $W$. Since $\alpha_i' = \pi(\alpha_i)$, $m + 1 \leq i \leq n$, form a basis in $V/W$, we may express $A': V/W \to V/W$ by writing

$$\begin{aligned} A'(\alpha_i') &= A'(\pi(\alpha_i)) = \pi(A(\alpha_i)) \\ &= \pi\left(\sum_{j=1}^{n} a_{ji}\alpha_j\right) = \sum_{j=m+1}^{n} a_{ji}\alpha_j', \qquad m + 1 \leq i \leq n. \end{aligned}$$

Thus the matrix representing $A'$ with respect to $\{\alpha_{m+1}', \ldots, \alpha_n'\}$ is equal to the matrix $\mathbf{A}_2$.

### EXERCISE 4.2

☆ **1.** Let $V_1$ and $V_2$ be vector spaces over $\mathfrak{F}$. Prove that there is a linear isomorphism $C: V_1 \oplus V_2 \to V_2 \oplus V_1$ such that $C(\alpha_1, \alpha_2) = (\alpha_2, \alpha_1)$ for all $(\alpha_1, \alpha_2) \in V_1 \oplus V_2$.

☆ **2.** Let $V_1$, $V_2$, and $V_3$ be vector spaces over $\mathfrak{F}$. Prove that there is a linear isomorphism $C: (V_1 \oplus V_2) \oplus V_3 \to V_1 \oplus V_2 \oplus V_3$ such that $C((\alpha_1, \alpha_2), \alpha_3) = (\alpha_1, \alpha_2, \alpha_3)$.

**3.** Let $W_i$ be a subspace of a vector space $V_i$, $i = 1, 2$ (both over $\mathfrak{F}$). Show that there is a natural isomorphism $C: (V_1 \oplus V_2)/(W_1 \oplus W_2) \to (V_1/W_1) \oplus (V_2/W_2)$ such that $C(\pi(\alpha_1, \alpha_2)) = (\pi_1(\alpha_1), \pi_2(\alpha_2))$ for $(\alpha_1, \alpha_2) \in V_1 \oplus V_2$.

›› **4.** Let $A$ be a linear transformation of $\mathfrak{R}^2$ defined by the matrix $\begin{bmatrix} 1 & -5 \\ 3 & 2 \end{bmatrix}$ with respect to the standard basis. Show that there is no proper subspace of $\mathfrak{R}^2$ which is invariant by $A$.

›› **5.** If we use the matrix in number 4 to define a linear transformation $A$ of $\mathfrak{C}^2$ with respect to the standard basis, does $\mathfrak{C}^2$ have a proper subspace invariant by $A$?

**6.** Let $A$ be a linear transformation of an $n$-dimensional vector space $V$. Suppose that $V$ has a sequence of subspaces $V = V_0 \supseteq V_1 \supseteq \cdots \supseteq V_n = (0)$ such that $\dim V_i = n - i$ and $A(V_i) \subseteq V_{i+1}$ for each $i$, $0 \leq i \leq n-1$. Show that we can find a basis of $V$ with respect to which $A$ can be represented by a matrix $[a_{ij}]$ such that $a_{ij} = 0$ whenever $i > j$.

☆ **7.** Let $V$ be a vector space over $\mathfrak{F}$ and let $V' = V - (0)$ be the set of all nonzero vectors in $V$. In $V'$ we write $\alpha \approx \beta$ if there exists $c \neq 0$ in $\mathfrak{F}$ such that $\beta = c\alpha$. Show that this is an equivalence relation in $V'$. The set of all equivalence classes is called the *projective space associated* with $V$ and is denoted by $P(V)$. If $\dim V = n + 1$, we define $\dim P(V) = n$. In particular, $P(\mathfrak{F}^{n+1})$ is called the *standard n-dimensional projective space over* $\mathfrak{F}$ and is denoted by $P^n(\mathfrak{F})$.

☆ **8.** (Continuation of 7) For any subspace $W$ of $V$ let $Q = \pi(W - (0))$, where $\pi$ is the natural projection of $V'$ onto $P(V)$. Prove that $Q$ is the projective space associated with $W$. It is called a *projective subspace* of $P(V)$.

☆ **9.** (Continuation of 7) For any linear isomorphism $A$ of $V$ onto itself, show that there is a unique transformation $f$ of $P(V)$ onto itself such that $f\pi = \pi f$. Such a transformation $f$ is called a *projective transformation* of $P(V)$.

☆ **10.** (Continuation of 7) Let $\dim V = n + 1$. Choose a basis $\{\alpha_0, \alpha_1, \ldots, \alpha_n\}$ in $V$. For any $p \in P(V)$, let $\alpha \in V'$ be such that $\pi(\alpha) = p$. Show that the components $(a_0, a_1, \ldots, a_n)$ of $\alpha$ with respect to the basis $\{\alpha_0, \alpha_1, \ldots, \alpha_n\}$ are determined by $p$ up to a nonzero scalar. Thus we obtain a rule which associates to each $p \in P(V)$ a set of $n + 1$ scalars which are not all zero, determined up to a nonzero scalar. This is called a system of *projective coordinates* (with respect to the basis $\{\alpha_0, \alpha_1, \ldots, \alpha_n\}$ in $V$). How are two systems of projective coordinates related to each other?

## 4.3 BILINEAR FUNCTIONS

We shall first make the following definition. Let $V_1, \ldots, V_r$ and $W$ be vector spaces over $\mathfrak{F}$.

### Definition 4.9

A mapping $L$ of the cartesian product $V_1 \times \cdots \times V_r$ into $W$ is said to be *r-linear* if the following condition is satisfied: For any $i$, $1 \leq i \leq r$, and for any fixed vectors $\alpha_1, \ldots, \alpha_{i-1}, \alpha_{i+1}, \ldots, \alpha_r$ in $V_1, \ldots, V_{i-1}, V_{i+1}, \ldots, V_r$, respectively, the mapping

$$\xi \in V_i \rightarrow L(\alpha_1, \ldots, \alpha_{i-1}, \xi, \alpha_{i+1}, \ldots, \alpha_r) \in W$$

is linear. When $r = 2$, we say that $L$ is *bilinear*.

The set of all $r$-linear mappings $E(V_1, \ldots, V_r; W)$ becomes a vector space over $\mathfrak{F}$ if we define addition and scalar multiplication as follows: For $L, M \in E(V_1, \ldots, V_r; W)$, $L + M$ is defined by

$$(L + M)(\alpha_1, \ldots, \alpha_r) = L(\alpha_1, \ldots, \alpha_r) + M(\alpha_1, \ldots, \alpha_r).$$

For $L \in E(V_1, \ldots, V_r; W)$ and $c \in \mathfrak{F}$, $cL$ is defined by

$$(cL)(\alpha_1, \ldots, \alpha_r) = cL(\alpha_1, \ldots, \alpha_r).$$

The verification that $L + M$ and $cL$ are $r$-linear and that $E(V_1, \ldots, V_r; W)$ is a vector space is straightforward.

We shall now mainly consider bilinear functions on a vector space, which are defined as follows.

**Definition 4.10**

By a *bilinear function* $f$ on a vector space $V$ over $\mathfrak{F}$ we mean a bilinear mapping $V \times V \to \mathfrak{F}$, namely, an element of $E(V,V;\mathfrak{F})$, where $\mathfrak{F}$ is considered as a one-dimensional vector space over itself.

***Example 4.7.*** Let $V$ be a vector space over $\mathfrak{F}$ and $V^*$ the dual space of $V$. Then $f(\alpha,\alpha^*) = \langle\alpha,\alpha^*\rangle$ defines a bilinear mapping of $V \times V^*$ into $\mathfrak{F}$.

***Example 4.8.*** Let $V = \mathfrak{F}^n$.

$$f(\xi,\eta) = \sum_{i=1}^{n} x_i y_i \qquad \text{for } \xi = (x_i) \text{ and } \eta = (y_i)$$

is a bilinear function on $V$.

We shall prove:

***Proposition 4.12.*** *Let $V$ be an $n$-dimensional vector space over $\mathfrak{F}$. Let $\{\alpha_1, \ldots, \alpha_n\}$ be a basis of $V$ and $\{\alpha_1^*, \ldots, \alpha_n^*\}$ the corresponding dual basis of $V^*$.*

**1.** *For each pair $(i,j)$, where $1 \leq i,j \leq n$, the function $f_{ij}(\alpha,\beta) = \langle\alpha,\alpha_i^*\rangle\langle\beta,\alpha_j^*\rangle$ is a bilinear function on $V \times V$.*

**2.** *The bilinear functions $f_{ij}$, $1 \leq i,j \leq n$, form a basis of the vector space $E(V,V;\mathfrak{F})$.*

*Proof.* 1. For a fixed $\beta \in V$, we have

$$\begin{aligned} f_{ij}(c\alpha + c'\alpha', \beta) &= \langle c\alpha + c'\alpha',\alpha_i^*\rangle\langle\beta,\alpha_j^*\rangle \\ &= (c\langle\alpha,\alpha_i^*\rangle + c'\langle\alpha',\alpha_i^*\rangle)\langle\beta,\alpha_j^*\rangle \\ &= cf_{ij}(\alpha,\beta) + c'f_{ij}(\alpha',\beta). \end{aligned}$$

2. We first show that $f_{ij}$, $1 \leq i, j \leq n$, are linearly independent. Suppose that $\sum_{i,j=1}^{n} c_{ij}f_{ij} = 0$. Noting that $f_{ij}(\alpha_k,\alpha_m) = \delta_{ki}\delta_{jm}$, we have

$$0 = \left(\sum_{i,j=1}^{n} c_{ij}f_{ij}\right)(\alpha_k,\alpha_m) = \sum_{i,j=1}^{n} c_{ij}\delta_{ki}\delta_{jm} = c_{km},$$

which proves our assertion. Now for any bilinear function $f$, set $a_{ij} = f(\alpha_i,\alpha_j)$ for $1 \leq i, j \leq n$. Then

$$\left(\sum_{i,j=1}^{n} a_{ij}f_{ij}\right)(\alpha_k,\alpha_m) = a_{km} = f(\alpha_k,\alpha_m).$$

Thus the bilinear functions $f$ and $g = \sum_{i,j=1}^{n} a_{ij}f_{ij}$ satisfy $f(\alpha_k,\alpha_m) = g(\alpha_k,\alpha_m)$ for $1 \leq k, m \leq n$. For the fixed vector $\alpha_m$, the linear functions $\xi \to f(\xi,\alpha_m)$ and $\xi \to g(\xi,\alpha_m)$ have the same value for each $\alpha_k$. Hence $f(\xi,\alpha_m) = g(\xi,\alpha_m)$ for all $\xi \in V$. By the same argument it follows that $f(\xi,\eta) = g(\xi,\eta)$ for all $\eta$. Thus $f = g$, showing that $f$ is a linear combination of $f_{ij}$, $1 \leq i, j \leq n$.

For a fixed basis $\Phi = \{\alpha_1, \ldots, \alpha_n\}$ of $V$, the matrix $\mathbf{A} = [a_{ij}]$, where $a_{ij} = f(\alpha_i,\alpha_j)$, $1 \leq i, j \leq n$, is called the *matrix associated* with the bilinear function $f$ with respect to $\Phi$.

***Corollary.*** *Let $\Phi = \{\alpha_1, \ldots, \alpha_n\}$ be a basis of a vector space $V$ over $\mathcal{F}$. For any $n \times n$ matrix $\mathbf{A} = [a_{ij}]$ over $\mathcal{F}$, there is one and only one bilinear function on $V$ which has $\mathbf{A}$ as the associated matrix with respect to $\Phi$.*

*Proof.* Using the notation of Proposition 4.12, it is obvious that $f = \sum_{i,j=1}^{n} a_{ij}f_{ij}$ is such a bilinear function.

We may also express the bilinear function $f$ in matrix notation. If

$$\xi = \sum_{i=1}^{n} x_i\alpha_i \qquad \text{and} \qquad \eta = \sum_{i=1}^{n} y_i\alpha_i,$$

then

$$f(\xi,\eta) = f\left(\sum_{i=1}^{n} x_i\alpha_i, \sum_{j=1}^{n} y_j\alpha_j\right)$$
$$= \sum_{i,j=1}^{n} x_iy_jf(\alpha_i,\alpha_j) = \sum_{i,j=1}^{n} a_{ij}x_iy_j,$$

which is equal to the matrix product

$$[x_1 \cdots x_n]\,\mathbf{A}\begin{bmatrix} y_1 \\ \cdot \\ \cdot \\ \cdot \\ y_n \end{bmatrix} \tag{4.7}$$

or, more briefly,

$${}^t\xi\mathbf{A}\eta,$$

where $\xi$ and $\eta$ stand for column vectors

$$\begin{bmatrix} x_1 \\ \cdot \\ \cdot \\ \cdot \\ x_n \end{bmatrix} \quad \text{and} \quad \begin{bmatrix} y_1 \\ \cdot \\ \cdot \\ \cdot \\ y_n \end{bmatrix},$$

respectively, and ${}^t\xi$ is the transpose of $\xi$.

The matrix associated with a bilinear function depends, of course, on a basis. In analogy to the corollary of Proposition 3.12, we have:

***Proposition 4.13.*** *Let $\Phi = \{\alpha_1, \ldots, \alpha_n\}$ and $\Phi' = \{\alpha'_1, \ldots, \alpha'_n\}$ be two bases in a vector space $V$, and let*

$$\alpha_i = \sum_{j=1}^{n} p_{ji}\alpha'_j \qquad \textit{and} \qquad \mathbf{P} = [p_{ji}].$$

*The matrix $\mathbf{A}$ associated with a bilinear function $f$ with respect to $\Phi$ and the matrix $\mathbf{A}'$ associated with $f$ with respect to $\Phi'$ are related by*

$$\mathbf{A} = {}^t\mathbf{P}\mathbf{A}'\mathbf{P}. \tag{4.8}$$

*Proof.* We have, setting $\mathbf{A} = [a_{ij}]$ and $\mathbf{A}' = [a'_{ij}]$,

$$\begin{aligned} a_{ij} = f(\alpha_i,\alpha_j) &= f\left(\sum_{k=1}^{n} p_{ki}\alpha'_k, \sum_{m=1}^{n} p_{mj}\alpha'_m\right) \\ &= \sum_{k,m=1}^{n} p_{ki}p_{mj}f(\alpha'_k,\alpha'_m) = \sum_{k,m=1}^{n} p_{ki}p_{mj}a'_{km} \\ &= \sum_{m=1}^{n}\left(\sum_{k=1}^{n} p_{ki}a'_{km}\right)p_{mj}, \end{aligned}$$

which shows that $\mathbf{A} = {}^t\mathbf{P}\mathbf{A}'\mathbf{P}$.

We now make the following definition:

**Definition 4.11**

A bilinear function $f$ on $V$ over $\mathfrak{F}$ is said to be *symmetric* [*skew-symmetric*] if

$$f(\eta,\xi) = f(\xi,\eta) \qquad [f(\eta,\xi) = -f(\xi,\eta)]$$

for all $\xi, \eta \in V$. Here $\mathfrak{F}$ is $\mathfrak{R}$ or $\mathfrak{C}$.† A symmetric bilinear function is also called a *bilinear form*.

***Proposition 4.14.*** *Let $\mathbf{A}$ be the matrix associated with a bilinear function $f$ on $V$ with respect to a basis $\{\alpha_1, \ldots, \alpha_n\}$ of $V$. Then $f$ is symmetric [skew-symmetric] if and only if $\mathbf{A}$ is symmetric, that is, ${}^t\mathbf{A} = \mathbf{A}$ [skew-symmetric, that is, ${}^t\mathbf{A} = -\mathbf{A}$].*

*Proof.* If $f$ is symmetric, then

$$a_{ij} = f(\alpha_i,\alpha_j) = f(\alpha_j,\alpha_i) = a_{ji},$$

that is, $\mathbf{A}$ is symmetric. Conversely, assume that $\mathbf{A}$ is symmetric. For any

$$\xi = \sum_{i=1}^{n} x_i\alpha_i \quad \text{and} \quad \eta = \sum_{i=1}^{n} y_i\alpha_i,$$

† Or more generally, $\mathfrak{F}$ can be a field of characteristic $\neq 2$. For characteristic 2, we have $a + a = 0$, that is, $a = -a$ for every $a \in \mathfrak{F}$ so that properties of symmetry and skew-symmetry coincide. (For the characteristic, see Sec. 5.1.)

we have

$$f(\xi,\eta) = \sum_{i=1}^{n} \sum_{j=1}^{n} x_i y_j f(\alpha_i,\alpha_j) = \sum_{i,j=1}^{n} a_{ij} x_i y_j$$
$$= \sum_{i,j=1}^{n} a_{ji} x_i y_j,$$

since $a_{ij} = a_{ji}$. By interchanging $i$ and $j$, the expression above is equal to

$$\sum_{i=1}^{n} \sum_{j=1}^{n} a_{ij} y_i x_j = \sum_{i=1}^{n} \sum_{j=1}^{n} y_i x_j f(\alpha_i,\alpha_j) = f(\eta,\xi),$$

that is, $f(\xi,\eta) = f(\eta,\xi)$. The proof for a skew-symmetric bilinear function is similar and hence is omitted.

***Proposition 4.15.*** *Let $f$ be a bilinear form on a vector space $V$. Then we have the polarization identity:*

$$f(\xi,\eta) = \tfrac{1}{2}[f(\xi + \eta, \xi + \eta) - f(\xi,\xi) - f(\eta,\eta)]$$

*for all $\xi$, $\eta \in V$.*

*Proof.* This is obvious from

$$f(\xi + \eta, \xi + \eta) = f(\xi,\xi) + f(\xi,\eta) + f(\eta,\xi) + f(\eta,\eta)$$
$$= f(\xi,\xi) + f(\eta,\eta) + 2f(\xi,\eta).$$

**Definition 4.12**

For a bilinear form $f$ on $V$, the function $g$ on $V$ defined by $g(\xi) = f(\xi,\xi)$ is called the *quadratic form* on $V$ associated with $f$.

Proposition 4.15 shows that the bilinear form $f$ is determined uniquely by the corresponding quadratic form $g$. Corresponding to matrix representations of bilinear functions with respect to a basis of $V$, we have

$$g(\xi) = \sum_{i,j=1}^{n} a_{ij} x_i x_j = [x_1, \ldots ,x_n]\mathbf{A} \begin{bmatrix} x_1 \\ \cdot \\ \cdot \\ \cdot \\ x_n \end{bmatrix} = {}^t\xi \mathbf{A} \xi,$$

where $\xi = \sum_{i=1}^{n} x_i \alpha_i$, which also stands for the column vector with components $x_1, \ldots , x_n$.

In order to study bilinear forms in detail, we make the following definition.

**Definition 4.13**

For a bilinear form $f$ on a vector space $V$, the *null space* $N_f$ of $f$ is $\{\alpha \in V; f(\alpha,\xi) = 0 \text{ for all } \xi \in V\}$, which is a subspace of $V$. $f$ is said to be *nondegenerate* (or *nonsingular*) if $N_f = (0)$. When $V$ is finite-dimensional, the *rank* of $f$ is defined to be $\dim V - \dim N_f$.

The bilinear form $f$ is nondegenerate if and only if the following condition is satisfied: If $f(\xi,\eta) = 0$ for all $\eta \in V$, then $\xi = 0$.

***Proposition 4.16.*** *Let $f$ be a bilinear form on an $n$-dimensional vector space $V$ and let $\mathbf{A}$ be the matrix associated with $f$ with respect to a basis $\{\alpha_1, \ldots, \alpha_n\}$. Then the rank of $f$ is equal to the rank of $\mathbf{A}$. In particular, $f$ is nondegenerate if and only if $\mathbf{A}$ is nonsingular.*

*Proof.* First take a basis $\Phi' = \{\alpha'_1, \ldots, \alpha'_n\}$ of $V$ such that $\alpha'_{r+1}, \ldots, \alpha'_n$ form a basis of $N_f$. Then

$$a'_{ij} = f(\alpha'_i,\alpha'_j) = 0 \qquad \text{for } 1 \le i \le n \text{ and } r+1 \le j \le n.$$

Since the matrix $\mathbf{A}' = [a'_{ij}]$ is symmetric, it follows that $\mathbf{A}'$ is of the form

$$\mathbf{A}' = \begin{bmatrix} * & \mathbf{O}_{r,n-r} \\ \mathbf{O}_{n-r,r} & \mathbf{O}_{n-r,n-r} \end{bmatrix}$$

where $*$ is an $r \times r$ matrix. In order to show that the first $r$ row vectors of $\mathbf{A}'$ are linearly independent, assume that $\sum_{i=1}^{r} c_i a'_{ij} = 0$ for every $j$, $1 \le j \le r$. This implies

$$f\left(\sum_{i=1}^{r} c_i\alpha'_i,\alpha'_j\right) = \sum_{i=1}^{r} c_i a'_{ij} = 0$$

for every $j$, $1 \le j \le r$. Since this holds for every $j$, $r+1 \le j \le n$, as well, we see that $\sum_{i=1}^{r} c_i\alpha'_i$ is in $N_f$. Hence $c_1 = \cdots = c_r = 0$. We have thus proved that the rank of $\mathbf{A}'$ is equal to $r$.

In order to conclude the proof of Proposition 4.16, let $\Phi$ be an arbitrary basis of $V$. The matrix $\mathbf{A}$ associated with $f$ with respect to $\Phi$ and the matrix $\mathbf{A}'$ above are related by $\mathbf{A} = {}^t\mathbf{P}\mathbf{A}'\mathbf{P}$, where $\mathbf{P}$ is a nonsingular matrix (Proposition 4.13). By Proposition 3.16 and its column analogue (Exercise 3.3, number 7), we see that the (row) rank of $\mathbf{A}'$ is equal to the (row) rank of ${}^t\mathbf{P}\mathbf{A}'$, which is equal to the (column) rank of ${}^t\mathbf{P}\mathbf{A}'\mathbf{P}$. Thus the rank of $\mathbf{A}'$ is equal to the rank of $\mathbf{A}$. Hence the rank of $\mathbf{A}$ is equal to $r$, namely, the rank of $f$.

We shall now consider the problem of finding a basis of $V$ which represents a given bilinear form $f$ in the simplest possible way. First we prove:

***Theorem 4.17.*** *Let $f$ be a bilinear form on an $n$-dimensional vector space $V$ over $\mathcal{F}$ and let $r$ be the rank of $f$. Then there exists a basis $\{\alpha_1, \ldots, \alpha_n\}$ of $V$ with the following properties:*
**1.** $f(\alpha_i,\alpha_j) = 0$ *for* $i \neq j$.
**2.** $c_i = f(\alpha_i,\alpha_i) \neq 0$ *for* $1 \leq i \leq r$.
**3.** $f(\alpha_j,\alpha_j) = 0$ *for* $r + 1 \leq j \leq n$.
*Hence the associated matrix* $\mathbf{A}$ *is a diagonal matrix of the form*

$$(4.9) \qquad \mathbf{A} = \begin{bmatrix} c_1 & & & & & & \\ & \cdot & & & & & \\ & & \cdot & & & & \\ & & & c_r & & & \\ & & & & 0 & & \\ & & & & & \cdot & \\ & & & & & & 0 \end{bmatrix}.$$

*Proof.* We shall first prove this result by assuming that the bilinear form $f$ is nondegenerate. Let $\alpha_1$ be an element of $V$ such that $c_1 = f(\alpha_1,\alpha_1)$ is not 0. There is such an element, because if $f(\xi,\xi) = 0$ for every $\xi \in V$, then the polarization identity (Proposition 4.15) shows that $f(\xi,\eta) = 0$ for all $\xi, \eta \in V$, contrary to the assumption. Let $W_1 = \mathrm{Sp}\{\alpha_1\}$ and $W_2 = \{\xi \in V; f(\xi,\eta) = 0 \text{ for all } \eta \in W_1\}$. We prove that $V = W_1 \oplus W_2$. To show that $W_1 \cap W_2 = (0)$, let $\xi \in W_1 \cap W_2$. Since $\xi \in W_1$, we have $\xi = a\alpha_1$ for some $a \in \mathcal{F}$. Since $\xi \in W_2$, we have

$$f(\xi,\alpha_1) = af(\alpha_1,\alpha_1) = 0,$$

which implies $a = 0$, because $f(\alpha_1,\alpha_1) \neq 0$. Hence $\xi = 0$. Next we shall show that $V = W_1 + W_2$. For any $\alpha \in V$, let $a = f(\alpha,\alpha_1)/c_1$ and let $\beta = \alpha - a\alpha_1$. Then

$$f(\beta,\alpha_1) = f(\alpha,\alpha_1) - af(\alpha_1,\alpha_1) = 0,$$

which implies that $\beta \in W_2$. Thus $\alpha = a\alpha_1 + \beta \in W_1 + W_2$, proving our assertion.

Now the restriction of $f$ to $W_2$ is clearly nondegenerate. In fact, assume that $\xi \in W_2$ and $f(\xi,\eta) = 0$ for all $\eta \in W_2$. Since $f(\xi,\eta) = 0$ for all $\eta \in W_1$ by definition of $W_2$, and since $V = W_1 \oplus W_2$, it follows that $f(\xi,\eta) = 0$ for all $\eta \in V$. Since $f$ is nondegenerate by assumption, we must have $\xi = 0$. Thus the restriction of $f$ to $W_2$ is nondegenerate. By choosing $\alpha_2 \in W_2$ such that $c_2 = f(\alpha_2,\alpha_2) \neq 0$, we see, just as before, that

$$W_2 = \mathrm{Sp}\{\alpha_2\} \oplus W_3, \qquad \text{where } W_3 = \{\xi \in W_2; f(\xi,\alpha_2) = 0\}.$$

Moreover, the restriction of $f$ to $W_3$ is nondegenerate. Continuing this process (or using induction on dim $V$ if we wish), we can find a basis $\{\alpha_1,\alpha_2, \ldots ,\alpha_n\}$ of $V$ such that $c_i = f(\alpha_i,\alpha_i) \neq 0$ for each $i$, $1 \leq i \leq n$, and such that $f(\alpha_i,\alpha_j) = 0$ for $i \neq j$, proving the desired result in the case where $f$ is nondegenerate.

Now, in the general case, let $N_f$ be the null space of $f$. We may choose a subspace $V_1$ of $V$ such that $V = V_1 \oplus N_f$. The restriction of $f$ to $V_1$ is a bilinear form on $V_1$ which is nondegenerate, as can be easily seen by the same kind of argument as before. Applying the theorem already proved for nondegenerate bilinear forms to this restriction of $f$ to $V_1$, we see that there exists a basis $\{\alpha_1, \ldots ,\alpha_r\}$ of $V_1$, where $r = \text{rank} f$, such that $c_i = f(\alpha_i,\alpha_i) \neq 0$ for each $i$ and such that $f(\alpha_i,\alpha_j) = 0$ for $i \neq j$. By adding a basis $\{\alpha_{r+1}, \ldots ,\alpha_n\}$ of $N_f$, we obtain a basis of $V$

$$\{\alpha_1, \ldots ,\alpha_r,\alpha_{r+1}, \ldots ,\alpha_n\},$$

which has properties 1, 2, and 3. We have thus proved Theorem 4.17.

***Corollary.*** *For any symmetric $n \times n$ matrix $\mathbf{A}$ over $\mathcal{F}$, there is a nonsingular matrix $\mathbf{P}$ over $\mathcal{F}$ such that ${}^t\mathbf{PAP}$ is of the form* (4.9), *where* $r =$ rank $\mathbf{A}$ *and* $c_i \neq 0$.

*Proof.* Consider the bilinear form on the vector space $\mathcal{F}^n$ which is represented by $\mathbf{A}$ with respect to the standard basis. Applying Theorem 4.17 to this bilinear form and recalling Proposition 4.13, we get the corollary.

We shall now specialize Theorem 4.17 to the cases $\mathcal{F} = \mathcal{R}$ and $\mathcal{C}$. We prove:

***Theorem 4.18.*** *Let $f$ be a bilinear form on a complex vector space $V$ of dimension $n$. Then there exists a basis $\{\alpha_1, \ldots ,\alpha_n\}$ of $V$ such that*

**1.** $f(\alpha_i,\alpha_j) = 0$ *for* $i \neq j$.

**2.** $f(\alpha_i,\alpha_i) = 1$ *for* $1 \leq i \leq r$, *where* $r =$ rank $f$.

**3.** $f(\alpha_j,\alpha_j) = 0$ *for* $r + 1 \leq j \leq n$.

*The associated matrix is of the form*

$$\begin{bmatrix} 1 & & & & & & \\ & \cdot & & & & & \\ & & \cdot & & & & \\ & & & 1 & & & \\ & & & & 0 & & \\ & & & & & \cdot & \\ & & & & & & 0 \end{bmatrix} \tag{4.10}$$

*Proof.* After choosing a basis $\{\alpha_1, \ldots, \alpha_n\}$ as in Theorem 4.17, we may find a complex number $d_i$ such that $d_i^2 = c_i$ for each $i$, $1 \leq i \leq r$. Then $f(\alpha_i/d_i, \alpha_i/d_i) = 1$, and $\{\alpha_1/d_1, \ldots, \alpha_r/d_r, \alpha_{r+1}, \ldots, \alpha_n\}$ is a desired basis.

**Corollary.** *For any symmetric complex matrix* $\mathbf{A}$, *there exists a complex nonsingular matrix* $\mathbf{P}$ *such that* ${}^t\mathbf{PAP}$ *is of the form* (4.10), *where the number of* 1*'s appearing equals the rank of* $\mathbf{A}$.

For a real vector space we prove

**Theorem 4.19.** *Let $f$ be a bilinear form on a real vector space $V$ of dimension $n$. Then there exists a basis $\{\alpha_1, \ldots, \alpha_n\}$ of $V$ such that:*

**1.** $f(\alpha_i, \alpha_j) = 0$ *for* $i \neq j$.
**2.** $f(\alpha_i, \alpha_i) = 1$ *for* $1 \leq i \leq p$.
**3.** $f(\alpha_k, \alpha_k) = -1$ *for* $p + 1 \leq k \leq r$.
**4.** $f(\alpha_j, \alpha_j) = 0$ *for* $r + 1 \leq j \leq n$.

*Here $r$ is the rank of $f$ and $p$ is a certain integer, $0 \leq p \leq r$, which is uniquely determined by $f$.*

*Proof.* Choose a basis $\{\alpha_1, \ldots, \alpha_n\}$ as in Theorem 4.17. For each $i$, $1 \leq i \leq r$, we modify the element $\alpha_i$ as follows. If $c_i > 0$, then there is a real number $d_i$ such that $d_i^2 = c_i$. Then

$$f\left(\frac{\alpha_i}{d_i}, \frac{\alpha_i}{d_i}\right) = \frac{1}{d_i^2} f(\alpha_i, \alpha_i) = \frac{c_i}{d_i^2} = 1.$$

If $c_i < 0$, then there is a real number $d_i$ such that $d_i^2 = -c_i$. Then

$$f\left(\frac{\alpha_i}{d_i}, \frac{\alpha_i}{d_i}\right) = \frac{1}{d_i^2} f(\alpha_i, \alpha_i) = \frac{c_i}{d_i^2} = -1.$$

Thus, by rearranging $\alpha_1/d_1, \ldots, \alpha_r/d_r, \alpha_{r+1}, \ldots, \alpha_n$, we obtain a basis of $V$ with properties 1, 2, 3, and 4, where $p$ is the number of positive $c_i$'s. We shall now prove the fact that $p$ is determined uniquely by $f$ (although a choice of a basis with properties 1, 2, 3, and 4 may not be unique).

First we make the following remark. For a basis $\{\alpha_1, \ldots, \alpha_n\}$ satisfying conditions 1, 2, 3, and 4, let

$$V^+ = \text{Sp}\,\{\alpha_1, \ldots, \alpha_p\} \quad \text{and} \quad V^- = \text{Sp}\,\{\alpha_{p+1}, \ldots, \alpha_n\}.$$

Then

$$f(\xi,\xi) > 0 \qquad \text{for any nonzero vector } \xi \text{ in } V^+,$$

and

$$f(\xi,\xi) < 0 \qquad \text{for any nonzero vector } \xi \text{ in } V^-.$$

In fact, if $\xi = \sum_{i=1}^{p} x_i\alpha_i$, then

$$f(\xi,\xi) = \sum_{i,j=1}^{p} x_ix_jf(\alpha_i,\alpha_j) = \sum_{i=1}^{p} x_i^2 > 0$$

unless all $x_i$'s are 0, proving the first assertion. The second assertion can be proved in the same way.

Let $\{\beta_1, \ldots ,\beta_n\}$ be a basis of $V$ such that:

1′. $f(\beta_i,\beta_j) = 0$ for $i \neq j$.
2′. $f(\beta_i,\beta_i) = 1$ for $1 \leq i \leq q$.
3′. $f(\beta_k,\beta_k) = -1$ for $q + 1 \leq k \leq r$.
4′. $f(\beta_j,\beta_j) = 0$ for $r + 1 \leq j \leq n$.

We wish to show that $p = q$. Assuming $p < q$, we shall obtain a contradiction. Let $\{\beta_1^*, \ldots ,\beta_n^*\}$ be the dual basis corresponding to $\{\beta_1, \ldots ,\beta_n\}$. Since $r - q < r - p$, the system of $r - q$ homogeneous linear equations on the subspace $\mathrm{Sp}\{\alpha_{p+1}, \ldots ,\alpha_r\}$ of dimension $r - p$:

$$\langle\xi,\beta_k^*\rangle = 0, \qquad q + 1 \leq k \leq r,$$

has a nonzero solution, say, $\alpha \in \mathrm{Sp}\,\{\alpha_{p+1}, \ldots ,\alpha_r\}$. This means that $\alpha$ is a linear combination of $\beta_1, \ldots ,\beta_q$ and $\beta_{r+1}, \ldots , \beta_n$ so that $\alpha = \beta + \gamma$, where $\beta \in \mathrm{Sp}\{\beta_1, \ldots ,\beta_q\}$ and $\gamma \in \mathrm{Sp}\{\beta_{r+1}, \ldots ,\beta_n\} = N_f$. Then we have

$$f(\alpha,\alpha) = f(\beta + \gamma, \beta + \gamma) = f(\beta,\beta)$$

because $\gamma \in N_f$. By the remark that we made on $V^+$ and $V^-$, $\beta \in \mathrm{Sp}\{\beta_1, \ldots ,\beta_q\}$ implies $f(\beta,\beta) \geq 0$, and $\alpha \in \mathrm{Sp}\{\alpha_{p+1}, \ldots ,\alpha_r\}$, $\alpha \neq 0$, implies that $f(\alpha,\alpha) < 0$. Thus we have a contradiction, proving that $p \geq q$. Similarly, we have $p \leq q$, proving that $p = q$. We have thus proved Theorem 4.19.

**Definition 4.14**

For a bilinear form $f$ on an $n$-dimensional real vector space $V$, the pair of integers $(p, r - p)$, where $r$ is the rank of $f$ and $p$ is the number in Theorem 4.18, is called the *signature* of $f$.

***Remark.*** Sometimes, the number $2p - r$ is called the signature of $f$.

***Example 4.9.*** In the vector space $\mathcal{R}^4$, the bilinear form $f$ defined by

$$f(\xi,\eta) = x_1y_1 + x_2y_2 + x_3y_3 - x_4y_4 \qquad \text{for } \xi = (x_i),\ \eta = (y_i)$$

has signature (3,1).

**Definition 4.15**

A bilinear form $f$ on a real vector space $V$ is said to be *positive-semidefinite* [*negative-semidefinite*] if $f(\xi,\xi) \geq 0$ [$f(\xi,\xi) \leq 0$] for all $\xi \in V$. If, moreover, $f(\xi,\xi) = 0$ implies $\xi = 0$, then $f$ is said to be *positive-definite* [*negative-definite*].

If $V$ is $n$-dimensional, $f$ is positive-semidefinite [negative-semidefinite] if and only if the signature of $f$ is $(r,0)$ [$(0,r)$], where $0 \leq r \leq n$. $f$ is positive-definite [negative-definite] if and only if its signature is $(n,0)$ [$(0,n)$].

*Example 4.10.* In the vector space $V = \mathcal{R}^\infty$ (Example 2.4), which is not finite-dimensional, define a bilinear form $f$ by

$$f(\xi,\eta) = \sum_i x_i y_i \qquad \text{for } \xi = (x_i),\ \eta = (y_i),$$

where we note that the summation actually extends over a finite set of indices $i$. The form $f$ is positive-definite.

*Example 4.11.* Consider the vector space $V$ of all real-valued continuous functions on the interval $[0,1]$ (Example 2.7). For two functions $f$ and $g$ in $V$, define

$$L(f,g) = \int_0^1 f(x)g(x)\,dx.$$

By elementary properties of the integral, it follows that $L$ is a bilinear function on $V$. It is positive-definite; in fact, $L(f,f) = \int_0^1 f(x)^2 \geq 0$ for any $f \in V$, and if the integral is 0, we must have $f(x) = 0$ for all $x$.

*Example 4.12.* In $V = \mathcal{R}^n$, define for each $m$, $1 \leq m \leq n$,

$$f(\xi,\eta) = \sum_{i=1}^{m} x_i y_i \qquad \text{for } \xi = (x_i) \text{ and } \eta = (y_i).$$

The bilinear form $f$ is positive-semidefinite. For $m = n$, it is positive-definite.

**Theorem 4.20.** *Let $f$ be a positive-definite bilinear form on an $n$-dimensional real vector space $V$. Then there exists a basis $\{\alpha_1, \ldots ,\alpha_n\}$ of $V$ such that*

$$f(\alpha_i,\alpha_j) = \delta_{ij}.$$

*Proof.* This is an immediate corollary to Theorem 4.19 in which we have $n = p$, since $f$ is positive-definite.

We shall give another proof of this theorem in Chap. 8.

**EXERCISE 4.3**

1. In $\mathcal{R}^2$, define $f(\xi,\eta) = x_1y_2 - x_2y_1$ for $\xi = (x_1,x_2)$ and $\eta = (y_1,y_2)$. Show that $f$ is bilinear and skew-symmetric.

›› 2. Prove that every skew-symmetric bilinear function on $\mathcal{R}^2$ is a scalar multiple of the function $f$ in number 1.

**3.** In $\mathcal{R}^2$, let $f(x,y) = x_1y_1 + x_2y_2$ for $\xi = (x_1,x_2)$ and $\eta = (y_1,y_2)$. What is the matrix associated with $f$ with respect to the basis $\{(2,1),(-1,1)\}$?

›› **4.** Prove that the bilinear form $f$ defined by

$$\mathrm{A} = \begin{bmatrix} 1 & 2 & 0 \\ 2 & 0 & -1 \\ 0 & -1 & 0 \end{bmatrix}$$

with respect to the standard basis in $\mathcal{R}^3$ is nondegenerate. Find its signature. Find a nonsingular matrix **P** such that ${}^t\mathbf{PAP}$ is diagonal.

**5.** In $\mathcal{R}^2$, let $f$ be the bilinear form which gives rise to the quadratic form $g(\xi) = ax_1^2 + bx_1x_2 + cx_2^2$ for $\xi = (x_1,x_2)$. Write down $f$. Prove that $f$ is positive-definite if and only if $a > 0$ and $b^2 - 4ac < 0$.

**6.** In the vector space $W = E(V,V;\mathfrak{F})$ of all bilinear functions on a vector space $V$ over $\mathfrak{F}$ (where $\mathfrak{F} = \mathcal{R}$, $\mathcal{C}$, or more generally, a field of characteristic $\neq 2$), let $W_1$ [$W_2$] be the set of all symmetric [skew-symmetric] bilinear functions on $V$. Prove that $W_1$ and $W_2$ are subspaces of $W$ and that $W = W_1 \oplus W_2$. Interpret this result in terms of $n \times n$ matrices.

›› **7.** Let $\dim V = n$ in number 6 and find the dimensions of $W$, $W_1$, and $W_2$.

**8.** Let $V$ be an $n$-dimensional vector space over $\mathcal{R}$ and let $V = N \oplus V' \oplus V''$ be the direct sum of three subspaces $N$, $V'$, and $V''$. Show that there exists a bilinear form $f$ on $V$ that has all the following properties:

(*a*) $f(\xi,\eta) = 0$ for $\xi \in N$ and $\eta \in V$.
(*b*) $f(\xi,\eta) = 0$ for $\xi \in V'$ and $\eta \in V''$.
(*c*) $f$ is positive-definite on $V'$.
(*d*) $f$ is negative-definite on $V''$.

☆ **9.** Let $f$ be a nondegenerate bilinear form on an $n$-dimensional vector space $V$ over $\mathfrak{F}$. Assume that $W$ is a subspace of $V$ on which the restriction of $f$ is nondegenerate. Prove that

››(*a*) $V = W \oplus W'$, where $W' = \{\xi \in V; f(\xi,\eta) = 0 \text{ for all } \eta \in W\}$.
(*b*) The restriction of $f$ to $W'$ is nondegenerate.

# 5 Some Algebraic Concepts

We introduce some algebraic concepts such as fields, polynomials, rings, ideals, modules, and extensions of fields and vector spaces. As soon as the concept of field is introduced, we are able to treat vector spaces over an arbitrary field instead of $\mathcal{R}$ or $\mathcal{C}$. The basic theory of vector spaces in Chaps. 2 to 4 goes over to vector spaces in general (with some exceptions for a field of characteristic 2 in the discussions of Sec. 4.3). Theorems 5.5 and 5.8 will be used in Chap. 7 for further study of linear transformations. On the other hand, we have an application of the vector space arguments to the discussion of finite extensions (Theorem 5.15).

## 5.1 FIELDS

In Sec. 2.2 we introduced the concept of additive group. We now define a field as follows.

**Definition 5.1**

A set $\mathcal{F}$ is called a *field* if it is an additive group (with respect to an operation denoted by $+$) and if there is defined a mapping $(x,y) \in \mathcal{F} \times \mathcal{F} \to xy \in \mathcal{F}$, called *multiplication*, which has the following properties:

**1.** $xy = yx$ for all $x, y \in \mathcal{F}$ (commutative law).

**2.** $(xy)z = x(yz)$ for all $x, y, z \in \mathcal{F}$ (associative law).

**3.** $x(y + z) = xy + xz$ for all $x, y, z \in \mathcal{F}$ (distributive law).

**4.** There is an element, denoted by 1, such that $x1 = x$ for every $x \in \mathcal{F}$.

**5.** For any $x \neq 0$ in $\mathcal{F}$, there is an element $x' \in \mathcal{F}$ such that $xx' = 1$. (Here 0 is the zero element of the additive group $\mathcal{F}$.)

We call $xy$ the *product* of $x$ and $y$.

What we proved for an additive group (Propositions 2.3 and 2.4) is valid for a field as far as its addition is concerned. We now prove other basic properties of a field.

***Proposition 5.1.*** *Let $\mathcal{F}$ be a field.*

**1.** *There is one and only one element, namely,* 1 *in condition* 4 *of Definition* 5.1, *such that $x1 = x$ for every $x \in \mathcal{F}$.*

**2.** *For any $x \neq 0$ in $\mathcal{F}$, there is one and only one element $x'$ such that* $xx' = 1$ (this element $x'$ is called the *inverse* of $x$ and is denoted by $x^{-1}$).

**3.** *$x0 = 0$ for every $x \in \mathcal{F}$.*

**4.** *If $a, b \in \mathcal{F}$ and $a \neq 0$, then there is one and only one element $x \in \mathcal{F}$ such that $ax = b$.*

*Proof.* 1. Let $1'$ be another element such that $x1' = x$ for every $x \in \mathcal{F}$. Then we have

$$1' = 1'1 = 11' = 1$$

by using commutativity and properties of 1 and $1'$.

2. Let $x''$ be another element such that $xx'' = 1$. By the associative law, we have $x'(xx'') = (x'x)x''$. But $x'(xx'') = x'1 = x'$ and $(x'x)x'' = 1x'' = x''$, proving $x' = x''$.

3. Since $0 + 0 = 0$, we have $x0 = x(0 + 0) = x0 + x0$ by the distributive law. By (4) of Proposition 2.3, we have $x0 = 0$.

4. Suppose $ax = b$; then $a^{-1}(ax) = a^{-1}b$. Since $a^{-1}(ax) = (a^{-1}a)x = 1x = x$, we have $x = a^{-1}b$. This proves the uniqueness of $x$ satisfying $ax = b$ for given $a$ and $b$. On the other hand, the element $a^{-1}b$ obviously satisfies the equation $ax = b$.

***Corollary.*** *If $ab = 0$, then $a = 0$ or $b = 0$.*

*Proof.* Suppose that $a \neq 0$. By (4) of the preceding proposition, the equation $ax = 0$ admits one and only one solution. Since $x = 0$ is obviously a solution by (3), we must have $b = 0$.

**Definition 5.2**

The element 1 is called the *identity* of $\mathcal{F}$.

In a field $\mathcal{F}$ we may define the product $x_1 \cdots x_k$ of a finite number of elements $x_1, \ldots, x_k$ inductively by

$$x_1 \cdots x_k = (x_1 \cdots x_{k-1})x_k.$$

For any elements $x_1, \ldots, x_k, x_{k+1}, \ldots, x_{k+m}$ in $\mathcal{F}$, we have

$$(x_1 \cdots x_k)(x_{k+1} \cdots x_{k+m}) = x_1 \cdots x_{k+m},$$

as can be proved in the same fashion as Proposition 2.4. We may also observe that the product $x_1 \cdots x_k$ is independent of the order of the elements.

We now mention a few examples of fields.

***Example 5.1.*** The real number system $\mathcal{R}$ is a field, and so is the complex number field $\mathcal{C}$. The set of all rational numbers is a field.

***Example 5.2.*** Let $\mathfrak{F}$ be the set of two elements denoted by 0 and 1. We define addition and multiplication in $\mathfrak{F}$ by the following rules:

$$0 + 0 = 0; \quad 0 + 1 = 1 + 0 = 1; \quad 1 + 1 = 0;$$
$$00 = 0; \quad 01 = 10 = 0; \quad 11 = 1.$$

$\mathfrak{F}$ is then a field with zero element 0 and identity 1. In a similar way we may construct a field having three elements, denoted by 0, 1, and 2, as follows:

$$0 + 0 = 0; \quad 0 + 1 = 1 + 0 = 1; \quad 0 + 2 = 2 + 0 = 2;$$
$$1 + 1 = 2; \quad 1 + 2 = 2 + 1 = 0; \quad 2 + 2 = 1;$$
$$00 = 0; \quad 01 = 10 = 0; \quad 02 = 20 = 0; \quad 11 = 1; \quad 12 = 21 = 2; \quad 22 = 1.$$

We shall discuss later how to construct a field having exactly $p$ elements, where $p$ is an arbitrary prime number (Example 5.21).

***Example 5.3.*** Let $\mathfrak{F}$ be the set of all real numbers of the form $a + b\sqrt{2}$, where $a$ and $b$ are arbitrary rational numbers. The usual sum and product of any two such numbers are again of the same form; in fact,

$$(a + b\sqrt{2}) + (c + d\sqrt{2}) = (a + c) + (b + d)\sqrt{2};$$
$$(a + b\sqrt{2})(c + d\sqrt{2}) = (ac + 2bd) + (bc + ad)\sqrt{2}.$$

Conditions 1 to 4 are obviously satisfied in Definition 5.1. By verifying (5), one can show that $\mathfrak{F}$ is a field.

**Definition 5.3**

Let $\mathcal{F}$ be a field. A subset $K$ of $\mathcal{F}$ is called a *subfield* of $\mathcal{F}$ if it has the following properties:

1. The zero element 0 and the identity 1 of $\mathcal{F}$ belong to $K$.
2. If $a, b \in K$, then $a + b$, $a - b$, $ab$ belong to $K$.
3. If $a \neq 0$ is in $K$, then $a^{-1}$ is in $K$.

Under these conditions, $K$ itself is a field (with addition and multiplication inherited from $\mathcal{F}$).

***Example 5.4.*** $\mathcal{R}$ is a subfield of $\mathcal{C}$. The rational number field $\mathcal{Q}$ is a subfield of $\mathcal{R}$. The field $\mathfrak{F}$ in Example 5.3 is a subfield of $\mathcal{R}$.

Having introduced the concept of a field, we can now define a vector space over an arbitrary field $\mathcal{F}$. The definition is exactly the same as Definition 2.3. All we have done in Chaps. 2 to 4 depends on the fact that $\mathcal{F}$ is a field and not on the particular properties of $\mathcal{R}$ or $\mathcal{C}$, except for those results where we specified $\mathcal{F}$ to be $\mathcal{R}$ or $\mathcal{C}$.

***Example 5.5.*** Let $\mathfrak{F} = \{0,1\}$ be the field in Example 5.2. If $V$ is a vector space over $\mathfrak{F}$ of dimension $n$, then the number of elements in $V$ is equal to $2^n$. In fact, let

$\{\alpha_1, \ldots, \alpha_n\}$ be an arbitrary basis of $V$. Every element of $V$ can be written uniquely in the form $\sum_{i=1}^{n} c_i\alpha_i$, where each $c_i$ is either 0 or 1. The number of such linear combinations is $2^n$.

***Example 5.6.*** Let $K$ be a subfield of a field $\mathfrak{F}$. We may regard $\mathfrak{F}$ as a vector space over $K$, where scalar multiplication $c\alpha$, $c \in K$ and $\alpha \in \mathfrak{F}$, is just the restriction of the multiplication in $\mathfrak{F}$ to elements of $K$ and elements of $\mathfrak{F}$.

***Example 5.7.*** Let $V$ be a vector space over $\mathfrak{R}$ of dimension 4. Denote by $\{1,i,j,k\}$ a fixed basis of $V$ so that every element of $V$ is uniquely written as $a1 + bi + cj + dk$, where $a, b, c, d \in \mathfrak{R}$. We define "multiplication" in $V$ by

$$\begin{aligned}(a1 + bi + cj + dk)\,(a'1 + b'i + c'j + d'k) &= (aa' - bb' - cc' - dd') + (ab' + ba' + cd' - dc')i \\ &\quad + (ac' - bd' + ca' + db')j + (ad' + bc' - cb' + da')k.\end{aligned}$$

In particular, we have

$$i^2 = j^2 = k^2 = -1; \qquad ij = -ji = k; \qquad jk = -kj = i; \qquad ki = -ik = j.$$

We may verify that addition in $V$ and the "multiplication" above satisfy all the conditions of a field except for the commutativity of multiplication. $V$ is called the *skew field of quaternions.* (Some authors use the term "field" without assuming the commutative law for multiplication and use the term "commutative field" for what we call a field.)

Every $a \in \mathfrak{R}$ may be identified with the element $a1$ and is still denoted by $a$ instead of $a1$ as an element of $V$. With this convention, $\mathfrak{R}$ is a subset of $V$, and every element of $V$ can be written as $a + bi + cj + dk$, which is the ordinary expression for a quaternion. Moreover, $bi$, $cj$, and $dk$ are exactly the products of $b$ and $i$, $c$ and $j$, and $d$ and $k$, respectively. We may further identify $\mathfrak{C}$ as a subset of $V$, namely, the set of all quaternions $a + bi + cj + dk$, where $c$ and $d$ are equal to 0.

We shall define the notion of characteristic of a field $\mathfrak{F}$. For any positive integer $n$, we denote by $n1$ the sum $1 + \cdots + 1$ ($n$ times). If $n1 \neq 0$ for any positive integer $n$, then we say that $\mathfrak{F}$ has *characteristic* 0. If $n1 = 0$ for some positive integer $n$, let $p$ be the smallest positive integer with the property $p1 = 0$. Then $p$ is a prime number. In fact, if $p = qr$, then

$$p1 = (q1)(r1)$$

as can be easily verified. By the corollary to Proposition 5.1 we must have $q1 = 0$ or $r1 = 0$, either of which is a contradiction. We call this prime number $p$ the *characteristic* of $\mathfrak{F}$.

The field $\{0,1\}$ in Example 5.2 has characteristic 2, and the field $\{0,1,2\}$ in the same example has characteristic 3.

**EXERCISE 5.1**

1. Show that the set $\mathfrak{F} = \{0,1,2\}$ in Example 1.2 is a field.
2. » Verify condition 5 of Definition 5.1 for Example 5.3.
3. Show that the set of all real numbers of the form $a + b\sqrt{3}$, where $a$ and $b$ are rational numbers, is a subfield of $\mathfrak{R}$.

**4.** Show that the set of all complex numbers of the form $a + bi$, where $a$ and $b$ are rational numbers, is a subfield of $\mathcal{C}$.

›› **5.** Is the following statement true for an arbitrary field? If $a + a = 0$, then $a = 0$.

**6.** Prove that for any elements $a$, $b$, $c$, and $d$ in an arbitrary field $\mathfrak{F}$, we have $cadb = abcd$.

☆ **7.** Let $\mathfrak{F}$ be the set of all $2 \times 2$ real matrices of the form $\mathbf{A} = \begin{bmatrix} a & -b \\ b & a \end{bmatrix}$. Prove:

(*a*) If $\mathbf{A}, \mathbf{B} \in \mathfrak{F}$, then $\mathbf{A} \pm \mathbf{B}, \mathbf{AB} \in \mathfrak{F}$.

(*b*) For any $\mathbf{A} \in \mathfrak{F}$ as above, let $f(\mathbf{A}) = a + bi \in \mathcal{C}$. Then $f$ is a one-to-one mapping of $\mathfrak{F}$ onto the complex number field $\mathcal{C}$ which preserves addition and multiplication, that is, $f(\mathbf{A} + \mathbf{B}) = f(\mathbf{A}) + f(\mathbf{B})$ and $f(\mathbf{AB}) = f(\mathbf{A})f(\mathbf{B})$ for all $\mathbf{A}, \mathbf{B} \in \mathfrak{F}$.

››(*c*) If $\mathbf{A} \in \mathfrak{F}$ and $\mathbf{A} \neq 0$, then $\mathbf{A}$ is nonsingular and $\mathbf{A}^{-1} \in \mathfrak{F}$.

☆ **8.** Let $Q$ be the set of all matrices of the form

$$\mathbf{A} = \begin{bmatrix} a & -b & -c & -d \\ b & a & -d & c \\ c & d & a & -b \\ d & -c & b & a \end{bmatrix}, \qquad \text{where } a, b, c, d \in \mathcal{R}.$$

Prove:

(*a*) If $\mathbf{A}, \mathbf{B} \in Q$, then $\mathbf{A} \pm \mathbf{B}, \mathbf{AB} \in Q$.

(*b*) For any $\mathbf{A} \in Q$ as above, let $f(\mathbf{A}) = a + bi + cj + dk$. Then $f$ is a one-to-one mapping of $Q$ onto the skew field of quaternions which preserves addition and multiplication.

››(*c*) If $\mathbf{A} \in Q$ and $\mathbf{A} \neq 0$, then $\mathbf{A}$ is nonsingular and $\mathbf{A}^{-1} \in Q$.

☆ **9.** In a field $\mathfrak{F}$, we define $na$, where $a \in \mathfrak{F}$ and $n$ is an arbitrary integer, as follows. For positive $n$, we define $na$ inductively by

$$1a = a, \qquad na = (n - 1)a + a.$$

For negative $n$, define $na = -((-n)a)$, where $-n > 0$ so that $(-n)a$ has already been defined. Finally, for $n = 0$, we set $0a = 0$ (the zero element of $\mathfrak{F}$). Prove the following:

(*a*) $(m + n)a = ma + na$.

(*b*) $(mn)a = m(na)$.

(*c*) $(na)b = n(ab) = a(nb)$.

(*d*) If $\mathfrak{F}$ has characteristic $p(\neq 0)$, then $pa = 0$ for every $a \in \mathfrak{F}$.

(*e*) If $na = 0$ for some $a \neq 0$ in $\mathfrak{F}$ and for some positive integer $n$, then the characteristic of $\mathfrak{F}$ is not 0.

## 5.2 POLYNOMIALS

A polynomial $f(x)$ over a field $\mathcal{F}$ with indeterminate $x$ is a formal expression of the form

$$a_0 + a_1x + a_2x^2 + \cdots + a_nx^n, \tag{5.1}$$

where $n$ is an integer $\geq 0$, $a_0, a_1, a_2, \ldots, a_n \in \mathcal{F}$ and $a_n \neq 0$, or, as a very special case, simply an element of $\mathcal{F}$

$$a, \tag{5.2}$$

which is a particular case of the expression (5.1) with $n = 0$ and no restriction on $a$. For the expression (5.1), $x$ is a symbol called the *indeterminate*, $n$ is called the *degree of the polynomial* $f(x)$, and $a_n$ is called the *leading coefficient.* The element $a$ in (5.2) is called a *constant polynomial* or, simply, a *constant.* Its degree is 0.

We denote by $\mathcal{F}[x]$ the set of all polynomials $f(x)$ over $\mathcal{F}$.

We shall now define polynomials in a different, more rigorous fashion. We consider the vector space $V = \mathcal{F}^\infty$ in Example 2.4. Each element $\alpha \in \mathcal{F}^\infty$ is a sequence

$$\alpha = (a_0, a_1, \ldots, a_n, 0, \ldots),$$

where $a_n$ is the last nonzero term of the sequence, unless $\alpha$ is the zero element $(0,0, \ldots)$ of $\mathcal{F}^\infty$. We already know how to add two elements in $\mathcal{F}^\infty$ and multiply an element of $\mathcal{F}^\infty$ by a scalar (element of the field $\mathcal{F}$). We shall now define multiplication in $\mathcal{F}^\infty$ as follows.

For $\alpha = (a_0, a_1, \ldots, a_n, 0, \ldots)$ and $\beta = (b_0, b_1, \ldots, b_m, 0, \ldots)$, we define $\alpha\beta$ to be the following element of $\mathcal{F}^\infty$:

$$\gamma = (c_0, c_1, \ldots, c_{n+m}, 0, \ldots),$$

where

$$c_k = \sum_{i+j=k} a_i b_j. \tag{5.3}$$

We may immediately observe that if $\alpha \neq 0$ and $\beta \neq 0$, then $\gamma = \alpha\beta$ is not 0. Indeed, if $a_n$ is the last nonzero term of $\alpha$ and $b_m$ is the last nonzero term of $\beta$, then we have, according to (5.3),

$$c_{m+n} = a_n b_m, \tag{5.4}$$

which is not 0 by the corollary to Proposition 5.1, and this is the last nonzero term of $\gamma$.

We now relate $\mathcal{F}^\infty$ to expressions of the form (5.1). First, we shall identify each element $a$ of $\mathcal{F}$ with the element of $\mathcal{F}^\infty$

$$(a, 0, 0, \ldots), \tag{5.5}$$

and still denote (5.5) by $a$. Next, we introduce a symbol $x$ to represent the element of $\mathcal{F}^\infty$

$$x = (0, 1, 0, \ldots), \tag{5.6}$$

where the second term 1 is the last nonzero term. According to the rule of multiplication (5.3), we have the square of $x$:

$$x^2 = (0, 0, 1, 0, \ldots)$$

and, more generally, the $n$th power of $x$

$$x^n = (0, 0, \ldots, 0, 1, 0, \ldots), \tag{5.7}$$

where 1 is the $(n+1)$th term, which is the last nonzero term. Finally, if we multiply $a$ of (5.5) with $x^n$ of (5.7) according to the rule of multiplication (5.3), we obtain

$$ax^n = (0,0, \ldots ,0,a,0, \ldots), \tag{5.8}$$

which is the same as the scalar multiple $ax^n$ in the vector space $\mathcal{F}^\infty$. This being said, an expression of the form (5.1) now makes sense as an element of $\mathcal{F}^\infty$, the addition $+$ being, of course, the addition in the vector space $\mathcal{F}^\infty$. Indeed, we have

$$a_0 + a_1x + \cdots + a_nx^n = (a_0,a_1, \ldots ,a_n,0, \ldots). \tag{5.9}$$

In this manner we have clarified the relationship between $\mathcal{F}^\infty$ and the set of all polynomials as defined in the beginning of this section. The rule of multiplication (5.3) in $\mathcal{F}^\infty$, if transferred to the set of polynomials $\mathcal{F}[x]$, will lead to the usual multiplication for polynomials, namely, for two polynomials

$$f(x) = \sum_{i=0}^{n} a_ix^i \qquad \text{and} \qquad g(x) = \sum_{j=0}^{m} b_jx^j$$

the product $f(x)g(x)$ is the polynomial

$$h(x) = \sum_{k=0}^{n+m} c_kx^k, \qquad \text{where } c_k = \sum_{i+j=k} a_ib_j.$$

It is now straightforward to verify the following properties of multiplication in $\mathcal{F}[x]$. By writing simply $f, g, fg$ for $f(x), g(x), f(x)g(x)$, we have

$$fg = gf; \qquad (fg)h = f(gh); \qquad f(g+h) = fg + fh; \qquad 1f = f,$$

where 1 is the constant polynomial equal to the identity of $\mathcal{F}$. We thus have the commutative, distributive, and associative laws for multiplication. These algebraic properties of $\mathcal{F}[x]$ can be expressed by saying that $\mathcal{F}[x]$ forms a *commutative ring*. We now give the precise definitions.

**Definition 5.4**

A set $A$ is called a *ring* if it is an additive group (with respect to an operation called addition and denoted by $+$) and if there is a mapping $(x,y) \in A \times A \to xy \in A$, to be called multiplication, which satisfies the following conditions:

**1.** $(xy)z = x(yz)$ (associative law).

**2.** $(x+y)z = xz + yz$ and $z(x+y) = zx + zy$ (distributive laws).

If $A$ has an element 1 such that $x1 = 1x = x$ for every $x \in A$, 1 is called the *identity* of $A$. [The uniqueness of such an element can be proved in the same way as (1) of Proposition 5.1.]

**Definition 5.5**

A ring is said to be *commutative* if

**3.** $xy = yx$ for all $x, y \in A$.

We shall mention some examples.

***Example 5.8.*** The set of all integers $I = \{0, \pm 1, \pm 2, \ldots\}$ forms a commutative ring with identity 1 with respect to usual addition and multiplication. The set of all even integers $\{0, \pm 2, \pm 4, \ldots\}$ is a commutative ring which does not have an identity.

***Example 5.9.*** A field is, of course, a commutative ring with identity. For a field $\mathfrak{F}$, the set of polynomials $\mathfrak{F}[x]$ is a commutative ring with identity, as we already stated.

***Example 5.10.*** For any field $\mathfrak{F}$, let $\mathfrak{F}^n{}_n$ be the set of all $n \times n$ matrices over $\mathfrak{F}$. With respect to addition and multiplication of matrices which we already introduced, $\mathfrak{F}^n{}_n$ forms a ring with identity (the unit matrix). The ring $\mathfrak{F}^n{}_n$, $n > 1$, is not commutative.

***Example 5.11.*** Let $A$ be the set of all continuous functions defined on the unit interval [0,1]. We know that $A$ is a vector space over $\mathfrak{R}$ (Example 2.7*b*). The usual product $f(x)g(x)$ of two continuous functions is again a continuous function. This multiplication in $A$ makes it a commutative ring with identity (the constant function 1).

We now go back to the ring of polynomials $\mathcal{F}[x]$. We have already proved the following.

***Proposition 5.2.*** *In $\mathcal{F}[x]$, we have*

**1.** $\deg(fg) = \deg f + \deg g$, *where* $f \neq 0$, $g \neq 0$ *and* $\deg f$ *means the degree of* $f$.

**2.** *If* $fg = 0$, *then* $f = 0$ *or* $g = 0$.

We shall now prove a basic result on division in $\mathcal{F}[x]$ which will be used in the next section. We start with:

***Lemma.*** *Let $f$ and $d$ be nonzero polynomials in $\mathcal{F}[x]$ such that* $\deg d \leq \deg f$. *Then there is a polynomial $g$ in $\mathcal{F}[x]$ such that*

$$\textit{either} \quad f - dg = 0 \quad \textit{or} \quad \deg(f - dg) < \deg f.$$

*Proof.* Let

$$f(x) = a_n x^n + \sum_{i=0}^{n-1} a_i x^i, \qquad \text{where } a_n \neq 0,$$

and

$$g(x) = b_k x^k + \sum_{i=0}^{k-1} b_i x^i, \qquad \text{where } b_k \neq 0 \text{ and } k \leq n.$$

Then

$$g(x) = (a_n b_k^{-1}) x^{n-k}$$

satisfies our requirement.

***Theorem 5.3.*** *Let $\mathcal{F}$ be a field and let $d \neq 0$ and $f$ be in $\mathcal{F}[x]$. Then there exist $q, r \in \mathcal{F}[x]$ such that:*
**1.** $f = dq + r$.
**2.** *Either $r = 0$ or $\deg r < \deg d$.*
*Moreover, such polynomials $q$ and $r$ are determined uniquely.*

*Proof.* We first prove the existence. If $\deg f < \deg d$, then we may take $q = 0$ and $r = f$. Assume $\deg f \geq \deg d$. By the preceding lemma, there exists $g \in \mathcal{F}[x]$ such that either $f - dg = 0$ or

$$\deg (f - dg) < \deg f.$$

If $f - dg = 0$ or $\deg(f - dg) < \deg d$, then we take $q = g$ and $r = f - dg$. If $f - dg \neq 0$ and $\deg(f - dg) \geq \deg d$, then we may apply the lemma to $f - dg$ and $d$, instead of $f$ and $d$, to find $h \in \mathcal{F}[x]$ such that either $f - dg - dh = 0$ or

$$\deg (f - dg - dh) < \deg (f - dg),$$

that is,
$$\deg (f - d(g + h)) < \deg (f - dg).$$

If $f - d(g + h) = 0$ or $\deg(f - d(g + h)) < \deg d$, then we take $q = g + h$ and $r = f - d(g + h)$. If $f - d(g + h) \neq 0$ and $\deg(f - d(g + h)) \geq \deg d$, we apply the lemma to $f - d(g + h)$ and $d$. By continuing this process, we may obtain $q$ and $r$ satisfying (1) and (2).

To prove the uniqueness of such polynomials, assume that $q'$, $r'$ are another pair of polynomials which satisfy (1) and (2). Then

$$f = dq + r = dq' + r',$$

so that
$$d(q - q') = r' - r.$$

If $q \neq q'$, then $\deg d(q - q')$ is equal to $\deg d + \deg (q - q')$ by Proposition 5.2. On the other hand, we have

$$\deg d + \deg (q - q') \geq \deg d > \deg (r' - r),$$

since $\deg r < \deg d$ and $\deg r' < \deg d$. This is a contradiction. Hence $q = q'$ and $r = r'$, completing the proof of Theorem 5.3.

**Definition 5.6**

In Theorem 5.3, $q$ is called the *quotient* and $r$ the *remainder* for the division of $f$ by $d$. If $r = 0$, that is, $f = dg$, we say that $f$ is a *multiple* of $g$ or that $f$ is *divisible* by $g$. $g$ is a *divisor* of $f$. In this case, we often write $g \mid f$.

***Example 5.12.*** In $\mathcal{R}[x]$ or $\mathcal{C}[x]$, the quotient and remainder can be found by the usual long division. For example,

$$3x^4 - 2x^3 + 5x^2 - 4x + 1 = (x^2 - x + 1)(3x^2 + x + 3) + (-2x - 2).$$

Thus, on dividing the left-hand-side polynomial by $x^2 - x + 1$, we obtain the quotient $3x^2 + x + 3$ and the remainder $-2x - 2$.

**EXERCISE 5.2**

» **1.** Let $\mathfrak{F}$ be the field $\{0,1\}$ in Example 5.2. How many polynomials of degree $n$ are there in $\mathfrak{F}[x]$?

**2.** Let $\mathfrak{F} = \{0,1\}$ be the field as above. How many elements does the ring $\mathfrak{F}^n{}_n$ have? In $\mathfrak{F}^2{}_2$, find two elements **A** and **B** such that $\mathbf{AB} \neq \mathbf{BA}$.

**3.** Let $X$ be an arbitrary set and $\mathfrak{F}$ a field. Show that the set of all functions on $X$ with values in $\mathfrak{F}$ forms a commutative ring with identity if we define

$$(f+g)(x) = f(x) + g(x) \qquad \text{and} \qquad (fg)(x) = f(x)g(x)$$

for two functions $f$ and $g$, where $x \in X$.

» **4.** In $\mathfrak{R}[x]$, divide $5x^5 - 3x^4 + 3x^2 - x + 6$ by $x^2 + x + 1$ and find the quotient and the remainder.

**5.** Let $a$ and $b$ be two elements in a ring such that $ab = ba$. Prove the formula

$$(a+b)^n = a^n + \sum_{k=1}^{n-1} \binom{n}{k} a^k b^{n-k} + b^n$$

where $n$ is any positive integer and $\displaystyle\binom{n}{k} = \frac{n!}{k!(n-k)!}$.

☆ **6.** A nonempty subset $S$ of a ring $A$ is called a *subring* of $A$ if $x, y \in S$ implies $x + y$, $x - y$, $xy \in S$. Show that a subring is a ring itself with respect to addition and multiplication in $A$ restricted to $S$.

»☆ **7.** Show that the set of all diagonal matrices in $\mathfrak{F}^n{}_n$ forms a commutative subring of the ring $\mathfrak{F}^n{}_n$. How about the set of all matrices $\mathbf{A} = [a_{ij}]$ in $\mathfrak{F}^n{}_n$, where $a_{ij} = 0$ whenever $i > j$? (See number 6.)

## 5.3 IDEALS AND FACTORIZATION IN $\mathcal{F}[x]$

We shall start with the following definitions.

### Definition 5.7

Let $A$ be a ring. A nonempty subset $I$ of $A$ is called a *left ideal* of $A$ if it has the following two properties:

**1.** For $a, b \in I$, we have $a + b$, $a - b \in I$.

**2.** For $a \in I$ and $x \in A$, we have $xa \in I$.

When (2) is replaced by

**2′.** For $a \in I$ and $x \in A$, we have $ax \in I$,

we have the notion of a *right ideal.* If $I$ satisfies (1), (2), and (2′), then it is called a *two-sided ideal* or, simply, an *ideal.*

Note that if $A$ is a commutative ring, (2) and (2′) amount to the same condition and there is no distinction between left and right ideals.

### Definition 5.8

Let $A$ be a commutative ring with identity. For any element $a \in A$, let $I(a) = \{xa; x \in A\}$. Then $I(a)$ is an ideal (in fact, the smallest ideal containing $a$), called the *principal ideal* generated by $a$.

It is easy to verify that $I(a)$ is an ideal. On the other hand, if $I$ is an ideal of $A$ containing $a$, then for any $x \in A$ we have $xa \in I$ and hence $I(a) \subseteq I$. The ideal $\{0\}$ consisting of 0 alone is called the *zero ideal.*

***Example 5.13.*** In the ring of all integers $\mathbb{Z}$, let $I(m)$ be the set of all multiples of a given integer $m$. Given two integers $m_1$ and $m_2$ which are not 0, $m_1$ is a multiple of $m_2$ if and only if $I(m_2) \supseteq I(m_1)$. We can also show that every ideal $I$ of the ring $\mathbb{Z}$ is a principal ideal. In fact, if $I = \{0\}$, then $I = I(0)$. If $I \neq \{0\}$, let $d$ be the smallest positive integer contained in $I$. All multiples of $d$ are in $I$, so that $I(d) \subseteq I$. Conversely, for any $m \in I$, $m > 0$, let $m = dq + r$, where $q \geq 0$ and $r$, $0 \leq r \leq d$, are integers which are uniquely determined. Then $r = m - dq \in I$. Since $d$ is the smallest positive integer in $I$, we must have $r = 0$, that is, $m = dq$, proving that $m \in I(d)$. If $m \in I$ and $m < 0$, then $-m > 0$ is in $I$ and hence $-m \in I(d)$. Thus $m \in I(d)$. We have thus shown that $I = I(d)$.

We have:

***Theorem 5.4.*** *Let $A$ be a commutative ring with identity. Then $A$ is a field if and only if it does not have any ideal other than $A$ itself and the zero ideal.*

*Proof.* Let $A$ be a field and $I$ an ideal of $A$. If $I \neq \{0\}$, let $a \neq 0$ be an element in $I$. Since $A$ is a field, there is $a^{-1}$ such that $aa^{-1} = 1$. Since $I$ is an ideal, it follows that $aa^{-1}$, namely, 1 is in the ideal $I$. For any element $x \in A$, we have then $x = 1x \in I$, proving that $I = A$.

Conversely, suppose that $A$ has no ideal other than $A$ and $\{0\}$. To show that $A$ is a field, we wish to prove that for any $a \neq 0$ in $A$ there is an $a'$ such that $aa' = 1$. Consider the principal ideal $I(a)$, which contains a nonzero element $a$, since $a = a1 \in I(a)$. By assumption, it follows that $I(a) = A$. Thus there is an element $a' \in A$ such that $1 = aa'$, proving our assertion.

We shall now go back to the polynomial ring $\mathcal{F}[x]$ and prove the following fundamental result.

***Theorem 5.5.*** *Let $\mathcal{F}$ be a field. Then every nonzero ideal $I$ of $\mathcal{F}[x]$ is a principal ideal. More precisely, there is a unique polynomial with leading coefficient* 1, *say*, $d(x)$, *such that* $I = I(d)$. (A polynomial with leading coefficient 1 is called a *monic polynomial.*)

*Proof.* If $I$ contains a nonzero constant $a$, then $I$ contains $a^{-1}a = 1$ and hence every polynomial $g = 1g$ in $\mathcal{F}[x]$. Thus we may assume that all the polynomials in $I$ have degree $\geq 1$. Let $d$ be a polynomial in $I$ with the smallest possible degree, say, $k$, among those polynomials that lie in $I$. If the leading coefficient $a$ of $d$ is not 1, then $a^{-1}d$ is a monic polynomial in $I$ of degree $k$. Therefore, let us assume that $d$ is monic. Since $d \in I$, it is obvious that $I(d) \subseteq I$. Conversely, let $f \in I$. Let

$$f = dq + r,$$

where $q$ is the quotient and $r$ the remainder. Since both $d$ and $f$ are in $I$, we have $dq \in I$ and hence $r = f - dq \in I$. If $r \neq 0$, then $\deg r < \deg d$, and this is a contradiction to the choice of $d$. Hence $r = 0$, that is, $f = dq$. This proves that $I \subseteq I(d)$, and hence $I = I(d)$.

We now show the uniqueness of a monic polynomial $d$ such that $I = I(d)$. Suppose that $d'$ is another such polynomial. Then $d = d'q'$ for some polynomial $q'$. If $q'$ is not a constant, then $\deg d > \deg d'$, which is a contradiction, since $d$ is a polynomial of smallest possible degree in $I$. Thus $q'$ is a constant. Since both $d$ and $d'$ are monic, we must have $q' = 1$, that is, $d = d'$.

As an application, we may prove:

***Theorem 5.6.*** *Let $\mathcal{F}$ be a field. For any polynomials $f_1, \ldots, f_n$ in $\mathcal{F}[x]$, there exists a unique monic polynomial $d$ with the following properties:*
**1.** *$d$ divides $f_1, \ldots, f_n$.*
**2.** *If $d' \in \mathcal{F}[x]$ divides $f_1, \ldots, f_n$, then $d'$ divides $d$.*
*Moreover, there exist $g_1, \ldots, g_n \in \mathcal{F}[x]$ such that*

$$d = f_1g_1 + \cdots + f_ng_n.$$

*Proof.* Consider the set $I$ of all polynomials of the form $\sum_{i=1}^{n} f_ig_i$, where $g_1, \ldots, g_n$ are arbitrary polynomials in $\mathcal{F}[x]$. It is easily verified that $I$ is an ideal of $\mathcal{F}[x]$. By Theorem 5.5 there exists a unique monic polynomial $d$ such that $I = I(d)$. Since $f_i \in I$, it follows that $d$ divides $f_i$ for each $i$. On the other hand, $d \in I$ is of the form $\sum_{i=1}^{n} f_ig_i$ with suitable $g_1, \ldots, g_n \in \mathcal{F}[x]$. If $d'$ is a polynomial which divides every $f_i$, then $d'$ divides $d$.

**Definition 5.9**

Given $f_1, \ldots, f_n \in \mathcal{F}[x]$, the unique monic polynomial $d \in \mathcal{F}[x]$ which has properties 1 and 2 in the preceding theorem is called the *greatest common divisor* of $f_1, \ldots, f_n$ and is denoted by $(f_1, \ldots, f_n)$.

**Definition 5.10**

$f_1, \ldots, f_n$ are said to be *relatively prime* if $(f_1, \ldots, f_n) = 1$.

***Corollary.*** *If $f_1, \ldots, f_n \in \mathcal{F}[x]$ are relatively prime, then there exist $g_1, \ldots, g_n \in \mathcal{F}[x]$ such that*

$$f_1g_1 + \cdots + f_ng_n = 1.$$

We shall now explain the so-called euclidean algorithm, which gives a practical method of finding the greatest common divisor of two polynomia's $f$ and $g$. Assume that $\deg f \geq \deg g$ and that $g$ is monic. (If $g$ is not monic, replace it by $a^{-1}g$, where $a$ is the leading coefficient of $g$.) Let

$$f = gq + r_1, \qquad \text{where } r_1 = 0 \text{ or } \deg r_1 < \deg g.$$

If $r_1 = 0$, then obviously $(f,g) = g$. If $r_1 \neq 0$, then

$$(f,g) = (g,r_1).$$

In fact, if a polynomial $d$ divides $f$ and $g$, then $d$ divides $f - gq$, that is, $r_1$. Conversely, if $d$ divides $g$ and $r_1$, then it divides $f$. Thus $(f,g) = (g,r_1)$. Now let

$$g = r_1q_1 + r_2, \qquad \text{where } r_2 = 0 \text{ or } \deg r_2 < \deg r_1.$$

If $r_2 = 0$, then $(f,g) = (g,r_1) = r_1$. If $r_2 \neq 0$, then

$$(f,g) = (g,r_1) = (r_1,r_2)$$

for the same reason as before. Continuing in this way, we obtain $r_1, \ldots, r_n$ with decreasing degrees such that

$$r_i = r_{i+1}q_{i+1} + r_{i+2} \qquad \text{for } i = 1, 2, \ldots, n-2$$

and

$$r_{n-1} = r_nq_n.$$

Thus we find the greatest common divisor of $f$ and $g$:

$$(f,g) = r_n.$$

*Example 5.14.* In $\mathcal{R}[x]$, let $f = x^4 + x - 2$ and $g = x^2 - 1$. Then

$$x^4 + x - 2 = (x^2 - 1)(x^2 + 1) + (x - 1),$$

$$x^2 - 1 = (x - 1)(x + 1),$$

so that

$$(f,g) = x - 1.$$

We now make the following definition.

**Definition 5.11**

A nonconstant polynomial $f(x) \in \mathcal{F}[x]$ is said to be *irreducible in* $\mathcal{F}[x]$ (or *irreducible over* $\mathcal{F}$) if $f = gh$, where $g, h \in \mathcal{F}[x]$, implies that one of $g$ and $h$ is a nonzero constant $a \in \mathcal{F}$ and the other is equal to $a^{-1}f$.

*Example 5.15.* Let $\mathcal{F}$ be the field $\{0,1\}$. Since

$$x^2 + 1 = (x + 1)^2 \qquad \text{in } \mathcal{F}[x],$$

$x^2 + 1$ is not irreducible. On the other hand, $x^2 + x + 1$ is irreducible over $\mathcal{F}$, because the only polynomials of degree 1 in $\mathcal{F}[x]$ are $x$ and $x + 1$ and none of the products $x^2$, $x(x + 1) = x^2 + x$, and $(x + 1)^2 = x^2 + 1$ is equal to $x^2 + x + 1$.

***Example 5.16.*** The polynomial $x^2 + 1$ is irreducible over $\mathcal{R}$ but not over $\mathcal{C}$. In fact, we have

$$x^2 + 1 = (x + i)(x - i) \qquad \text{in } \mathcal{C}[x].$$

We shall now prove:

***Theorem 5.7.*** *Let $f, g \in \mathcal{F}[x]$. If the product $fg$ is divisible by a polynomial $p$ and if $p$ is irreducible over $\mathcal{F}$, then $f$ or $g$ is divisible by $p$.*

*Proof.* Without loss of generality we may assume that $p$ is monic. Let $d = (p,f)$. Since $p$ is irreducible, we have $d = p$ or $d = 1$ (note that both $p$ and $d$ are monic). In the first case, $p$ divides $f$. In the second case, there exist, by virtue of the corollary to Theorem 5.6, two polynomials $h, k \in \mathcal{F}[x]$ such that

$$fh + pk = 1,$$

so that

$$fhg + pkg = g.$$

Since $p$ divides $fg$ by assumption, $p$ divides $fhg$ and, of course, $pkg$. Thus $p$ divides their sum, namely, $g$.

***Corollary.*** *If $p$ is irreducible in $\mathcal{F}[x]$ and divides the product $f_1 \cdots f_k$, where $f_i \in \mathcal{F}[x]$, then $p$ divides at least one $f_i$.*

We are now in position to prove the following main theorem.

***Theorem 5.8.*** *Let $\mathcal{F}$ be a field. A nonconstant monic polynomial $f$ in $\mathcal{F}[x]$ can be expressed as a product of monic irreducible polynomials in $\mathcal{F}[x]$. Such a factorization is unique up to order. More precisely, if*

$$f = p_1 \cdots p_m = q_1 \cdots q_k$$

*where $p_i$'s and $q_j$'s are irreducible monic polynomials in $\mathcal{F}[x]$, then $m = k$ and, with a suitable rearrangement of indices, $p_i = q_i$ for each $i$, $1 \leq i \leq m$.*

*Proof.* We shall use induction on $\deg f$. When $\deg f = 1$, $f$ is irreducible and the assertion is obvious. Assume that the theorem is valid for polynomials of degree less than $n$ and let $f$ be a given polynomial of degree $n$. If $f$ is not irreducible, then we may write $f$ in the form $f = gh$ in $\mathcal{F}[x]$, where $\deg g < n$ and $\deg h < n$. By the inductive assumption, we may express $g$ and $h$ in the form

$$g = g_1 \cdots g_k \qquad \text{and} \qquad h = h_1 \cdots h_m,$$

where $g_i$'s and $h_j$'s are irreducible monic polynomials in $\mathcal{F}[x]$. Then, obviously, $f$ is the product of these irreducible monic polynomials.

In order to prove the uniqueness, let

$$f = p_1 \cdots p_m = q_1 \cdots q_k$$

as in the statement of the theorem. Since $p_1$ divides the product $q_1 \cdots q_k$, it follows that $p_1$ divides one of $q_1, \ldots, q_k$ by virtue of the corollary to Theorem 5.7. By changing the indices if necessary, we may assume that $p_1$ divides $q_1$. Since they are both monic and irreducible, we must have $p_1 = q_1$. We then have

$$p_2 \cdots p_m = q_2 \cdots q_k.$$

[In fact, we have, more generally, the following cancellation law for polynomials: If $fg = fh$ with nonzero $f$, then $g = h$. This follows from (2) of Proposition 5.2 by considering $f(g - h) = 0$.] Since $p_2 \cdots p_m = q_2 \cdots q_k$ is a polynomial of degree $< n$, the inductive assumption shows that $m = k$ and $p_2 = q_2, \ldots, p_m = q_m$ by a suitable rearrangement of indices, thus proving our assertion for $f$.

***Example 5.17.*** Let $a_1, \ldots, a_n$ be distinct elements in a field $\mathfrak{F}$. For any positive integers $k_1, \ldots, k_n$, the $n$ polynomials

$$(x - a_1)^{k_1}, \ldots, (x - a_n)^{k_n}$$

are relatively prime. In fact, suppose $d$ is the greatest common divisor of these $n$ polynomials. If $d \neq 1$, let $p$ be one of the irreducible factors of $d$. Since $p$ divides $(x - a_1)^{k_1}$, $p$ divides $(x - a_1)$. Hence $p = x - a_1$, and it does not divide $x - a_2$, because $a_2 \neq a_1$. This is a contradiction, proving that $d = 1$.

We shall conclude this section by stating the following basic properties of the real number field $\mathfrak{R}$ and the complex number field $\mathfrak{C}$ from the point of view of factorization.

**Theorem 5.9.** *In $\mathfrak{R}[x]$, every irreducible polynomial is of degree* 1 *or* 2.

**Theorem 5.10.** *In $\mathfrak{C}[x]$, every irreducible polynomial is of degree* 1.

We shall not prove these theorems here, but simply indicate their consequences.

Theorem 5.9 implies that a monic irreducible polynomial in $\mathfrak{R}[x]$ is either of the form $x - a$ or of the form $x^2 + bx + c$, where $b^2 - 4c < 0$. In the second case, one can also write it in the form $(x - a)^2 + b^2$, $b \neq 0$, by suitable choice of $a, b \in \mathfrak{R}$.

Theorem 5.10 is equivalent to the so-called *fundamental theorem of algebra*, which states that every algebraic equation with coefficients in $\mathfrak{C}$ has a root in $\mathfrak{C}$. We have:

**Corollary.** *In $\mathfrak{C}[x]$, every polynomial can be factored in the form*

$$c(x - a_1)^{k_1} \cdots (x - a_m)^{k_m},$$

*where $a_1, \ldots, a_m$ are distinct elements in $\mathfrak{C}$, $c \in \mathfrak{C}$, and $k_1, \ldots, k_m$ are certain positive integers.*

**EXERCISE 5.3**

**1.** Let $A$ be a ring. For any $a \in A$, show that

$$L(a) = \{xa; x \in A\}$$

is a left ideal and

$$R(a) = \{ax; x \in A\}$$

is a right ideal of $A$.

**2.** Let $A$ be a ring. For any $a_1, \ldots, a_k \in A$, show that

$$I(a_1, \ldots, a_k) = \left\{ \sum_{i=1}^{n} x_i a_i; x_1, \ldots, x_k \in A \right\}$$

is a left ideal of $A$. State and prove the analogue for a right ideal.

›› **3.** Find the greatest common divisor of

$$f = x^4 - 2x^3 + 2x^2 - 2x + 1 \qquad \text{and} \qquad g = x^4 - 2x^3 - 2x - 1$$

by the euclidean algorithm.

**4.** Find the factorization for $f$ and $g$ in number 3 in $\mathcal{R}[x]$ and in $\mathcal{C}[x]$.

›› **5.** Find all irreducible polynomials of degree 3 in $\mathfrak{F}[x]$, where $\mathfrak{F}$ is the field $\{0,1\}$.

››☆ **6.** Prove that the ring $\mathfrak{F}^n{}_n$ has no proper two-sided ideal.

## 5.4 EXTENSION OF FIELDS

Let $\mathcal{F}$ be a field and consider the polynomial ring $\mathcal{F}[x]$. For each polynomial

$$f(x) = \sum_{i=0}^{n} a_i x^i \in \mathcal{F}[x],$$

we shall consider a mapping of $\mathcal{F}$ into $\mathcal{F}$ given by

$$c \in \mathcal{F} \to \sum_{i=0}^{n} a_i c^i \in \mathcal{F}.$$

This mapping is called the *polynomial function* corresponding to the polynomial $f$. It is convenient to denote by $f(c)$ the value $\sum_{i=0}^{n} a_i c^i$ of the polynomial function for $c \in \mathcal{F}$.

A polynomial and the corresponding polynomial function should not be confused.

***Example 5.18.*** Let $\mathfrak{F}$ be the field $\{0,1\}$. Let

$$f(x) = x^2 + 1 \qquad \text{and} \qquad g(x) = x + 1 \qquad \text{in } \mathfrak{F}[x].$$

These are two distinct polynomials, but the corresponding polynomial functions coincide, that is, they have the same value for each element of $\mathfrak{F}$:

$$f(0) = 0 + 1 = 1, \qquad g(0) = 0 + 1 = 1;$$

$$f(1) = 1^2 + 1 = 0, \qquad g(1) = 1 + 1 = 0.$$

***Proposition 5.11.*** *Let $c$ be a fixed element of $\mathcal{F}$. Then the mapping*

$$f(x) \in \mathcal{F}[x] \to f(c) \in \mathcal{F}$$

*preserves addition and multiplication, that is,*

$$(f+g)(c) = f(c) + g(c) \quad \text{and} \quad (fg)(c) = f(c)g(c)$$

*for all $f, g \in \mathcal{F}[x]$.*

*Proof.* Let

$$f(x) = \sum_{i=0}^{n} a_i x^i \quad \text{and} \quad g(x) = \sum_{j=0}^{m} b_j x^j.$$

Then $$f(x)g(x) = \sum_{k=0}^{n+m} d_k x^k, \quad \text{where } d_k = \sum_{i+j=k} a_i b_j.$$

We have

$$(fg)(c) = \sum_{k=0}^{n+m} d_k c^k = \sum_{i=0}^{n} \sum_{j=0}^{m} a_i b_j c^{i+j}$$

$$= \left(\sum_{i=0}^{n} a_i c^i\right)\left(\sum_{j=0}^{m} b_j c^j\right) = f(c)g(c).$$

The proof for $(f+g)(c) = f(c) + g(c)$ is much simpler and is omitted.

***Proposition 5.12.*** *Let $f(x) \in \mathcal{F}[x]$ and $c \in \mathcal{F}$. Then $f$ is divisible by $x - c$ if and only if $f(c) = 0$.*

*Proof.* Let

$$f(x) = (x - c)q(x) + r(x),$$

where $r(x)$ is a constant polynomial $r$. By Proposition 5.11, we have

$$f(c) = (c - c)q(c) + r = r.$$

Thus $r = 0$ if and only if $f(c) = 0$.

If $f(c) = 0$, we say that $c \in \mathcal{F}$ is a *root* of $f$, or a *root* of the equation $f(x) = 0$. A given polynomial $f$ in $\mathcal{F}[x]$ may have no root in $\mathcal{F}$. We shall later see that we may always find a larger field $\mathcal{F}'$, namely, a field $\mathcal{F}'$ containing $\mathcal{F}$ as a subfield, in which $f$ has a root.

We shall here introduce the following definitions.

**Definition 5.12**

A mapping $\phi$ of a ring $A$ into a ring $B$ is called a *homomorphism* if it satisfies the following conditions for all $a, b \in A$:

**1.** $\phi(a + b) = \phi(a) + \phi(b)$.

**2.** $\phi(ab) = \phi(a)\phi(b)$.

If $\phi$ is, moreover, one-to-one, it is called an *isomorphism* of $A$ into $B$.

Proposition 5.11 may be expressed by saying that the mapping

$$f(x) \in \mathcal{F}[x] \rightarrow f(c) \in \mathcal{F}$$

is a homomorphism of the ring $\mathcal{F}[x]$ into $\mathcal{F}$.

**Definition 5.13**

For a homomorphism $\phi$ of a ring $A$ into a ring $B$, the *kernel* of $\phi$ is the set of all $a \in A$ such that $\phi(a) = 0$.

We shall prove the basic properties of homomorphisms.

***Proposition 5.13.*** *Let $\phi$ be a homomorphism of a ring $A$ into a ring $B$.*
**1.** $\phi(0) = 0$, *where* 0 *on the left-hand side denotes the zero element of $A$ and* 0 *on the right-hand side denotes that of $B$.*
**2.** $\phi(-a) = -\phi(a)$ *for every* $a \in A$.
**3.** *The kernel of $\phi$ is a two-sided ideal of $A$.*
**4.** $\phi$ *is one-to-one if and only if* $\phi(a) = 0$ *implies* $a = 0$, *that is, the kernel is* $\{0\}$.

*Proof.* 1. From $\phi(0) = \phi(0 + 0) = \phi(0) + \phi(0)$, we have $\phi(0) = 0$. [This is essentially the same argument as that for (2) of Proposition 3.1.]

2. From $\phi(a) + \phi(-a) = \phi(a + (-a)) = \phi(0) = 0$, we have $\phi(-a) = -\phi(a)$.

3. Let $I$ be the kernel of $\phi$. If $a, b \in I$, then

$$\phi(a + b) = \phi(a) + \phi(b) = 0$$

and

$$\phi(a - b) = \phi(a + (-b)) = \phi(a) + \phi(-b) = \phi(a) - \phi(b) = 0$$

so that $a + b$, $a - b \in I$. If $a \in I$ and $x \in A$, then

$$\phi(xa) = \phi(x)\phi(a) = \phi(x)0 = 0$$

and

$$\phi(ax) = \phi(a)\phi(x) = 0\phi(x) = 0$$

so that $xa$ and $ax$ are in $I$, proving that $I$ is a two-sided ideal.

4. If $\phi$ is one-to-one, then $\phi(a) = 0$ implies that $a = 0$, since we know that $\phi(0) = 0$. Conversely, assume that the kernel $I$ is $\{0\}$. If $\phi(a) = \phi(b)$ for $a, b \in A$, then

$$\phi(b - a) = \phi(b) - \phi(a) = 0,$$

which means that $b - a \in I$, that is, $b = a$.

It is necessary to consider a more general situation than that of Proposition 5.11. First we make the following definition.

**Definition 5.14**

By an *extension* of a field $\mathcal{F}$ we mean a field $\mathcal{F}'$ which contains $\mathcal{F}$ as a subfield.

Let $\mathcal{F}'$ be an extension of a field $\mathcal{F}$. For any element $c' \in \mathcal{F}'$, the mapping

$$f(x) \in \mathcal{F}[x] \to f(c') \in \mathcal{F}'$$

can be defined by

$$f(c') = \sum_{i=0}^{n} a_i c'^i \qquad \text{if } f(x) = \sum_{i=0}^{n} a_i x^i.$$

In the same way as Proposition 5.11, one can prove that this mapping is a homomorphism of the ring $\mathcal{F}[x]$ into $\mathcal{F}'$.

**Definition 5.15**

Let $\mathcal{F}'$ be an extension of a field $\mathcal{F}$. An element $c' \in \mathcal{F}'$ is said to be *transcendental* over $\mathcal{F}$ if the kernel $I$ of the homomorphism $f(x) \in \mathcal{F}[x] \to f(c') \in \mathcal{F}'$ is $\{0\}$. Otherwise, $c'$ is said to be *algebraic* over $\mathcal{F}$.

Thus $c'$ is algebraic over $\mathcal{F}$ if there exists a nonzero polynomial $f \in \mathcal{F}[x]$ such that $f(c) = 0$. If $f_0$ is a monic polynomial of the smallest possible degree such that $f_0(c') = 0$ (that is, a monic polynomial of the smallest degree in $I$), then we know that $I = I(f_0)$ by virtue of Theorem 5.5. We shall call $f_0$ the *minimal polynomial* of $c'$.

***Proposition 5.14.*** *If $c'$ is algebraic over $\mathcal{F}$, then its minimal polynomial $f_0 \in \mathcal{F}[x]$ is irreducible over $\mathcal{F}$.*

*Proof.* Assume that $f_0 = gh$, where $g$ and $h$ are nonconstant polynomials in $\mathcal{F}[x]$. Then we have $f_0(c') = g(c')h(c') = 0$. Since $\deg g$ and $\deg h$ are smaller than $\deg f$, $g$ and $h$ are not in the ideal $I$, so that $g(c') \neq 0$, and $h(c') \neq 0$. This is a contradiction, since $\mathcal{F}'$ is a field.

***Example 5.19.*** $\mathcal{R}$ is an extension of the rational number field $\mathcal{Q}$. The number $\sqrt{2}$ is algebraic over $\mathcal{Q}$, since it satisfies the equation $x^2 - 2 = 0$. Its minimal polynomial is $x^2 - 2$. On the other hand, it is known that the numbers $\pi$ and $e$ are transcendental over $\mathcal{Q}$.

As in Example 5.6, an extension $\mathcal{F}'$ of a field $\mathcal{F}$ can be considered as a vector space over $\mathcal{F}$.

**Definition 5.16**

An extension $\mathcal{F}'$ of $\mathcal{F}$ is called a *finite extension* of $\mathcal{F}$ if the vector space $\mathcal{F}'$ over $\mathcal{F}$ is finite-dimensional. Its dimension is called the *degree* of $\mathcal{F}'$ over $\mathcal{F}$ and is denoted by $(\mathcal{F}' : \mathcal{F})$.

We shall apply the vector space argument to obtain the following result.

***Theorem 5.15.*** *If $\mathcal{F}'$ is a finite extension of a field $\mathcal{F}$, then every element of $\mathcal{F}'$ is algebraic over $\mathcal{F}$.*

*Proof.* Let $c' \in \mathcal{F}'$ and consider the elements $1, c', c'^2, \ldots, c'^n$, where $n = (\mathcal{F}' : \mathcal{F})$. These $n + 1$ elements in the vector space $\mathcal{F}'$ over $\mathcal{F}$ are linearly dependent since $\dim \mathcal{F}' = n$. Thus there exist $a_0, a_1, \ldots, a_n \in \mathcal{F}$, which are not all 0, such that $\sum_{i=0}^{n} a_i c'^i = 0$. Let $a_m$, $m \leq n$, be the last nonzero element among $a_0, a_1, \ldots, a_n$, and let

$$f(x) = \sum_{i=0}^{n} a_i x^i \in \mathcal{F}[x].$$

We have then $f(c') = 0$, showing that $c'$ is algebraic over $\mathcal{F}$.

We shall say that an extension $\mathcal{F}'$ of $\mathcal{F}$ is *algebraic* if every element of $\mathcal{F}'$ is algebraic over $\mathcal{F}$. The preceding theorem says that a finite extension is an algebraic extension.

Let $\mathcal{F}$ be a field and $S$ a subset of $\mathcal{F}$. Consider all subfields of $\mathcal{F}$ which contain $S$ (for example, $\mathcal{F}$ itself). The intersection of all such subfields is obviously a subfield of $\mathcal{F}$ and, indeed, the smallest subfield of $\mathcal{F}$ containing $S$. We shall call it the *subfield* of $\mathcal{F}$ *generated by* $S$.

We shall apply this construction to the following situation. Let $\mathcal{F}'$ be an extension of a field $\mathcal{F}$ and let $c' \in \mathcal{F}'$. The subfield of $\mathcal{F}'$ generated by the subset $\mathcal{F} \cup \{c'\}$ is denoted by $\mathcal{F}(c')$. We say that $\mathcal{F}(c')$ is obtained from $\mathcal{F}$ by *adjunction* of $c'$.

**Definition 5.17**

An extension $\mathcal{F}'$ of $\mathcal{F}$ is called a *simple extension* if there exists an element $c' \in \mathcal{F}'$ such that $\mathcal{F}' = \mathcal{F}(c')$. If $c'$ is algebraic, then $\mathcal{F}'$ is called a *simple algebraic extension.* If $c'$ is transcendental, then $\mathcal{F}'$ is called a *simple transcendental extension.*

If $\mathcal{F}' = \mathcal{F}(c')$ and if $c'$ is algebraic, then we can show that $\mathcal{F}'$ is algebraic, that is, every element of $\mathcal{F}'$ is algebraic over $\mathcal{F}$. Thus the expression "simple algebraic extension" means "simple extension $\mathcal{F}(c')$ with $c'$ algebraic" as well as "simple extension which is algebraic," both being mathematically equivalent. This result can be stated more precisely as follows.

***Theorem 5.16.*** *Let $\mathcal{F}' = \mathcal{F}(c')$, where $c'$ is algebraic over $\mathcal{F}$. Then every element of $\mathcal{F}'$ is of the form $f(c')$, where $f$ is a polynomial in $\mathcal{F}[x]$ whose degree is less than the degree $n$ of the minimal polynomial of $c'$. Moreover, $\mathcal{F}'$ is a finite, hence algebraic, extension with $(\mathcal{F}' : \mathcal{F}) = n$.*

*Proof.* Let

$$f_0(x) = \sum_{i=0}^{n} a_i x^i \qquad \text{with } a_n = 1$$

be the minimal polynomial of $c'$. Consider the following subset $\mathcal{F}^*$ of $\mathcal{F}'$:

$$\mathcal{F}^* = \{f(c'); f \in \mathcal{F}[x], \deg f \leq n - 1\}.$$

$\mathcal{F}^*$ contains $\mathcal{F}$ (take constant polynomials $f$) and $c'$. We show that $\mathcal{F}^* = \mathcal{F}'$ by proving that $\mathcal{F}^*$ is a subfield of $\mathcal{F}'$.

If $f$ and $g$ are in $\mathcal{F}[x]$ with degrees $\leq n - 1$, the same holds for $f + g$ and $f - g$. Since $(f + g)(c') = f(c') + g(c')$ and $(f - g)(c') = f(c') - g(c')$ by the remark following Definition 5.14, we see that $\mathcal{F}^*$ contains the sum and difference of any two elements of $\mathcal{F}^*$. As for the product $f(c')g(c')$ of two elements $f(c')$ and $g(c')$ in $\mathcal{F}^*$, we write

$$f(x)g(x) = f_0(x)q(x) + r(x),$$

where $r(x) = 0$ or $\deg r < \deg f_0$. Then

$$f(c')g(c') = f_0(c')q(c') + r(c') = r(c')$$

because $f_0(c') = 0$. If $r(c') = 0$, then $f(c')g(c') = 0$. If $r(c') \neq 0$, then $f(c')g(c')$ is still in $\mathcal{F}^*$, because $\deg r < n$. We shall now prove that the inverse of a nonzero element in $\mathcal{F}^*$ lies in $\mathcal{F}^*$. Assume that $f(c') \neq 0$, where $\deg f \leq n - 1$. Since the minimal polynomial $f_0$ of $c'$ is irreducible by Proposition 5.14, we have $(f, f_0) = 1$. By the corollary to Theorem 5.6, there exist $g, g_0 \in \mathcal{F}[x]$ such that

$$fg + f_0 g_0 = 1.$$

Thus

$$f(c')g(c') = 1.$$

If $\deg g \leq n - 1$, it is obvious that $g(c') \in \mathcal{F}^*$. If $\deg g \geq n$, then let

$$g = f_0 q + r \qquad \text{in } \mathcal{F}[x],$$

where $r \neq 0$ and $\deg r < \deg f_0 = n$. Then $g(c') = r(c')$ and this belongs to $\mathcal{F}^*$. At any rate, we have shown that there is an element $c'' \in \mathcal{F}^*$ such that $f(c')c'' = 1$, thus completing the proof that $\mathcal{F}^*$ is a subfield of $\mathcal{F}'$.

The second assertion of Theorem 5.16 follows easily from the first. Since every element of $\mathcal{F}' = \mathcal{F}(c')$ is of the form $f(c')$, where $f \in \mathcal{F}[x]$ and $\deg f \leq n - 1$, it follows that the elements $1, c', c'^2, \ldots, c'^{n-1}$ span the vector space $\mathcal{F}'$ over $\mathcal{F}$. On the other hand, they are linearly independent, because otherwise there exist $a_0, a_1, \ldots, a_{n-1}$ in $\mathcal{F}$ which are not all 0 such that

$$a_0 + a_1 c' + \cdots + a_{n-1} c'^{n-1} = 0.$$

This means that the minimal polynomial of $c'$ must be of degree $\leq n - 1$, contrary to the assumption. Thus $1, c', \ldots, c'^{n-1}$ form a basis of the vector space $\mathcal{F}'$ over $\mathcal{F}$ and $(\mathcal{F}' : \mathcal{F}) = n$.

EXERCISE 5.4

» 1. Let $\mathfrak{F}$ be the field $\{0,1,2\}$ as in Example 5.2. Determine the polynomial function corresponding to the polynomial $x^3 + 2x \in \mathfrak{F}[x]$. Obtain also a factorization of this polynomial into irreducible factors.

2. Show that $\sqrt{3}$ is algebraic over $\mathbb{Q}$. What is the degree of $\mathbb{Q}(\sqrt{3})$ over $\mathbb{Q}$? ($\mathbb{Q}$: the rational number field.)

» 3. Show that $\sqrt{2} + \sqrt{3}$ is algebraic over $\mathbb{Q}$ and find its minimal polynomial, where $\mathbb{Q}$ is the rational number field.

»☆ 4. Prove that if $\mathfrak{F}''$ is a finite extension of $\mathfrak{F}'$ and if $\mathfrak{F}'$ is a finite extension of $\mathfrak{F}$, then $\mathfrak{F}''$ is a finite extension of $\mathfrak{F}$ and $(\mathfrak{F}'':\mathfrak{F}) = (\mathfrak{F}'':\mathfrak{F}')\,(\mathfrak{F}':\mathfrak{F})$.

☆ 5. Let $\mathfrak{F}'$ be an extension of $\mathfrak{F}$. Prove that the set of all elements in $\mathfrak{F}'$ which are algebraic over $\mathfrak{F}$ is a subfield of $\mathfrak{F}'$. (A complex number which is algebraic over $\mathbb{Q}$ is called an *algebraic number*. A real number which is algebraic over $\mathbb{Q}$ is called a *real algebraic number.*)

»☆ 6. Prove that if $\mathfrak{F}''$ is an algebraic extension of $\mathfrak{F}'$ and if $\mathfrak{F}'$ is an algebraic extension of $\mathfrak{F}$, then $\mathfrak{F}''$ is an algebraic extension of $\mathfrak{F}$.

☆ 7. Let $\mathfrak{F}'$ be an extension of a field $\mathfrak{F}$. Prove that if a system of linear equations over $\mathfrak{F}$:

$$\sum_{j=1}^{n} a_{ij}x_j = b_i, \qquad 1 \leq i \leq m,$$

has a solution in $\mathfrak{F}'$, then it has a solution in $\mathfrak{F}$.

## 5.5 CONSTRUCTION OF EXTENSIONS

We shall now discuss the method of constructing simple algebraic extensions of a given field. Let $\mathcal{F}$ be a field. A given polynomial $f \in \mathcal{F}[x]$ may not have a root in $\mathcal{F}$, but it is possible to construct an extension $\mathcal{F}'$ in which $f$ has a root, say, $c'$. In order to do this, we may assume that $f$ is irreducible over $\mathcal{F}$. For, if $f$ is not irreducible, let $f = f_1 \cdots f_k$ be a factorization into irreducible factors. If we can find an extension $\mathcal{F}'$ and a root $c' \in \mathcal{F}'$ of $f_1$, then $c'$ is also a root of $f$.

We shall start with a general discussion. Let $A$ be a commutative ring with identity. Given an ideal $I$ of $A$, we introduce an equivalence relation in $A$. We shall say that *$a$ is congruent to $b$* mod $I$ if $a - b \in I$ and write $a \equiv b \bmod I$, or simply, $a \equiv b$; this is an equivalence relation, as can be easily verified. (This relation is analogous to what we considered for a vector space and a subspace in Sec. 4.2.)

Let $A/I$ denote the set of equivalence classes with respect to this relation, and let $\pi$ be the mapping $A \to A/I$ which associates to $a \in A$ the equivalence class $\pi(a) \in A/I$ which contains $a$. Just in the same way as we did in Sec. 4.2, we can make $A/I$ into an additive group by defining addition in $A/I$ in such a way that

$$\pi(a + b) = \pi(a) + \pi(b) \qquad \text{for all } a, b \in A.$$

Before we define multiplication in $A/I$, we observe:

***Lemma.*** *If $\pi(a) = \pi(b)$, then $\pi(ac) = \pi(bc)$ for any $c \in A$.*

*Proof.* $\pi(a) = \pi(b)$ implies that $b - a \in I$. Since $I$ is an ideal, we have $c(b - a) = cb - ca \in I$, which means $\pi(ac) = \pi(bc)$, where $c$ is arbitrary.

Now for any elements $a^*, b^* \in A/I$, choose $a, b \in A$ such that $\pi(a) = a^*$ and $\pi(b) = b^*$. We shall define $a^*b^*$ to be $\pi(ab)$ after showing that $\pi(ab)$ does not depend on the choice of $a$ and $b$. Suppose that

$$\pi(a) = \pi(a') = a^* \qquad \text{and} \qquad \pi(b) = \pi(b') = b^*.$$

Then the lemma above implies

$$\pi(ab) = \pi(a'b) = \pi(a'b'),$$

proving our assertion.

We can thus define multiplication in $A/I$ in such a way that

$$\pi(ab) = \pi(a)\pi(b) \qquad \text{for all } a, b \in A.$$

It is straightforward to show that $A/I$ now becomes a ring and that $\pi$ is a homomorphism of the ring $A$ onto the ring $A/I$. The kernel is precisely the ideal $I$.

**Definition 5.18**

The ring $A/I$ is called the *quotient ring* of $A$ over $I$. The mapping $\pi$ is called the *projection* of $A$ onto $A/I$.

We note that $A/I$ is commutative together with $A$. The element $\pi(1)$ is the identity for $A/I$.

***Example 5.20.*** Let $A$ be the ring of integers $\mathbb{Z}$ and let $I$ be the ideal $I(m)$, that is, the set of all multiples of a given integer $m$. The quotient ring $\mathbb{Z}/I(m)$ consists of the residue classes modulo $m$, that is, $0^*, 1^*, \ldots, (m-1)^*$. For $a^*, b^*$, we can find $a^* + b^*$ and $a^*b^*$ by taking $a + b$ and $ab$ and taking their residue classes modulo $m$.

**Definition 5.19**

An ideal $I \neq A$ of a ring $A$ is said to be *maximal* if there is no ideal, other than $A$ and $I$, which contains $I$.

We shall prove:

***Theorem 5.17.*** *Let $A$ be a commutative ring with identity, and let $I$ be an ideal of $A$. The quotient ring $A/I$ is a field if and only if $I$ is maximal.*

*Proof.* Assume that $A/I$ is a field. By Theorem 5.4, $A/I$ has no ideal other than $A/I$ and $\{0\}$. But if $I$ is not maximal, there is an ideal $J$ such that $I \subset J \subset A$. The image $\pi(J)$ of $J$ by the projection $\pi\colon A \to A/I$ is obviously an ideal of $A/I$. Hence $\pi(J) = A/I$ or $\pi(J) = \{0\}$, which will imply $J = A$ or $J = I$, contrary to the remark above.

Conversely, assume that $I$ is maximal. If $A/I$ is not a field, then, again by Theorem 5.4, there exists an ideal $J^*$ other than $A/I$ and $\{0\}$. Then $\pi^{-1}(J^*)$ is obviously an ideal of $A$ which contains $I$. Since $I$ is maximal, we have $\pi^{-1}(J^*) = A$, which implies $J^* = A/I$, contrary to the assumption.

***Example 5.21.*** In Example 5.20, the ideal $I(m)$ is maximal if and only if $m$ is a prime number. In fact, if $m$ is not prime and if $m = pq$, where $p$ and $q$ are integers $>1$, then $I(m) \subset I(p) \subset \mathbb{Z}$, so that $I(m)$ is not maximal. Conversely, if $p$ is a prime, then $I(p)$ is maximal, for if $J$ is an ideal such that $I(p) \subset J \subset \mathbb{Z}$, then there is $d \in \mathbb{Z}$ such that $d \neq 1$ and $J = I(d)$ (cf. Example 5.13). This means $d|p$, contrary to the assumption. By Theorem 5.17 we see that the quotient ring $\mathbb{Z}/I(m)$ is a field if and only if $m$ is a prime. In particular, for any prime number $p$, $\mathbb{Z}/I(p)$ is a field which has exactly $p$ elements. One can verify that for $p = 2$ and 3 we have the fields we defined in Example 5.2.

In order to apply Theorem 5.17 to the polynomial ring, we first establish:

***Lemma.*** *Let $\mathfrak{F}$ be a field. An ideal $I(f)$, $f \in \mathfrak{F}[x]$ with* $\deg f > 1$, *is maximal if and only if $f$ is irreducible over $\mathfrak{F}$.*

*Proof.* If $J$ is an ideal containing $I(f)$, then there is a monic polynomial $g \in \mathfrak{F}[x]$, by virtue of Theorem 5.5, such that $J = I(g)$, and it follows that $g$ divides $f$. Hence if $f$ is irreducible, $I(f)$ is a maximal ideal. Conversely, if $I(f)$ is maximal, then for any proper factor $g$ of $f$ we have $I(f) \subset I(g)$ so that $I(g) = \mathfrak{F}[x]$, that is, $g$ is a nonzero constant. Hence $f$ is irreducible over $\mathfrak{F}$.

We shall now prove the main result of this section.

***Theorem 5.18.*** *Let $f$ be an irreducible polynomial in $\mathfrak{F}[x]$. Then $\mathfrak{F}^* = \mathfrak{F}[x]/I(f)$ is a field. Moreover, $\mathfrak{F}^*$ is a simple algebraic extension of $\mathfrak{F}$, in which the equation $f(x) = 0$ has a root $c'$ and $\mathfrak{F}^* = \mathfrak{F}(c')$.*

*Proof.* By the preceding lemma, $I(f)$ is a maximal ideal of $\mathfrak{F}[x]$. By Theorem 5.17 it follows that $\mathfrak{F}^*$ is a field. Let $\pi$ be the projection of $\mathfrak{F}[x]$ onto $\mathfrak{F}^*$ and let $c' = \pi(x)$ be the image of the polynomial $x$. Since $I(f)$ does not contain any nonzero constant, $\pi$ maps $\mathfrak{F}$, regarded as a field contained in the ring $\mathfrak{F}[x]$, isomorphically into $\mathfrak{F}^*$. Hence we shall identify every $a \in \mathfrak{F}$ with its image $\pi(a) \in \mathfrak{F}^*$. Now if

$$f(x) = \sum_{i=0}^{n} a_i x^i \qquad \text{in } \mathfrak{F}[x],$$

then

$$\pi(f) = \sum_{i=0}^{n} a_i c'^i = 0 \qquad \text{in } \mathfrak{F}^*,$$

since $f \in I(f)$, which is the kernel of $\pi$. This means that $c'$ satisfies $f(c') = 0$. Hence $c'$ is algebraic over $\mathcal{F}$. The minimal polynomial of $c'$ is indeed $f$; in fact, if $g(c') = 0$ for a polynomial $g$ over $\mathcal{F}$ with smaller degree than $f$, then $\pi(g) = g(c')$ as above, and this is 0. It follows that $g \in I(f)$ and $f$ divides $g$, contrary to $\deg g < \deg f$. We now prove $\mathcal{F}^* = \mathcal{F}(c')$. Since $\pi: \mathcal{F}[x] \to \mathcal{F}^*$ is onto, every element of $\mathcal{F}^*$ is of the form $\pi(h)$, where $h \in \mathcal{F}[x]$. But, as above, $\pi(h) = h(c') \in \mathcal{F}(c')$. Hence $\mathcal{F}^* = \mathcal{F}(c')$.

***Example 5.22.*** Let $f(x) = x^2 + 1$ in $\mathcal{R}[x]$. Then the field $\mathcal{R}[x]/I(f)$ is isomorphic with the complex number field $\mathcal{C}$. In fact, the element $c' = \pi(x)$ in the notation of the preceding theorem satisfies $c'^2 + 1 = 0$ and corresponds to the imaginary unit $i = \sqrt{-1}$.

**Definition 5.20**

A field $\mathcal{F}$ is said to be *algebraically closed* if every polynomial $f(x) \in \mathcal{F}[x]$ has a root in $\mathcal{F}$.

Thus $\mathcal{F}$ is algebraically closed if and only if every irreducible polynomial in $\mathcal{F}[x]$ is of degree 1. Theorem 5.10 says that $\mathcal{C}$ is algebraically closed. We shall state one important theorem without proof.

***Theorem 5.19.*** *Every field $\mathcal{F}$ has an algebraic extension $\mathcal{F}^*$ which is algebraically closed.*

Such an extension $\mathcal{F}^*$ is unique up to isomorphism and is called the *algebraic closure* of $\mathcal{F}$.

**EXERCISE 5.5**

**1.** Let $A$ be a commutative ring with identity. Assume that an equivalence relation $x \sim y$ in $A$ satisfies the following conditions:

(*a*) If $x_1 \sim y_1$ and $x_2 \sim y_2$, then $x_1 + x_2 \sim y_1 + y_2$.

(*b*) If $x \sim y$, then $xz \sim yz$ for every $z \in A$.

Prove that $I = \{x \in A;\ x \sim 0\}$ is an ideal of $A$ and that $x \sim y$ if and only if $y - x \in I$.

☆ **2.** Let $A$ and $B$ be commutative rings with identity. Prove that if $\phi$ is a homomorphism of $A$ onto $B$, then there is an isomorphism $\psi$ of the quotient ring $A/I$, where $I$ is the kernel of $\phi$, onto $B$ such that the diagram

$$\begin{array}{ccc} A & \xrightarrow{\phi} & B \\ & {\scriptstyle\pi}\searrow \quad \swarrow{\scriptstyle\psi} & \\ & A/I & \end{array}$$

is commutative.

›› **3.** In Example 5.20, let $m = 6$. Show that in $\mathcal{Z}/I(6)$ there are two nonzero elements whose product is 0.

›› **4.** Let $\mathcal{F}$ be the field $\mathcal{Z}/I(2) = \{0,1\}$. How many elements does the field $\mathcal{F}[x]/I(f)$ have, where $f(x) = x^2 + x + 1$?

☆ **5.** The *prime subfield* $\mathfrak{F}_0$ of a field $\mathfrak{F}$ is defined as the subfield generated by the subset $\{0,1\}$ (that is, the smallest subfield of $\mathfrak{F}$, since any subfield contains 0 and 1).

(*a*) Prove that if $\mathfrak{F}$ has characteristic 0, then $\mathfrak{F}_0$ is isomorphic with the field of rational numbers.

(*b*) Prove that if $\mathfrak{F}$ has characteristic $p$, then $\mathfrak{F}_0$ is isomorphic with the field $\mathbb{Z}/I(p)$ in Example 5.21.

☆ **6.** Let $\mathfrak{F}$ be a finite field.

(*a*) Prove that the characteristic of $\mathfrak{F}$ is not 0.

(*b*) If the characteristic of $\mathfrak{F}$ is $p$, show that the number of elements in $\mathfrak{F}$ is $p^n$, where $n$ is a certain positive integer. (Show that $\mathfrak{F}$ is a finite extension of its prime subfield.)

## 5.6 QUOTIENT FIELDS

We shall first discuss the construction of the quotient field of an integral domain and apply it to the construction of a simple transcendental extension of a given field.

### Definition 5.21

A commutative ring with identity $A$ is called an *integral domain* if $ab = 0$ in $A$ implies that $a = 0$ or $b = 0$.

***Example 5.23.*** A field $\mathfrak{F}$ is an integral domain (corollary to Proposition 5.1). The polynomial ring $\mathfrak{F}[x]$ over $\mathfrak{F}$ is an integral domain [(2) of Proposition 5.2].

***Example 5.24.*** The quotient ring $\mathbb{Z}/I(m)$, where $\mathbb{Z}$ is the ring of integers and $m \in \mathbb{Z}$ (Example 5.20), is an integral domain if $m$ is prime (in fact, it is then a field). If $m = pq$, $p > 1$, $q > 1$, then the product $p^*q^*$ of the residue classes $p^*$ and $q^*$ of $p$ and $q$ is 0, while $p^* \neq 0$ and $q^* \neq 0$, so that $\mathbb{Z}/I(m)$ is not an integral domain.

Given an integral domain $A$ which is not a field, we may construct a field $\mathfrak{F}$ containing $A$ as a subset. This construction is patterned after the construction of the field of rational numbers from the integral domain $\mathbb{Z}$.

For this purpose we consider the set $S$ of all pairs $(a,b)$, where $a, b \in A$ and $b \neq 0$. We define a relation $(a,b) \approx (c,d)$ in $S$ as follows: We say that $(a,b)$ is *equivalent to* $(c,d)$ and write $(a,b) \approx (c,d)$ if $ad = bc$.

***Lemma.*** *$(a,b) \approx (c,d)$ is an equivalence relation.*

*Proof.* First, $(a,b) \approx (a,b)$ follows from $ab = ba$. Second, $(a,b) \approx (c,d)$ implies $(c,d) \approx (a,b)$, because $ad = bc$ implies $cb = da$, since $A$ is commutative. Finally, we show that $(a,b) \approx (c,d)$ and $(c,d) \approx (e,f)$ imply $(a,b) \approx (e,f)$. Multiplying $ad = bc$ and $cf = de$, we obtain $afcd = becd$. If $c \neq 0$, then $cd \neq 0$, since $d \neq 0$ and since $A$ is an integral domain. Then $afcd = becd$, that is, $(af - be)cd = 0$, implies $af = be$, proving that $(a,b) \approx (e,f)$. If $c = 0$, then $ad = bc = 0$, and hence $a = 0$

because $d \neq 0$. Similarly, $cf = de = 0$, and hence $e = 0$ because $d \neq 0$. Then $af = 0 = be$, so that $(a,b) \approx (e,f)$. This completes the proof of the lemma.

Now let $\mathcal{F}$ be the set of all equivalence classes of $S$ with respect to the equivalence relation $\approx$. We wish to make $\mathcal{F}$ into a field by defining addition and multiplication as follows: Denoting the equivalence class of $(a,b) \in S$ by $(a,b)^*$, we wish to define

$$(a,b)^* + (c,d)^* = (ad + bc,\ bd)^* \tag{5.10}$$

and

$$(a,b)^*(c,d)^* = (ac,bd)^*. \tag{5.11}$$

First, note that $b \neq 0$ and $d \neq 0$ imply that $bd \neq 0$. We verify that the right-hand sides of (5.10) and (5.11) are independent of the choice of representatives of the equivalence classes $(a,b)^*$ and $(c,d)^*$. Suppose that $(a,b) \approx (a',b')$ and $(c,d) \approx (c',d')$. Then $ab' = ba'$ and $cd' = dc'$. These imply

$$ab'dd' + bb'cd' = a'bdd' + bb'c'd,$$

namely, $(ad + bc)b'd' = (a'd' + b'c')bd$, that is,

$$(ad + bc,\ bd) \approx (a'd' + b'c',\ b'd').$$

The assertion for (5.11) can be verified in a similar way.

Now all conditions for $\mathcal{F}$ to satisfy as a field can be verified easily; we shall only note that:

1. The zero element is $(0,1)^*$.
2. The additive inverse of $(a,b)^*$ is $(-a,b)^*$.
3. The identity element is $(1,1)^*$.
4. If $(a,b)^* \neq (0,1)^*$ (which amounts to $a \neq 0$), then the inverse of $(a,b)^*$ is $(b,a)^*$.

In order to identify $A$ with a certain subset of $\mathcal{F}$, we consider a mapping $\phi$ of $A$ into $\mathcal{F}$ defined by

$$\phi(a) = (a,1)^* \qquad \text{for every } a \in A. \tag{5.12}$$

We have obviously

1. $\phi(a + b) = \phi(a) + \phi(b)$ for all $a, b \in A$.
2. $\phi(ab) = \phi(a)\phi(b)$ for all $a, b \in A$.
3. $\phi$ is one-to-one.

Furthermore:

4. Every element of $\mathcal{F}$ can be written in the form $\phi(a)/\phi(b)$ [that is, $\phi(a)(\phi(b))^{-1}$], where $a$ and $b \neq 0$ are certain elements of $A$.

We shall verify (4). Let $(a,b)^*$ be an arbitrary element of $\mathcal{F}$. Then, since

$$(a,1)^*(1,b)^* = (a,b)^* \qquad \text{and} \qquad (1,b)^* = ((b,1)^*)^{-1},$$

we have

$$\phi(a)(\phi(b))^{-1} = (a,b)^*.$$

The mapping $a \in A \to \phi(a) \in \mathcal{F}$ is thus an isomorphism of $A$ into the field $\mathcal{F}$, and we may identify $a \in A$ with $\phi(a) \in \mathcal{F}$ and denote this element still by $a$. Property 4 means that every element of $\mathcal{F}$ is of the form $a/b$, where $a, b \in A$ and $b \neq 0$.

**Definition 5.22**

The field $\mathcal{F}$ constructed above is called the *quotient field* of the integral domain $A$.

***Example 5.25.*** The quotient field of the ring of integers $\mathbb{Z}$ is the rational number field.

**Definition 5.23**

For a field $\mathcal{F}$, the quotient field of the polynomial ring $\mathcal{F}[x]$ is called the *field of rational functions over* $\mathcal{F}$, and it is denoted by $\mathcal{F}(x)$. Elements of $\mathcal{F}(x)$ are called *rational functions* over $\mathcal{F}$ (with indeterminate $x$).

Thus a rational function over $\mathcal{F}$ is of the form $f(x)/g(x)$, where $f$ and $g \neq 0$ are polynomials in $\mathcal{F}[x]$. The field $\mathcal{F}(x)$ is a simple extension of $\mathcal{F}$ obtained by adjunction of the polynomial $x \in \mathcal{F}[x]$. The element $x$ is transcendental over $\mathcal{F}$, for if $f$ is a polynomial over $\mathcal{F}$ such that $f(x) = 0$ in $\mathcal{F}(x)$, then, by definition of polynomials, all the coefficients of $f$ must be 0.

***Theorem 5.20.*** *A simple transcendental extension $\mathcal{F}' = \mathcal{F}(c')$ of a field $\mathcal{F}$ is isomorphic with the field of rational functions $\mathcal{F}(x)$ over $\mathcal{F}$; more precisely, there is an isomorphism $\psi$ of $\mathcal{F}'$ onto $\mathcal{F}(x)$ such that $\psi(c') = x$ and $\psi(a) = a$ for every $a \in \mathcal{F}$.*

*Proof.* Since $c'$ is transcendental, the mapping $\phi: f(x) \in \mathcal{F}[x] \to f(c') \in \mathcal{F}(c')$ is an isomorphism of the ring $\mathcal{F}[x]$ into $\mathcal{F}'$. For an arbitrary element $r(x) \in \mathcal{F}(x)$, we may write $r = f/g$ with $f, g \in \mathcal{F}[x]$. If $d = (f,g)$, we may divide $f$ and $g$ by $d$ and so assume that $r = f/g$ with $(f,g) = 1$. Such a representation of $r$ is unique if we assume furthermore that $g$ is monic. Now we associate to $r$ the element $f(c')/g(c') \in \mathcal{F}'$, where we know that $g(c') \neq 0$ because $g \neq 0$. It is easy to verify that this mapping $\phi$ of $\mathcal{F}(x)$ into $\mathcal{F}'$ is actually an isomorphism of $\mathcal{F}(x)$ onto $\mathcal{F}(c')$ which maps $x$ upon $c'$ and every $a \in \mathcal{F}$ upon itself. By taking the inverse $\psi$ of this isomorphism, we have an isomorphism of $\mathcal{F}'$ onto $\mathcal{F}(x)$ with the desired properties.

EXERCISE 5.6

» 1. Let $A$ be the ring of all continuous functions on the unit interval (Example 5.11). Is $A$ an integral domain?

2. In an integral domain, prove that $ax = bx$ and $x \neq 0$ imply $a = b$.

☆ 3. Let $K$ be a commutative ring with identity. An ideal $I$ of $K$ is called a *prime ideal* if $ab \in I$ implies $a \in I$ or $b \in I$. Prove that an ideal $I$ is prime if and only if the quotient ring $K/I$ is an integral domain.

☆ 4. Prove that in the ring $\mathfrak{F}[x]$, where $\mathfrak{F}$ is a field, every prime ideal is maximal.

## 5.7 EXTENSION OF VECTOR SPACES

Let $V$ be a vector space over $\mathcal{F}$. Given an extension $\mathcal{F}'$ of the field $\mathcal{F}$, we shall construct a vector space $V'$ over $\mathcal{F}'$ which contains $V$ in a certain natural manner.

We first remark that any vector space $W'$ over $\mathcal{F}'$ can be considered as a vector space over $\mathcal{F}$. If $f$ is a mapping of a vector space $V$ over $\mathcal{F}$ into a vector space $W'$ over $\mathcal{F}'$ such that

$$f(\alpha + \beta) = f(\alpha) + f(\beta), \qquad \text{where } \alpha, \beta \in V,$$
$$f(c\alpha) = cf(\alpha), \qquad \text{where } \alpha \in V \text{ and } c \in \mathcal{F},$$

then we say that $f$ is $\mathcal{F}$-*linear*. In other words, $f$ is $\mathcal{F}$-linear if it is a linear mapping of $V$ into $W'$ when we regard $W'$ as a vector space over $\mathcal{F}$.

### Definition 5.24

A vector space $V'$ over $\mathcal{F}'$ is called an *extension* of $V$ corresponding to the extension $\mathcal{F}'$ of $\mathcal{F}$ if there exists an $\mathcal{F}$-linear mapping $\theta$ of $V$ into $V'$ having the following properties:

1. The image $\theta(V)$ spans $V'$, that is, the smallest subspace of $V'$ (over $\mathcal{F}'$) containing $\theta(V)$ is $V'$.

2. For any $\mathcal{F}$-linear mapping $f$ of $V$ into a vector space $W'$ over $\mathcal{F}'$, there exists a linear mapping $f'$ of $V'$ into $W'$ such that $f = f'\theta$, that is, the diagram

$$\begin{array}{ccc} V & \xrightarrow{\theta} & V' \\ & {\scriptstyle f}\searrow & \swarrow{\scriptstyle f'} \\ & W' & \end{array}$$

is commutative.

***Example 5.26.*** Let $\mathfrak{F}'$ be an extension of a field $\mathfrak{F}$. The vector space $\mathfrak{F}'^n$ is an extension of the vector space $\mathfrak{F}^n$. In fact, any $\alpha = (a_1, \ldots, a_n) \in \mathfrak{F}^n$ can be regarded as an element of $\mathfrak{F}'^n$. The mapping $\theta_0$ of $\mathfrak{F}^n$ into $\mathfrak{F}'^n$ which maps $\alpha \in \mathfrak{F}^n$ into $\alpha \in \mathfrak{F}'^n$ is obviously $\mathfrak{F}$-linear. We show that $\theta_0$ has the properties 1 and 2 in the preceding definition. Let

$$\epsilon_1 = (1,0, \ldots ,0), \qquad \epsilon_2 = (0,1,0, \quad \ldots ,0), \ldots , \qquad \epsilon_n = (0, \ldots ,0,1).$$

Then $\{\epsilon_1, \ldots ,\epsilon_n\}$ is the standard basis for $\mathfrak{F}^n$ as well as for $\mathfrak{F}'^n$ [as elements of $\mathfrak{F}'^n$ we express them by $\theta_0(\epsilon_i)$]. Then it is obvious that $\theta_0(\epsilon_1), \ldots , \theta_0(\epsilon_n)$ span $\mathfrak{F}'^n$, so that condition 1 is satisfied. Let $f$ be any $\mathfrak{F}$-linear mapping of $\mathfrak{F}^n$ into a vector space $W'$ over $\mathfrak{F}'$. By Proposition 3.2, there exists a unique linear mapping $f'\colon \mathfrak{F}'^n \to W'$ such that

$$f'(\theta_0(\epsilon_1)) = f(\epsilon_i) \qquad \text{for } 1 \leq i \leq n.$$

It follows that $f(\alpha) = f'(\theta_0(\alpha))$ for every $\alpha \in \mathfrak{F}^n$. Condition 2 is thus satisfied.

We shall prove:

***Proposition 5.21.*** *Let $V$ be a finite-dimensional vector space over a field $\mathfrak{F}$. For any extension $\mathfrak{F}'$ of $\mathfrak{F}$, there is a corresponding extension $V'$ with $\theta\colon V \to V'$. Such an extension is unique up to an isomorphism; more precisely, if $V''$ with $\theta'\colon V \to V''$ is another extension, then there exists a linear isomorphism $\phi$ of $V'$ onto $V''$ such that*

$$\begin{array}{ccc} & \overset{\theta}{\nearrow} & V' \\ V & & \downarrow \phi \\ & \underset{\theta'}{\searrow} & V'' \end{array}$$

*is commutative.*

*Proof.* The existence is almost immediate from Example 5.22. In fact, let $\Phi$ be a linear isomorphism of $V$ onto $\mathfrak{F}^n$ corresponding to a basis $\{\alpha_i\}$ of $V$, where $n = \dim V$. For the extension $\theta_0\colon \mathfrak{F}^n \to \mathfrak{F}'^n$ in Example 5.22, we have the extension $\theta = \theta_0\Phi\colon V \to \mathfrak{F}'^n$. Moreover, this shows that $V$ has an $n$-dimensional vector space over $\mathfrak{F}'$ as an extension.

We now prove the uniqueness. Let $\theta\colon V \to V'$ and $\theta'\colon V \to V''$ be two extensions. By property 2 in Definition 5.24, there is a linear mapping $\phi\colon V' \to V''$ such that $z' = \phi\theta$ and a linear mapping $\psi\colon V'' \to V'$ such that $\theta = \psi\theta'$. Thus we have

$$\theta' = \phi\psi\theta' \qquad \text{and} \qquad \theta = \psi\phi\theta,$$

so that $\psi\phi$ is the identity on $\theta(V)$ and $\phi\psi$ is the identity on $\theta'(V)$. By property 1 of Definition 5.24, $\theta(V)$ and $\theta'(V)$ span $V'$ and $V''$, respectively. Hence $\phi\psi$ and $\psi\phi$ are the identity mappings of $V''$ and $V'$, respectively. Hence $\phi$ is a linear isomorphism of $V'$ onto $V''$ such that $\theta' = \phi\theta$, proving the uniqueness part of Proposition 5.21.

Incidentally, we have proved:

***Corollary.*** *If $V$ is an $n$-dimensional vector space over $\mathfrak{F}$, then an extension $V'$ of $V$ corresponding to an extension $\mathfrak{F}'$ of $\mathfrak{F}$ is $n$-dimensional.*

***Proposition 5.22.*** *Let $V'$ with $\theta$: $V \to V'$ be an extension corresponding to an extension $\mathfrak{F}'$ of $\mathfrak{F}$. For any linear transformation $A$ of $V$, there is a unique linear transformation $A'$ of $V'$ such that*

$$\begin{array}{ccc} V & \xrightarrow{\theta} & V' \\ {\scriptstyle A}\downarrow & & \downarrow{\scriptstyle A'} \\ V & \xrightarrow{\theta} & V' \end{array}$$

*is commutative.* ($A'$ is called the *extension* of $A$ to $V'$.)

*Proof.* Consider the mapping $f = \theta A$ of $V$ into $V'$, which is $\mathfrak{F}$-linear. By property 2 of Definition 5.24 there exists a linear mapping $A'$ of $V'$ into itself such that $\theta A = A'\theta$. If $A''$ is another linear transformation of $V'$ satisfying $\theta A = A''\theta$, then $A'$ and $A''$ coincide on $\theta(V)$. Since $\theta(V)$ spans $V'$, it follows that $A' = A''$, showing the uniqueness of $A'$.

***Remark.*** Let $\{\alpha_1, \ldots, \alpha_n\}$ be a basis of $V$ so that $\{\theta(\alpha_1), \ldots, \theta(\alpha_n)\}$ is a basis of $V'$. The matrix representing the extension $A'$ with respect to $\{\theta(\alpha_i)\}$ is the same as the matrix representing $A$ with respect to $\{\alpha_i\}$. This corresponds to a rather obvious fact that an $n \times n$ matrix **A** over $\mathfrak{F}$ can be regarded as a linear transformation of $\mathfrak{F}^n$ as well as a linear transformation of $\mathfrak{F}'^n$.

We shall now consider the special case where $\mathfrak{F} = \mathfrak{R}$ and $\mathfrak{F}' = \mathfrak{C}$. For an $n$-dimensional real vector space $V$, an extension $V'$ corresponding to the extension $\mathfrak{C}$ of $\mathfrak{R}$ is called the *complexification* of $V$. We shall denote it by $V^c$ and now identify $\alpha \in V$ with $\theta(\alpha) \in V^c$.

***Proposition 5.23.*** *Let $V^c$ be the complexification of an n-dimensional real vector space $V$.*

**1.** *Every $\gamma \in V^c$ can be expressed uniquely in the form $\gamma = \alpha + i\beta$, where $\alpha, \beta \in V$.*

**2.** *For any $\gamma \in V^c$, write $\gamma = \alpha + i\beta$ with $\alpha, \beta \in V$ as in* (1) *and define $\bar{\gamma} = \alpha - i\beta$.* ($\bar{\gamma}$ is called the *conjugate* of $\gamma$.) *Then*

*a.* $\overline{(\gamma_1 + \gamma_2)} = \bar{\gamma}_1 + \bar{\gamma}_2$.

*b.* $\overline{c\gamma} = \bar{c}\,\bar{\gamma}$, *where $\bar{c}$ is the conjugate of a complex number c.*

*c.* $\bar{\bar{\gamma}} = \gamma$.

*d.* $\gamma \in V^c$ *belongs to $V$ if and only if* $\bar{\gamma} = \gamma$.

*Proof.* 1. Since $V$ spans $V^c$, every element $\gamma$ of $V^c$ is of the form $\sum_{k=1}^{m} c_k\alpha_k$, where $\alpha_k \in V$ and $c_k$'s are complex numbers. Write $c_k = a_k + ib_k$, where $a_k$'s and $b_k$'s are real numbers. Then

$$\gamma = \alpha + i\beta,$$

where $$\alpha = \sum_{k=1}^{m} a_k\alpha_k \quad \text{and} \quad \beta = \sum_{k=1}^{m} b_k\alpha_k$$

are in $V$. In order to show that such a representation is unique, we show that if $\alpha = i\beta$, $\alpha, \beta \in V$, then $\alpha = \beta = 0$. If $\alpha$ and $\beta$ are linearly independent in $V$, then they are part of a basis of $V$, which is also a basis in $V^c$. Thus $\alpha$ and $\beta$ are linearly independent in $V^c$, contrary to $\alpha = i\beta$. Assume that $\alpha \neq 0$, so that $\beta = a\alpha$ with $a \in \mathfrak{R}$. Then from $\beta = a\alpha$ and $\alpha = i\beta$ we get $\beta = ia\beta$, so that $\beta = 0$. This shows that $\alpha = \beta = 0$.

2. The verification of ($a$), ($b$), and ($c$) is easy. To prove ($d$), let $\gamma = \alpha + i\beta \in V^c$ with $\alpha, \beta \in V$. If $\gamma \in V$, then by the uniqueness part of (1) we must have $\gamma = \alpha$ and $\beta = 0$, and hence $\bar{\gamma} = \gamma$. Conversely, assume that $\bar{\gamma} = \gamma$, so that $\alpha - i\beta = \alpha + i\beta$. Again by the uniqueness part of (1), we have $-\beta = \beta$, so that $\beta = 0$, and hence $\gamma = \alpha$ is in $V$.

**EXERCISE 5.7**

**1.** Let $\theta_1: V_1 \to V_1'$ and $\theta_2: V_2 \to V_2'$ be extensions of vector spaces corresponding to an extension $\mathfrak{F}'$ of $\mathfrak{F}$. Prove that for any linear mapping $A: V_1 \to V_2$ there is a unique linear mapping $A': V_1' \to V_2'$ such that $\theta_2 A = A'\theta_1$. Prove that $A'$ is nonsingular if and only if $A$ is nonsingular.

**2.** Interpret the statement in number 1 in terms of matrices by considering $V_1 = \mathfrak{F}^n$ and $V_2 = \mathfrak{F}^m$.

» **3.** An $n \times n$ complex matrix $\mathbf{A}$ is a real matrix if and only if $\mathbf{A}\,\xi \in \mathfrak{R}^n$ for every $\xi \in \mathfrak{R}^n$.

☆ **4.** In the complex vector space $\mathfrak{C}^n$, define

$$\bar{\alpha} = (\bar{c}_1, \ldots, \bar{c}_n) \qquad \text{for } \alpha = (c_1, \ldots, c_n) \in \mathfrak{C}^n,$$

where $\bar{c}_k$ denotes the complex conjugate of the complex number $c_k$. Prove that $\bar{\alpha}$ is the conjugate of $\alpha$ in the sense of Proposition 5.23 when $\mathfrak{C}^n$ is regarded as the complexification of the real vector space $\mathfrak{R}^n$.

☆ **5** In the vector space $\mathfrak{C}^2$, let $\alpha = (c_1, c_2)$, where $c_2 \neq 0$, and let $W = \text{Sp}\,\{\alpha\}$. Prove that $\mathfrak{C}^2 = W \oplus \overline{W}$ if and only if $c_1/c_2$ is not a real number. (For the conjugation see number 4.)

## 5.8 MODULES

In this section we intend to indicate how the notion of a vector space can be generalized.

**Definition 5.25**

Let $K$ be a commutative ring with identity. By a *module* $M$ over $K$ (or a *K-module*) we mean an additive group $M$ provided with a mapping $(c,\alpha) \in K \times M \to c\alpha \in M$ satisfying the following conditions:

**1.** $c(\alpha + \beta) = c\alpha + c\beta$.
**2.** $(a + b)\alpha = a\alpha + b\alpha$.
**3.** $(ab)\alpha = a(b\alpha)$.
**4.** $1\alpha = \alpha$.

Here $a, b \in K$ and $\alpha, \beta \in M$.

***Remark.*** More generally, we may define the concept of left (or right) module over a ring which is not commutative. We shall not, however, discuss these more general concepts.

***Example 5.27.*** An additive group $M$ can be regarded as a module over the ring of integers $\mathbb{Z}$ in the following manner. For positive integers $n$ we define $n\alpha$, $\alpha \in M$, inductively by

$$1\alpha = \alpha \qquad \text{and} \qquad n\alpha = (n-1)\alpha + \alpha.$$

For a negative integer $n$, we define

$$n\alpha = -((-n)\alpha),$$

where on the right-hand side $-n$ is positive and hence $(-n)\alpha$ is already defined. Finally, for $n = 0$, we define

$$0\alpha = 0,$$

where 0 on the right-hand side is, of course, the zero element of the additive group $M$.

***Example 5.28.*** Let $K$ be a commutative ring with identity 1. Denote by $K^n$ ($n$: positive integer) the set of all $n$-tuples of elements of $K$

$$\alpha = (a_1, \ldots, a_n), \qquad a_i \in K.$$

The set $K^n$ can be made into a $K$-module by defining

$$(a_1, \ldots, a_n) + (b_1, \ldots, b_n) = (a_1 + b_1, \ldots, a_n + b_n),$$
$$c(a_1, \ldots, a_n) = (ca_1, \ldots, ca_n).$$

If $K$ is a field $\mathfrak{F}$, then $K^n = \mathfrak{F}^n$ is a vector space over $\mathfrak{F}$ as we already know.

***Example 5.29.*** Let $K$ be as above. Denote by $K^m{}_n$ the set of all $m \times n$ matrices over $K$, namely,

$$[a_{ij}] = \begin{bmatrix} a_{11} & a_{12} & \cdots & a_{1n} \\ a_{21} & a_{22} & \cdots & a_{2n} \\ \cdots & \cdots & \cdots & \cdots \\ a_{m1} & a_{m2} & \cdots & a_{mn} \end{bmatrix}, \qquad \text{where } a_{ij} \in K.$$

The set $K^m{}_n$ can be made into a $K$-module by defining

$$[a_{ij}] + [b_{ij}] = [a_{ij} + b_{ij}],$$
$$c[a_{ij}] = [ca_{ij}], \qquad c \in K.$$

Compare this with Example 2.8.

***Example 5.30.*** We may define multiplication for matrices over a commutative ring $K$ with 1 as in the case of matrices over a field. For an $m \times n$ matrix $\mathbf{A} = [a_{ij}]$ and an $n \times p$ matrix $\mathbf{B} = [b_{jk}]$, $\mathbf{AB}$ is an $m \times p$ matrix whose $(i,k)$ component $c_{ik}$ is given by

$$c_{ik} = \sum_{j=1}^{n} a_{ij}b_{jk}.$$

In particular, in the module $K^n{}_n$ we have multiplication:

$$(\mathbf{A},\mathbf{B}) \in K^n{}_n \times K^n{}_n \to \mathbf{AB} \in K^n{}_n.$$

It is easy to verify that $K^n{}_n$ is a ring (not commutative for $n > 1$) with identity element $\mathbf{I}_n$ (unit matrix $[\delta_{ij}]$).

***Example 5.31.*** Let $K$ be a commutative ring with identity 1. A polynomial $f(x)$ over $K$ with indeterminate $x$ is a formal expression of the form

$$a_0 + a_1x + \cdots + a_nx^n, \qquad \text{where } a_0, a_1, \ldots, a_n \in K,$$

as in the case where $K$ is a field. The set of all polynomials $K[x]$ over $K$ can be made into a module over $K$ by defining $f + g$ and $cf$, $c \in K$, as in Sec. 5.2. Moreover, the product $fg$ can be defined in the same fashion, so that $K[x]$ becomes a commutative ring with identity.

***Example 5.32.*** Let $\mathfrak{F}$ be a field and let $K = \mathfrak{F}[x]$ be the polynomial ring over $\mathfrak{F}$ with indeterminate $x$. With another indeterminate $y$ we may form the polynomial ring $K[y] = \mathfrak{F}[x][y]$ over $K = \mathfrak{F}[x]$. Each element $g$ in $K[y]$ is of the form

$$f_0(x) + f_1(x)y + \cdots + f_m(x)y^m,$$

where $f_0(x), f_1(x), \ldots, f_m(x)$ are polynomials in $\mathfrak{F}[x]$. We may rewrite $g$ as a usual polynomial in $x$ and $y$:

$$g(x,y) = \sum_{i=0}^{n} \sum_{j=0}^{m} a_{ij}x^i y^j, \qquad a_{ij} \in \mathfrak{F}.$$

We shall denote $\mathfrak{F}[x][y]$ by $\mathfrak{F}[x,y]$. It is called the *ring of polynomials over $\mathfrak{F}$ in two indeterminates $x$ and $y$.*

Many notions which we defined for vector spaces have analogues for modules. For example, we make the following definitions.

Let $M$ be a $K$-module, where $K$ is a ring. By a *submodule* $N$ of $M$ we mean a subset $N$ of $M$ such that:

1. If $\alpha, \beta \in N$, then $\alpha + \beta$, $\alpha - \beta \in N$.
2. If $\alpha \in N$ and $c \in K$, then $c\alpha \in N$.

A submodule $N$ is itself a module. Given a subset $S$ of $M$, the set

$$N = \{\sum c_i\alpha_i \text{ (finite sum)}; c_i \in K, \alpha_i \in S\}$$

is a submodule, called the *submodule generated* by $S$.

A subset $S$ of $M$ is said to be a *basis* of $M$ if:

1. The submodule generated by $S$ is $M$.
2. When $\sum_{i=1}^{n} c_i\alpha_i = 0$, where $\alpha_1, \ldots, \alpha_k$ are distinct elements of $S$ and $c_1, \ldots, c_k \in K$, then $c_1 = \cdots = c_k = 0$.

We can express condition 2 by saying that $S$ is linearly independent, just as in the case of a vector space.

A module $M$ is said to be *of finite type* if there exist a finite number of elements $\alpha_1, \ldots, \alpha_r$ which generate $M$. A module $M$ is said to be *free* if there exist a finite number of elements $\alpha_1, \ldots, \alpha_r$ which form a basis of $M$.

For a vector space $V$ over a field, $V$ is free if and only if $V$ is finite-dimensional. If $V$ is of finite type in the above sense, then it follows that it is free, that is, finite-dimensional (Theorem 2.12); moreover, the number of elements in any basis of $V$ is uniquely determined (Theorem 2.11). But a module of finite type over a commutative ring with identity may not be free (see the example below). For a free module, the number of elements in any basis is still uniquely determined, although one has to prove it in a different manner from the case of a vector space (cf. Exercise 6.2, number 8).

***Example 5.33.*** Consider the additive group $\mathbb{Z}_2 = \mathbb{Z}/(2)$ (Example 5.20). Regarding $\mathbb{Z}_2$ as a module over $\mathbb{Z}$, we see that it is of finite type (in fact, $1^*$ generates $\mathbb{Z}_2$). But it is not free (note that $2\,1^* = 0^*$).

The notion of direct sum can be defined for two modules $M_1$ and $M_2$ over the same commutative ring $K$ with identity. In the cartesian product $M_1 \times M_2$, one defines

$$(\alpha_1,\alpha_2) + (\beta_1,\beta_2) = (\alpha_1 + \beta_1, \alpha_2 + \beta_2)$$

and

$$a(\alpha_1,\alpha_2) = (a\alpha_1,a\alpha_2), \qquad \text{where } a \in K.$$

The module so obtained is called the *direct sum* of $M_1$ and $M_2$ and is denoted by $M_1 \oplus M_2$. Obviously, the direct sum $M_1 \oplus \cdot \cdot \cdot \oplus M_k$ can be defined for any number of modules over $K$.

Let $M$ be a $K$-module and $N$ a submodule of $M$. The notion of the quotient module $M/N$ can be defined in the same way as the quotient space of a vector space. We shall omit the details.

For two modules $M$ and $N$ over $K$, a $K$-*homomorphism* $f$ of $M$ into $N$ is a mapping of $M$ into $N$ such that

$$f(\alpha + \beta) = f(\alpha) + f(\beta), \qquad \text{where } \alpha, \beta \in M,$$

and

$$f(a\alpha) = af(\alpha), \qquad \text{where } \alpha \in M \text{ and } a \in K.$$

A one-to-one $K$-homomorphism of $M$ into $N$ is called a $K$-isomorphism. Let $\text{Hom}_K(M,N)$ denote the set of all $K$-homomorphisms of $M$ into $N$. For $f, g \in \text{Hom}_K(M,N)$, we define

$$(f + g)(\alpha) = f(\alpha) + g(\alpha), \qquad \text{where } \alpha \in M,$$

and

$$(af)(\alpha) = af(\alpha), \qquad \text{where } \alpha \in M \text{ and } a \in K.$$

It is easy to verify that both the mappings $f + g$ and $af$ are $K$-homomorphisms of $M$ into $N$. For example,

$$(af)(b\alpha) = a(f(b\alpha)) = abf(\alpha) = baf(\alpha) = b(af)(\alpha),$$

by using the commutativity of $K$. One can verify that $\text{Hom}_K(M,N)$ becomes a module over $K$.

***Example 5.34.*** Let $M$ be a free module over $K$ with a basis $\alpha_1, \ldots, \alpha_n$ and $N$ a free module over $K$ with a basis $\beta_1, \ldots, \beta_m$. With any $K$-homomorphism $f$ of $M$ into $N$, we associate a matrix $\mathbf{A} = [a_{ij}]$ over $K$, where $a_{ij}$'s are determined by

$$f(\alpha_j) = \sum_{i=1}^{m} a_{ij}\beta_i, \qquad 1 \leq j \leq n.$$

In this way, we get a $K$-isomorphism of the $K$-module $\text{Hom}_K(M,N)$ onto the $K$-module $K^m{}_n$ (Example 5.29). The proof is left to the reader (cf. Exercise 5.8, number 3).

Let $K$ be a commutative ring with identity. $K$ itself can be considered as a $K$-module. For any $K$-module $M$ an element $f \in \text{Hom}_K(M,K)$ is called a *linear function* on $M$. The module $\text{Hom}_K(M,K)$ is called the *dual module* of $K$ and is denoted by $K^*$.

**EXERCISE 5.8**

**1.** Let $K$ be a commutative ring with identity. When $K$ is regarded as a module over $K$, prove that an ideal of $K$ is a submodule and vice versa.

**2.** Show that the module $\mathbb{Z}/(m)$ over $\mathbb{Z}$ (Example 5.20), where $m$ is an arbitrary integer $\neq 0$, is of finite type but not free.

**3.** Let $M$ be a free $K$-module with a basis $\{\alpha_1, \ldots, \alpha_n\}$. Let $N$ be a $K$-module and let $\beta_1, \ldots, \beta_n$ be arbitrary elements in $N$. Prove that there is a unique $K$-homomorphism $f$ of $M$ into $N$ such that $f(\alpha_i) = \beta_i$ for $1 \leq i \leq n$. Using this, prove the assertion in Example 5.31.

☆ **4.** Let $f$ be a $K$-homomorphism of a $K$-module $M$ into a $K$-module $N$. Prove that if $M_1$ is a submodule of $M$, then $f(M_1)$ is a submodule of $N$. If $N_1$ is a submodule of $N$, then $f^{-1}(N_1)$ is a submodule of $M$. In particular, the *kernel* of $f$, that is, $f^{-1}(0)$, is a submodule of $M$.

☆ **5.** Continuing number 4, prove that if the kernel of $f$ and $f(M)$ are of finite type, then $M$ is of finite type.

☆ **6.** A commutative ring $K$ with identity is said to be *Noetherian* if every proper ideal of $K$ is of finite type (as a submodule of the module $K$ as in number 1). If $\mathfrak{F}$ is a field, then $\mathfrak{F}$ and the polynomial ring $\mathfrak{F}[x]$ are Noetherian. The polynomial ring $\mathfrak{F}[x,y]$ is also Noetherian.

# 6 Determinants

We first discuss permutations together with the concept of group and then proceed to the definition of alternating $n$-linear functions on a vector space and that of determinants for $n \times n$ matrices over an arbitrary field $\mathcal{F}$ (or, more generally, a commutative ring $K$ with identity). The basic properties of determinants are proved, and applications to linear transformations and systems of linear equations (in particular, Cramer's rule) are given.

## 6.1 PERMUTATIONS

**Definition 6.1**

Let $M$ be an arbitrary set. By a *permutation* (or *transformation*) of $M$ we mean a one-to-one mapping of $M$ onto itself.

For two permutations $\sigma$ and $\tau$ of $M$, the *composite* (or *product*) defined by

$$\sigma\tau(x) = \sigma(\tau(x)), \qquad x \in M,$$

is again a permutation. For a permutation $\sigma$ the *inverse mapping* $\sigma^{-1}$, defined by

$$\sigma^{-1}(x) = y \qquad \text{if } \sigma(y) = x,$$

is also a permutation. The identity mapping $\epsilon$ which maps every $x \in M$ into itself is called the *identity permutation* of $M$. The set of all permutations of $M$ forms a group in the sense of the following definition.

**Definition 6.2**

A set $G$ is called a *group* if there is a mapping $(x,y) \in G \times G \to xy \in G$, called *multiplication*, which satisfies the following conditions:

**1.** $(xy)z = x(yz)$ for all $x, y, z \in G$.

**2.** There exists an element, denoted by $e$, such that $xe = ex = x$ for every $x \in G$.

**3.** For each $x \in G$ there exists an element $x' \in G$ such that $xx' = x'x = e$.

If we assume furthermore

**4.** $xy = yx$ for all $x, y \in G$,

then we say that $G$ is an *abelian group* (or *commutative group*).

Recalling Definition 2.2, we see that an abelian group is essentially the same as an additive group, the only difference being how we denote the operation $(x,y) \in G \times G \to xy \in G$; when this is denoted by $+$ and is called addition, we have an additive group.

***Proposition 6.1.*** *Let $G$ be a group.*

**1.** *There is one and only one element $e \in G$ such that $xe = ex = x$ for every $x \in G$.* (This element is called the *identity* of $G$.)

**2.** *For each $x \in G$, there is one and only one element $x'$ such that $xx' = x'x = e$.* (This element $x'$ is called the *inverse* of $x$ and is denoted by $x^{-1}$.)

**3.** *For $a, b \in G$, there is one and only one element $x \in G$ such that $ax = b$ (or $xa = b$).*

**4.** *For $a, b \in G$, we have $(ab)^{-1} = b^{-1}a^{-1}$.*

*Proof.* 1. We know that there is one such element $e$. Suppose that $e'$ is another such element. Then

$$e = ee' = e'e = e'.$$

2. We know that there is one such element $x'$. Suppose that $x''$ is another such element. Then

$$x'' = ex'' = (x'x)x'' = x'(xx'') = x'e = x'.$$

3. If we take $x = a^{-1}b$, then

$$ax = a(a^{-1}b) = (aa^{-1})b = eb = b.$$

Conversely, if $ax = b$, then, multiplying both sides by $a^{-1}$ on the left, we have $a^{-1}b = a^{-1}(ax) = (a^{-1}a)x = ex = x$. The argument for the equation $xa = b$ is similar.

4. We have

$$(ab)(b^{-1}a^{-1}) = a(b(b^{-1}a^{-1})) = a((bb^{-1})a^{-1}) = a(ea^{-1}) = aa^{-1} = e,$$

showing that

$$(ab)^{-1} = b^{-1}a^{-1}.$$

***Example 6.1.*** Let $V$ be a vector space over $\mathfrak{F}$. The set $GL(V)$ of all nonsingular linear transformations of $V$ onto $V$ forms a group. In particular, taking $V = \mathfrak{F}^n$, the set of all nonsingular $n \times n$ matrices over $\mathfrak{F}$ forms a group, which we denote by $GL(n,\mathfrak{F})$ and call the *general linear group of degree n over* $\mathfrak{F}$. This group is not commutative for $n > 1$.

***Example 6.2.*** The set of all positive real numbers forms a group with respect to the usual multiplication. This group is abelian.

We are particularly interested in permutations of a finite set $M$. When $M$ has $n$ elements, we denote the elements by the letters $1, 2, \ldots, n$. A permutation of $M$ (or a *permutation on n letters*) can be expressed by

$$\sigma = \begin{pmatrix} 1 & 2 & \cdots & n \\ i_1 & i_2 & \cdots & i_n \end{pmatrix},$$

where $i_k$ is the image $\sigma(k)$ of the letter $k$ by $\sigma$. We may think of $\sigma$ as a way of rearranging the letters $1, 2, \ldots, n$ in the order $i_1, i_2, \ldots, i_n$. The number of all permutations on $n$ letters is therefore $n!$.

**Definition 6.3**

The group consisting of all permutations on $n$ letters is called the *permutation group* (or *symmetric group*) of *degree n* and is denoted by $S_n$.

***Example 6.3.*** We illustrate some products and inverses in $S_n$, $n = 4$. If

$$\sigma = \begin{pmatrix} 1 & 2 & 3 & 4 \\ 3 & 1 & 2 & 4 \end{pmatrix}, \qquad \tau = \begin{pmatrix} 1 & 2 & 3 & 4 \\ 2 & 4 & 1 & 3 \end{pmatrix},$$

then

$$\sigma\tau = \begin{pmatrix} 1 & 2 & 3 & 4 \\ 1 & 4 & 3 & 2 \end{pmatrix}, \qquad \tau\sigma = \begin{pmatrix} 1 & 2 & 3 & 4 \\ 1 & 2 & 4 & 3 \end{pmatrix}.$$

Thus $\sigma\tau \neq \tau\sigma$. We also have

$$\sigma^{-1} = \begin{pmatrix} 1 & 2 & 3 & 4 \\ 2 & 3 & 1 & 4 \end{pmatrix}, \qquad \tau^{-1} = \begin{pmatrix} 1 & 2 & 3 & 4 \\ 3 & 1 & 4 & 2 \end{pmatrix}.$$

We now define special types of permutations. For any distinct letters $i_1, i_2, \ldots, i_k$ among $1, 2, \ldots, n$, the symbol $(i_1, i_2, \ldots, i_k)$, which may be written without commas, will denote a permutation on $n$ letters which maps $i_1$ into $i_2$, $i_2$ into $i_3$, $\ldots$, $i_{k-1}$ into $i_k$, and $i_k$ into $i_1$, while leaving the other letters invariant. Such a permutation is called a *cyclic permutation* (or *cycle*).

In particular, for two distinct letters $i$ and $j$, the cycle $(i,j)$ is called a *transposition*. It interchanges $i$ and $j$ and leaves all the other letters invariant.

***Example 6.4.*** Let $n = 6$. We have

$$(1\ \ 2\ \ 4)\,(3\ \ 5\ \ 6) = \begin{pmatrix} 1 & 2 & 3 & 4 & 5 & 6 \\ 2 & 4 & 5 & 1 & 6 & 3 \end{pmatrix} = (3\ \ 5\ \ 6)\,(1\ \ 2\ \ 4),$$

$$(1\ \ 3)^2 = (1\ \ 3)\,(1\ \ 3) = \epsilon, \qquad \text{identity permutation,}$$

and

$$(1\ \ 2\ \ 3)^3 = \epsilon.$$

We shall prove:

***Proposition 6.2***

**1.** *If $\sigma$ and $\tau$ are two cycles which do not have a common letter, then $\sigma\tau = \tau\sigma$.*

**2.** *If $\sigma$ is a cycle of $k$ letters, then $\sigma^k = \epsilon$.*

**3.** *Every permutation is a product of a finite number of cycles any two of which have no common letter.*

**4.** *A cycle is a product of transpositions.*

*Proof.* Parts 1 and 2 are easy. To prove (3), we use induction on the number $n$ of letters. For $n = 1$ there is nothing to prove. Assume that (3) is valid for permutations on $m$ letters, where $m < n$, and let $\sigma$ be a permutation on $n$ letters. Consider $\sigma(1)$, $\sigma^2(1)$, . . . ; there must be positive integers $p$ and $q$, $p < q$, such that $\sigma^p(1) = \sigma^q(1)$. Then $\sigma^{q-p}(1) = 1$. Thus there is a positive integer $s$ such that $\sigma^s(1) = 1$. Now let $s$ be the smallest positive integer with this property. Then $1, \sigma(1), \sigma^2(1), \ldots, \sigma^{s-1}(1)$ are all distinct; for otherwise there would be $p$ and $q$, $0 \leq p < q \leqq s - 1$, such that $\sigma^p(1) = \sigma^q(1)$ and hence $\sigma^{q-p}(1) = 1$, where $q - p \leqq s - 1$, contrary to the choice of $s$. Now consider the cycle $\tau_1 = (1,\sigma(1),\sigma^2(1), \ldots, \sigma^{s-1}(1))$. Let $\sigma' = \sigma\tau_1^{-1}$. Then

$$\sigma'(1) = \sigma\tau_1^{-1}(1) = \sigma\sigma^{s-1}(1) = \sigma^s(1) = 1$$

and
$$\sigma'(\sigma^k(1)) = \sigma\tau_1^{-1}(\sigma^k(1)) = \sigma(\sigma^{k-1}(1)) = \sigma^k(1)$$

for $1 \leqq k \leqq s - 1$. In other words, $\sigma'$ leaves the letters $1, \sigma(1), \ldots, \sigma^{s-1}(1)$ invariant and permutes the other letters among themselves. Considering $\sigma'$ as a permutation of these $n - s$ letters other than $1, \sigma(1), \ldots, \sigma^{s-1}(1)$, we may express $\sigma'$, by inductive assumption, as a product of cycles:

$$\sigma' = \tau_k \cdots \tau_2,$$

any two of which have no common letter. It is clear that we have

$$\sigma = \tau_k \cdots \tau_2\tau_1,$$

where any two cycles have no common letter, proving (3).

In order to prove (4), we may assume that $\sigma = (1 \quad 2 \quad \cdots \quad r)$. Then

$$\sigma = (1 \quad r)\,(1 \quad r-1) \quad \cdots \quad (1 \quad 3)\,(1 \quad 2),$$

as can be easily verified.

It follows from (3) and (4) of Proposition 6.2 that every permutation is a product of transpositions. Such a representation is not unique; for example, we have

$$(1 \quad 2) = (1 \quad 2)\,(3 \quad 4)\,(3 \quad 4) = (1 \quad 2)\,(5 \quad 6)\,(5 \quad 6).$$

We have, however, one important result: If a permutation is a product of transpositions in two ways, say,

$$\sigma_1 \cdots \sigma_p = \tau_1 \cdots \tau_q,$$

then the number $q - p$ is even. In other words, a given permutation is a product of either an odd number or an even number of transpositions.

In order to prove this result, we consider a polynomial

$$f = \prod_{i<j} (x_i - x_j) = \begin{array}{llll} (x_1 - x_2) & (x_1 - x_3) & \cdots & (x_1 - x_n) \\ & (x_2 - x_3) & \cdots & (x_2 - x_n) \\ & & \cdots & \cdots \\ & & & (x_{n-1} - x_n) \end{array}$$

in $n$ indeterminates $x_1, x_2, \ldots, x_n$ over the ring of integers $\mathbb{Z}$. For any permutation $\sigma$ on $n$ letters, we consider the polynomial $f^\sigma(x_1, \ldots, x_n)$ given by

$$f^\sigma(x_1, \ldots, x_n) = f(x_{\sigma(1)}, \ldots, x_{\sigma(n)}),$$

which is equal to either $f(x_1, \ldots, x_n)$ or $-f(x_1, \ldots, x_n)$, since $f^\sigma(x_1, \ldots, x_n)$ is the product of all differences $(x_{\sigma(i)} - x_{\sigma(j)}), i < j$, except possibly for a sign of each term. Now consider the transposition $\sigma = (j,k)$, $j < k$. Those factors of $f$ which are changed by $\sigma$ are divided into the following categories:

1. $(x_j - x_k) \to (x_k - x_j)$.
2. $(x_m - x_j) \to (x_m - x_k)$, $(x_m - x_k) \to (x_m - x_j)$ for $m$ such that $m < j < k$.
3. $(x_j - x_m) \to (x_k - x_m)$, $(x_m - x_k) \to (x_m - x_j)$ for $m$ such that $j < m < k$.
4. $(x_j - x_m) \to (x_k - x_m)$, $(x_k - x_m) \to (x_j - x_m)$ for $m$ such that $j < k < m$.

We see that the product of the two terms in (2), (3), and (4), for the same value of $m$, is unchanged by $\sigma$. Since (1) changes the sign of the factor $x_j - x_k$, we may conclude that $f^\sigma = -f$ for the transposition $\sigma$.

Now every permutation is a product of transpositions, as we know. If $\sigma$ is a product of an even number of transpositions, then $f^\sigma = f$, and if $\sigma$ is a product of an odd number of transpositions, then $f^\sigma = -f$. We have thus concluded the proof of the following:

***Theorem 6.3.*** *Every permutation can be expressed as a product of transpositions. The number of transpositions which appear for such a representation of a given permutation $\sigma$ is either even or odd depending on $\sigma$.*

**Definition 6.4**

The *sign* $s(\sigma)$ of a permutation $\sigma$ is defined as follows:

$s(\sigma) = 1$ if $\sigma$ is a product of an even number of transpositions,
$s(\sigma) = -1$ if $\sigma$ is a product of an odd number of transpositions.

$\sigma$ is called an *even permutation* or an *odd permutation* according as $s(\sigma) = 1$ or $s(\sigma) = -1$.

The following is almost obvious.

***Proposition 6.4.*** *For any $\sigma$, $\tau$ in $S_n$ we have*

$$s(\sigma\tau) = s(\sigma)s(\tau) \qquad \text{and} \qquad s(\sigma^{-1}) = s(\sigma).$$

***Example 6.5.*** For a cycle $\sigma$ having $r$ letters, we have

$$s(\sigma) = \begin{cases} 1 & \text{if } r \text{ is odd,} \\ -1 & \text{if } r \text{ is even,} \end{cases}$$

as follows from the expression

$$(1 \quad 2 \quad \cdots \quad r) = (1 \quad r)\,(1 \quad r-1) \cdots (1 \quad 3)\,(1 \quad 2).$$

***Example 6.6.*** Let

$$\sigma = \begin{pmatrix} 1 & 2 & 3 & 4 & 5 & 6 & 7 & 8 \\ 3 & 7 & 5 & 6 & 1 & 2 & 8 & 4 \end{pmatrix}.$$

Then

$$\sigma = (1 \quad 3 \quad 5)\,(2 \quad 7 \quad 8 \quad 4 \quad 6)$$

so that

$$s(\sigma) = s(1 \quad 3 \quad 5)s(2 \quad 7 \quad 8 \quad 4 \quad 6) = 1 \cdot 1 = 1,$$

by Proposition 6.4 and Example 6.5.

Let $A_n$ be the set of all even permutations on $n$ letters. If $\sigma, \tau \in A_n$, then obviously $\sigma\tau \in A_n$ and $\sigma^{-1} \in A_n$. We express this fact by saying that $A_n$ is a subgroup of $S_n$. More precisely:

**Definition 6.5**

Let $G$ be a group. A nonempty subset $H$ of $G$ is called a *subgroup* of $G$ if it has the following properties:

1. If $a, b \in H$, then $ab \in H$.
2. If $a \in H$, then $a^{-1} \in H$.

Let $H$ be a subgroup. For any $a \in H$ we have $a^{-1} \in H$ by condition 2 and $e = aa^{-1} \in H$ by condition 1. Thus $H$ contains the identity of $G$. The associativity $(ab)c = a(bc)$ is satisfied in $H$, since it is satisfied in $G$. Now it is clear that $H$ itself is a group.

**Definition 6.6**

The subgroup $A_n$ of the symmetric group $S_n$ is called the *alternating group* of degree $n$.

The number of elements in $A_n$ is $n!/2$. Indeed, if $\tau$ is an arbitrary odd permutation, say, $\tau = (1 \quad 2)$, then

$$\sigma \longrightarrow \sigma' = \sigma\tau$$

is a one-to-one mapping of $A_n$ onto the set of all odd permutations; hence the number of even permutations is equal to the number of odd permutations.

### EXERCISE 6.1

**1.** Express each of the following permutations as a product of transpositions and find its sign:

$$\text{»}(a) \begin{pmatrix} 1 & 2 & 3 & 4 & 5 & 6 & 7 \\ 3 & 5 & 6 & 7 & 4 & 1 & 2 \end{pmatrix}; \quad (b) \begin{pmatrix} 1 & 2 & 3 & 4 & 5 & 6 & 7 & 8 \\ 8 & 7 & 4 & 1 & 2 & 5 & 6 & 3 \end{pmatrix}.$$

**2.** Let $\sigma = (1 \ \ 2)(3 \ \ 4 \ \ 5)$. Find the smallest possible integer $k$ such that $\sigma^k = \epsilon$. Also write down $\sigma^{-1}$ as a product of cycles.

☆ **3.** Let $G$ be a group and $a \in G$. Show that each of the following mappings is a one-to-one mapping of $G$ onto itself: (a) $x \to ax$; (b) $x \to xa$; (c) $x \to axa^{-1}$; (d) $x \to x^{-1}$.

☆ **4.** Let $\sigma = \begin{pmatrix} 1 & 2 & \cdots & n \\ i_1 & i_2 & \cdots & i_n \end{pmatrix}$. For each $r$, $1 \leq r \leq n$, let $k_r$ be the number of $s$, $s < r$, such that $i_s > i_r$. (Such a pair $i_s > i_r$ with $s < r$ is called an *inversion* in $\sigma$.) Show that $\sigma$ is even [odd] if and only if the total number of inversions $\sum_{r=1}^{n} k_r$ is even [odd].

☆ **5.** Let $G_1$ and $G_2$ be two groups. In the cartesian product $G = G_1 \times G_2$, we define

$$(a_1,a_2) \cdot (b_1,b_2) = (a_1b_1,a_2b_2).$$

Show that $G$ is a group with respect to this operation. (It is called the *direct product* of $G_1$ and $G_2$.)

☆' **6.** A mapping $f$ of a group $G$ into another group $H$ is called a *homomorphism* if

$$f(ab) = f(a)f(b) \qquad \text{for all } a, b \in G.$$

Prove that for a homomorphism $f: G \to H$:

(a) $f(a^{-1}) = (f(a))^{-1}$ for every $a \in G$.

(b) The subset $N = \{a \in G; f(a) = e'\}$, where $e'$ denotes the identity element of $H$, is a subgroup of $G$. (It is called the *kernel* of $f$.)

(c) The kernel $N$ has the following property: If $a \in N$ and $x \in G$, then $xax^{-1} \in N$. (Any subgroup of $G$ having this property is called a *normal subgroup* of $G$.)

☆ **7.** Let $G$ be a group and $H$ a subgroup of $G$. We define a relation $a \sim b$ by $a^{-1}b \in H$. Show that this is an equivalence relation. Denote by $G/H$ the set of all equivalence classes and by $\pi$ the mapping of $G$ onto $G/H$ which maps $a \in G$ into its equivalence class $a^* = \pi(a)$. (The set $G/H$ is called the *quotient set* of $G$ over $H$.) For each $a \in G$, we define a permutation $f_a$ of $G/H$ by

$$f_a(b^*) = \pi(ab), \qquad \text{where } b^* \in G/H \text{ and } b^* = \pi(b),\ b \in G.$$

Show that this definition is independent of the choice of $b \in G$ such that $\pi(b) = b^*$. Show that $f_{ab} = f_af_b$ and that $\{f_a; a \in G\}$ is a subgroup of the permutation group of the set $G/H$.

☆ **8.** Let $N$ be a normal subgroup of a group $G$. In the quotient set $G/N$, we define

$$a^*b^* = \pi(ab), \qquad \text{where } a,b \in G \text{ and } \pi(a) = a^*,\ \pi(b) = b^*.$$

Show that $a^*b^*$ is defined independently of the choice of $a$ and $b$. Prove then that $G/N$ becomes a group and that $\pi: G \to G/N$ is a homomorphism. (The group $G/N$ is called the *quotient group* of $G$ over $N$.)

☆ **9.** Let $f$ be a homomorphism of a group $G$ onto a group $G'$ and let $N$ be the kernel of $f$. Prove that there exists an isomorphism (that is, a one-to-one homomorphism) $g$ of $G/N$ onto $G'$ such that the diagram

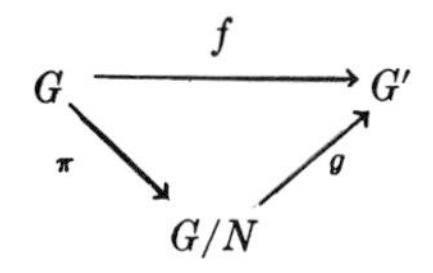


is commutative.

☆**10.** Let $G$ be a subgroup of the permutation group $S_n$. Prove that if $\sigma = (1,2, \ldots ,n-1)$ and $\tau = (1,n)$ are contained in $G$, then $G = S_n$.

»☆**11.** Prove that the product of two transpositions is either a cycle with three letters or a product of two such cycles. Prove that every even permutation is a product of cycles each of which has three letters.

## 6.2 ALTERNATING $n$-LINEAR FUNCTIONS

For a vector space $V$ over a field $\mathcal{F}$ we have already defined the notion of an $r$-linear mapping of $V \times \cdots \times V$ ($r$-fold cartesian product) into $\mathcal{F}$ (Definition 4.9). We now observe that we can define this notion more generally for a module over a commutative ring $K$ with identity.

**Definition 6.7**

Let $M$ be a module over $K$. By an *r-linear function* on $M$, we mean a mapping

$$f: M \times \cdots \times M \text{ ($r$-fold cartesian product)} \to K$$

having the following property: For each $i$, $1 \leq i \leq r$, and for any $r-1$ elements $\alpha_1, \ldots, \alpha_{i-1}, \alpha_{i+1}, \ldots, \alpha_r$ in $M$, the mapping

$$\xi \in M \to f(\alpha_1, \ldots ,\alpha_{i-1},\xi,\alpha_{i+1}, \ldots ,\alpha_r) \in K$$

is a $K$-homomorphism.

**Definition 6.8**

An $r$-linear function $f$ on $M$ is said to be *alternating* if

$$f(\alpha_1, \ldots ,\alpha_r) = 0 \qquad \text{whenever two of the } \alpha_i\text{'s are equal.}$$

***Proposition 6.5.*** *An alternating r-linear function f on M is skew-symmetric, that is,*

$$f(\alpha_1, \ldots ,\alpha_j, \ldots ,\alpha_i, \ldots ,\alpha_r) = -f(\alpha_1, \ldots ,\alpha_i, \ldots ,\alpha_j, \ldots ,\alpha_r)$$

*for $\alpha_1, \ldots, \alpha_r \in M$. If $K$ is a field of characteristic $\neq 2$ and $M$ a vector space over $K$, then a skew-symmetric r-linear function is alternating.*

*Proof.* Suppose $f$ is alternating. Then we have

$$\begin{aligned} 0 &= f(\alpha_1, \ldots, \alpha_i + \alpha_j, \ldots, \alpha_i + \alpha_j, \ldots, \alpha_r) \\ &= f(\alpha_1, \ldots, \alpha_i, \ldots, \alpha_i, \ldots, \alpha_r) \\ &\quad + f(\alpha_1, \ldots, \alpha_i, \ldots, \alpha_j, \ldots, \alpha_r) \\ &\quad + f(\alpha_1, \ldots, \alpha_j, \ldots, \alpha_i, \ldots, \alpha_r) \\ &\quad + f(\alpha_1, \ldots, \alpha_j, \ldots, \alpha_j, \ldots, \alpha_r). \end{aligned}$$

Since the first and fourth terms are 0, we have

$$\begin{aligned} &f(\alpha_1, \ldots, \alpha_j, \ldots, \alpha_i, \ldots, \alpha_r) \\ &= -f(\alpha_1, \ldots, \alpha_i, \ldots, \alpha_j, \ldots, \alpha_r). \end{aligned}$$

Suppose that $K$ is a field of characteristic $\neq 2$ and $M$ a vector space over $K$. If $f$ is skew-symmetric, then we have

$$\begin{aligned} f(\alpha_1, \ldots, \alpha_i, \ldots, \alpha_i, \ldots, \alpha_r) \\ = -f(\alpha_1, \ldots, \alpha_i, \ldots, \alpha_i, \ldots, \alpha_r), \end{aligned}$$

that is, $$2f(\alpha_1, \ldots, \alpha_i, \ldots, \alpha_i, \ldots, \alpha_r) = 0$$

and hence $$f(\alpha_1, \ldots, \alpha_i, \ldots, \alpha_i, \ldots, \alpha_r) = 0.$$

We shall now prove the following basic result.

***Theorem 6.6.*** *Let $K$ be a commutative ring with identity* 1 *and let $M = K^n$* (Example 5.28). *There exists a unique alternating $n$-linear function $f$ on $M$ such that*

$$f(\epsilon_1, \ldots, \epsilon_n) = 1,$$

*where $\{\epsilon_1, \ldots, \epsilon_n\}$ is the standard basis of $M$, that is,*

$$\epsilon_1 = (1,0,0, \ldots, 0), \ \epsilon_2 = (0,1,0, \ldots, 0), \ \ldots, \ \epsilon_n = (0,0, \ldots, 0,1).$$

*Proof.* We first prove the uniqueness of $f$. Assume that $f$ is an alternating $n$-linear function on $M$. For any $n$ elements $\alpha_i$, $1 \leq i \leq n$, given by

$$\alpha_i = \sum_{j=1}^{n} a_{ji}\epsilon_j,$$

we have

$$\begin{aligned} f(\alpha_1, \ldots, \alpha_n) &= f\left(\sum_{j=1}^{n} a_{j1}\epsilon_j, \ldots, \sum_{j=1}^{n} a_{jn}\epsilon_j\right) \\ &= \sum_{j_1, \ldots, j_n = 1}^{n} a_{j_11} \cdots a_{j_nn} f(\epsilon_{j_1}, \ldots, \epsilon_{j_n}). \end{aligned}$$

If $j_1, \ldots, j_n$ are not all distinct, then $f(\epsilon_{j_1}, \ldots, \epsilon_{j_n}) = 0$, since $f$ is alternating. If $j_1, \ldots, j_n$ are all distinct, then

$$\sigma = \begin{pmatrix} 1 & 2 & \cdots & n \\ j_1 & j_2 & \cdots & j_n \end{pmatrix}$$

is a permutation. We show that

$$f(\epsilon_{j_1}, \ldots, \epsilon_{j_n}) = s(\sigma) f(\epsilon_1, \ldots, \epsilon_n).$$

Indeed, if $s(\sigma) = 1$, then $\sigma$ is a product of an even number of transpositions. Since each transposition, say $(i,j)$, changes the sign of $f$, we see that $\sigma$ does not change the sign of $f$. If $s(\sigma) = -1$, then $\sigma$ changes the sign of $f$, proving the assertion above. We thus obtain

$$(*) \qquad f(\alpha_1, \ldots, \alpha_n) = \sum_{\sigma \in S_n} s(\sigma) a_{\sigma(1)1} a_{\sigma(2)2} \cdots a_{\sigma(n)n} f(\epsilon_1, \ldots, \epsilon_n).$$

This shows that an alternating $n$-linear function $f$ is completely determined by its value $f(\epsilon_1, \ldots, \epsilon_n)$. When this is 1 as assumed in Theorem 6.6, we have

$$(**) \qquad f(\alpha_1, \ldots, \alpha_n) = \sum_{\sigma \in S_n} s(\sigma) a_{\sigma(1)1} a_{\sigma(2)2} \cdots a_{\sigma(n)n},$$

proving the uniqueness of $f$ in Theorem 6.6.

We shall now prove that (**) actually defines an alternating $n$-linear function such that $f(\epsilon_1, \ldots, \epsilon_n) = 1$. First, $f$ is $n$-linear. Indeed, let

$$\alpha_i = \sum_{j=1}^{n} a_{ji} \epsilon_j, \qquad 1 \leq i \leq n, \qquad \text{and} \qquad \beta = \sum_{j=1}^{n} b_j \epsilon_j.$$

Then

$$\begin{aligned} & f(\alpha_1, \ldots, \alpha_i + \beta, \ldots, \alpha_n) \\ &= f\left(\sum_{j=1}^{n} a_{j1} \epsilon_j, \ldots, \sum_{j=1}^{n} (a_{ji} + b_j) \epsilon_j, \ldots, \sum_{j=1}^{n} a_{jn} \epsilon_j\right) \\ &= \sum_{\sigma \in S_n} s(\sigma) a_{\sigma(1)1} \cdots (a_{\sigma(i)i} + b_{\sigma(i)}) \cdots a_{\sigma(n)n} \\ &= f(\alpha_1, \ldots, \alpha_i, \ldots, \alpha_n) + f(\alpha_1, \ldots, \beta, \ldots, \alpha_n). \end{aligned}$$

Similarly,

$$f(\alpha_1, \ldots, c\alpha_i, \ldots, \alpha_n) = c f(\alpha_1, \ldots, \alpha_i, \ldots, \alpha_n).$$

Thus $f$ is $n$-linear. We also have

$$f(\epsilon_1, \ldots, \epsilon_n) = 1$$

for the following reason. If $\sigma$ is not the identity permutation, there exists $i$, $1 \leq i \leq n$, such that $\sigma(i) \neq i$. If $\alpha_i = \epsilon_i$ on the left-hand side of (**), then on the right-hand side we have

$$a_{\sigma(i)i} = \delta_{\sigma(i)i} = 0,$$

where $\delta_{ji}$ is the Kronecker delta. Thus the product $a_{\sigma(1)1} \cdots a_{\sigma(n)n}$ is equal to 0. This being the case for every permutation other than the identity permutation, the right-hand side of (**) reduces to one single term corresponding to the identity permutation, namely,

$$\delta_{11} \cdots \delta_{nn} = 1.$$

We have thus shown that $f(\epsilon_1, \ldots, \epsilon_n) = 1$.

The essential part of the proof is to show that $f$ defined by (**) is alternating. We shall first show that it is skew-symmetric and then that it is alternating.

Let $\sigma$ be any permutation; we want to prove that

$$(***) \qquad f(\alpha_{\sigma(1)}, \ldots, \alpha_{\sigma(n)}) = s(\sigma)f(\alpha_1, \ldots, \alpha_n).$$

It is sufficient to prove (***) for the case of a transposition, say $(i,j)$, because every permutation is a product of transpositions. In fact, (***) for any transposition means that $f$ is skew-symmetric.

Let $\tau = (i,j)$, where $i < j$. Then

$$f(\alpha_1, \ldots, \alpha_j, \ldots, \alpha_i, \ldots, \alpha_n)$$
$$= \sum_{\sigma \in S_n} s(\sigma) a_{\sigma(1)1} \cdots a_{\sigma(i)j} \cdots a_{\sigma(j)i} \cdots a_{\sigma(n)n}.$$

If we set $\sigma' = \sigma\tau$, then we have

$$\sigma'(j) = \sigma\tau(j) = \sigma(i),$$
$$\sigma'(i) = \sigma\tau(j) = \sigma(j),$$

and

$$\sigma'(k) = \sigma\tau(k) = \sigma(k) \qquad \text{for } k \neq i, j.$$

Thus

$$a_{\sigma(1)1} \cdots a_{\sigma(i)j} \cdots a_{\sigma(j)i} \cdots a_{\sigma(n)n}$$
$$= a_{\sigma'(1)1} \cdots a_{\sigma'(j)j} \cdots a_{\sigma'(i)i} \cdots a_{\sigma'(n)n}.$$

We also have

$$s(\sigma) = s(\sigma'\tau^{-1}) = s(\sigma')s(\tau) = -s(\sigma').$$

Since the mapping $\sigma \to \sigma' = \sigma\tau$ is a one-to-one mapping of $S_n$ onto itself, we have

$$f(\alpha_1, \ldots, \alpha_j, \ldots, \alpha_i, \ldots, \alpha_n)$$
$$= - \sum_{\sigma' \in S_n} s(\sigma') a_{\sigma'(1)1} \cdots a_{\sigma'(j)j} \cdots a_{\sigma'(i)i} \cdots a_{\sigma'(n)n}$$
$$= -f(\alpha_1, \ldots, \alpha_i, \ldots, \alpha_j, \ldots, \alpha_n),$$

as we wanted to show. This proves that $f$ is skew-symmetric.

We now want to show that $f$ is alternating. Let $\tau = (i,j)$, where $i < j$; we shall prove that

$$f(\alpha_1, \ldots, \alpha_i, \ldots, \alpha_j, \ldots, \alpha_n) = 0 \qquad \text{if } \alpha_i = \alpha_j.$$

Let $S'$ be the set of all permutations $\sigma \in S_n$ such that $\sigma(i) < \sigma(j)$ and let $S''$ be the set of permutations $\sigma \in S_n$ such that $\sigma(i) > \sigma(j)$. Obviously,

$S_n = S' \cup S''$ and $S' \cap S''$ is the empty set. We observe that $\sigma' \to \sigma'' = \sigma'\tau$ is a one-to-one mapping of $S'$ onto $S''$. Let us write

$$\begin{aligned} f(\alpha_1, \ldots, \alpha_i, \ldots, \alpha_j, \ldots, \alpha_n) \\ = \sum_{\sigma' \in S'} s(\sigma') a_{\sigma'(1)1} \cdots a_{\sigma'(i)i} \cdots a_{\sigma'(j)j} \cdots a_{\sigma'(n)n} \\ + \sum_{\sigma'' \in S''} s(\sigma'') a_{\sigma''(1)1} \cdots a_{\sigma''(i)i} \cdots a_{\sigma''(j)j} \cdots a_{\sigma''(n)n}. \end{aligned}$$

For each $\sigma' \in S'$ take the corresponding $\sigma'' = \sigma'\tau$. We have

$\sigma''(i) = \sigma'(j)$, $\sigma''(j) = \sigma'(i)$, and $\sigma''(k) = \sigma'(k)$ for $k \neq i,j$.

Since $\alpha_i = \alpha_j$, we have $a_{mi} = a_{mj}$ for every $m$ and hence

$$a_{\sigma'(i)i} a_{\sigma'(j)j} = a_{\sigma''(j)i} a_{\sigma''(i)j} = a_{\sigma''(j)j} a_{\sigma''(i)i}.$$

We have therefore

$$a_{\sigma'(1)1} \cdots a_{\sigma'(n)n} = a_{\sigma''(1)1} \cdots a_{\sigma''(n)n}$$

and

$$s(\sigma') a_{\sigma'(1)1} \cdots a_{\sigma'(n)n} + s(\sigma'') a_{\sigma''(1)1} \cdots a_{\sigma''(n)n} = 0,$$

because $s(\sigma'') = s(\sigma'\tau) = s(\sigma')s(\tau) = -s(\sigma')$. Taking the sum of the left-hand side for all $\sigma' \in S'$, we obtain the expression for $f(\alpha_1, \ldots, \alpha_n)$, proving that this value is 0. This completes the proof of Theorem 6.6.

***Corollary.*** *Let $f$ be the alternating $n$-linear function on the module $K^n$. For any alternating $n$-linear function $g$ there exists $c \in K$ such that $g = cf$.*

*Proof.* This follows from (*) in the proof of Theorem 6.6. We let $c = g(\epsilon_1, \ldots, \epsilon_n)$.

***Example 6.7.*** Let $K$ be a commutative ring with 1. For $M = K^2$, the function $f$ in Theorem 6.6 is given by

$$f\left(\begin{bmatrix} a_1 \\ a_2 \end{bmatrix}, \begin{bmatrix} b_1 \\ b_2 \end{bmatrix}\right) = a_1b_2 - a_2b_1,$$

where we have written elements of $K^2$ as $\begin{bmatrix} a_1 \\ a_2 \end{bmatrix}$.

***Example 6.8.*** Similarly, for $M = K^3$, the function $f$ is given by

$$f\left(\begin{bmatrix} a_1 \\ a_2 \\ a_3 \end{bmatrix}, \begin{bmatrix} b_1 \\ b_2 \\ b_3 \end{bmatrix}, \begin{bmatrix} c_1 \\ c_2 \\ c_3 \end{bmatrix}\right) = a_1b_2c_3 + a_3b_1c_2 + a_2b_3c_1 - a_1b_3c_2 - a_2b_1c_3 - a_3b_2c_1.$$

We obtain this expression by writing out (**) for the following six elements of $S_3$:

$$\begin{pmatrix} 1 & 2 & 3 \\ 1 & 2 & 3 \end{pmatrix}, \quad \begin{pmatrix} 1 & 2 & 3 \\ 3 & 1 & 2 \end{pmatrix}, \quad \begin{pmatrix} 1 & 2 & 3 \\ 2 & 3 & 1 \end{pmatrix}, \quad \text{even,}$$

$$\begin{pmatrix} 1 & 2 & 3 \\ 1 & 3 & 2 \end{pmatrix}, \quad \begin{pmatrix} 1 & 2 & 3 \\ 2 & 1 & 3 \end{pmatrix}, \quad \begin{pmatrix} 1 & 2 & 3 \\ 3 & 2 & 1 \end{pmatrix}, \quad \text{odd.}$$

***Proposition 6.7.*** *Let $V$ be an $n$-dimensional vector space over a field $\mathcal{F}$. The set of all alternating $n$-linear functions on $V$ forms a vector space $E_n(V)$ over $\mathcal{F}$ under addition and scalar multiplication:*

$$(f+g)(\alpha_1, \ldots, \alpha_n) = f(\alpha_1, \ldots, \alpha_n) + g(\alpha_1, \ldots, \alpha_n)$$

$$(cf)(\alpha_1, \ldots, \alpha_n) = cf(\alpha_1, \ldots, \alpha_n), \qquad \text{where } c \in \mathcal{F}.$$

*The dimension of this vector space is* 1.

*Proof.* We know that all $n$-linear functions on $V$ form a vector space over $\mathcal{F}$ (Sec. 4.3). The alternating $n$-linear functions form a subspace. Now since $V$ is $n$-dimensional and hence isomorphic with the vector space $\mathcal{F}^n$, there is a natural isomorphism between the vector space $E_n(V)$ and $E_n(\mathcal{F}^n)$. By the corollary above, we know that $\dim E_n(\mathcal{F}^n) = 1$. Hence $\dim E_n(V) = 1$.

EXERCISE 6.2

In numbers 1 to 6 one may take any field $\mathcal{F}$ instead of a commutative ring $K$ (if Sec. 5.8 was omitted).

» **1.** Let $K$ be a commutative ring with 1. Among the following functions on $K^2 \times K^2$, determine those which are ($a$) two-linear; ($b$) two-linear and alternating:

1. $f(\alpha,\beta) = a_1b_1 + a_2b_2$.
2. $f(\alpha,\beta) = a_1a_2 + b_1b_2$.
3. $f(\alpha,\beta) = a_1b_2$.
4. $f(\alpha,\beta) = a_2b_1$.
5. $f(\alpha,\beta) = a_1b_2 - a_2b_1$.

Here $\alpha = (a_1,a_2)$ and $\beta = (b_1,b_2)$.

**2.** Among the following functions on $\mathcal{R}^3 \times \mathcal{R}^3 \times \mathcal{R}^3$, determine those which are ($a$) three-linear; ($b$) three-linear and alternating:

1. $f(\alpha,\beta,\gamma) = a_1b_2 - a_2b_1$.
2. $f(\alpha,\beta,\gamma) = a_1a_2a_3 + b_1b_2b_3 + c_1c_2c_3$.
3. $f(\alpha,\beta,\gamma) = a_1b_2c_3$.
4. $f(\alpha,\beta,\gamma) = a_1b_2c_3 - a_2b_1c_3$.
5. $f(\alpha,\beta,\gamma) = a_1b_2c_3 + 2a_2b_3c_1 + a_3b_1c_2 - a_1b_3c_2 - 2a_2b_1c_3 - a_3b_2c_1$.

☆ **3.** Let $h$ be an arbitrary $n$-linear function on a module $M$ over a commutative ring with 1. Define

$$g(\alpha_1, \ldots, \alpha_n) = \sum_{\sigma \in S_n} s(\sigma)h(\alpha_{\sigma(1)}, \ldots, \alpha_{\sigma(n)})$$

for all $\alpha_1, \ldots, \alpha_n \in M$. Prove that $g$ is an alternating $n$-linear function on $M$.

☆ **4.** Let $M = K^n$, where $K$ is a commutative ring with 1, and let

$$h(\alpha_1, \ldots, \alpha_n) = a_{11}a_{22} \cdots a_{nn},$$

where $\alpha_i = (a_{1i}, a_{2i}, \ldots, a_{ni}) \in K^n$, $1 \le i \le n$. Show that $h$ is $n$-linear. Then construct $g$ as in number 3. Prove that $g$ is the function in Theorem 6.6.

**5.** Let $M$ be a module over a commutative ring with 1. For any $n$-linear function $h$, define

$$g(\alpha_1, \ldots, \alpha_n) = \sum_{\sigma \in S_n} h(\alpha_{\sigma(1)}, \ldots, \alpha_{\sigma(n)}).$$

Show that $g$ is $n$-linear and *symmetric*, that is, $g$ does not change its value when we interchange any two of $\alpha_1, \ldots, \alpha_n$ [or, $g$ is symmetric if

$$g(\alpha_{\sigma(1)}, \ldots, \alpha_{\sigma(n)}) = g(\alpha_1, \ldots, \alpha_n)$$

for any $\sigma \in S_n$].

**6.** For $M = K^n$, where $K$ is a commutative ring with identity, prove that an alternating $r$-linear function on $M$ is identically 0 if $r > n$.

☆ **7.** Assume that a module $M$ over a commutative ring with identity has a basis with $n$ elements. Prove that there is a nonzero alternating $n$-linear function on $M$ (cf. Theorem 6.6). Show also that an alternating $r$-linear function on $M$ is identically equal to 0 for $r > n$ (cf. number 6).

☆ **8.** Using number 7, prove that, for a free module $M$ over a commutative ring with identity, the number of elements in any basis of $M$ is uniquely determined (cf. Sec. 5.8).

## 6.3 DETERMINANTS

We shall apply the results in the preceding section and define the determinant function for $n \times n$ matrices over a field $\mathcal{F}$ or, more generally, over a commutative ring $K$ with 1. We shall denote by $K^n{}_n$ the set of all $n \times n$ matrices over $K$ (Example 5.29). For each matrix $\mathbf{A} = [a_{ij}]$ in $K^n{}_n$ consider the columns of $\mathbf{A}$,

$$\alpha_j = \begin{bmatrix} a_{1j} \\ \cdot \\ \cdot \\ \cdot \\ a_{nj} \end{bmatrix}, \qquad 1 \leq j \leq n,$$

as elements of the module $K^n$ (written vertically).

**Definition 6.9**

Let $f$ be the alternating $n$-linear function on $K^n$ such that

$$f(\epsilon_1, \ldots, \epsilon_n) = 1,$$

where $\{\epsilon_1, \ldots, \epsilon_n\}$ is the standard basis of $K^n$ (cf. Theorem 6.6). For any $\mathbf{A} \in K^n{}_n$, we call

$$\det \mathbf{A} = f(\alpha_1, \ldots, \alpha_n)$$

the *determinant* of $\mathbf{A}$. The function $f$, regarded as a function on $K^n{}_n$, is called the *determinant function.*

Thus the determinant function is an alternating $n$-linear function of the $n$ column vectors of matrices in $K^n{}_n$ such that $\det \mathbf{I}_n = 1$, where $\mathbf{I}_n$ is the unit matrix. det **A** is the value of this function for the matrix **A**, and is denoted also by $|\mathbf{A}|$ or by the symbol

$$\begin{vmatrix} a_{11} & a_{12} & \cdots & a_{1n} \\ a_{21} & a_{22} & \cdots & a_{2n} \\ \cdots & \cdots & \cdots & \cdots \\ a_{n1} & a_{n2} & \cdots & a_{nn} \end{vmatrix}.$$

From the formula for the function $f$ in the proof of Theorem 6.6, we have

$$|\mathbf{A}| = \sum_{\sigma \in S_n} s(\sigma) a_{\sigma(1)1} a_{\sigma(2)2} \cdots a_{\sigma(n)n}.$$

*Example 6.9.* For $\mathcal{R}^2{}_2$, we have

$$\begin{vmatrix} a & b \\ c & d \end{vmatrix} = ad - bc.$$

*Example 6.10.* Let $K = \mathcal{R}[x]$ be the polynomial ring over $\mathcal{R}$ and consider $K^2{}_2$.

$$\begin{vmatrix} x-1 & 3 \\ x+1 & x-2 \end{vmatrix} = (x-1)(x-2) - 3(x+1) = x^2 - 6x - 1.$$

*Example 6.11.* For $\mathcal{R}^3{}_3$, we have

$$\begin{vmatrix} a_1 & b_1 & c_1 \\ a_2 & b_2 & c_2 \\ a_3 & b_3 & c_3 \end{vmatrix} = a_1 b_2 c_3 + b_1 c_2 a_3 + c_1 b_3 a_2 - a_3 b_2 c_1 - b_3 c_2 a_1 - c_3 b_1 a_2,$$

as is seen from Example 6.8.

We shall now prove the basic properties of the determinant function.

***Theorem 6.8.*** det $({}^t\mathbf{A})$ = det **A**, *where* ${}^t\mathbf{A}$ *is the transpose of* **A** (cf. Definition 4.6).

*Proof.* Let $\mathbf{A} = [a_{ij}]$, $\mathbf{B} = {}^t\mathbf{A} = [b_{ij}]$, so that $b_{ij} = a_{ji}$. We have

$$\det \mathbf{B} = \sum_{\sigma \in S_n} s(\sigma) b_{\sigma(1)1} b_{\sigma(2)2} \cdots b_{\sigma(n)n}.$$

For each $\sigma \in S_n$, consider $\sigma^{-1}$. If $\sigma(i) = j$, then $\sigma^{-1}(j) = i$, so that we may rewrite $b_{\sigma(1)1} \cdots b_{\sigma(n)n}$ in the form $b_{1\sigma^{-1}(1)} \cdots b_{n\sigma^{-1}(n)}$ (by rearranging the factors). Since $s(\sigma^{-1}) = s(\sigma)$, we have

$$\det \mathbf{B} = \sum_{\sigma \in S_n} s(\sigma^{-1}) b_{1\sigma^{-1}(1)} \cdots b_{n\sigma^{-1}(n)}.$$

Since $\sigma \to \sigma^{-1}$ is a one-to-one mapping of $S_n$ onto itself, we may replace $\sigma^{-1}$ by $\sigma$ in the expression above. Hence

$$\det \mathbf{B} = \sum_{\sigma \in S_n} s(\sigma) b_{1\sigma(1)} \cdots b_{n\sigma(n)}$$

$$= \sum_{\sigma \in S_n} s(\sigma) a_{\sigma(1)1} \cdots a_{\sigma(n)n} = \det \mathbf{A},$$

completing the proof of Theorem 6.8.

By virtue of Theorem 6.8 we may consider det **A** as an alternating $n$-linear function of the $n$ row vectors of **A**. In other words, if

$$\mathbf{A} = \begin{bmatrix} \beta_1 \\ \cdot \\ \cdot \\ \cdot \\ \beta_n \end{bmatrix}$$

denotes an $n \times n$ matrix over $K$, where each $\beta_i$ is an element of $K^n$ written horizontally as the $i$th row of **A**, then det **A** is linear in each variable $\beta_i$ when other rows are kept fixed, and det **A** is 0 if two of the rows are equal.

We shall sum up in:

***Theorem 6.9.*** *Regarding* det **A** *as a function of n column vectors of* **A**, *we have*

**1.** $\det [\alpha_1, \ldots, \alpha_i + \alpha', \ldots, \alpha_n]$
$$= \det [\alpha_1, \ldots, \alpha_i, \ldots, \alpha_n] + \det [\alpha_1, \ldots, \alpha', \ldots, \alpha_n].$$
**2.** $\det [\alpha_1, \ldots, c\alpha_i, \ldots, \alpha_n] = c \det [\alpha_1, \ldots, \alpha_i, \ldots, \alpha_n].$
**3.** $\det [\alpha_1, \ldots, \alpha_i, \ldots, \alpha_i, \ldots, \alpha_n] = 0.$
**4.** $\det [\alpha_1, \ldots, \alpha_j, \ldots, \alpha_i, \ldots, \alpha_n]$
$$= -\det [\alpha_1, \ldots, \alpha_i, \ldots, \alpha_j, \ldots, \alpha_n].$$
**5.** $\det [\alpha_1, \ldots, \alpha_i, \ldots, \alpha_j + c\alpha_i, \ldots, \alpha_n]$
$$= \det [\alpha_1, \ldots, \alpha_i, \ldots, \alpha_j, \ldots, \alpha_n].$$

*Similar properties hold for* det **A** *as a function of n row vectors of* **A**.

*Proof.* Properties 1 and 2 say that det **A** is $n$-linear, and (3) says that it is alternating. Property 4 is a consequence of (3), as we have seen in Proposition 6.5. Property 5 follows from (1), (2), and (3).

Property 5 is useful for practical computation.

***Theorem 6.10.*** det (**AB**) = det **A** · det **B**.

*Proof.* Let

$$\mathbf{A} = [a_{ij}] \quad \text{and} \quad \mathbf{B} = [\beta_1, \ldots, \beta_n] = [b_{ij}].$$

Then we have

$$(*) \qquad \mathbf{AB} = [\mathbf{A}\beta_1, \ldots, \mathbf{A}\beta_n],$$

where $\mathbf{A}\beta_i$ is the product of **A** with the $i$th column of **B**. In fact, we have

$$(\mathbf{AB})_{ik} = \sum_{j=1}^{n} a_{ij}b_{jk},$$

so that the $k$th column of **AB** has the $i$th component equal to the $i$th component of $\mathbf{A}\beta_k$.

Now we fix a matrix $\mathbf{A}$ and define a function $D$ on $K^n{}_n$ (or a function of $n$ columns of $\mathbf{B} \in K^n{}_n$) by

$$(**) \qquad D(\mathbf{B}) = D[\beta_1, \ldots, \beta_n] = \det(\mathbf{AB}).$$

By virtue of (*) we see that $D$ is an alternating $n$-linear function of $n$ columns of $B$, because $\beta \to \mathbf{A}\beta$ is $K$-linear. By the corollary to Theorem 6.6 combined with the definition of the determinant function, we have

$$D(\mathbf{B}) = D(\mathbf{I}_n) \det \mathbf{B},$$

where

$$D(\mathbf{I}_n) = \det(\mathbf{AI}_n) = \det \mathbf{A}.$$

Thus

$$\det(\mathbf{AB}) = \det \mathbf{A} \cdot \det \mathbf{B}.$$

***Theorem 6.11***

$$\det \begin{bmatrix} \mathbf{A} & \mathbf{B} \\ \mathbf{O} & \mathbf{C} \end{bmatrix} = \det \mathbf{A} \cdot \det \mathbf{C},$$

*where* $\mathbf{A}$ *is an* $m \times m$ *matrix,* $\mathbf{B}$ *is an* $m \times (n - m)$ *matrix,* $\mathbf{C}$ *is an* $(n - m) \times (n - m)$ *matrix, and* $\mathbf{O}$ *is the* $(n - m) \times m$ *zero matrix.*

*Proof.* Fixing $\mathbf{A}$ and $\mathbf{B}$, we consider a function $f$ on $K^{n-m}{}_{n-m}$ defined by

$$f(\mathbf{C}) = \det \begin{bmatrix} \mathbf{A} & \mathbf{B} \\ \mathbf{O} & \mathbf{C} \end{bmatrix}.$$

Regarded as a function of $n - m$ rows of $\mathbf{C}$, $f$ is $(n - m)$-linear and alternating. By the corollary of Theorem 6.6 applied to the determinant function as an alternating $(n - m)$-linear function of the rows, we have

$$f(\mathbf{C}) = f(\mathbf{I}_{n-m}) \cdot \det \mathbf{C},$$

where

$$f(\mathbf{I}_{n-m}) = \det \begin{bmatrix} \mathbf{A} & \mathbf{B} \\ \mathbf{O} & \mathbf{I}_{n-m} \end{bmatrix}, \qquad \mathbf{I}_{n-m}\text{: identity matrix.}$$

We shall now prove that the right-hand side is equal to $\det \mathbf{A}$. Fixing $\mathbf{B}$, we define a function $g$ on $K^m{}_m$ by

$$g(\mathbf{A}) = \det \begin{bmatrix} \mathbf{A} & \mathbf{B} \\ \mathbf{O} & \mathbf{I}_{n-m} \end{bmatrix}.$$

Regarded as a function of the column vectors of $\mathbf{A}$, the function $g$ is alternating and $m$-linear. As before, we have

$$g(\mathbf{A}) = g(\mathbf{I}_m) \cdot \det \mathbf{A} = \det \begin{bmatrix} \mathbf{I}_m & \mathbf{B} \\ \mathbf{O} & \mathbf{I}_{n-m} \end{bmatrix} \cdot \det \mathbf{A}.$$

Finally, we shall prove

$$\det \begin{bmatrix} \mathbf{I}_m & \mathbf{B} \\ \mathbf{O} & \mathbf{I}_{n-m} \end{bmatrix} = 1, \qquad \text{where } \mathbf{B} = \begin{bmatrix} b_{11} & \cdots & b_{1,n-m} \\ \cdots & \cdots & \cdots \\ b_{m1} & \cdots & b_{m,n-m} \end{bmatrix}.$$

In the matrix $\begin{bmatrix} \mathbf{I}_m & \mathbf{B} \\ \mathbf{O} & \mathbf{I}_{n-m} \end{bmatrix}$, we subtract $b_{11}$ times the $(m+1)$st row from the first, $b_{12}$ times the $(m+2)$nd row from the first, and so on until we get the zero vector for the first row of the block **B**. Likewise, we may reduce all rows of the block **B** to the zero vectors by the same kind of operations until the whole matrix is reduced to

$$\mathbf{I}_n = \begin{bmatrix} \mathbf{I}_m & \mathbf{O} \\ \mathbf{O} & \mathbf{I}_{n-m} \end{bmatrix}.$$

Since these operations do not change the determinant by virtue of (5) of Theorem 6.9, we see

$$\det \begin{bmatrix} \mathbf{I}_m & \mathbf{B} \\ \mathbf{O} & \mathbf{I}_{n-m} \end{bmatrix} = \det \mathbf{I}_n = 1.$$

We have therefore proved

$$\det \begin{bmatrix} \mathbf{A} & \mathbf{B} \\ \mathbf{O} & \mathbf{C} \end{bmatrix} = \det \mathbf{A} \cdot \det \mathbf{C}.$$

***Corollary***

$$\det \begin{bmatrix} 1 & b_1 & \cdots & b_m \\ 0 & & & \\ \cdot & & & \\ \cdot & & \mathbf{B} & \\ \cdot & & & \\ 0 & & & \end{bmatrix} = \det \mathbf{B},$$

*where* **B** *is an* $m \times m$ *matrix.*

***Example 6.12***

$$\begin{vmatrix} 3 & 1 & 7 & 5 \\ 0 & 2 & -1 & 6 \\ 0 & 0 & 3 & 2 \\ 0 & 0 & 1 & -1 \end{vmatrix} = \begin{vmatrix} 3 & 1 \\ 0 & 2 \end{vmatrix} \begin{vmatrix} 3 & 2 \\ 1 & -1 \end{vmatrix} = 6 \cdot (-5) = -30.$$

***Example 6.13***

$$\begin{vmatrix} 0 & 0 & 1 & 2 \\ 0 & 0 & 3 & 4 \\ 3 & -1 & 1 & 6 \\ 2 & 4 & 1 & 5 \end{vmatrix} = \begin{vmatrix} 3 & -1 & 1 & 6 \\ 2 & 4 & 1 & 5 \\ 0 & 0 & 1 & 2 \\ 0 & 0 & 3 & 4 \end{vmatrix} = \begin{vmatrix} 3 & -1 \\ 2 & 4 \end{vmatrix} \begin{vmatrix} 1 & 2 \\ 3 & 4 \end{vmatrix} = -28,$$

where we first interchanged the first and third rows and then the second and fourth rows.

We shall give some more examples.

***Example 6.14***

$$\begin{vmatrix} a+b+2c & a & b \\ c & b+c+2a & b \\ c & a & c+a+2b \end{vmatrix}$$

$$= \begin{vmatrix} a+b+c & -(a+b+c) & 0 \\ 0 & a+b+c & -(a+b+c) \\ c & a & c+a+2b \end{vmatrix}$$ (by subtracting the second row from the first and the third row from the second)

$$= (a+b+c)^2 \begin{vmatrix} 1 & -1 & 0 \\ 0 & 1 & -1 \\ c & a & c+a+2b \end{vmatrix}$$

$$= (a+b+c)^2(c+a+2b+c+a) = 2(a+b+c)^3.$$

***Example 6.15***

$$\begin{vmatrix} a_{33} & a_{23} & a_{13} \\ a_{32} & a_{22} & a_{12} \\ a_{31} & a_{21} & a_{11} \end{vmatrix} = \begin{vmatrix} a_{11} & a_{12} & a_{13} \\ a_{21} & a_{22} & a_{23} \\ a_{31} & a_{32} & a_{33} \end{vmatrix},$$

as can be seen as follows: Interchange the first and third rows and then interchange the first and third columns, getting

$$\begin{vmatrix} a_{11} & a_{21} & a_{31} \\ a_{12} & a_{22} & a_{32} \\ a_{13} & a_{23} & a_{33} \end{vmatrix}.$$

Finally, we take the transpose.

***Example 6.16.*** We show

$$\begin{vmatrix} 1 & 1 & 1 \\ x & y & z \\ x^2 & y^2 & z^2 \end{vmatrix} = (x-y)(y-z)(z-x).$$

In fact,

$$\begin{vmatrix} 1 & 1 & 1 \\ x & y & z \\ x^2 & y^2 & z^2 \end{vmatrix} = \begin{vmatrix} 1 & 0 & 0 \\ x & y-x & z-x \\ x^2 & y^2-x^2 & z^2-x^2 \end{vmatrix}$$ (subtracting the first column from the second and the third)

$$= \begin{vmatrix} y-x & y^2-x^2 \\ z-x & z^2-x^2 \end{vmatrix}$$ (taking the transpose and using the corollary to Theorem 6.11)

$$= (y-x)(z-x)\begin{vmatrix} 1 & 1 \\ y+x & z+x \end{vmatrix}$$

$$= (y-x)(z-x)(z-y) = (x-y)(y-z)(z-x).$$

## EXERCISE 6.3

1. Compute the following determinants (all matrices are over $\mathcal{R}$).

›(a) $\begin{vmatrix} a & -b \\ b & a \end{vmatrix}$; (b) $\begin{vmatrix} \cos\theta & -\sin\theta \\ \sin\theta & \cos\theta \end{vmatrix}$; ›(c) $\begin{vmatrix} 0 & -1 \\ 0 & 0 \end{vmatrix}$;

(d) $\begin{vmatrix} \cos\theta & -\sin\theta & a \\ \sin\theta & \cos\theta & b \\ 0 & 0 & 1 \end{vmatrix}$;

»(*e*) $\begin{vmatrix} \cos\theta\cos\varphi & \sin\theta\cos\varphi & -\sin\varphi \\ \cos\theta\sin\varphi & \sin\theta\sin\varphi & \cos\varphi \\ -\sin\theta & \cos\theta & 0 \end{vmatrix}$;

(*f*) $\begin{vmatrix} 0 & 0 & 1 \\ 0 & 1 & 0 \\ 1 & 0 & 0 \end{vmatrix}$; »(*g*) $\begin{vmatrix} 1 & 0 & 1 \\ 0 & 1 & 0 \\ 1 & 0 & 1 \end{vmatrix}$.

**2.** Compute

»(*a*) $\begin{vmatrix} 2 & 1 & 3 & 4 \\ 0 & 1 & -2 & 5 \\ 0 & 0 & 4 & 1 \\ 0 & 0 & 3 & -1 \end{vmatrix}$; (*b*) $\begin{vmatrix} 2 & 3 & 0 & 0 \\ 1 & 1 & 0 & 0 \\ 4 & 5 & 4 & 2 \\ -2 & 3 & 3 & 2 \end{vmatrix}$;

»(*c*) $\begin{vmatrix} 1 & 1 & 1 & 1 & -1 & -1 \\ 0 & 1 & 1 & 1 & 0 & 1 \\ 0 & 0 & 1 & 0 & 0 & 2 \\ 0 & 0 & 0 & 1 & 0 & 0 \\ 0 & 0 & 0 & 1 & 1 & 0 \\ 0 & 0 & 0 & 1 & 1 & 1 \end{vmatrix}$; (*d*) $\begin{vmatrix} 0 & 1 & 0 & 1 \\ -1 & 0 & -1 & 0 \\ 0 & 1 & 0 & 1 \\ -1 & 0 & -1 & 0 \end{vmatrix}$.

» **3.** Prove that

$$\begin{vmatrix} \mathbf{O} & \mathbf{C} \\ \mathbf{A} & \mathbf{B} \end{vmatrix} = (-1)^m|\mathbf{A}|\,|\mathbf{C}|,$$

where **O** is the $m \times m$ zero matrix and **A**, **B**, and **C** are $m \times m$ matrices.

**4.** Let $\mathbf{A} = [a_{ij}]$ be an $n \times n$ matrix, where $a_{ij} = 0$ whenever $i < j$. Prove that

$$\det \mathbf{A} = a_{11}a_{22} \ldots a_{nn}.$$

» **5.** Compute $\begin{vmatrix} x & 1 & 1 & 1 \\ 1 & x & 1 & 1 \\ 1 & 1 & x & 1 \\ 1 & 1 & 1 & x \end{vmatrix}$.

☆ **6.** Let $\mathbf{A} = [a_{ij}]$ be an $n \times n$ matrix, where $a_{ij} = 1 - \delta_{ij}$ ($\delta_{ij}$: Kronecker's delta). Prove that

$$\det \mathbf{A} = (n-1)(-1)^{n-1}.$$

»☆ **7.** Prove

$$\begin{vmatrix} a_{11} & a_{12} & \cdots & a_{1n} \\ a_{21} & a_{22} & \cdots & a_{2n} \\ \cdots & \cdots & \cdots & \cdots \\ a_{n1} & a_{n2} & \cdots & a_{nn} \end{vmatrix} = \begin{vmatrix} a_{nn} & a_{n-1,n} & \cdots & a_{1n} \\ a_{n,n-1} & a_{n-1,n-1} & \cdots & a_{1,n-1} \\ \cdots & \cdots & \cdots & \cdots \\ a_{n1} & a_{n-1,1} & \cdots & a_{11} \end{vmatrix}.$$

**8.** Compute

$$f(x) = \begin{vmatrix} x & 1 & 0 & x \\ 0 & x & x & 1 \\ 1 & x & x & 0 \\ x & 0 & 1 & x \end{vmatrix}$$

and find all $a \in \mathfrak{R}$ such that $f(a) = 0$.

☆ **9.** Show that

$$\begin{vmatrix} a & -b & -c & -d \\ b & a & -d & c \\ c & d & a & -b \\ d & -c & b & a \end{vmatrix} = (a^2 + b^2 + c^2 + d^2)^2.$$

»☆ **10.** Let **A** and **B** be real $n \times n$ matrices. Prove that

$$\det \begin{bmatrix} \mathbf{A} & -\mathbf{B} \\ \mathbf{B} & \mathbf{A} \end{bmatrix} = |\det(\mathbf{A} + i\mathbf{B})|^2,$$

where $\mathbf{A} + i\mathbf{B}$ is the $n \times n$ complex matrix whose $(i,j)$ component is equal to $\mathbf{A}_{ij} + i\mathbf{B}_{ij}$. ($|\det|^2$ is the square of the absolute value of det.)

☆ **11.** Prove

$$\begin{vmatrix} 1 & 1 & \cdots & 1 \\ x_1 & x_2 & \cdots & x_n \\ x_1^2 & x_2^2 & \cdots & x_n^2 \\ \cdots & \cdots & \cdots & \cdots \\ x_1^{n-1} & x_2^{n-1} & \cdots & x_n^{n-1} \end{vmatrix} = \prod_{1 \le i < j \le n} (x_j - x_i).$$

(This is called the *Vandermonde determinant.*)

☆ **12.** (*a*) Let $e$ be an elementary row operation and $\mathbf{E} = e(\mathbf{I}_n)$ the corresponding elementary matrix. Prove that

$$\det e(\mathbf{A}) = \det \mathbf{E} \cdot \det \mathbf{A}$$

for any $n \times n$ matrix **A**.

(*b*) Using (*a*), show that $\det(\mathbf{E}_1 \cdots \mathbf{E}_k) = \det\mathbf{E}_1 \cdots \det\mathbf{E}_k$ for any elementary matrices $\mathbf{E}_1, \ldots, \mathbf{E}_k$. (Do not appeal to Theorem 6.10.)

(*c*) Using (*b*) and Theorem 3.23, prove that $\det(\mathbf{AB}) = \det\mathbf{A} \cdot \det\mathbf{B}$ for nonsingular matrices **A** and **B**.

(*d*) For a singular matrix **A**, show that $\det \mathbf{A} = 0$.

(*e*) Using (*c*) and (*d*), prove $\det(\mathbf{AB}) = \det\mathbf{A} \cdot \det\mathbf{B}$ for any two matrices. (We thus obtain an alternative proof of Theorem 6.10.)

» **13.** Let $\sigma \in S_n$ and let $A_\sigma$ be the linear transformation of $\mathfrak{F}^n$ which maps $\epsilon_i$ upon $\epsilon_{\sigma(i)}$ for each $1 \le i \le n$, where $\{\epsilon_i\}$ is the standard basis of $\mathfrak{F}^n$. Express this transformation by a matrix (using $\{\epsilon_i\}$) and find its determinant.

☆ **14.** Prove that under matrix multiplication

(*a*) $GL^+(n,\mathfrak{R}) = \{\mathbf{A} \in GL(n,\mathfrak{R}); \det \mathbf{A} > 0\}$ is a group [subgroup of $GL(n,\mathfrak{R})$ defined in Example 6.1].

(*b*) $SL(n,\mathfrak{R}) = \{\mathbf{A} \in GL(n,\mathfrak{R}); \det \mathbf{A} = 1\}$ is a group [subgroup of $GL^+(n,\mathfrak{R})$].

(*c*) $SL(n,\mathfrak{C}) = \{\mathbf{A} \in GL(n,\mathfrak{C}); \det \mathbf{A} = 1\}$ is a group [subgroup of $GL(n,\mathfrak{C})$ defined in Example 6.1].

$SL(n,\mathfrak{R})$ and $SL(n,\mathfrak{C})$ are called the *special linear groups* over $\mathfrak{R}$ and $\mathfrak{C}$, respectively.

☆ **15.** A set $\mathfrak{G}$ of $n \times n$ real or complex matrices is said to be *arcwise-connected* if for any two matrices $\mathbf{A}_0$ and $\mathbf{A}_1$ in $\mathfrak{G}$ there is a continuous family of matrices $\mathbf{A}(t) \in \mathfrak{G}$, $0 \le t \le 1$, such that $\mathbf{A}(0) = \mathbf{A}_0$ and $\mathbf{A}(1) = \mathbf{A}_1$. [A family of matrices $\mathbf{A}(t)$, $0 \le t \le 1$, is continuous if every component $a_{ij}(t)$ of $\mathbf{A}(t)$, $1 \le i, j \le n$, is a continuous function of $t$, $0 \le t \le 1$.]

(*a*) Is $GL(n,\mathfrak{R})$ arcwise-connected?

(*b*) Prove that both $GL^+(n,\mathfrak{R})$ and $SL(n,\mathfrak{R})$ are arcwise-connected.

(*c*) Prove that $GL(n,\mathfrak{C})$ is arcwise-connected.

(*d*) Is $SL(n,\mathfrak{C})$ arcwise-connected?

☆ **16.** Let $V$ be an $n$-dimensional real vector space and let $\Phi = \{\alpha_1, \ldots, \alpha_n\}$ and $\Psi = \{\beta_1, \ldots, \beta_n\}$ be two ordered bases of $V$. We say that $\Phi$ is *positively related* to $\Psi$ if the matrix of transition $\mathbf{P} = [p_{ij}]$ given by

$$\alpha_i = \sum_{j=1}^{n} p_{ji}\beta_j, \qquad 1 \le i \le n,$$

has a positive determinant. Prove:

(*a*) $\Phi$ is positively related to $\Phi$ itself.

(*b*) If $\Phi$ is positively related to $\Psi$, then $\Psi$ is positively related to $\Phi$.

(*c*) If $\Phi$ is positively related to $\Psi$ and if $\Psi$ is positively related to $\Lambda$, then $\Phi$ is positively related to $\Lambda$.

Thus we have an equivalence relation in the set $\mathfrak{D}$ of all ordered bases of $V$; $\mathfrak{D}$ is divided into two nonempty disjoint subsets $\mathfrak{D}_1$ and $\mathfrak{D}_2$ such that, for each $i = 1, 2$:

1. If $\Phi, \Psi \in \mathfrak{D}_i$, then $\Phi$ is positively related to $\Psi$.
2. If $\Phi \in \mathfrak{D}_i$ and if $\Phi$ is positively related to $\Psi$, then $\Psi \in \mathfrak{D}_i$.

By an *orientation* of $V$, we mean $\mathfrak{D}_1$ or $\mathfrak{D}_2$. Thus $V$ has two orientations; one is said to be the opposite of the other. With an orientation chosen on $V$, we say that $V$ is an *oriented* vector space.

☆**17.** (Continuation of number 16) Prove that $\Phi = \{\alpha_1, \ldots, \alpha_n\}$ is positively related to $\Psi = \{\beta_1, \ldots, \beta_n\}$ if and only if there is a continuous family of ordered bases $\Phi(t) = \{\alpha_1(t), \ldots, \alpha_n(t)\}$, $0 \leq t \leq 1$, such that $\Phi(0) = \Phi$ and $\Phi(1) = \Psi$. [$\Phi(t)$ is continuous if, for each $i$, $\alpha_i(t)$ is continuous in $t$; cf. Exercise 3.3, number 9.]

## 6.4 EXPANSIONS OF DETERMINANTS

In this section we prove the expansion theorem for determinants and give its applications.

### Definition 6.10

Let $\mathbf{A}$ be an $n \times n$ matrix. The *(i,j) minor* of $\mathbf{A}$, where $1 \leq i, j \leq n$, is the determinant of the $(n-1) \times (n-1)$ matrix which results by deleting the $i$th row and the $j$th column of $\mathbf{A}$. The *(i,j) cofactor* $\Delta_{ij}$ of $\mathbf{A}$ is $(-1)^{i+j}$ times the $(i,j)$ minor of $\mathbf{A}$.

***Example 6.17.*** For

$$\mathbf{A} = \begin{bmatrix} 4 & 3 & -1 & 0 \\ 2 & 1 & 3 & 1 \\ 1 & 0 & 2 & 3 \\ 3 & 2 & 1 & 4 \end{bmatrix}$$

the (2,3) cofactor is

$$(-1)^{2+3} \cdot \begin{vmatrix} 4 & 3 & 0 \\ 1 & 0 & 3 \\ 3 & 2 & 4 \end{vmatrix} = (-1)(-9) = 9.$$

We now prove:

***Theorem 6.12.*** *Let* $\mathbf{A} = [a_{ij}]$ *be an* $n \times n$ *matrix over a commutative ring with* 1 *and let* $\Delta_{ij}$ *be the* $(i,j)$ *cofactor of* $\mathbf{A}$.

**1.** *For any fixed* $j$, *we have the expansion along the* $j$*th column, namely,*

$$\det \mathbf{A} = a_{1j}\Delta_{1j} + a_{2j}\Delta_{2j} + \cdots + a_{nj}\Delta_{nj}.$$

**2.** *For any fixed i, we have the expansion along the ith row, namely,*

$$\det \mathbf{A} = a_{i1}\Delta_{i1} + a_{i2}\Delta_{i2} + \cdots + a_{in}\Delta_{in}.$$

*Proof.* We shall prove only (1) here. Part 2 can be proved in a similar way, or it can be obtained from (1) by taking the transpose.

Let $\alpha_1, \ldots, \alpha_n$ be the columns of $\mathbf{A}$, so that $\mathbf{A} = [\alpha_1, \ldots, \alpha_n]$. Using the standard basis $\{\epsilon_1, \ldots, \epsilon_n\}$ of $K^n$ (whose elements are written vertically), we have

$$\alpha_j = \sum_{i=1}^{n} a_{ij}\epsilon_i.$$

We shall first consider the case where $j = 1$. Then

$$\det \mathbf{A} = \det\left[\sum_{i=1}^{n} a_{i1}\epsilon_i, \alpha_2, \ldots, \alpha_n\right]$$

$$= \sum_{i=1}^{n} a_{i1} \det [\epsilon_i, \alpha_2, \ldots, \alpha_n].$$

In order to obtain the expansion 1 for $j = 1$, it is sufficient to prove

$$\det [\epsilon_i, \alpha_2, \ldots, \alpha_n] = \Delta_{i1}, \qquad 1 \le i \le n.$$

We have

$$\begin{vmatrix} 0 & a_{12} & \cdots & a_{1n} \\ 0 & a_{22} & \cdots & a_{2n} \\ \cdot & \cdot & \cdots & \cdot \\ 1 & a_{i2} & \cdots & a_{in} \\ \cdot & \cdot & \cdots & \cdot \\ 0 & a_{n2} & \cdots & a_{2n} \end{vmatrix} = (-1)^{i-1} \begin{vmatrix} 1 & a_{i2} & \cdots & a_{in} \\ 0 & a_{12} & \cdots & a_{1n} \\ \cdot & \cdot & \cdots & \cdot \\ \cdot & \cdot & \cdots & \cdot \\ \cdot & \cdot & \cdots & \cdot \\ 0 & a_{n2} & \cdots & a_{nn} \end{vmatrix}$$

$$= (-1)^{i-1} \begin{vmatrix} a_{12} & \cdots & a_{1n} \\ \cdot & \cdots & \cdot \\ a_{n2} & \cdots & a_{nn} \end{vmatrix} = \Delta_{i1},$$

by moving the $i$th row to the top and then using the corollary to Theorem 6.11.

In order to obtain the expansion 1 for any $j$, we have

$$\det \mathbf{A} = \det\left[\alpha_1, \ldots, \alpha_{j-1}, \sum_{i=1}^{n} a_{ij}\epsilon_i, \alpha_{j+1}, \ldots, \alpha_n\right]$$

$$= \sum_{i=1}^{n} a_{ij} \det [\alpha_1, \ldots, \alpha_{j-1}, \epsilon_i, \alpha_{j+1}, \ldots, \alpha_n],$$

so that it is sufficient to show

$$\det [\alpha_1, \ldots, \alpha_{j-1}, \epsilon_i, \alpha_{j+1}, \ldots, \alpha_n] = \Delta_{ij}, \qquad 1 \le i \le n.$$

This can be done by moving the $j$th column to the left end and using the result for $j = 1$ which we have already proved:

$$\begin{vmatrix} a_{11} & \cdots & a_{1,j-1} & 0 & a_{1,j+1} & \cdots & a_{1n} \\ \cdots & \cdots & \cdots & \cdots & \cdots & \cdots & \cdots \\ a_{i1} & \cdots & a_{i,j-1} & 1 & a_{i,j+1} & \cdots & a_{in} \\ \cdots & \cdots & \cdots & \cdots & \cdots & \cdots & \cdots \\ a_{n1} & \cdots & a_{n,j-1} & 0 & a_{n,j+1} & \cdots & a_{nn} \end{vmatrix}$$

$$= (-1)^{j-1} \begin{vmatrix} 0 & a_{11} & \cdots & a_{1,j-1} & a_{1,j+1} & \cdots & a_{1n} \\ \cdots & \cdots & \cdots & \cdots & \cdots & \cdots & \cdots \\ 1 & a_{i1} & \cdots & a_{i,j-1} & a_{i,j+1} & \cdots & a_{in} \\ \cdots & \cdots & \cdots & \cdots & \cdots & \cdots & \cdots \\ 0 & a_{n1} & \cdots & a_{n,j-1} & a_{n,j+1} & \cdots & a_{nn} \end{vmatrix}$$

$$= (-1)^{j-1}(-1)^{i-1} \begin{vmatrix} a_{11} & \cdots & a_{1,j-1} & a_{1,j+1} & \cdots & a_{1n} \\ \cdots & \cdots & \cdots & \cdots & \cdots & \cdots \\ a_{n1} & \cdots & a_{n,j-1} & a_{n,j+1} & \cdots & a_{nn} \end{vmatrix} = \Delta_{ij}.$$

***Example 6.18.*** Using the expansion along the second column, we have

$$\begin{vmatrix} 3 & 0 & 5 & -2 \\ 1 & 5 & 1 & -1 \\ 2 & -1 & 2 & 3 \\ 4 & 0 & 1 & 3 \end{vmatrix} = (-1)^{2+2}5 \begin{vmatrix} 3 & 5 & -2 \\ 2 & 2 & 3 \\ 4 & 1 & 3 \end{vmatrix} + (-1)^{2+3}(-1) \begin{vmatrix} 3 & 5 & -2 \\ 1 & 1 & -1 \\ 4 & 1 & 3 \end{vmatrix}.$$

In order to derive an important consequence of Theorem 6.12, we make the following definition.

**Definition 6.11**

For an $n \times n$ matrix $\mathbf{A} = [a_{ij}]$, the *adjoint* of $\mathbf{A}$, denoted by $\tilde{\mathbf{A}}$, is the matrix whose $(i,j)$ component is equal to $\Delta_{ji}$ for $1 \leq i, j \leq n$.

***Theorem 6.13.*** *For any $n \times n$ matrix $\mathbf{A}$ over a commutative ring $K$ with 1, we have*

$$\mathbf{A}\tilde{\mathbf{A}} = \tilde{\mathbf{A}}\mathbf{A} = (\det \mathbf{A})\mathbf{I}_n.$$

*Proof.* We shall prove for $\mathbf{A} = [a_{ij}]$

1. $\displaystyle\sum_{j=1}^{n} a_{ij}\tilde{\mathbf{A}}_{ji} = \sum_{j=1}^{n} a_{ij}\Delta_{ij} = \det \mathbf{A}$ for each $i$

and

2. $\displaystyle\sum_{j=1}^{n} a_{ij}\tilde{\mathbf{A}}_{jk} = \sum_{j=1}^{n} a_{ij}\Delta_{kj} = 0$ for $i \neq k$,

which together imply that $\mathbf{A}\tilde{\mathbf{A}} = (\det \mathbf{A})\mathbf{I}_n$. The proof of $\tilde{\mathbf{A}}\mathbf{A} = (\det \mathbf{A})\mathbf{I}_n$ is similar.

Now, (1) is nothing but the expansion of det $\mathbf{A}$ along the $i$th row. In order to prove (2), consider the matrix $\mathbf{B} = [b_{ij}]$, which is obtained by replacing the $k$th row of $\mathbf{A}$ by the $i$th row of $\mathbf{A}$, that is, $b_{kj} = a_{ij}$ and $b_{mj} = a_{mj}$ for $m \neq k$, where $1 \leq j \leq n$. Obviously, $\det \mathbf{B} = 0$. On the other hand, the expansion of $\det \mathbf{B}$ along the $k$th row gives

$$\sum_{j=1}^{n} b_{kj}\Delta'_{kj} = \sum_{j=1}^{n} a_{ij}\Delta_{kj} = 0,$$

since the $(k,j)$ cofactor $\Delta'_{kj}$ of $\mathbf{B}$ is equal to the $(k,j)$ cofactor $\Delta_{kj}$ of $\mathbf{A}$. We have thus proved (2).

***Theorem 6.14.*** *An $n \times n$ matrix $\mathbf{A}$ over a commutative ring $K$ with 1 is invertible if and only if $\det \mathbf{A}$ has an inverse in $K$, that is, an element $b \in K$ such that $(\det \mathbf{A})b = b(\det \mathbf{A}) = 1$. In this case,*

$$\mathbf{A}^{-1} = b\tilde{\mathbf{A}}, \qquad \textit{that is,} \qquad \mathbf{A}^{-1} = (\det \mathbf{A})^{-1}\tilde{\mathbf{A}}.$$

*Proof.* Suppose that $\mathbf{A}$ is invertible in $K^n{}_n$, that is, there exists $\mathbf{B} \in K^n{}_n$ such that $\mathbf{AB} = \mathbf{BA} = \mathbf{I}_n$. By Theorem 6.10 we have

$$\det \mathbf{A} \cdot \det \mathbf{B} = \det \mathbf{I}_n = 1,$$

so that the element $\det \mathbf{A} \in K$ has an inverse (namely, $\det \mathbf{B}$) in $K$. Conversely, suppose that $\det \mathbf{A}$ has an inverse $b \in K$. Then let $\mathbf{B} = b\tilde{\mathbf{A}}$, where $\tilde{\mathbf{A}}$ is the adjoint of $\mathbf{A}$. By Theorem 6.13 we have

$$\mathbf{AB} = \mathbf{BA} = b(\mathbf{A}\tilde{\mathbf{A}}) = b(\det \mathbf{A})\mathbf{I}_n = \mathbf{I}_n,$$

showing that $\mathbf{B}$ is an inverse of $\mathbf{A}$.

***Corollary.*** *Let $\mathbf{A}$ be an $n \times n$ matrix over a field $\mathcal{F}$. Then $\mathbf{A}$ is invertible if and only if $\det \mathbf{A} \neq 0$. In this case,*

$$\mathbf{A}^{-1} = \frac{\tilde{\mathbf{A}}}{\det \mathbf{A}} \qquad \textit{and} \qquad \det(\mathbf{A}^{-1}) = (\det \mathbf{A})^{-1}.$$

*Proof.* We have only to observe that in a field $\mathcal{F}$ an element $a$ is invertible if and only if $a \neq 0$.

***Example 6.19.*** Over the ring of integers $\mathbb{Z}$,

$$\begin{bmatrix} 2 & 1 \\ 1 & 1 \end{bmatrix} \text{ has the inverse } \begin{bmatrix} 1 & -1 \\ -1 & 2 \end{bmatrix},$$

but

$$\begin{bmatrix} 3 & 1 \\ 1 & 1 \end{bmatrix} \text{ is not invertible, since } \begin{vmatrix} 3 & 1 \\ 1 & 1 \end{vmatrix} = 2.$$

***Corollary.*** *Let $\mathbf{A}$ be an $n \times n$ matrix over a commutative ring $K$ with 1. If there is a left [right] inverse $\mathbf{B}$ of $\mathbf{A}$, that is, a matrix $\mathbf{B}$ over $K$ such that $\mathbf{BA} = \mathbf{I}_n$ [$\mathbf{AB} = \mathbf{I}_n$], then $\mathbf{A}$ is invertible and $\mathbf{B} = \mathbf{A}^{-1}$.*

*Proof.* From $\mathbf{BA} = \mathbf{I}_n$ we have $\det \mathbf{B} \cdot \det \mathbf{A} = 1$ by Theorem 6.10. Since $K$ is commutative, $\det \mathbf{A} \cdot \det \mathbf{B} = 1$, that is, $\det \mathbf{A}$ has an inverse, namely, $\det \mathbf{B}$. Let $\mathbf{A}^{-1} \in K^n{}_n$ be the inverse of $\mathbf{A}$. Then

$$\mathbf{B} = \mathbf{B}(\mathbf{AA}^{-1}) = (\mathbf{BA})\mathbf{A}^{-1} = \mathbf{I}_n\mathbf{A}^{-1} = \mathbf{A}^{-1}.$$

The case of a right inverse is similar.

We now consider matrices over a field $\mathfrak{F}$.

**Definition 6.12**

For an $m \times n$ matrix $\mathbf{A} = [a_{ij}]$, a *submatrix* of degree $r$ is a matrix of the form

$$\begin{bmatrix} a_{i_1j_1} & a_{i_1j_2} & \cdots & a_{i_1j_r} \\ a_{i_2j_1} & a_{i_2j_2} & \cdots & a_{i_2j_r} \\ \cdots & \cdots & \cdots & \cdots \\ a_{i_rj_1} & a_{i_rj_2} & \cdots & a_{i_rj_r} \end{bmatrix},$$

where $1 \leq i_1 < i_2 < \cdots < i_r \leq m$ and $1 \leq j_1 < j_2 < \cdots < j_r \leq n$. The *determinant rank of* $\mathbf{A}$ is the largest possible integer $r$ such that there is a submatrix of degree $r$ with nonzero determinant.

We shall prove:

***Theorem 6.15.*** *For any $m \times n$ matrix $\mathbf{A}$ over a field $\mathfrak{F}$, the determinant rank of $\mathbf{A}$ is equal to the rank of $\mathbf{A}$.*

*Proof.* Recall that the row rank and the column rank coincide (Theorem 3.17). First, assume that there is a submatrix $\mathbf{B}$ of degree $s$ such that $\det \mathbf{B} \neq 0$. Then $\mathbf{B}$ is nonsingular (the first corollary to Theorem 6.14) and hence the row rank is $s$, that is, the $s$ row vectors of $\mathbf{B}$ are linearly independent. Then it follows that the $s$ row vectors of $\mathbf{A}$ which correspond to the row vectors of $\mathbf{B}$ are linearly independent, proving that the row rank of $\mathbf{A}$ is greater than or equal to $s$. Hence,

$$\text{row rank of } \mathbf{A} \geq \text{determinant rank of } \mathbf{A}.$$

Now assume that the row rank of $\mathbf{A}$ is $r$ and let $\alpha_{i_1}, \ldots, \alpha_{i_r}$, $1 \leq i_1 < i_2 < \cdots < i_r \leq m$, be a linearly independent set of row vectors of $\mathbf{A}$. Let $\mathbf{B}$ be the $r \times n$ matrix with $\alpha_{i_1}, \ldots, \alpha_{i_r}$ as the row vectors. Since the column rank of $\mathbf{B}$ is equal to $r$, we may choose a linearly independent set of $r$ column vectors. Let $\mathbf{C}$ be the submatrix of $\mathbf{B}$ with these $r$ column vectors. $\mathbf{C}$ is then a submatrix of degree $r$ of the original matrix $\mathbf{A}$. The rank of $\mathbf{C}$ is $r$ and $\mathbf{C}$ is nonsingular, that is, $\det \mathbf{C} \neq 0$. This proves that the determinant rank of $\mathbf{A}$ is at least $r$. Hence

$$\text{row rank of } \mathbf{A} \leq \text{determinant rank of } \mathbf{A},$$

which, together with the inequality above, proves our assertion.

***Corollary.*** *A set of $n$ vectors $\alpha_1, \ldots, \alpha_n$ in $\mathfrak{F}^n$ are linearly independent if and only if*

$$\det [\alpha_1, \ldots, \alpha_n] \neq 0.$$

*Proof.* $\alpha_1, \ldots, \alpha_n$ are linearly independent if and only if the (column) rank of the matrix $[\alpha_1, \ldots, \alpha_n]$ is $n$, that is, $[\alpha_1, \ldots, \alpha_n]$ is nonsingular.

### EXERCISE 6.4

›› **1.** Find the (2,3) cofactors of

$$\begin{bmatrix} 3 & -1 & 1 \\ 4 & 5 & -2 \\ 2 & 3 & 4 \end{bmatrix} \quad \text{and} \quad \begin{bmatrix} 3 & -1 & 1 \\ 3 & 1 & 0 \\ 2 & 3 & 5 \end{bmatrix}.$$

**2.** Write down the expansion of

(a) $\begin{vmatrix} 3 & -1 & 1 \\ 4 & 5 & -2 \\ 2 & 3 & 4 \end{vmatrix}$ along the second row;

(b) $\begin{vmatrix} 3 & -1 & 1 \\ 3 & 1 & 0 \\ 2 & 3 & 5 \end{vmatrix}$ along the third column.

**3.** Give a proof of $\tilde{\mathbf{A}}\mathbf{A} = (\det \mathbf{A})\mathbf{I}_n$ in Theorem 6.13.

›› **4.** Decide whether each of the following matrices is invertible and find its inverse if it exists:

(a) $\begin{bmatrix} 3 & -1 \\ -5 & 2 \end{bmatrix}$ over $\mathbb{Z}$; (b) $\begin{bmatrix} 3 & -2 \\ 1 & -1 \end{bmatrix}$ over $\mathbb{Z}$; (c) $\begin{bmatrix} 2 & 4 \\ 1 & 3 \end{bmatrix}$ over $\mathbb{Z}$;

(d) $\begin{bmatrix} x+1 & x \\ x & x-1 \end{bmatrix}$ over $\mathbb{Z}[x]$; (e) $\begin{bmatrix} x^2+1 & x^2 \\ x+1 & x \end{bmatrix}$ over $\mathbb{R}[x]$.

**5.** Find whether each of the following sets of vectors is linearly independent:
(a) $\{(1,-1,2),(3,2,-1),(-3,1,8)\}$ in $\mathbb{R}^3$.
(b) $\{(1,1,0,0),(0,1,1,0),(0,0,1,1),(1,0,0,1)\}$ in $\mathbb{R}^4$.

**6.** Prove $\begin{vmatrix} a & b & c \\ c & a & b \\ b & c & a \end{vmatrix} = a^3 + b^3 + c^3 - 3abc.$

››**7.** Find all $x \in \mathbb{R}$ so that rank $\begin{bmatrix} 1 & x & x & x \\ x & 1 & x & x \\ x & x & 1 & x \\ x & x & x & 1 \end{bmatrix} < 4.$

For those values of $x$, find the rank of the corresponding matrix.

**8.** Find the inverse of

$$(a)\ \begin{bmatrix} & & a_1 \\ & \mathbf{I}_n & \vdots \\ & & a_n \\ 0 & \cdots\ 0 & 1 \end{bmatrix}; \qquad (b)\ \begin{bmatrix} 1 & -1 & & & \\ & 1 & -1 & & \\ & & 1 & \ddots & \\ & & & \ddots & -1 \\ & & & & 1 \end{bmatrix}$$

(all other components 0).

» **9.** Let $\Delta_{ij}$ be the $(i,j)$ cofactor of an $n \times n$ matrix $\mathbf{A} = [a_{ij}]$. Prove

$$\begin{vmatrix} \Delta_{11} & \Delta_{12} & \cdots & \Delta_{1n} \\ \Delta_{21} & \Delta_{22} & \cdots & \Delta_{2n} \\ \cdots & \cdots & \cdots & \cdots \\ \Delta_{n1} & \Delta_{n2} & \cdots & \Delta_{nn} \end{vmatrix} = |\mathbf{A}|^{n-1}.$$

☆**10.** Using the same notation as in number 9, prove

$$\begin{vmatrix} \Delta_{11} & \cdots & \Delta_{1r} \\ \cdots & \cdots & \cdots \\ \Delta_{r1} & \cdots & \Delta_{rr} \end{vmatrix} = |\mathbf{A}|^{r-1} \begin{vmatrix} a_{r+1,r+1} & \cdots & a_{r+1,n} \\ \cdots & \cdots & \cdots \\ a_{n,r+1} & \cdots & a_{nn} \end{vmatrix}$$

for $1 \le r < n$.

☆**11.** Let $\tilde{\mathbf{A}}$ be the adjoint of an $n \times n$ matrix $\mathbf{A}$. Prove
(*a*) If rank $\mathbf{A} = n$, then rank $\tilde{\mathbf{A}} = n$.
(*b*) If rank $\mathbf{A} = n - 1$, then rank $\tilde{\mathbf{A}} = 1$.
(*c*) If rank $\mathbf{A} < n - 1$, then rank $\tilde{\mathbf{A}} = 0$.

☆**12.** A mapping $t \to \mathbf{A}(t) = [a_{ij}(t)]$ of the interval $I_0 = (0,1)$ into $\mathcal{R}^n_n$ is said to be *differentiable* if each $a_{ij}(t)$ is differentiable, and $d\mathbf{A}/dt$ is defined to be the matrix $[da_{ij}/dt]$ (cf. Exercise 3.3, number 10).

(*a*) For two mappings $\mathbf{A}(t)$ and $\mathbf{B}(t)$ both of which are differentiable, prove that

$$\frac{d(\mathbf{A}+\mathbf{B})}{dt} = \frac{d\mathbf{A}}{dt} + \frac{d\mathbf{B}}{dt} \qquad \text{and} \qquad \frac{d(\mathbf{AB})}{dt} = \frac{d\mathbf{A}}{dt}\mathbf{B} + \mathbf{A}\frac{d\mathbf{B}}{dt}.$$

(*b*) If $\mathbf{A}(t) = [\alpha_1(t),\alpha_2(t), \ldots ,\alpha_n(t)]$, where $\alpha_i(t)$ are column vectors, is differentiable, then the function det $\mathbf{A}(t)$ is differentiable and its derivative is equal to

$$\det\left[\frac{d\alpha_1}{dt},\alpha_2, \ldots ,\alpha_n\right] + \det\left[\alpha_1, \frac{d\alpha_2}{dt}, \ldots ,\alpha_n\right] + \cdots + \det\left[\alpha_1,\alpha_2, \ldots , \frac{d\alpha_n}{dt}\right].$$

## 6.5 DETERMINANT OF A LINEAR TRANSFORMATION

Let $V$ be an $n$-dimensional vector space over a field $\mathcal{F}$. For any linear transformation of $V$ we shall define its determinant. We choose a basis $\Phi = \{\alpha_1, \ldots ,\alpha_n\}$ of $V$ and represent the given linear transformation $A$ by a matrix

$$A_\Phi = [a_{ij}], \qquad \text{where } A(\alpha_j) = \sum_{i=1}^n a_{ij}\alpha_i, \qquad 1 \le j \le n.$$

We wish to define det $A$ to be det $A_\Phi$; for this purpose we have to verify that this definition is independent of the choice of a basis $\Phi$ we are using. Let $\Psi = \{\beta_1, \ldots, \beta_n\}$ be any other basis of $V$; the matrix $A_\Psi$ representing $A$ with respect to $\Psi$ is given by

$$A_\Psi = \mathbf{P} A_\Phi \mathbf{P}^{-1},$$

where $\mathbf{P}$ is a nonsingular matrix (corollary to Proposition 3.12). By Theorem 6.10, we have

$$\det A_\Psi = \det \mathbf{P} \cdot \det A_\Phi \cdot \det \mathbf{P}^{-1} = \det A_\Phi,$$

since $\det \mathbf{P} \cdot \det \mathbf{P}^{-1} = 1$.

When we regard an $n \times n$ matrix $\mathbf{A}$ over $\mathcal{F}$ as defining a linear transformation of $\mathcal{F}^n$ (with respect to the standard basis), the determinant of the transformation $\mathbf{A}$ is, of course, equal to the determinant of the matrix $\mathbf{A}$.

We have:

***Proposition 6.16***

1. *The determinant of the identity transformation is* 1.
2. *For linear transformations $A$ and $B$,* $\det(AB) = (\det A)(\det B)$.
3. *A linear transformation $A$ is invertible if and only if* $\det A$ *is not* 0.

*Proof.* These assertions follow from the correspondence between linear transformations and matrices (with respect to any fixed basis) and the properties of determinants for matrices (cf. Theorem 6.10 and the first corollary to Theorem 6.14).

***Proposition 6.17.*** *Let $f$ be any nonzero alternating $n$-linear function on an $n$-dimensional vector space $V$. For any linear transformation $A$ of $V$, we have*

$$f(A(\alpha_1), \ldots, A(\alpha_n)) = (\det A) f(\alpha_1, \ldots, \alpha_n)$$

*for all* $\alpha_1, \ldots, \alpha_n \in V$.

*Proof.* We first observe that the function $g$ defined by

$$g(\alpha_1, \ldots, \alpha_n) = f(A(\alpha_1), \ldots, A(\alpha_n))$$

is alternating and $n$-linear. By Proposition 6.7 we have $g = cf$, where $c$ is a scalar. We shall prove that $c = \det A$. It is sufficient to prove

$$g(\alpha_1, \ldots, \alpha_n) = (\det A) f(\alpha_1, \ldots, \alpha_n)$$

for some basis $\Phi = \{\alpha_1, \ldots, \alpha_n\}$ of $V$. Let $\mathbf{A} = [a_{ij}]$ be the matrix which represents $A$ with respect to $\Phi$:

$$A(\alpha_j) = \sum_{i=1}^{n} a_{ij}\alpha_i, \qquad 1 \leq j \leq n.$$

Then

$$\begin{aligned} g(\alpha_1, \ldots, \alpha_n) &= f(A(\alpha_1), \ldots, A(\alpha_n)) \\ &= f\left(\sum_{j=1}^{n} a_{j1}\alpha_j, \ldots, \sum_{j=1}^{n} a_{jn}\alpha_j\right) \\ &= \left(\sum_{\sigma \in S_n} s(\sigma) a_{\sigma(1)1} \cdots a_{\sigma(n)n}\right) f(\alpha_1, \ldots, \alpha_n) \\ &= (\det A) f(\alpha_1, \ldots, \alpha_n). \end{aligned}$$

***Remark.*** If we assume that the function $f$ has, furthermore, the property

$$f(\alpha_1, \ldots, \alpha_n) = 1$$

for a certain fixed basis $\{\alpha_1, \ldots, \alpha_n\}$ of $V$, then

$$\det A = f(A(\alpha_1), \ldots, A(\alpha_n)).$$

We could have defined det $A$ by this formula. As a special case, let $V = \mathfrak{F}^n$ and let $\{\alpha_i\}$ be the standard basis $\{\epsilon_i\}$. Consider any $n \times n$ matrix $\mathbf{A}$ over $\mathfrak{F}$ as a linear transformation $\mathfrak{F}^n \to \mathfrak{F}^n$. Then

$$\mathbf{A} = [\mathbf{A}(\epsilon_1), \ldots, \mathbf{A}(\epsilon_n)],$$

that is, the $i$th column vector of $\mathbf{A}$ is $\mathbf{A}(\epsilon_i)$. We have

$$\det \mathbf{A} = f(\mathbf{A}(\epsilon_1), \ldots, \mathbf{A}(\epsilon_n))$$

as in Definition 6.9.

**EXERCISE 6.5**

**1.** Let $A$ be a linear transformation of an $n$-dimensional vector space $V$ and let ${}^tA: V^* \to V^*$ be the transpose of $A$. Prove that $\det {}^tA = \det A$.

»☆ **2.** Let $V = \mathfrak{F}^n_n$ be the vector space of all $n \times n$ matrices over a field $\mathfrak{F}$. For any $\mathbf{A} \in V$ let $f_{\mathbf{A}}$ be the linear transformation of $V$ defined by $f_{\mathbf{A}}(\mathbf{B}) = \mathbf{AB}$. Express $\det f_{\mathbf{A}}$ in terms of $\det \mathbf{A}$. (Compare with Exercise 3.2, number 17.)

**3.** Let $A$ be a linear transformation of an $n$-dimensional vector space $V$ and let $U$ be a subspace of $V$ invariant by $A$. Let $A_U$ be the restriction of $A$ to $U$ and let $A_{V/U}$ be the linear transformation induced by $A$ on the quotient space $V/U$ (cf. corollary to Proposition 4.11). Prove that

$$\det A = (\det A_U)(\det A_{V/U}).$$

»☆ **4.** Let $A$ be a linear transformation of an $n$-dimensional vector space $V$ over a field $\mathfrak{F}$ of characteristic $\neq 2$. For any alternating $n$-linear function $f$, define

$$(Df)(\alpha_1, \ldots, \alpha_n) = f(A(\alpha_1), \alpha_2, \ldots, \alpha_n) + f(\alpha_1, A(\alpha_2), \ldots, \alpha_n) + \cdots + f(\alpha_1, \alpha_2, \ldots, \alpha_{n-1}, A(\alpha_n))$$

for all $\alpha_1, \ldots, \alpha_n \in V$.

(*a*) Show that $Df$ is an alternating $n$-linear function.

(*b*) If $f \neq 0$, show that $Df = cf$, where $c = \text{trace } A$.

☆ **5.** Let $V$ be an $n$-dimensional real vector space with an orientation $\mathfrak{O}_1$ (cf. Exercise 6.3, number 16). A nonsingular linear transformation $A$ of $V$ is said to be *orientation-preserving* if $\Phi = \{\alpha_1, \ldots, \alpha_n\} \in \mathfrak{O}_1$ implies $\{A(\alpha_1), \ldots, A(\alpha_n)\} \in \mathfrak{O}_1$. Prove that this condition is equivalent to the condition $\det A > 0$.

## 6.6 CRAMER'S FORMULA

Let $\mathcal{F}$ be a field and consider a system of $n$ linear equations with $n$ unknowns:

$$\begin{aligned} a_{11}x_1 + \cdots + a_{1n}x_n &= b_1, \\ \cdots\cdots\cdots\cdots &\ , \\ a_{n1}x_1 + \cdots + a_{nn}x_n &= b_n, \end{aligned}$$

where $a_{ij} \in \mathcal{F}, b_i \in \mathcal{F}$. Using matrix notation, we have

$$\mathbf{Ax} = \mathbf{b},$$

where $\mathbf{A} = [a_{ij}]$ is an $n \times n$ matrix, $\mathbf{x} = [x_i] \in \mathcal{F}^n$, and $\mathbf{b} = [b_i] \in \mathcal{F}^n$.

We discussed in Sec. 3.5 a method of solving such a system. If $\mathbf{A}$ is nonsingular, then the system has a unique solution

$$\mathbf{x} = \mathbf{A}^{-1}\mathbf{b}.$$

$\mathbf{A}$ is nonsingular if and only if $\det \mathbf{A} \neq 0$ (corollary to Theorem 6.14); we can prove the following explicit formula.

***Theorem 6.18.*** *If* $\det \mathbf{A} \neq 0$, *the system has a unique solution given by*

$$x_1 = \frac{\begin{vmatrix} b_1 & a_{12} & \cdots & a_{1n} \\ \cdots & \cdots & \cdots & \cdots \\ b_n & a_{1n} & \cdots & a_{nn} \end{vmatrix}}{\det \mathbf{A}}, \ \ldots, \ x_n = \frac{\begin{vmatrix} a_{11} & \cdots & a_{1,n-1} & b_1 \\ \cdots & \cdots & \cdots & \cdots \\ a_{n1} & \cdots & a_{n,n-1} & b_n \end{vmatrix}}{\det \mathbf{A}}.$$

*Proof.* Let $\tilde{\mathbf{A}}$ be the adjoint of $\mathbf{A}$, so that

$$\tilde{\mathbf{A}}\mathbf{A} = (\det \mathbf{A})\mathbf{I}_n$$

by Theorem 6.13. Then we have

$$\tilde{\mathbf{A}}\mathbf{A}\mathbf{x} = (\det \mathbf{A})\mathbf{I}_n\mathbf{x} = (\det \mathbf{A})\mathbf{x}.$$

On the other hand, if $\mathbf{Ax} = \mathbf{b}$, then

$$\tilde{\mathbf{A}}\mathbf{A}\mathbf{x} = \tilde{\mathbf{A}}\mathbf{b},$$

so that

$$\mathbf{x} = \frac{\tilde{\mathbf{A}}\mathbf{b}}{\det \mathbf{A}}.$$

The $i$th component of $\tilde{\mathbf{A}}\mathbf{b}$ is equal to

$$\sum_{j=1}^{n} \tilde{\mathbf{A}}_{ij}b_j = \sum_{j=1}^{n} \Delta_{ji}b_j,$$

which in turn is nothing but the expansion along the $i$th column of the determinant

$$\begin{vmatrix} a_{11} & \cdots & b_1 & \cdots & a_{1n} \\ \cdot & \cdot & \cdot & \cdot & \cdot \\ a_{n1} & \cdots & b_n & \cdots & a_{nn} \end{vmatrix}.$$

***Example 6.20.*** Consider the system of linear equations

$$\begin{aligned} (m+1)x + \quad y + \quad z &= 2 - m, \\ x + (m+1)y + \quad z &= -2, \\ x + \quad y + (m+1)z &= m, \end{aligned}$$

over the real number field, where $m$ is a real number. The solution set will vary depending on the value of $m$. In order to describe the solution sets for all values of $m$, we shall first find the values of $m$ so that the coefficient matrix

$$\mathbf{A} = \begin{bmatrix} m+1 & 1 & 1 \\ 1 & m+1 & 1 \\ 1 & 1 & m+1 \end{bmatrix}$$

is nonsingular. We have

$$\det \mathbf{A} = m^2(m+3).$$

Thus, if $m \neq 0, -3$, then $\det \mathbf{A} \neq 0$ and the system has a unique solution which is given by Cramer's formula:

$$x = \frac{\begin{vmatrix} 2-m & 1 & 1 \\ -2 & m+1 & 1 \\ m & 1 & m+1 \end{vmatrix}}{m^2(m+3)}, \qquad y = \frac{\begin{vmatrix} m+1 & 2-m & 1 \\ 1 & -2 & 1 \\ 1 & m & m+1 \end{vmatrix}}{m^2(m+3)},$$

$$z = \frac{\begin{vmatrix} m+1 & 1 & 2-m \\ 1 & m+1 & -2 \\ 1 & 1 & m \end{vmatrix}}{m^2(m+3)}.$$

For $m = 0$, the system reduces to

$$\begin{aligned} x + y + z &= 2, \\ x + y + z &= -2, \\ x + y + z &= 0, \end{aligned}$$

which has apparently no solution.

For $m = -3$, the system reduces to

$$\begin{aligned} -2x + y + z &= 5, \\ x - 2y + z &= -2, \\ x + y - 2z &= -3, \end{aligned}$$

whose solution is given by

$$x = c - 8/3, \qquad y = c - 1/3, \qquad z = c,$$

where $c$ is arbitrary.

**EXERCISE 6.6**

1. Write down Cramer's formula for the following system of linear equations:

»(a)
$$\begin{aligned} 3x_1 - 2x_2 + x_3 &= 1, \\ x_1 + 4x_2 - 5x_3 &= 3, \\ -x_1 + 3x_2 + 2x_3 &= -4. \end{aligned}$$

(b)
$$\begin{aligned} x_1 - 3x_2 + x_3 &= 1, \\ 2x_1 \qquad - 3x_3 &= 4, \\ x_2 + 2x_3 &= 0. \end{aligned}$$

2. Describe the solution set of the system

$$\begin{aligned} x + 2y + z &= m, \\ mx + y + (m+1)z &= 1, \\ x + my + (m-1)z &= 0, \end{aligned}$$

for all possible real values of $m$.

# 7 Minimal Polynomials

We introduce the notions of eigenvalue, eigenvector, characteristic polynomial, and minimal polynomial, which are important for the study of linear transformations. We prove a basic result (Theorem 7.5) which will reduce the study of arbitrary linear transformations to that of linear transformations of a special kind (namely, those whose minimal polynomials are powers of irreducible polynomials). In the case of a vector space over an algebraically closed field such as $\mathcal{C}$, it follows that an arbitrary linear transformation is decomposed into the sum of linear transformations of the form $cI + A$, where $I$ is the identity transformation and $A$ a nilpotent transformation. We also prove the famous theorem of Cayley and Hamilton (Theorem 7.7). We then study nilpotent transformations in detail and proceed to the discussion of the Jordan normal forms. In the last section, we discuss some results based on irreducibility of a family of linear transformations, getting Schur's lemma as a special case. The notion of complex structure is also studied here.

## 7.1 EIGENVALUES AND CHARACTERISTIC POLYNOMIALS

Let $A$ be a linear transformation of an $n$-dimensional vector space $V$ over a field $\mathfrak{F}$.

**Definition 7.1**

An element $c \in \mathfrak{F}$ is called an *eigenvalue* (or *proper value, characteristic value*) of $A$ if there exists a nonzero vector $\alpha \in V$ such that

$$A(\alpha) = c\alpha.$$

Any nonzero vector $\alpha$ satisfying this condition is called an *eigenvector* (or *proper vector, characteristic vector*) of $A$ corresponding to the eigenvalue $c$.

***Example 7.1.*** Let $A = cI$, where $I$ is the identity transformation and $c \in \mathfrak{F}$. Then $c$ is an eigenvalue and any nonzero vector $\alpha \in V$ is an eigenvector corresponding to $c$. On the other hand, there is no other eigenvalue. Suppose $c' \in \mathfrak{F}$ is an eigenvalue and let $A(\alpha) = c'\alpha$ for some $\alpha \neq 0$. Then $c\alpha = c'\alpha$, which implies $c = c'$ because $\alpha \neq 0$.

***Example 7.2.*** Let $A$ be the linear transformation of $\mathcal{R}^2$ defined by

$$\begin{bmatrix} x_1 \\ x_2 \end{bmatrix} \to \begin{bmatrix} 1 & -1 \\ 2 & -1 \end{bmatrix} \begin{bmatrix} x_1 \\ x_2 \end{bmatrix}.$$

We show that $A$ has no eigenvalue. Suppose $c \in \mathcal{R}$ is an eigenvalue and let

$$\alpha = \begin{bmatrix} a_1 \\ a_2 \end{bmatrix}$$

be an eigenvector such that $A(\alpha) = c\alpha$. We have then

$$a_1 - a_2 = ca_1 \qquad \text{and} \qquad 2a_1 - a_2 = ca_2,$$

that is,

$$\begin{aligned} (1 - c)a_1 - \qquad a_2 &= 0, \\ 2a_1 - (1 + c)a_2 &= 0. \end{aligned}$$

Since this system of linear equations has a nontrivial solution $(a_1,a_2) \neq (0,0)$, we must have

$$\begin{vmatrix} 1 - c & -1 \\ 2 & -(1 + c) \end{vmatrix} = 0, \qquad \text{that is,} \qquad c^2 + 1 = 0,$$

which is impossible, since $c$ is a real number.

***Example 7.3.*** Consider the linear transformation $A$ of the complex vector space $\mathcal{C}^2$ defined by the same matrix as that in Example 7.2. If $c \in \mathcal{C}$ is an eigenvalue of $A$, then the same computation shows that $c^2 + 1 = 0$, that is, $c = \pm i$. Actually, $i$ is an eigenvalue for which

$$\alpha_1 = \begin{bmatrix} (1 + i)/2 \\ 1 \end{bmatrix}$$

is an eigenvector, and $-i$ is an eigenvalue for which

$$\alpha_2 = \begin{bmatrix} (1 - i)/2 \\ 1 \end{bmatrix}$$

is an eigenvector. We note that $\alpha_1$ and $\alpha_2$ form a basis of $\mathcal{C}^2$, with respect to which $A$ can be represented by the matrix $\begin{bmatrix} i & 0 \\ 0 & -i \end{bmatrix}$.

In order to find all possible eigenvalues of a given linear transformation, we introduce the notion of characteristic polynomial. We shall first define the characteristic polynomial of an $n \times n$ matrix $\mathbf{A} = [a_{ij}]$ over $\mathcal{F}$. Let us consider

$$x\mathbf{I}_n - \mathbf{A} = \begin{bmatrix} x - a_{11} & -a_{12} & \cdots & -a_{1n} \\ -a_{21} & x - a_{22} & \cdots & -a_{2n} \\ \cdots & \cdots & \cdots & \cdots \\ -a_{n1} & -a_{n2} & \cdots & x - a_{nn} \end{bmatrix},$$

which is an $n \times n$ matrix over the polynomial ring $\mathcal{F}[x]$ (with indeterminate $x$). Its determinant is an element of $\mathcal{F}[x]$ whose degree is obviously $n$.

**Definition 7.2**

For an $n \times n$ matrix $\mathbf{A}$ over $\mathcal{F}$,

$$f_{\mathbf{A}}(x) = \det(x\mathbf{I}_n - \mathbf{A}) \in \mathcal{F}[x]$$

is called the *characteristic polynomial* of $\mathbf{A}$.

***Proposition 7.1.*** *If* $\mathbf{B} = \mathbf{P}^{-1}\mathbf{A}\mathbf{P}$ *with a nonsingular matrix* $\mathbf{P}$, *then*

$$f_{\mathbf{B}}(x) = f_{\mathbf{A}}(x).$$

*Proof.* We have

$$x\mathbf{I}_n - \mathbf{B} = x\mathbf{I}_n - \mathbf{P}^{-1}\mathbf{A}\mathbf{P} = \mathbf{P}^{-1}(x\mathbf{I}_n - \mathbf{A})\mathbf{P},$$

and hence

$$\det(x\mathbf{I}_n - \mathbf{B}) = \det(x\mathbf{I}_n - \mathbf{A}),$$

that is,

$$f_{\mathbf{B}}(x) = f_{\mathbf{A}}(x).$$

**Definition 7.3**

For a linear transformation $A$ of an $n$-dimensional vector space $V$ over $\mathcal{F}$, the *characteristic polynomial* $f_A(x)$ is defined to be that of the matrix $A_\Phi$ which represents $A$ with respect to an arbitrary basis $\Phi$ of $V$. It is independent of the choice of a basis; if $\Psi$ is any other basis, then $A_\Phi$ and $A_\Psi$ satisfy the assumption of Proposition 7.1 (cf. corollary to Proposition 3.12) and hence have the same characteristic polynomial.

***Example 7.4.*** For $\mathbf{A} = c\mathbf{I}_n$, the characteristic polynomial is $(x - c)^n$.

***Example 7.5.*** For the matrix

$$\mathbf{A} = \begin{bmatrix} 1 & -1 \\ 2 & -1 \end{bmatrix}$$

in Examples 7.2 and 7.3, the characteristic polynomial is

$$\begin{vmatrix} x - 1 & 1 \\ -2 & x + 1 \end{vmatrix} = x^2 + 1.$$

We shall now relate eigenvalues to the characteristic polynomial. For a linear transformation $A$, suppose that $c \in \mathcal{F}$ is an eigenvalue. Then there exists $\alpha \neq 0$ in $V$ such that $A(\alpha) = c\alpha$, that is,

$$(cI - A)(\alpha) = 0.$$

Since $\alpha \neq 0$, this shows that the transformation $cI - A$ is singular, so that

$$\det(cI - A) = 0;$$

in other words,

$$f_A(c) = 0.$$

Thus an eigenvalue $c$ is a root of the characteristic polynomial. Conversely, suppose that $c \in \mathfrak{F}$ is a root of the characteristic polynomial $f_A(x)$. Then $cI - A$ with zero determinant is singular, that is, there exists an $\alpha \neq 0$ in $V$ such that

$$(cI - A)(\alpha) = 0, \qquad \text{that is,} \qquad A(\alpha) = c\alpha.$$

In other words, $c$ is an eigenvalue with $\alpha$ as an eigenvector.

Summarizing our results, we have:

***Proposition 7.2.*** *An eigenvalue of $A$ is a root of $f_A(x)$. Conversely, a root of $f_A(x)$ in $\mathfrak{F}$ is an eigenvalue of $A$.*

Note that $f_A(x)$ may not have a root in $\mathfrak{F}$. It may have roots in some extension $\mathfrak{F}'$ of $\mathfrak{F}$ [in fact, there is always an extension $\mathfrak{F}'$ of $\mathfrak{F}$ in which $f_A(x)$ has a root, since any irreducible factor of $f_A(x)$ has a root in some extension by virtue of Theorem 5.18], but those roots which are not in $\mathfrak{F}$ are not eigenvalues. This explains the difference between Examples 7.2 and 7.3.

It is obvious that the problems involving eigenvalues are easier to study in the case where $\mathfrak{F}$ is algebraically closed (for example, $\mathfrak{F} = \mathfrak{C}$). We shall prove:

***Theorem 7.3.*** *Let $A$ be a linear transformation of an n-dimensional vector space $V$ over an algebraically closed field $\mathfrak{F}$. Then there exists a basis $\{\alpha_1, \ldots, \alpha_n\}$ in $V$ with respect to which $A$ is represented by a matrix of the form (superdiagonal matrix)*

$$\begin{bmatrix} c_1 & & & & * \\ & c_2 & & & \\ & & \cdot & & \\ & & & \cdot & \\ & & & & \cdot \\ 0 & & & & c_n \end{bmatrix},$$

*(that is, all entries below the diagonal are* 0*). The diagonal elements $c_1, \ldots, c_n$ are all of the eigenvalues of $A$.*

*Proof.* We shall use induction on dim $V$. The theorem is trivial for dim $V = 1$. Assume that it is valid for vector spaces of dimension $< n$. Since $f_A(x)$ has a root in $\mathfrak{F}$, $A$ has an eigenvalue, say, $c_1$, by Proposition 7.2. Let $\alpha_1$ be an eigenvector corresponding to $c_1$. If we denote by $W$ the subspace spanned by $\alpha_1$, then $A(W) \subseteq W$. By Proposition 4.11 there exists a linear transformation $A'$ of the quotient space $V' = V/W$ such that $A'\pi = \pi A$, where $\pi \colon V \to V'$ is the projection. Since dim $V' = n - 1$, the inductive assumption implies that $A'$ can be represented by a matrix of the form

$$\begin{bmatrix} c_2 & & & & * \\ & c_3 & & & \\ & & \cdot & & \\ & & & \cdot & \\ & & & & \cdot \\ 0 & & & & c_n \end{bmatrix}$$

with respect to a certain basis $\{\alpha_2', \ldots, \alpha_n'\}$ of $V'$. We now choose $\alpha_2, \ldots, \alpha_n$ in $V$ such that $\pi(\alpha_i) = \alpha_i'$ for $2 \leq i \leq n$. Then $\{\alpha_1, \ldots, \alpha_n\}$ is a basis of $V$ with respect to which $A$ can be represented by a matrix

$$\begin{bmatrix} c_1 & & & & * \\ & c_2 & & & \\ & & \cdot & & \\ & & & \cdot & \\ & & & & \cdot \\ 0 & & & & c_n \end{bmatrix}.$$

The characteristic polynomial $f_A(x)$ is given by

$$\det \begin{bmatrix} x - c_1 & & & & * \\ & x - c_2 & & & \\ & & \cdot & & \\ & & & \cdot & \\ & & & & \cdot \\ 0 & & & & x - c_n \end{bmatrix} = (x - c_1)(x - c_2) \cdots (x - c_n)$$

by using the corollary to Theorem 6.11. This shows that $c_1, \ldots, c_n$ exhaust all eigenvalues of $A$, completing the proof of Theorem 7.3.

***Remark.*** For any $n \times n$ matrix $\mathbf{A}$ over a field $\mathfrak{F}$ we may consider the characteristic polynomial $f_{\mathbf{A}}(x)$ as a polynomial over an algebraic closure $\mathfrak{F}'$ of $\mathfrak{F}$. Then $f_{\mathbf{A}}(x)$ has all the roots $c_1, c_2, \ldots, c_n$ in $\mathfrak{F}'$, so that

$$f_{\mathbf{A}}(x) = (x - c_1)(x - c_2) \cdots (x - c_n).$$

These roots are called the *characteristic roots* of $\mathbf{A}$. (Sometimes, a characteristic root is used in the same sense as an eigenvalue; according to our definition, an eigenvalue is a characteristic root that lies in the field $\mathfrak{F}$.)

In this connection, we prove:

***Proposition 7.4.*** *Let $A$ be a linear transformation of an $n$-dimensional vector space $V$ over $\mathfrak{F}$. If $V'$ is an extension of $V$ corresponding to an extension $\mathfrak{F}'$ of $\mathfrak{F}$* (Sec. 5.7), *the extension $A'$ of $A$ to $V'$ has the same characteristic polynomial as $A$.*

*Proof.* By the remark following Proposition 5.22 we know the following. If $\{\alpha_1, \ldots, \alpha_n\}$ is a basis of $\mathfrak{F}$, then $\{\theta(\alpha_1), \ldots, \theta(\alpha_n)\}$ is a basis of $V'$, where $\theta\colon V \to V'$ is the mapping in Definition 5.24; the matrix representing $A'$ with respect to $\{\theta(\alpha_1), \ldots, \theta(\alpha_n)\}$ is the same as the matrix representing $A$ with respect to $\{\alpha_1, \ldots, \alpha_n\}$. Thus $f_{A'}(x) = f_A(x)$.

***Corollary.*** *If* $\mathbf{A}$ *is an* $n \times n$ *matrix over a field* $\mathfrak{F}$, *there exists a nonsingular matrix* $\mathbf{P}$ *over an algebraic closure* $\mathfrak{F}'$ *of* $\mathfrak{F}$ *such that* $\mathbf{PAP}^{-1}$ *is a superdiagonal matrix with all the characteristic roots of* $\mathbf{A}$ *on the diagonal. In particular, this is valid for* $\mathfrak{F} = \mathcal{R}$ *and* $\mathfrak{F}' = \mathcal{C}$.

*Proof.* Consider $\mathbf{A}$ as a linear transformation $A$ of the vector space $\mathfrak{F}'^n$ over $\mathfrak{F}'$ and apply Theorem 7.3. With respect to a new basis, $A$ can be expressed by a matrix of the form $\mathbf{PAP}^{-1}$, where $\mathbf{P}$ is a nonsingular matrix over $\mathfrak{F}'$ (cf. corollary to Proposition 3.12).

**EXERCISE 7.1**

**1.** Find the characteristic polynomials of the following matrices over $\mathcal{R}$:

$$\text{»}(a)\ \begin{bmatrix} 2 & 1 \\ -1 & 3 \end{bmatrix}; \quad (b)\ \begin{bmatrix} 3 & -1 & 1 \\ 2 & 4 & 0 \\ 1 & 0 & 2 \end{bmatrix}; \quad \text{»}(c)\ \begin{bmatrix} 0 & 1 & 3 \\ 0 & 0 & 4 \\ 0 & 0 & -3 \end{bmatrix};$$

$$(d)\ \begin{bmatrix} 0 & 1 & 0 \\ 0 & 0 & 1 \\ 1 & 0 & 0 \end{bmatrix}; \quad \text{»}(e)\ \begin{bmatrix} 0 & & & & 1 \\ & & & 1 & \\ & & 1 & & \\ & 1 & & & \\ 1 & & & & 0 \end{bmatrix}; \quad (f)\ \begin{bmatrix} 1 & 2 & -1 \\ 2 & 0 & 0 \\ -1 & 0 & -1 \end{bmatrix}.$$

**2.** Let $V$ be an $n$-dimensional vector space over $\mathfrak{F}$ with a basis $\{\alpha_1, \ldots, \alpha_n\}$. Let $A$ be a linear transformation of $V$ such that

$$A(\alpha_i) = \alpha_{i+1} \qquad \text{for } 1 \le i \le n-1$$

and

$$A(\alpha_n) = -c_n\alpha_1 - c_{n-1}\alpha_2 - \cdots - c_1\alpha_n,$$

where $c_1, \ldots, c_n \in \mathfrak{F}$.

(*a*) Find the matrix representing $A$ with respect to $\{\alpha_1, \ldots, \alpha_n\}$.

(*b*) Find $f_A(x)$.

»☆ **3.** Prove that if all the characteristic roots of an $n \times n$ real matrix $\mathbf{A}$ are real, then there exists an $n \times n$ nonsingular real matrix $\mathbf{P}$ such that $\mathbf{PAP}^{-1}$ is a superdiagonal matrix with all the characteristic roots of $\mathbf{A}$ on the diagonal.

**4.** For an $n \times n$ matrix $\mathbf{A}$ of the block form

$$\mathbf{A} = \begin{bmatrix} \mathbf{A}_1 & \mathbf{B} \\ \mathbf{O} & \mathbf{A}_2 \end{bmatrix}$$

show that $f_{\mathbf{A}}(x) = f_{\mathbf{A}_1}(x) f_{\mathbf{A}_2}(x)$.

» **5.** Assume that an $n$-dimensional vector space $V$ is the direct sum of subspaces $W_1$ and $W_2$, where $\dim V = n$ and $\dim W_1 = m$. Find the characteristic polynomial of the projection $P$ of $V$ onto $W_1$.

»☆ **6.** For any $n \times n$ matrix $\mathbf{A}$, show that the transpose ${}^t\mathbf{A}$ has the same characteristic polynomial as $\mathbf{A}$.

**7.** For any $n \times n$ matrix $\mathbf{A}$ over $\mathfrak{F}$, let

$$f_{\mathbf{A}}(x) = x^n + s_1x^{n-1} + \cdots + s_n.$$

Prove that:

(*a*) $s_1 = -\text{trace } \mathbf{A}$, where trace $\mathbf{A} = \sum_{i=1}^{n} a_{ii}$ for $\mathbf{A} = [a_{ij}]$.

(*b*) $s_n = (-1)^n \det \mathbf{A}$.

(*c*) trace $\mathbf{A}^k = c_1^k \ . \ . \ . \ c_n^k$, where $c_1, \ldots, c_n$ are the characteristic roots of $\mathbf{A}$ and $k$ is an arbitrary positive integer.

☆ **8.** Let $\mathbf{A}$ be an $n \times n$ real matrix, where $n$ is odd. Show that $\mathbf{A}$ has at least one real characteristic root. Using this, show that any linear transformation $A$ of an odd-dimensional real vector space has at least one eigenvalue.

»☆ **9.** Let $\mathbf{A}$ be an $n \times n$ real matrix, where $n$ is even.

(*a*) Give an example of such an $\mathbf{A}$ which has no real characteristic root.

(*b*) Prove that if $\det \mathbf{A} < 0$, then $\mathbf{A}$ has at least two distinct real characteristic roots (cf. number 7, part *b*).

**10.** Let $c$ be an eigenvalue of a linear transformation $A$ of a vector space $V$ over $\mathfrak{F}$. Show that, for any polynomial $g \in \mathfrak{F}[x]$, $g(c)$ is an eigenvalue for the transformation $g(A)$. [If $g(x) = \sum_{i=0}^{n} a_ix^i$, then $g(A)$ is defined to be the linear transformation $\sum_{i=0}^{n} a_iA^i$. For the details, see the beginning of the next section.]

»☆ **11.** Let $\mathbf{A}$ and $\mathbf{B}$ be $n \times n$ real matrices. Prove that if $\mathbf{A}$ is nonsingular, then $\mathbf{AB}$ and $\mathbf{BA}$ have the same characteristic polynomial. (As a matter of fact, this result holds without assuming that $\mathbf{A}$ is nonsingular; observe that the coefficients of $f_{\mathbf{A}}$ depend continuously on the components of $\mathbf{A}$ and that a singular matrix can be approximated by nonsingular matrices.)

## 7.2 MINIMAL POLYNOMIALS

Let $V$ be an $n$-dimensional vector space over a field $\mathcal{F}$ and let $A$ be a linear transformation of $V$. For any polynomial

$$g(x) = a_0 + a_1x + \cdots + a_nx^n \qquad \text{in } \mathcal{F}[x],$$

we consider the linear transformation

$$a_0I + a_1A + \cdots + a_nA^n, \qquad I\text{: identity transformation,}$$

which we denote by $g(A)$. We have

$$(g + h)(A) = g(A) + h(A), \qquad (gh)(A) = g(A)h(A)$$

for any two polynomials $g$ and $h$ in $\mathcal{F}[x]$. In other words, the mapping

$$g(x) \in \mathcal{F}[x] \rightarrow g(A) \in \text{Hom}(V,V)$$

is a homomorphism of the ring $\mathcal{F}[x]$ into the ring $\text{Hom}(V,V)$ (recall Definition 5.12). We observe that all linear transformations of the form $g(A)$, $g \in \mathcal{F}[x]$, commute with each other (in particular, with $A$).

Let $K$ be the kernel of this homomorphism. $K$ is an ideal of $\mathcal{F}[x]$. By Theorem 5.5 there exists a unique monic polynomial $\phi(x) \in \mathcal{F}[x]$ such that $K = (\phi)$, the principal ideal generated by $\phi$; $\phi(x)$ is the unique monic polynomial of the smallest degree such that $\phi(A) = 0$.

### Definition 7.4

$\phi(x)$ is called the *minimal polynomial* of $A$ and is denoted by $\phi_A(x)$.

***Example 7.6.*** If $V = \mathfrak{F}^n$, we have $\text{Hom}(V,V) = \mathfrak{F}^n{}_n$. The minimal polynomial of a matrix $\mathbf{A} \in \mathfrak{F}^n{}_n$ is a unique monic polynomial of the smallest degree such that $\phi(\mathbf{A}) = 0$.

***Example 7.7.*** Let

$$\mathbf{A} = \begin{bmatrix} 2 & 1 \\ 0 & 1 \end{bmatrix} \quad \text{over } \mathfrak{R}.$$

For $\phi(x) = x^2 - 3x + 2 = (x - 2)(x - 1)$, we have

$$\phi(\mathbf{A}) = \begin{bmatrix} 2 & 1 \\ 0 & 1 \end{bmatrix}^2 - 3\begin{bmatrix} 2 & 1 \\ 0 & 1 \end{bmatrix} + 2\begin{bmatrix} 1 & 0 \\ 0 & 1 \end{bmatrix} = \begin{bmatrix} 0 & 0 \\ 0 & 0 \end{bmatrix}.$$

On the other hand, it is obvious that, for any polynomial $g$ of degree 1, $g(\mathbf{A})$ is not the zero matrix. We conclude that $\phi(x)$ is the minimal polynomial of $\mathbf{A}$.

***Example 7.8.*** Over any field $\mathfrak{F}$, consider

$$\mathbf{A} = \begin{bmatrix} a & 0 \\ 0 & a \end{bmatrix} \quad \text{and} \quad \mathbf{B} = \begin{bmatrix} a & b \\ 0 & a \end{bmatrix}, \qquad \text{where } b \neq 0.$$

Then the characteristic polynomials of $\mathbf{A}$ and $\mathbf{B}$ coincide:

$$f_{\mathbf{A}}(x) = f_{\mathbf{B}}(x) = (x - a)^2,$$

while the minimal polynomials are different:

$$\phi_{\mathbf{A}}(x) = x - a, \qquad \phi_{\mathbf{B}}(x) = (x - a)^2.$$

We shall now prove the following basic result.

***Theorem 7.5.*** *Let $A$ be a linear transformation of an $n$-dimensional vector space $V$ over a field $\mathfrak{F}$. Assume that the minimal polynomial $\phi_A(x)$ of $A$ is a product of two polynomials $\phi_1(x)$ and $\phi_2(x)$ which are relatively prime over $\mathfrak{F}$. Set*

$$V_1 = \{\alpha \in V;\ \phi_1(A)(\alpha) = 0\}, \qquad V_2 = \{\alpha \in V;\ \phi_2(A)(\alpha) = 0\}.$$

*Then*

**1.** $V = V_1 \oplus V_2$, *and* $A(V_1) \subseteq V_1$, $A(V_2) \subseteq V_2$.

**2.** $V_1 = \{\phi_2(A)(\alpha);\ \alpha \in V\}$, $V_2 = \{\phi_1(A)(\alpha);\ \alpha \in V\}$.

**3.** *The restriction $A_1$ [$A_2$] of $A$ to $V_1$ [$V_2$] has $\phi_1(x)$ [$\phi_2(x)$] as its minimal polynomial.*

For the proof, we first prove:

***Lemma.*** *Assume that $\phi_1(x)$ and $\phi_2(x)$ are relatively prime in $\mathfrak{F}[x]$. Then*

**1.** *For any $\alpha \in V$, there exist $\beta, \gamma \in V$ such that*

$$\alpha = \phi_1(A)(\beta) + \phi_2(A)(\gamma).$$

**2.** *If* $\phi_1(A)(\alpha) = \phi_2(A)(\alpha) = 0$, *then* $\alpha = 0$.

*Proof.* By the corollary to Theorem 5.6, there exist $g(x)$ and $h(x)$ in $\mathcal{F}[x]$ such that

$$\phi_1(x)g(x) + \phi_2(x)h(x) = 1.$$

For any $\alpha \in V$, let $\beta = g(A)(\alpha)$ and $\gamma = h(A)(\alpha)$. Then

$$\phi_1(A)(\beta) + \phi_2(A)(\gamma) = \alpha,$$

proving (1). Now assume that $\phi_1(A)(\alpha) = \phi_2(A)(\alpha) = 0$. Then

$$\alpha = g(A)\phi_1(A)(\alpha) + h(A)\phi_2(A)(\alpha) = 0,$$

proving (2).

*Proof of the Theorem.* 1. We first observe that $V_1$ and $V_2$ are subspaces of $V$, because they are the null spaces of $\phi_1(A)$ and $\phi_2(A)$, respectively. Since $A\phi_1(A) = \phi_1(A)A$, we see that if $\phi_1(A)(\alpha) = 0$, then

$$\phi_1(A)(A(\alpha)) = A(\phi_1(A)(\alpha)) = 0,$$

that is, $\alpha \in V_1$ implies $A(\alpha) \in V_1$. Thus $A(V_1) \subseteq V_1$. Similarly, $A(V_2) \subseteq V_2$. By (2) of the lemma above, we have $V_1 \cap V_2 = (0)$. For any $\alpha \in V$ we have

$$\alpha = \phi_1(A)(\beta) + \phi_2(A)(\gamma)$$

for some $\beta, \gamma \in V$ by (1) of the preceding lemma. Since $\phi_1(x)\phi_2(x)$ is the minimal polynomial of $A$, we have

$$\phi_1(A)\phi_2(A)(\gamma) = \phi_A(A)(\gamma) = 0,$$

so that

$$\phi_2(A)(\gamma) \in V_1.$$

Similarly, we have

$$\phi_1(A)(\beta) \in V_2.$$

Thus $V = V_1 + V_2$, concluding the proof of $V = V_1 \oplus V_2$.

2. Using $g$ and $h$ in the proof of the lemma, we see for any $\alpha \in V_1$ that

$$\alpha = g(A)\phi_1(A)(\alpha) + h(A)\phi_2(A)(\alpha) = \phi_2(A)(h(A)(\alpha)),$$

which shows that $\alpha$ is in the range of $\phi_2(A)$. Conversely, if $\alpha \in V$, then $\phi_1(A)\phi_2(A)(\alpha) = \phi_A(A)(\alpha) = 0$, which shows $\phi_2(A)(\alpha) \in V_1$. Thus $V_1$ is equal to the range of $\phi_2(A)$. The case of $V_2$ is similar.

3. Let $\psi_i(x)$ be the minimal polynomial of $A_i$, $i = 1, 2$. Since $\psi_1(A)(\alpha) = 0$ for every $\alpha \in V_1$, we have

$$\psi_2(A)\psi_1(A)(\alpha) = 0 \qquad \text{for every } \alpha \in V_1.$$

Similarly,

$$\psi_1(A)\psi_2(A)(\alpha) = 0 \qquad \text{for every } \alpha \in V_2.$$

This means that $\psi_1(A)\psi_2(A) = 0$, and hence $\psi_1(x)\psi_2(x)$ is divisible by $\phi_A(x) = \phi_1(x)\phi_2(x)$. On the other hand, it is clear that $\psi_1$ divides $\phi_1$ and $\psi_2$ divides $\phi_2$. Considering the degrees of these polynomials, we have $\psi_1 = \phi_1$ and $\psi_2 = \phi_2$.

By a repeated application of Theorem 7.5, we have:

***Corollary.*** *Assume that the minimal polynomial $\phi_A(x)$ of $A$ is a product $\phi_1^{p_1}\phi_2^{p_2} \cdots \phi_k^{p_k}$, where $\phi_1, \ldots, \phi_k$ are distinct irreducible polynomials in $\mathfrak{F}[x]$. Then*

$$V = V_1 \oplus V_2 \oplus \cdots \oplus V_k,$$

*where*

$$V_i = \{\alpha \in V;\ (\phi_i(A))^{p_i}(\alpha) = 0\}, \qquad A(V_i) \subseteq V_i, \qquad 1 \leq i \leq k,$$

*and the minimal polynomial of the restriction of $A$ to $V_i$ is equal to $\phi_i^{p_i}$ for each $i$.*

In the case where $\mathfrak{F}$ is algebraically closed (in particular, if $\mathfrak{F} = \mathbb{C}$), every polynomial $\phi$ is of the form

$$\phi(x) = (x - c_1)^{p_1} \cdots (x - c_k)^{p_k},$$

where $c_1, \ldots, c_k$ are distinct elements in $\mathfrak{F}$ (corollary to Theorem 5.10 and Definition 5.20). We obtain:

***Theorem 7.6.*** *Let $A$ be a linear transformation of an $n$-dimensional vector space over an algebraically closed field $\mathfrak{F}$ (in particular, $\mathfrak{F} = \mathbb{C}$). Let*

$$\phi_A(x) = (x - c_1)^{p_1} \cdots (x - c_k)^{p_k}$$

*be the factorization of the minimal polynomial of $A$ into a product of powers of distinct linear factors. Then*

$$V = V_1 \oplus \cdots \oplus V_k,$$

*where*

$$V_i = \{\alpha \in V;\ (A - c_iI)^{p_i}(\alpha) = 0\}, \qquad A(V_i) \subseteq V_i,$$

*and the minimal polynomial of the restriction of $A$ to $V_i$ is $(x - c_i)^{p_i}$ for each $i$, $1 \leq i \leq k$. Moreover, $c_1, \ldots, c_k$ exhaust the eigenvalues of $A$ and* dim $V_i$ *is equal to the multiplicity of $c_i$ in the characteristic polynomial $f_A(x)$, that is,*

$$f_A(x) = (x - c_1)^{m_1} \cdots (x - c_k)^{m_k}, \qquad \textit{where } m_i = \dim V_i.$$

*Proof.* We have only to prove the assertion on the eigenvalues and their multiplicities. Suppose $c$ is an eigenvalue of $A$ which is different from

$c_1, \ldots, c_k$ and let $\alpha \neq 0$ be an eigenvector: $A(\alpha) = c\alpha$. We may write $\alpha$ in the form

$$\alpha = \alpha_1 + \cdots + \alpha_k, \qquad \text{where } \alpha_i \in V_i, \qquad 1 \leq i \leq k.$$

For the polynomial $g(x) = x - c$ we have $g(A)\alpha = 0$, and hence

$$0 = g(A)\alpha = \sum_{i=1}^{k} g(A)(\alpha_i), \qquad \text{where } g(A)(\alpha_i) \in V_i,$$

because $A(V_i) \subseteq V_i$ and $g(A)(V_i) \subseteq V_i$ for each $i$. We see that $g(A)(\alpha_i) = 0$ for each $i$. Since $c$ is different from $c_i$, $g(x)$ and $(x - c_i)^{p_i}$ are relatively prime. By (2) of the lemma for Theorem 7.5, we see that $\alpha_i = 0$ for each $i$, that is, $\alpha = 0$, which is a contradiction. This shows that there are no eigenvalues other than $c_1, \ldots, c_k$.

In order to show that the multiplicity of $c_i$ in $f_A(x)$ is equal to $m_i = \dim V_i$, we first observe that $f_A(x)$ is the product of the characteristic polynomials $f_i(x)$ of the restriction $A_i$ of $A$ to $V_i$. This can be shown as follows. If we take a basis of $V$ made up of bases of $V_i$'s, then the matrix representing $A$ is of the block form

$$\mathbf{A} = \begin{bmatrix} \mathbf{A}_1 & & & & \\ & \cdot & & & \\ & & \cdot & & \\ & & & \cdot & \\ & & & & \mathbf{A}_k \end{bmatrix},$$

where each $\mathbf{A}_i$ is the matrix representing $A_i$. Thus

$$f_{\mathbf{A}}(x) = f_{\mathbf{A}_1}(x) \cdots f_{\mathbf{A}_k}(x),$$

and hence our assertion.

We shall now show that $f_i(x) = (x - c_i)^{n_i}$. By applying the preceding argument to $A_i$ on $V_i$, we see that $A_i$ has no eigenvalue other than $c_i$. Since $\mathcal{F}$ is algebraically closed, $f_i$ is a power of $(x - c_i)$—the exponent being $m_i$ since $f_i$ is of degree $m_i$. Thus we have $f_i(x) = (x - c_i)^{m_i}$, concluding our proof.

One consequence of Theorem 7.6 is that $A$ is the direct sum of linear transformations $A_i$ (cf. Example 4.5), where $(A_i - c_iI)^{p_i} = 0$. In other words, $A_i = c_iI + B_i$, where $B_i^{p_i} = 0$. We say that $B_i$ is *nilpotent*. Thus the study of an arbitrary linear transformation of a vector space over an algebraically closed field is reduced to that of nilpotent transformations.

Another consequence of Theorem 7.6 is that $\phi_A(x)$ divides $f_A(x)$ and that they have the same roots in $\mathcal{F}$. It also follows that there is a positive integer $d \leq n$ such that $f_A(x)$ divides $(\phi_A(x))^d$. These facts will be valid even when $\mathcal{F}$ is not algebraically closed. Namely, we have:

***Theorem 7.7.*** *For any linear transformation $A$ of an $n$-dimensional vector space $V$ over a field $\mathfrak{F}$, the minimal polynomial $\phi_A$ divides the characteristic polynomial $f_A$ and there exists a positive integer $d \leq n$ such that $f_A$ divides $\phi_A{}^d$.*

***Corollary* (*Cayley-Hamilton*).** $f_A(A) = 0$.

*Proof.* Let $\mathfrak{F}'$ be the algebraic closure of $\mathfrak{F}$ and let $V'$ be the corresponding extension of $V$. For the extension $A'$ of $A$ to $V'$ we know that $f_{A'}$ and $\phi_{A'}$ have the properties stated in Theorem 7.7. We also know by Proposition 7.4 that $f_A = f_{A'}$. Hence $f_A(A') = 0$. Since $A$ is the restriction of $A'$ to $V$, we have $f_A(A) = 0$. Thus $\phi_A$ divides $f_A$. On the other hand, since $\phi_A(A) = 0$, it follows that $\phi_A(A') = 0$. In fact, if $\{\alpha_1, \ldots, \alpha_n\}$ is a basis of $V$ and $\theta$ is the extension mapping of $V$ into $V'$, then

$$\{\theta(\alpha_1), \ldots, \theta(\alpha_n)\}$$

is a basis of $V'$ and we have

$$A'(\theta(\alpha_i)) = \theta(A(\alpha_i))$$

so that

$$\phi_A(A')(\theta(\alpha_i)) = \theta(\phi_A(A)(\alpha_i)) = 0,$$

which implies that $\phi_A(A') = 0$. This means that $\phi_{A'}$ divides $\phi_A$ (over $\mathfrak{F}'$). Now if $f_{A'}$ divides $\phi_{A'}{}^d$, it follows that $f_A$ divides $\phi_A{}^d$ (over $\mathfrak{F}$).

***Remark.*** As a matter of fact, Theorem 7.7 implies $\phi_A = \phi_{A'}$ (a fact which is not obvious). We can, however, prove this fact directly and use it for the proof of Theorem 7.7. For another proof of the corollary, see Exercise 7.2, number 12.

Applying Theorem 7.6, we shall now prove:

***Theorem 7.8.*** *Let $V$ be an $n$-dimensional vector space over a field $\mathfrak{F}$. A linear transformation $A$ of $V$ can be represented by a diagonal matrix with respect to a suitable basis of $V$ if and only if $\phi_A(x)$ is a product of distinct linear factors in $\mathfrak{F}$, that is,*

$$\phi_A(x) = (x - c_1) \cdots (x - c_k),$$

*where $c_1, \ldots, c_k$ are all distinct elements in $\mathfrak{F}$.*

*Proof.* Assume first that there exists a basis of $V$ with respect to which $A$ can be represented by a diagonal matrix. By rearranging the order of basis elements if necessary, we may assume that the diagonal matrix $\mathbf{A}$ is of the form

$$\mathbf{A} = \begin{bmatrix} c_1\mathbf{I}_{m_1} & & & \\ & c_2\mathbf{I}_{m_2} & & \\ & & \ddots & \\ & & & c_k\mathbf{I}_{m_k} \end{bmatrix},$$

where $c_1, \ldots, c_k$ are distinct, $\mathbf{I}_{m_i}$ denotes the $m_i \times m_i$ identity matrix, and $m_1 + \cdots + m_k = n$. Then we have

$$c_1\mathbf{I}_n - \mathbf{A} = \begin{bmatrix} \mathbf{O}_{m_1} & & & & \\ & (c_1 - c_2)\mathbf{I}_{m_2} & & & \\ & & \cdot & & \\ & & & \cdot & \\ & & & & \cdot \\ & & & & (c_1 - c_k)\mathbf{I}_{m_k} \end{bmatrix},$$

where $\mathbf{O}_{m_1}$ is the zero matrix of degree $m_1$, and similarly for $c_i\mathbf{I}_n - \mathbf{A}$, $2 \leq i \leq k$. It follows that

$$(c_1\mathbf{I}_n - \mathbf{A})\,(c_2\mathbf{I}_n - \mathbf{A}) \cdots (c_k\mathbf{I}_n - \mathbf{A}) = \mathbf{O}.$$

It is also clear that no proper factor of the left-hand side is equal to the zero matrix. This shows that

$$\phi_{\mathbf{A}}(x) = (x - c_1)\,(x - c_2) \cdots (x - c_k),$$

where $c_1, \ldots, c_k$ are distinct. Conversely, assume that $\phi_A(x)$ is of this form. By Theorem 7.6, we have

$$V = V_1 \oplus \cdots \oplus V_k,$$

where $\qquad A(\alpha) = c_i\alpha \qquad$ for every $\alpha \in V_i, \qquad 1 \leq i \leq k.$

If we take a basis of $V$ made up by bases of $V_1, \ldots, V_k$, then $A$ can be represented by the diagonal matrix

$$\begin{bmatrix} c_1\mathbf{I}_{m_1} & & & \\ & \cdot & & \\ & & \cdot & \\ & & & \cdot \\ & & & c_k\mathbf{I}_{m_k} \end{bmatrix},$$

where $m_i = \dim V_i$, $1 \leq i \leq k$, concluding the proof.

***Corollary.*** *Let* $\mathbf{A}$ *be an* $n \times n$ *matrix over a field* $\mathfrak{F}$. *Then there exists a nonsingular* $n \times n$ *matrix* $\mathbf{P}$ *such that* $\mathbf{PAP}^{-1}$ *is a diagonal matrix if and only if the minimal polynomial* $\phi_{\mathbf{A}}$ *is a product of distinct linear factors in* $\mathfrak{F}[x]$.

***Remark.*** If a polynomial $g(x)$ over a field $\mathfrak{F}$ is a product of distinct linear factors in $\mathfrak{F}$:

$$g(x) = (x - c_1) \cdots (x - c_k), \qquad c_1, \ldots, c_k \in \mathfrak{F},$$

we say that $g$ has *simple roots* in $\mathfrak{F}$.

***Example 7.9.*** Over $\mathfrak{R}$, consider

$$\mathbf{A} = \begin{bmatrix} 5 & -6 & -6 \\ -1 & 4 & 2 \\ 3 & -6 & -4 \end{bmatrix}.$$

By computation we find

$$f_{\mathbf{A}}(x) = (x - 1)(x - 2)^2.$$

By Theorem 7.7 we know that $\phi_{\mathbf{A}}(x)$ is either $(x - 1)(x - 2)$ or $f_{\mathbf{A}}(x)$. Since we can verify that

$$(\mathbf{A} - \mathbf{I}_3)(\mathbf{A} - 2\mathbf{I}_3) = 0,$$

we have

$$\phi_{\mathbf{A}}(x) = (x - 1)(x - 2).$$

By Theorem 7.8, we see that there exists a nonsingular matrix $\mathbf{P}$ such that

$$\mathbf{P}^{-1}\mathbf{A}\mathbf{P} = \begin{bmatrix} 1 & 0 & 0 \\ 0 & 2 & 0 \\ 0 & 0 & 2 \end{bmatrix}.$$

In order to find $\mathbf{P}$, let

$$V_1 = \{\alpha \in \mathcal{R}^3;\ (\mathbf{A} - \mathbf{I})(\alpha) = 0\} = \{(\mathbf{A} - 2\mathbf{I})(\alpha);\ \alpha \in \mathcal{R}^3\},$$

$$V_2 = \{\alpha \in \mathcal{R}^3;\ (\mathbf{A} - 2\mathbf{I})(\alpha) = 0\} = \{(\mathbf{A} - \mathbf{I})(\alpha);\ \alpha \in \mathcal{R}^3\}.$$

Thus $V_1$ is the range of

$$\mathbf{A} - 2\mathbf{I} = \begin{bmatrix} 3 & -6 & -6 \\ -1 & 2 & 2 \\ 3 & -6 & -6 \end{bmatrix}$$

and $V_2$ is the range of

$$\mathbf{A} - \mathbf{I} = \begin{bmatrix} 4 & -6 & -6 \\ -1 & 3 & 2 \\ 3 & -6 & -5 \end{bmatrix}.$$

Apparently, $V_1$ is spanned by

$$\alpha_1 = \begin{bmatrix} 3 \\ -1 \\ 3 \end{bmatrix}$$

and $V_2$ is spanned by

$$\alpha_2 = \begin{bmatrix} 4 \\ -1 \\ 3 \end{bmatrix} \quad \text{and} \quad \alpha_3 = \begin{bmatrix} -6 \\ 3 \\ -6 \end{bmatrix}.$$

(Recall that the range of the mapping defined by a matrix is nothing but the column space.) We know that $A$ can be represented by the diagonal matrix (with 1, 2, 2 on the diagonal) with respect to $\{\alpha_1,\alpha_2,\alpha_3\}$. Thus if we take

$$\mathbf{P} = \begin{bmatrix} 3 & 4 & -6 \\ -1 & -1 & 3 \\ 3 & 3 & -6 \end{bmatrix},$$

then

$$\mathbf{P}^{-1}\mathbf{A}\mathbf{P} = \begin{bmatrix} 1 & 0 & 0 \\ 0 & 2 & 0 \\ 0 & 0 & 2 \end{bmatrix}.$$

It goes without saying that such $\mathbf{P}$ is not unique.

***Example 7.10.*** Let $\mathbf{A}$ be a $2 \times 2$ real symmetric matrix. We show that $\mathbf{A}$ can be transformed into a diagonal matrix, that is, there exists a nonsingular matrix $\mathbf{P}$ such that $\mathbf{P}\mathbf{A}\mathbf{P}^{-1}$ is diagonal. Let

$$\mathbf{A} = \begin{bmatrix} a & b \\ b & c \end{bmatrix}.$$

We have

$$f_{\mathbf{A}}(x) = \begin{vmatrix} x-a & -b \\ -b & x-c \end{vmatrix} = x^2 - (a+c)x + (ac - b^2).$$

Since $(a+c)^2 - 4(ac - b^2) = (a-c)^2 + 4b^2 \geq 0$, we have the following two cases.

1. $f_{\mathbf{A}}$ has two distinct real roots, say, $t$, $s$. Then

$$f_{\mathbf{A}}(x) = (x-t)\,(x-s)$$

so that

$$\phi_{\mathbf{A}}(x) = (x-t)\,(x-s).$$

Thus $\mathbf{A}$ can be transformed into $\begin{bmatrix} t & 0 \\ 0 & s \end{bmatrix}$.

2. $f_{\mathbf{A}}$ has a double root, say, $t$. In this case, $(a-c)^2 + 4b^2 = 0$, that is, $a = c$ and $b = 0$. The given matrix $\mathbf{A}$ itself is diagonal.

## EXERCISE 7.2

**1.** Find the minimal polynomial of each of the following matrices:

»(*a*) $\begin{bmatrix} 2 & -1 \\ 1 & 0 \end{bmatrix}$; (*b*) $\begin{bmatrix} 1 & 0 & 1 \\ 0 & 1 & 0 \\ 1 & 0 & 1 \end{bmatrix}$;

»(*c*) $\begin{bmatrix} 1 & 1 & 1 \\ 0 & 1 & 1 \\ 0 & 0 & 1 \end{bmatrix}$; (*d*) $\begin{bmatrix} a & b & c \\ 0 & a & b \\ 0 & 0 & a \end{bmatrix}$.

**2.** For each of the following matrices, find a nonsingular (complex) matrix $\mathbf{P}$ such that $\mathbf{PAP}^{-1}$ is diagonal, if any:

»(*a*) $\begin{bmatrix} 1 & 1 \\ 0 & 1 \end{bmatrix}$; (*b*) $\begin{bmatrix} 2 & 1 \\ 0 & 1 \end{bmatrix}$; »(*c*) $\begin{bmatrix} 0 & -1 \\ 1 & 0 \end{bmatrix}$;

(*d*) $\begin{bmatrix} 1 & 1 & -1 \\ 0 & 2 & 2 \\ 0 & 0 & -1 \end{bmatrix}$; (*e*) $\begin{bmatrix} 2 & 0 & 0 \\ 0 & 1 & 2 \\ 0 & 2 & 0 \end{bmatrix}$.

»☆ **3.** Prove that any $n \times n$ matrix $\mathbf{A}$ and its transpose ${}^t\mathbf{A}$ have the same minimal polynomial.

»☆ **4.** Do $\mathbf{AB}$ and $\mathbf{BA}$ have the same minimal polynomial for all $n \times n$ matrices $\mathbf{A}$ and $\mathbf{B}$? What about the case where one of them is nonsingular? (Compare with Exercise 7.1, number 11.)

☆ **5.** Let $A$ be a linear transformation of an $n$-dimensional vector space $V$ over $\mathfrak{F}$. For any $\alpha \in V$ let

$$W(\alpha) = \{g(A)(\alpha);\, g \in \mathfrak{F}[x]\}.$$

(*a*) Show that $W(\alpha)$ is a subspace invariant by $A$.

(*b*) Let $\phi(x;\alpha)$ be the minimal polynomial of the restriction of $A$ to $W(\alpha)$. [We call $\phi(x;\alpha)$ the *minimal polynomial for* $\alpha$ (*with respect to* $A$).] Prove that, for each $\alpha \in V$, $\phi(x;\alpha)$ divides $\phi_A(x)$.

(*c*) If a polynomial $\psi(x) \in \mathfrak{F}[x]$ is divisible by $\phi(x;\alpha)$ for every $\alpha \in V$, then $\psi(x)$ is divisible by $\phi_A(x)$.

(*d*) Prove that there is an $\alpha \in V$ such that $\phi(x;\alpha) = \phi_A(x)$.

☆ **6.** In the notation of Theorem 7.5, prove that if $B$ is a linear transformation of $V$ such that $AB = BA$, then $B(V_i) \subseteq V_i$ for each $i = 1, 2$. State and prove the analogous result for the corollary to Theorem 7.5.

»☆ **7.** Let $A$ and $B$ be linear transformations of a finite-dimensional vector space $V$ over $\mathfrak{F}$. If $AB = BA$ and if $\phi_A$ and $\phi_B$ have simple roots in $\mathfrak{F}$, prove that there exists a basis $\Phi = \{\alpha_1, \ldots, \alpha_n\}$ such that both $A_\Phi$ and $B_\Phi$ are diagonal.

☆ **8.** Let $A$ be a linear transformation of a finite-dimensional vector space $V$ over $\mathfrak{F}$. Assume that $f_A(x)$ has all its roots in $\mathfrak{F}$. Prove that there exist a linear transformation $B$ which is representable by a diagonal matrix (with respect to a certain basis) and a nilpotent linear transformation $N$ such that $A = B + N$ and $BN = NB$. Prove that such $B$ and $N$ are uniquely determined by $A$ and that there exist polynomials $g, h \in \mathfrak{F}[x]$ such that $B = g(A)$ and $N = h(A)$.

**9.** In the notation of Theorem 7.5, prove that for any subspace $U$ of $V$ such that $A(U) \subseteq U$, we have

$$U = U_1 \oplus U_2, \qquad \text{where } U_i = U \cap V_i \text{ for } i = 1, 2.$$

State the analogous result for the corollary to Theorem 7.5.

☆ **10.** Let $A$ be a linear transformation of an $n$-dimensional vector space $V$ over an algebraically closed field $\mathfrak{F}$. Prove that the following two conditions are equivalent:

(*a*) $A$ can be represented by a diagonal matrix (with respect to a suitable basis).

(*b*) For any subspace $U$ of $V$ such that $A(U) \subseteq U$ there exists a subspace $W$ of $V$ such that

$$V = U \oplus W \qquad \text{and} \qquad A(W) \subseteq W.$$

(When a linear transformation $A$ satisfies condition $b$, $A$ is said to be *semisimple.*)

☆ **11.** (Generalization of 10) Let $A$ be a linear transformation of an $n$-dimensional vector space $V$ over a field $\mathfrak{F}$ which is *perfect.* (A field $\mathfrak{F}$ is said to be *perfect* if any irreducible polynomial $f(x)$ over $\mathfrak{F}$ has simple roots in the algebraic closure $\mathfrak{F}'$ of $\mathfrak{F}$.) Prove that the following two conditions are equivalent:

(*a*) $A$ is semisimple.

(*b*) The extension $A'$ of $A$ to the extension $V'$ corresponding to the algebraic closure $\mathfrak{F}'$ of $\mathfrak{F}$ can be represented by a diagonal matrix (with respect to a suitable basis in $V'$).

☆ **12.** Let $\mathbf{A}$ be an $n \times n$ matrix over a field $\mathfrak{F}$. Prove the Cayley-Hamilton theorem [that is, $f_{\mathbf{A}}(\mathbf{A}) = 0$] along the following lines:

(*a*) Show that the $(i,j)$ cofactor $\Delta_{ij}$ of the matrix $x\mathbf{I}_n - \mathbf{A}$ over $\mathfrak{F}[x]$ is a polynomial of degree $\leq n - 1$ and set

$$\Delta_{ij} = b_{ij,0}x^{n-1} + b_{ij,1}x^{n-2} + \cdots + b_{ij,n-1}.$$

(*b*) If $\mathbf{B}_k$ is the transpose of the matrix $[b_{ij,k}]$, then we have, by Theorem 6.13 applied to the matrix $x\mathbf{I}_n - \mathbf{A}$ over $\mathfrak{F}[x]$,

$$\begin{aligned} f_{\mathbf{A}}(x)\mathbf{I}_n &= \det(x\mathbf{I}_n - \mathbf{A})\mathbf{I}_n \\ &= (x\mathbf{I}_n - \mathbf{A})\,(x^{n-1}\mathbf{B}_0 + x^{n-2}\mathbf{B}_1 + \cdots + \mathbf{B}_{n-1}) \\ &= (x^{n-1}\mathbf{B}_0 + x^{n-2}\mathbf{B}_1 + \cdots + \mathbf{B}_{n-1})\,(x\mathbf{I}_n - \mathbf{A}). \end{aligned}$$

From this it follows that $\mathbf{B}_i\mathbf{A} = \mathbf{A}\mathbf{B}_i$ for $0 \leq i \leq n - 1$.

(*c*) We may substitute $\mathbf{A}$ for $x$ in the preceding equation (why?) and obtain

$$f_{\mathbf{A}}(\mathbf{A})\mathbf{I}_n = (\mathbf{A} - \mathbf{A})\,(\mathbf{A}^{n-1}\mathbf{B}_0 + \mathbf{A}^{n-2}\mathbf{B}_1 + \cdots + \mathbf{B}_{n-1}) = \mathbf{O}$$

that is, $f_{\mathbf{A}}(\mathbf{A}) = \mathbf{O}$.

## 7.3 NILPOTENT TRANSFORMATIONS

We begin with:

**Definition 7.5**

A linear transformation $A$ of a vector space $V$ over $\mathcal{F}$ is said to be *nilpotent* if its minimal polynomial is $x^k$, where $k$ is a certain positive integer which will be called the *index* of $A$. This definition also applies to a matrix.

In other words, $A$ is nilpotent if some power of $A$ is 0. If $A$ has index $k$, then $A^k = 0$ but $A^{k-1} \neq 0$, that is, there exists an $\alpha \in V$ such that $A^{k-1}(\alpha) \neq 0$. We have:

***Proposition 7.9.*** *If $A$ is a nilpotent transformation of index $k$ of an $n$-dimensional vector space $V$, then*

$$f_A(x) = x^n \qquad \textit{and} \qquad k \leq n.$$

*Proof.* By Theorem 7.7 we know that $\phi_A = x^k$ divides $f_A$ and $f_A$ divides some power of $\phi_A$. Hence, $f_A$ is a power of $x$. Since its degree is $n$, we have $f_A(x) = x^n$. Since $x^k$ divides $x^n$, we must have $k \leq n$.

In Theorem 7.6 we saw the importance of studying nilpotent transformations. We shall now proceed to the study of arbitrary nilpotent transformations.

The following definition is convenient for our purpose.

**Definition 7.6**

Let $W$ be a subspace of a vector space $V$. We shall say that $\alpha_1, \ldots, \alpha_k$ in $V$ are *linearly independent modulo* $W$ if $\pi(\alpha_1), \ldots, \pi(\alpha_k)$ are linearly independent in the quotient space $V/W$, where $\pi: V \to V/W$ is the projection.

$\alpha_1, \ldots, \alpha_k$ are linearly independent modulo $W$ if and only if $\sum_{i=1}^{k} c_i\alpha_i \in W$ implies $c_1 = \cdots = c_k = 0$.

We establish:

***Lemma 1.*** *Let $A$ be a nilpotent transformation of index $k$ of an $n$-dimensional vector space $V$. Let*

$$W = \{\alpha \in V;\, A^{k-1}(\alpha) = 0\},$$

*and let* $\alpha_1, \ldots, \alpha_m \in V$ *be linearly independent modulo* $W$. *Then the vectors*

$$(*) \qquad \begin{array}{l} \alpha_1, A(\alpha_1), \ldots, A^{k-1}(\alpha_1), \\ \alpha_2, A(\alpha_2) \ldots, A^{k-1}(\alpha_2), \\ \ldots\ldots\ldots\ldots\ldots, \\ \alpha_m, A(\alpha_m), \ldots, A^{k-1}(\alpha_m) \end{array}$$

*are linearly independent.*

*Proof.* Suppose we have

$$(**) \qquad \begin{array}{l} c_{10}\alpha_1 + c_{11}A(\alpha_1) + \cdots + c_{1,k-1}A^{k-1}(\alpha_1) \\ +c_{20}\alpha_2 + c_{21}A(\alpha_2) + \cdots + c_{2,k-1}A^{k-1}(\alpha_2) \\ + \cdots\cdots\cdots\cdots\cdots\cdots \\ +c_{m0}\alpha_m + c_{m1}A(\alpha_m) + \cdots + c_{m,k-1}A^{k-1}(\alpha_m) = 0. \end{array}$$

By applying $A^{k-1}$, we obtain

$$A^{k-1}(c_{10}\alpha_1 + c_{20}\alpha_2 + \cdots + c_{m0}\alpha_m) = 0,$$

so that $c_{10}\alpha_1 + c_{20}\alpha_2 + \cdots + c_{m0}\alpha_m \in W$. Since $\alpha_1, \ldots, \alpha_m$ are linearly independent modulo $W$ by assumption, we conclude that $c_{10} = c_{20} = \cdots = c_{m0} = 0$. After eliminating the terms with these zero coefficients in (**) and applying $A^{k-2}$, we obtain

$$A^{k-1}(c_{11}\alpha_1 + \cdots + c_{m1}\alpha_m) = 0.$$

By an argument similar to that we have already used, we conclude that $c_{11} = \cdots = c_{m1} = 0$. Now it is obvious that we may continue this process and see that all the coefficients $c_{ij}$, $1 \leq i \leq m, 0 \leq j \leq k - 1$, are 0.

***Lemma 2.*** *Under the same assumptions as Lemma* 1, *let* $U$ *be the span of the vectors* (*). *Then there exists a subspace* $S$ *of* $V$ *such that*

$$V = U \oplus S \qquad \textit{and} \qquad A(S) \subseteq S.$$

*Proof.* We shall use induction on $\dim V$. If $\dim V = 1$, there is nothing to prove. Assume that Lemma 2 is valid for vector spaces $V$ of dimension $< n$.

We obviously have $A(W) \subseteq W$, in fact, $A(V) \subseteq W$, because $\alpha \in V$ implies $A^{k-1}(A(\alpha)) = A^k(\alpha) = 0$ and hence $A(\alpha) \in W$. For the given vectors $\alpha_1, \ldots, \alpha_m$, which are linearly independent modulo $W$, we may choose $\alpha_{m+1}, \ldots, \alpha_{m+p} \in V$ such that $\alpha_1, \ldots, \alpha_m, \alpha_{m+1}, \ldots, \alpha_{m+p}$ are linearly independent modulo $W$, where $m + p = \dim(V/W)$; it follows that $\pi(\alpha_1), \ldots, \pi(\alpha_{m+p})$ form a basis of $V/W$. We have clearly

$$V = W \oplus \mathrm{Sp}\,\{\alpha_1, \ldots, \alpha_{m+p}\}.$$

We now consider the restriction of $A$ to $W$, which is nilpotent of index $k - 1$. The elements $A(\alpha_1), \ldots, A(\alpha_{m+p})$ are in $W$, since $A(V) \subseteq W$ as we have seen before. We show that they are linearly independent modulo the subspace $W_1 = \{\alpha \in W; A^{k-2}(\alpha) = 0\}$. Indeed, suppose

$$\sum_{i=1}^{m+p} c_i A(\alpha_i) \in W_1.$$

By applying $A^{k-2}$, we have

$$\sum_{i=1}^{m+p} c_i(A^{k-1}(\alpha_i)) = 0.$$

Since $a_1, \ldots, a_{m+p}$ are linearly independent modulo $W$, Lemma 1 implies that $A^{k-1}(\alpha_1), \ldots, A^{k-1}(\alpha_{m+p})$ are linearly independent. Hence $c_1 = \cdots = c_{m+p} = 0$.

Since dim $W < n$, we may apply the inductive assumption to the restriction of $A$ to $W$ and the elements $A(\alpha_1), \ldots, A(\alpha_{m+p})$ (which we have proved to be linearly independent modulo $W_1$); there exists a subspace $T$ of $W$ such that

$$W = T \oplus \mathrm{Sp}\{A(\alpha_i), \ldots, A^{k-1}(\alpha_i),\ 1 \leq i \leq m+p\}$$

and

$$A(T) \subseteq T.$$

Then we obtain

$$V = T \oplus \mathrm{Sp}\{\alpha_1, \ldots, \alpha_{m+p}\} \oplus \mathrm{Sp}\{A(\alpha_i), \ldots, A^{k-1}(\alpha_i),\ 1 \leq i \leq m+p\}.$$

Now let

$$U = \mathrm{Sp}\{\alpha_i, A(\alpha_i), \ldots, A^{k-1}(\alpha_i),\ 1 \leq i \leq m\}$$

and $S = T \oplus \mathrm{Sp}\{\alpha_i, A(\alpha_i), \ldots, A^{k-1}(\alpha_i),\ m+1 \leq i \leq m+p\}$.

Since $\alpha_1, A\alpha_1, \ldots, A^{k-1}(\alpha_1); \ldots; \alpha_{m+p}, A(\alpha_{m+p}), \ldots, A^{k-1}(\alpha_{m+p})$ are linearly independent by Lemma 1, it follows that

$$V = U \oplus S, \qquad \text{where } A(S) \subseteq S,$$

completing the proof of Lemma 2.

We shall now prove the basic results concerning nilpotent transformations.

***Theorem 7.10.*** *Let $A$ be a nilpotent transformation of index $k$ of an $n$-dimensional vector space $V$ over $\mathcal{F}$. Then there exist a positive integer $p$, a set of $p$ positive integers $k = k_1 \geq k_2 \geq \cdots \geq k_p$ with $\sum_{i=1}^{p} k_i = n$, and elements $\alpha_1, \ldots, \alpha_p$ in $V$ such that*

$$A^{k_1}(\alpha_1) = 0, \ldots, A^{k_p}(\alpha_p) = 0$$

*and such that*

$$\Phi = \{\alpha_1, A\alpha_1, \ldots, A^{k_1-1}(\alpha_1); \ldots; \alpha_p, A(\alpha_p), \ldots, A^{k_p-1}(\alpha_p)\}$$

*is a basis of $V$. Moreover, $p$ and the system of integers $\{k_1, \ldots, k_p\}$ are uniquely determined (although elements $\alpha_1, \ldots, \alpha_p$ are not unique).*

*Proof.* We shall use induction on $\dim V$. If $\dim V = 1$, there is nothing to prove. Assume that Theorem 7.10 is valid for vector spaces of dimension $< n$. Let

$$W = \{\alpha \in V;\ A^{k-1}(\alpha) = 0\},$$

and let $\alpha_1, \ldots, \alpha_m$ be elements in $V$ such that $\pi(\alpha_1), \ldots, \pi(\alpha_m)$ form a basis of the quotient space $V/W$, where $m > 0$. By Lemmas 1 and 2, we have

$$V = U \oplus S, \qquad A(U) \subseteq U \qquad \text{and} \qquad A(S) \subseteq S,$$

where

$$U = \mathrm{Sp}\{\alpha_1, A(\alpha_1), \ldots, A^{k-1}(\alpha_1); \ldots; \alpha_m, A(\alpha_m), \ldots, A^{k-1}(\alpha_m)\}.$$

Since $\dim S < n$, we may apply the inductive assumption to the restriction of $A$ to the subspace $S$, which is nilpotent of index, say, $l$, where $l \leq k$. Then there exist a positive integer $q$, a set of integers $l = l_1 \geq l_2 \geq \cdots \geq l_q$ with $\sum_{i=1}^{q} l_i = \dim S$, and elements $\beta_1, \ldots, \beta_q$ in $S$ such that

$$A^{l_1}(\beta_1) = 0, \ldots, A^{l_q}(\beta_q) = 0$$

and such that

$$\beta_1, A(\beta_1), \ldots, A^{l_1-1}(\beta_1); \ldots; \beta_q, A(\beta_q), \ldots, A^{l_q-1}(\beta_q)$$

form a basis of $S$. We now set

$$p = m + q, \qquad k_1 = k_2 = \cdots = k_m = k,$$
$$k_{m+1} = l_1, \ldots, k_{m+q} = l_q,$$
$$\alpha_{m+1} = \beta_1, \ldots, \alpha_{m+q} = \beta_q;$$

then $p, k_1, \ldots, k_p, \alpha_1, \ldots, \alpha_p$ satisfy the conditions of Theorem 7.10 for the given transformation $A$.

We shall now prove the uniqueness of $p, k_1, \ldots, k_p$. First of all, it is obvious that $A$ is nilpotent of index $k_1$ and hence $k_1 = k$. If $k = 1$, then obviously $p = n$ and $k_1 = \cdots = k_n = 1$. We assume that $k \geq 2$ and also that the uniqueness is valid for vector spaces of dimension $< n$, since the case where $n = 1$ is trivial.

Again, let $W$ be the subspace $\{\alpha \in V;\ A^{k-1}(\alpha) = 0\}$. We show that the number of $k_i$'s which are equal to $k_1$ (namely, the integer $m$ such that $k_1 = k_2 = \cdots = k_m > k_{m+1}$) is equal to $\dim (V/W)$. It is sufficient to prove that $\pi(\alpha_1), \ldots, \pi(\alpha_m)$ form a basis of $V/W$, where $\pi$ is the natural projection of $V$ onto $V/W$. First, $\pi(A^i(\alpha_j))$, where $1 \leq j \leq p$ and $0 \leq i \leq k_j - 1$, span $V/W$. Among these elements, all are 0 except $\pi(\alpha_1), \ldots, \pi(\alpha_m)$; in fact,

$$A^{k-1}(A^i(\alpha_j)) = A^{k+i-1}(\alpha_j) = 0$$

if $1 \leq j \leq m$ and $1 \leq i \leq k - 1$ or if $m + 1 \leq j \leq p$ and $0 \leq i \leq k_j - 1$ (since $k > k_{m+1} \geq \cdots \geq k_p$). Second, we show that $\pi(\alpha_1), \ldots, \pi(\alpha_m)$ are linearly independent in $V/W$. Indeed, if $\sum_{i=1}^{m} c_i\pi(\alpha_i) = 0$, that is, $\sum_{i=1}^{m} c_i\alpha_i \in W$, then

$$A^{k-1}\left(\sum_{i=1}^{m} c_i\alpha_i\right) = \sum_{i=1}^{m} c_iA^{k-1}(\alpha_i) = 0.$$

Since $A^{k-1}(\alpha_1), \ldots, A^{k-1}(\alpha_m)$ are part of a basis, we have $c_1 = \cdots = c_m = 0$, proving our assertion. Thus we have shown that $m = \dim(V/W)$.

Next, we consider the restriction $A_W$ of $A$ to $W$, which is nilpotent of index $k_1 - 1$. The elements

$$\beta_1 = A(\alpha_1), \ldots, \beta_m = A(\alpha_m), \alpha_{m+1}, \ldots, \alpha_p$$

are in $W$ and have the property that

$$A_W{}^{k_1-1}(\beta_1) = 0, \ldots, A_W{}^{k_m-1}(\beta_m) = 0,$$
$$A_W{}^{k_{m+1}}(\alpha_{m+1}) = 0, \ldots, A_W{}^{k_p}(\alpha_p) = 0,$$

where $k_1 - 1 \geqq k_2 - 1 \geqq \cdots \geqq k_m - 1 \geqq k_{m+1} \geqq \cdots \geqq k_p$ and their sum is $n - m = \dim W$. Also, the elements

$$\beta_1, A_W(\beta_1), \ldots, A_W{}^{k_1-2}(\beta_1); \ldots; \beta_m, A_W(\beta_m), \ldots, A_W{}^{k_m-2}(\beta_m);$$
$$\alpha_{m+i}, A_W(\alpha_{m+i}), \ldots, A_W{}^{k_{m+i}-1}(\alpha_{m+i}), \qquad 1 \leq i \leq p - m,$$

form a basis of $W$. By inductive assumption we see that these integers $p$, $k_1 - 1, k_2 - 1, \ldots, k_m - 1, k_{m+1}, \ldots, k_p$ are uniquely determined by $A_W$. Since $W$, $m = \dim(V/W)$, and $A_W$ are uniquely determined by the given transformation $A$ and do not depend on the choice of $\alpha_1, \ldots, \alpha_p$, it follows that $p, k_1, k_2, \ldots, k_p$ are also uniquely determined by $A$.

**Definition 7.7**

The system of integers $\{k_1, \ldots, k_p\}$ in Theorem 7.10 is called the *invariant system* of the nilpotent transformation $A$.

We shall now show that given $n$, $p$, and a system of integers $k_1 \geq k_2 \geq \cdots \geq k_p$ such that $\sum_{i=1}^{p} k_i = n$, there is essentially one nilpotent transformation, with $\{k_1, \ldots, k_p\}$ as its invariant system, on an $n$-dimensional vector space. More precisely, we have:

***Theorem 7.11.*** *Let $n > 0$, $p > 0$, and $k_1, \ldots, k_p$ be given integers such that*

$$k_1 \geq k_2 \geq \cdots \geq k_p \geq 1 \qquad \textit{and} \qquad \sum_{i=1}^{p} k_i = n.$$

**1.** *For any vector space $V$ of dimension $n$ over a field $\mathfrak{F}$, there exists a nilpotent transformation of $V$ with $\{k_1, \ldots, k_p\}$ as its invariant system.*
**2.** *If a nilpotent transformation $A$ of a vector space $V$ and a nilpotent transformation $B$ of a vector space $W$, where both $V$ and $W$ are $n$-dimensional vector spaces over $\mathfrak{F}$, have the same invariant system, there exists a linear isomorphism $C$ of $V$ onto $W$ such that*

$$\begin{array}{ccc} V & \xrightarrow{C} & W \\ {\scriptstyle A}\downarrow & & \downarrow{\scriptstyle B} \\ V & \xrightarrow{C} & W \end{array}$$

*is commutative.*
*Proof.* 1. Let

$$\{\alpha_{11}, \ldots, \alpha_{1,k_1-1}, \alpha_{21}, \ldots, \alpha_{2,k_2-1}, \ldots, \alpha_{p1}, \ldots, \alpha_{p,k_p-1}\}$$

be a basis of $V$ and define $A$ to be the linear transformation of $V$ with the following properties:

$$\begin{array}{ll} A(\alpha_{ij}) = \alpha_{i,j+1} & \text{for each } i,\ 1 \leq i \leq p, \text{ and each } j,\ 1 \leq j < k_i - 1, \\ A(\alpha_{i,k_i-1}) = 0 & \text{for each } i,\ 1 \leq i \leq p. \end{array}$$

Then $A$ is a nilpotent transformation of index $k_1$ whose invariant system is $\{k_1, k_2, \ldots, k_p\}$.

2. Let $\alpha_1, \ldots, \alpha_p$ be elements in $V$ such that

$$A^{k_1}(\alpha_1) = 0, \ldots, A^{k_p}(\alpha_p) = 0$$

and such that $\{\alpha_1, A(\alpha_1), \ldots, A^{k_1-1}(\alpha_1); \ldots; \alpha_p, A(\alpha_p), \ldots, A^{k_p-1}(\alpha_p)\}$ is a basis of $V$. Similarly, let $\beta_1, \ldots, \beta_p$ be elements in $W$ with the same properties with respect to $B$. We define $C$ to be a linear isomorphism of $V$ onto $W$ which maps $A^i(\alpha_j)$ upon $B^i(\beta_j)$ for each $i$ and $j$, where $1 \leq j \leq p$ and $0 \leq i \leq k_j - 1$. Then $BC = CA$.

We state the following corollary to Theorem 7.10.

**Corollary.** *Let $A$ be a nilpotent transformation of an $n$-dimensional vector space $V$. If $\{k_1, \ldots, k_p\}$ is its invariant system, then there exists a basis of $V$ with respect to which $A$ can be represented by a matrix of the block form*

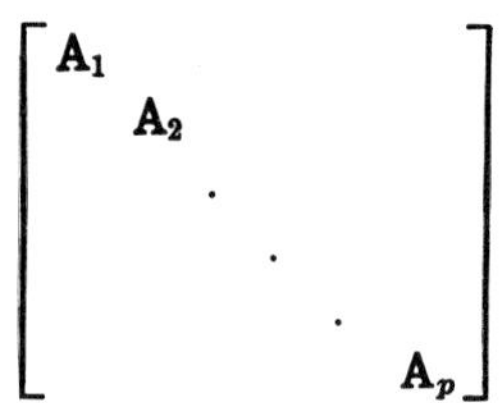

$$\begin{bmatrix} \mathbf{A}_1 & & & \\ & \mathbf{A}_2 & & \\ & & \ddots & \\ & & & \mathbf{A}_p \end{bmatrix}$$

*where each* $\mathbf{A}_i$ *is a* $k_i \times k_i$ *matrix of the form*

$$\mathbf{A}_i = \begin{bmatrix} 0 & & & & & \\ 1 & 0 & & & & \\ & 1 & 0 & & & \\ & & \cdot & \cdot & & \\ & & & \cdot & \cdot & \\ & & & & \cdot & \cdot \\ & & & & 1 & 0 \end{bmatrix}$$

(*that is,* 1 *directly below each diagonal element and* 0 *everywhere else*).

***Example 7.11.*** Let $A$ be a nilpotent transformation of $\mathcal{R}^3$. By Proposition 7.9 we have $A^3 = 0$. The following cases are possible.

Case 1. The index $k$ of $A$ is 3. There exists $\alpha \in V$ such that $\alpha$, $A(\alpha)$, $A^2(\alpha)$ form a basis of $V$. With respect to this basis, $A$ can be represented by

$$\begin{bmatrix} 0 & 0 & 0 \\ 1 & 0 & 0 \\ 0 & 1 & 0 \end{bmatrix}.$$

Case 2. The index $k$ of $A$ is 2. The invariant system is $\{2,1\}$. There exist $\alpha_1$, $\alpha_2$ such that $\alpha_1$, $A(\alpha_1)$, $\alpha_2$ form a basis, where $A(\alpha_2) = 0$. With respect to this basis, $A$ can be represented by

$$\begin{bmatrix} 0 & 0 & 0 \\ 1 & 0 & 0 \\ 0 & 0 & 0 \end{bmatrix}.$$

Case 3. The index $k$ of $A$ is 1. This means $A = 0$.

***Example 7.12.*** Let $A$ be a nilpotent transformation of $\mathcal{R}^5$ of index 3. The following cases arise.

Case 1. The invariant system is $\{3,2\}$. $A$ can be represented by the matrix

$$\begin{bmatrix} 0 & 0 & 0 & 0 & 0 \\ 1 & 0 & 0 & 0 & 0 \\ 0 & 1 & 0 & 0 & 0 \\ 0 & 0 & 0 & 0 & 0 \\ 0 & 0 & 0 & 1 & 0 \end{bmatrix}$$

with respect to a suitable basis.

Case 2. The invariant system is $\{3,1,1\}$. $A$ can be represented by

$$\begin{bmatrix} 0 & 0 & 0 & 0 & 0 \\ 1 & 0 & 0 & 0 & 0 \\ 0 & 1 & 0 & 0 & 0 \\ 0 & 0 & 0 & 0 & 0 \\ 0 & 0 & 0 & 0 & 0 \end{bmatrix}$$

with respect to a suitable basis.

EXERCISE 7.3

» 1. Prove that if an $n \times n$ complex matrix $\mathbf{A}$ has no eigenvalue other than 0, then $\mathbf{A}$ is nilpotent.

2. Suppose that $A$ and $B$ are nilpotent transformations of a vector space $V$. If $AB = BA$, then show that both $A + B$ and $AB$ are nilpotent.

☆ 3. Let $A$ be a nilpotent transformation of an $n$-dimensional vector space over a field $\mathfrak{F}$ of characteristic 0. Define

$$\exp A = \sum_{m=0}^{k-1} \frac{A^m}{m!},$$

where $k$ is the index of $A$ and $A^0 = I$. Prove that $\exp A$ is nonsingular and find its inverse. Show also that, in the case where $\mathfrak{F} = \mathfrak{R}$ or $\mathfrak{C}$,

$$\det(\exp A) = \exp(\text{trace } A).$$

» 4. Classify all nilpotent transformations on a vector space of dimension 6.

» 5. Prove that the trace of a nilpotent matrix is 0.

## 7.4 JORDAN CANONICAL FORM

Combining Theorems 7.6 and 7.10, we shall now obtain a classical theorem on Jordan canonical forms. First we give the following notation and definition.

For any $c \in \mathcal{F}$ and for any positive integer $k$, we shall denote by $\mathbf{J}(c;k)$ the following $k \times k$ matrix over $\mathcal{F}$:

$$\mathbf{J}(c;k) = \begin{bmatrix} c & & & & & \\ 1 & c & & & & \\ & 1 & c & & & \\ & & \ddots & \ddots & & \\ & & & \ddots & \ddots & \\ & & & & 1 & c \end{bmatrix}$$

where all the other entries are 0.

For any $k_1 \geq k_2 \geq \cdots \geq k_p$ we shall denote by $\mathbf{J}(c;k_1, \ldots, k_p)$ the following matrix

$$\mathbf{J}(c;k_1, \ldots, k_p) = \begin{bmatrix} \mathbf{J}(c;k_1) & & & \\ & \mathbf{J}(c;k_2) & & \\ & & \ddots & \\ & & & \mathbf{J}(c;k_p) \end{bmatrix},$$

whose degree is $\sum_{i=1}^{p} k_i$.

**Definition 7.8**

An $n \times n$ matrix $\mathbf{J}$ over $\mathcal{F}$ is called a *Jordan matrix* if it is of the form

$$\begin{bmatrix} \mathbf{J}(c_1;k_1^{(1)}, \ldots, k_{p_1}^{(1)}) & & \\ & \ddots & \\ & & \mathbf{J}(c_m;k_1^{(m)}, \ldots, k_{p_m}^{(m)}) \end{bmatrix},$$

where $c_1, \ldots, c_m \in \mathcal{F}$ and for each $i$, $1 \le i \le m$,

$$k_1^{(i)} \ge k_2^{(i)} \ge \cdots \ge k_{p_i}^{(i)}.$$

Now we have:

***Theorem 7.12.*** *Let $A$ be a linear transformation of an $n$-dimensional vector space $V$ over an algebraically closed field $\mathcal{F}$ (in particular, $\mathcal{F} = \mathcal{C}$). Let $c_1, \ldots, c_m$ be all the distinct eigenvalues of $A$ with multiplicities $n_1, \ldots, n_m$, respectively. Then there exists a basis with respect to which $A$ can be represented by a Jordan matrix*

$$\mathbf{J} = \begin{bmatrix} \mathbf{J}(c_1;k_1^{(1)}, \ldots, k_{p_1}^{(1)}) & & \\ & \ddots & \\ & & \mathbf{J}(c_m;k_1^{(m)}, \ldots, k_{p_m}^{(m)}) \end{bmatrix},$$

*where $c_1, \ldots, c_m$ are the eigenvalues and*

$$n_i = \sum_{j=1}^{p_i} k_j^{(i)}, \qquad 1 \le i \le m,$$

*and* $$k_1^{(i)} \ge k_2^{(i)} \ge \cdots \ge k_{p_i}^{(i)}, \qquad 1 \le i \le m.$$

*A Jordan matrix representing $A$ is uniquely determined up to the order of the blocks $\mathbf{J}(c_i;k_1^{(i)}, \ldots, k_{p_i}^{(i)})$, $1 \le i \le m$.*

*Proof.* Let

$$\phi_A(x) = (x - c_1)^{r_1} \cdots (x - c_m)^{r_m}$$

be the decomposition of the minimal polynomial $\phi_A(x)$ of $A$ into irreducible factors, where $c_1, \ldots, c_m$ are all the distinct eigenvalues of $A$. By Theorem 7.6 we have

$$V = V_1 \oplus V_2 \oplus \cdots \oplus V_m, \qquad A(V_i) \subseteq V_i,$$

where

$$V_i = \{\alpha \in V;\ (A - c_iI)^{r_i}(\alpha) = 0\} \qquad \text{for } 1 \le i \le m,$$

and the restriction of $A$ to $V_i$ has $(x - c_iI)^{r_i}$ as the minimal polynomial, while $\dim V_i = n_i$.

On each $V_i$, the linear transformation $B_i = A - c_iI$ is nilpotent of index $r_i$. By Theorem 7.10 there exist a positive integer $p_i$, a set of integers $r_i = k_1^{(i)} \geq k_2^{(i)} \geq \cdots \geq k_{p_i}^{(i)}$ with $\sum_{j=1}^{p_i} k_j^{(i)} = n_i$, and a basis of $V_i$, so that $B_i$ can be represented by a matrix of the form

$$\mathrm{J}(0;k_1^{(i)}, \ldots, k_{p_i}^{(i)}).$$

Then $A$ on $V_i$ can be represented by the matrix $\mathrm{J}(c_i;k_1^{(i)}, \ldots, k_{p_i}^{(i)})$. By putting these bases for $V_i$'s together, we obtain a basis of $V$ with respect to which $A$ can be represented by the Jordan matrix stated in the theorem.

The uniqueness of a Jordan matrix representing $A$ can be seen as follows. If there exists a basis of $V$ with respect to which $A$ can be represented by a Jordan matrix $\mathrm{J}$ of the form in the theorem, then, first of all, $V$ is the direct sum of invariant subspaces, say, $U_1, \ldots, U_m$. On each $U_i$, the transformation $A - c_iI$ is nilpotent of index $k_1^{(i)}$ and $\dim U_i = \sum_{j=1}^{p_i} k_j^{(i)} = n_i$. Thus the minimal polynomial of $A$ on $U_i$ is $(x - c_i)^{k_1^{(i)}}$. It follows that the minimal polynomial of $A$ on $V$ is

$$(x - c_1)^{k_1^{(1)}} \cdots (x - c_m)^{k_1^{(m)}}.$$

Since the minimal polynomial of $A$ is independent of the choice of a basis in $V$, it follows that $c_1, \ldots, c_m$ are indeed the distinct eigenvalues of $A$ and that the integers $k_1^{(i)}$ are uniquely determined by $A$. We see also that $U_i$ is equal to the null space of $(A - c_iI)^{k_1^{(i)}}$. $\{k_1^{(i)}, \ldots, k_{p_i}^{(i)}\}$ being the invariant system of the transformation $A - c_iI$ on $V^i$, those integers are also uniquely determined by $A$. Thus a Jordan matrix representing $A$ is unique except for the order of the blocks $\mathrm{J}(c_i;k_1^{(i)}, \ldots, k_{p_i}^{(i)})$, $1 \leq i \leq n$.

**Corollary 1.** *Let $A$ be a linear transformation of a vector space of dimension $n$ over an algebraically closed field. Then the Jordan canonical form of $A$ is diagonal if and only if the minimal polynomial $\phi_A(x)$ has simple roots.*

**Corollary 2.** *Let $A$ and $B$ be linear transformations of vector spaces $V$ and $W$, respectively, where both $V$ and $W$ are $n$-dimensional vector spaces over an algebraically closed field. Then the Jordan canonical forms of $A$ and $B$ coincide if and only if there is a linear isomorphism $C$ of $V$ onto $W$ such that*

$$\begin{array}{ccc} V & \xrightarrow{C} & W \\ {\scriptstyle A}\downarrow & & \downarrow{\scriptstyle B} \\ V & \xrightarrow{C} & W \end{array}$$

*is commutative.*

*Proof of Corollary* 2. Assume that the Jordan canonical forms of $A$ and $B$ coincide. Take a basis $\{\alpha_i\}$ in $V$ and a basis $\{\beta_i\}$ in $V$ with respect to which $A$ and $B$ can be represented by the same Jordan matrix. Then the linear mapping $C: V \to W$ such that $C(\alpha_i) = \beta_i$ for every $i$ satisfies the condition. The converse is almost obvious.

Two $n \times n$ matrices $\mathbf{A}$ and $\mathbf{B}$ over a field $\mathcal{F}$ are said to be *similar* if there is a nonsingular matrix $\mathbf{C}$ over $\mathcal{F}$ such that $\mathbf{A} = \mathbf{C}^{-1}\mathbf{BC}$. The matrix version of Corollary 2 can be stated as follows.

***Corollary 3.*** *Let* $\mathbf{A}$ *and* $\mathbf{B}$ *be* $n \times n$ *matrices over an algebraically closed field* $\mathcal{F}$. *Then the Jordan canonical forms of* $\mathbf{A}$ *and* $\mathbf{B}$ *coincide if and only if* $\mathbf{A}$ *and* $\mathbf{B}$ *are similar.*

***Remark.*** Theorem 7.12 and its corollaries are valid without the assumption that the field $\mathcal{F}$ is algebraically closed, provided that the linear transformations (or matrices) involved have all their characteristic roots in $\mathcal{F}$ (cf. Exercise 7.4, number 5).

***Example 7.13.*** Let

$$\mathbf{A} = \begin{bmatrix} 3 & 0 & 0 \\ a & 3 & 0 \\ b & c & -2 \end{bmatrix} \quad \text{over } \mathcal{R}.$$

Then we have

$$f_{\mathbf{A}}(x) = (x-3)^2(x+2).$$

Case 1. $\phi_{\mathbf{A}}(x) = (x-3)(x+2)$. This is so if and only if

$$(\mathbf{A} - 3\mathbf{I}_3)(\mathbf{A} + 2\mathbf{I}_3) = \begin{bmatrix} 0 & 0 & 0 \\ 5a & 0 & 0 \\ ac & 0 & 0 \end{bmatrix} = 0,$$

that is, if and only if $a = 0$. The Jordan form of $\mathbf{A}$ is

$$\begin{bmatrix} 3 & 0 & 0 \\ 0 & 3 & 0 \\ 0 & 0 & -2 \end{bmatrix}.$$

Case 2. $\phi_{\mathbf{A}}(x) = (x-3)^2(x+2)$. The Jordan form of $\mathbf{A}$ is

$$\begin{bmatrix} 3 & 0 & 0 \\ 1 & 3 & 0 \\ 0 & 0 & -2 \end{bmatrix}.$$

As a consequence, all matrices of the given form with $a \neq 0$ are similar to each other, since they have the same Jordan canonical form.

***Example 7.14.*** We may classify all $2 \times 2$ complex matrices up to similarity. For such a matrix $\mathbf{A}$ we have either

$$f_{\mathbf{A}}(x) = (x-a)(x-b), \qquad a \neq b,$$

or

$$f_{\mathbf{A}}(x) = (x-a)^2.$$

In the first case we have

$$\phi_{\mathbf{A}}(x) = (x - a)(x - b)$$

and hence **A** is similar to

$$\begin{bmatrix} a & 0 \\ 0 & b \end{bmatrix}.$$

In the second case we have either

$$(a) \quad \phi_{\mathbf{A}}(x) = x - a \qquad \text{or} \qquad (b) \quad \phi_{\mathbf{A}}(x) = (x - a)^2.$$

In case (*a*), **A** is equal to

$$\begin{bmatrix} a & 0 \\ 0 & a \end{bmatrix}.$$

In case (*b*), **A** is similar to

$$\begin{bmatrix} a & 0 \\ 1 & a \end{bmatrix}.$$

**EXERCISE 7.4**

**1.** Write down in full the following matrices over $\mathcal{C}$: ››(*a*) $\mathbf{J}(2;3)$; (*b*) $\mathbf{J}(1 + i; 4)$; ››(*c*) $\mathbf{J}(i;3,2)$; (*d*) $\mathbf{J}(-1;3,2,1)$; ››(*e*) a Jordan matrix with blocks $\mathbf{J}(2;3)$ and $\mathbf{J}(1;4,3)$.

**2.** Find the characteristic polynomials and minimal polynomials of the matrices in number 1.

**3.** Find the Jordan canonical form of each of the following matrices:

$$››(a) \begin{bmatrix} 3 & -2 \\ 1 & -2 \end{bmatrix}; \quad ››(b) \begin{bmatrix} 0 & 1 \\ -1 & 2 \end{bmatrix}; \quad (c) \begin{bmatrix} 1 & -1 \\ 2 & -1 \end{bmatrix};$$

$$››(d) \begin{bmatrix} 1 & 0 & 0 \\ 1 & 2 & 0 \\ 3 & 0 & -3 \end{bmatrix}; \quad ››(e) \begin{bmatrix} 1 & 0 & 0 \\ 0 & 1 & 0 \\ 1 & 0 & 1 \end{bmatrix}; \quad (f) \begin{bmatrix} 3 & -4 & 0 \\ 3 & -2 & 1 \\ 0 & 0 & 2 \end{bmatrix}.$$

**4.** Classify all $3 \times 3$ complex matrices up to similarity.

☆ **5.** Let $A$ be a linear transformation of an $n$-dimensional vector space over $\mathfrak{F}$, where $\mathfrak{F}$ is not necessarily algebraically closed (for example, $\mathfrak{F} = \mathcal{R}$). Prove that if all the characteristic roots of $A$ are in $\mathfrak{F}$, then there is a basis of $V$ with respect to which $A$ can be represented by a Jordan matrix over $\mathfrak{F}$.

››☆ **6.** Prove that any $n \times n$ complex matrix **A** is similar to its transpose ${}^t\mathbf{A}$.

**7.** Classify all $n \times n$ complex matrices **J** such that $\mathbf{J}^2 = -\mathbf{I}_n$ up to similarity.

☆ **8.** Prove the following theorem along the lines indicated further below.

*Theorem (rational form). Let $A$ be a linear transformation of an $n$-dimensional vector space $V$ over a field $\mathfrak{F}$. Then there exist $s$ nonzero vectors $\beta_1, \ldots, \beta_s$ such that:*

**1.** *$V = W_1 \oplus \cdots \oplus W_s$, where $W_i$ is the smallest invariant subspace containing $\beta_i$, $1 \leq i \leq s$.*

**2.** *If $\psi_i$ is the minimal polynomial of $A$ restricted to $W_i$ for each $i$, then $\psi_{i+1}$ divides $\psi_i$, $1 \leq i \leq s - 1$.*

*Furthermore, the integers $s$ and $\psi_1, \ldots, \psi_s$ are uniquely determined.*

*Hints for the Proof:*

1. Let

$$\phi_A = \phi_1^{p_1} \cdots \phi_k^{p_k},$$

where $\phi_1, \ldots, \phi_k$ are irreducible polynomials which are relatively prime. By Theorem 7.5 one has

$$V = V_1 \oplus \cdots \oplus V_k,$$

where each $V_i$ is invariant by $A$ and the restriction of $A$ to $V_i$ has $\phi_i^{p_i}$ as its minimal polynomial.

2. Let $B_i = \phi_i(A)$ on $V_i$. Since $B_i$ is nilpotent of index $p_i$, let $\{p_i^{(1)}, \ldots, p_i^{(r_i)}\}$ be its invariant system. By rearranging the indices $i = 1, 2, \ldots, k$, one assumes that

$$r_1 \geq r_2 \geq \cdots \geq r_k.$$

3. In each $V_i$ let $\alpha_i^{(1)}, \ldots, \alpha_i^{(r_i)}$ be elements such that

$$(B_i)^t(\alpha_i^{(j)}), \qquad 0 \leq t \leq p_i^{(j)} - 1, \qquad 1 \leq j \leq r_i,$$

form a basis of $V_i$.

4. Let $\beta_1 = \alpha_1^{(1)} + \alpha_2^{(1)} + \cdots + \alpha_k^{(1)}$. Show that the minimal polynomial of $\beta_1$ (with respect to $A$, see Exercise 7.2, number 5) is equal to $\phi_A$; call it $\psi_1$.

5. Let $\beta_2 = \alpha_1^{(2)} + \alpha_2^{(2)} + \cdots + \alpha_k^{(2)}$. Show that the minimal polynomial of $\beta_2$ is the product of $\phi_i^{p_i^{(2)}}$, $1 \leq i \leq k$. Thus this polynomial, which we call $\psi_2$, divides $\psi_1$.

6. In general, for each $j$, $1 \leq j \leq r_1$, let $\beta_j$ be the sum of $\alpha_i^{(j)}$'s, where we set $\alpha_i^{(j)} = 0$ if $r_i < j$. Let $\psi_j$ be the minimal polynomial of $\beta_j$.

7. Show that $\beta_1, \ldots, \beta_s$, with $s = r_1$, satisfy the conditions of the theorem. Then show the uniqueness of $s$ and $\psi_1, \ldots, \psi_s$.

## 7.5 SCHUR'S LEMMA AND COMPLEX STRUCTURES

Let $V$ be an $n$-dimensional vector space over a field $\mathfrak{F}$ $(n > 0)$.

### Definition 7.9

Given a family $\mathfrak{G}$ of linear transformations of $V$, we say that a subspace $U$ of $V$ is *$\mathfrak{G}$-invariant* if $A(U) \subseteq U$ for every $A \in \mathfrak{G}$. We say that the pair $\{V,\mathfrak{G}\}$ is *simple* (or *irreducible*) if there is no proper subspace $U$ [that is, a subspace $U$ different from $V$ and $(0)$] which is $\mathfrak{G}$-invariant. When $\{V,\mathfrak{G}\}$ is simple, we shall also say that $\mathfrak{G}$ *acts irreducibly* on $V$.

We shall prove the following theorem, which is often called *Schur's lemma.*

***Theorem 7.13.*** *Let $\{V,\mathfrak{G}\}$ and $\{W,\mathfrak{H}\}$ be finite-dimensional vector spaces over $\mathfrak{F}$ with a family of linear transformations $\mathfrak{G}$ and $\mathfrak{H}$, respectively. Let $C$ be a linear mapping of $V$ into $W$ satisfying the following conditions:*

1. *For any $A \in \mathfrak{G}$, there exists a $B \in \mathfrak{H}$ such that $CA = BC$.*
2. *For any $B \in \mathfrak{H}$, there exists an $A \in \mathfrak{G}$ such that $CA = BC$.*

*Assume that both $\{V,\mathfrak{G}\}$ and $\{W,\mathfrak{H}\}$ are simple. Then either $C$ is the zero mapping or $C$ is a linear isomorphism of $V$ onto $W$ (and hence $\mathfrak{H} = \{CAC^{-1}; A \in \mathfrak{G}\}$).*

*Proof.* Let $U = \{\alpha \in V; C(\alpha) = 0\}$, namely, the null space of $C$. We show that $U$ is $\mathfrak{G}$-invariant. Let $A \in \mathfrak{G}$ and take $B \in \mathfrak{H}$ such that $CA = BC$. Then

$$C(A(\alpha)) = B(C(\alpha)) = B(0) = 0 \qquad \text{for every } \alpha \in U,$$

showing that $A(U) \subseteq U$. Since $\{V,\mathfrak{G}\}$ is simple, we have either $U = V$ or $U = (0)$. In the first case, $C$ is the zero mapping. In the second case, $C$ is one-to-one. We shall now see that $C$ is onto. Let $S = C(V)$ be the range of $C$. For any $B \in \mathfrak{H}$, let $A \in \mathfrak{G}$ be such that $CA = BC$. Then

$$B(C(\alpha)) = C(A(\alpha)) \in C(V), \qquad \text{hence } B(S) \subseteq S,$$

showing that $S$ is $\mathfrak{H}$-invariant. Since $\{W,\mathfrak{H}\}$ is simple, we have either $S = (0)$ or $S = W$. Excluding the case where $C$ is the zero mapping, we have $S = W$, that is, $C$ is a linear isomorphism of $V$ onto $W$.

**Corollary.** *Let $V$ be a finite-dimensional vector space over an algebraically closed field $\mathfrak{F}$ (in particular, $\mathfrak{F} = \mathfrak{C}$). If a family of transformations $\mathfrak{G}$ acts irreducibly on $V$, then every linear transformation $C$ which commutes with every $A \in \mathfrak{G}$ is of the form $C = cI$, where $c \in \mathfrak{F}$ and $I$ is the identity transformation.*

*Proof.* Since $\mathfrak{F}$ is algebraically closed, $C$ has at least one eigenvalue, say, $c$. The linear transformation $C - cI$ commutes with every $A \in \mathfrak{G}$. By Theorem 7.13, $C - cI$ has to be the zero mapping, since it is singular [consider an eigenvector $\alpha \neq 0$ for which $(C - cI)(\alpha) = 0$]. Thus $C = cI$.

The corollary plays an important role in representation theory of groups.

More generally, we have, without assuming that $\mathfrak{F}$ is algebraically closed:

**Theorem 7.14.** *Let $V$ be a finite-dimensional vector space over $\mathfrak{F}$. If a family of transformations $\mathfrak{G}$ acts irreducibly on $V$ and if a linear transformation $C$ commutes with every $A \in \mathfrak{G}$, then the minimal polynomial of $C$ is irreducible over $\mathfrak{F}$.*

*Proof.* Assume that

$$\phi_C = \phi_1 \cdot \phi_2,$$

where $\phi_1$ and $\phi_2$ are nonconstant polynomials in $\mathfrak{F}[x]$ which are relatively prime. By Theorem 7.5 we have

$$V = V_1 \oplus V_2,$$

where

$$V_1 = \{\alpha \in V; \phi_1(C)(\alpha) = 0\} \quad \text{and} \quad V_2 = \{\alpha \in V; \phi_2(C)(\alpha) = 0\}.$$

We note that $V_1 \neq (0)$ and $V_2 \neq (0)$. We now show that $V_1$ is $\mathfrak{G}$-invariant. For any $A \in \mathfrak{G}$, we have

$$\phi_1(A)(C\alpha) = C(\phi_1(A))\alpha = C(0) = 0,$$

since $\phi_1(A)$ and $C$ commute because $A$ and $C$ commute. Thus $V_1$ is $\mathfrak{G}$-invariant, and similarly for $V_2$. Since $\{V,\mathfrak{G}\}$ is simple, we must have $V_1 = (0)$ or $V_1 = V$. Since $V_1 \neq (0)$, we must have $V_1 = V$ and, consequently, $V_2 = (0)$, which is a contradiction. Thus $\phi_C$ is irreducible over $\mathcal{F}$.

Let us remark that the corollary to Theorem 7.13 follows from Theorem 7.14 as well. In fact, if $\mathcal{F}$ is algebraically closed, any irreducible polynomial over $\mathcal{F}$ is of the form $x - c$. In particular, $\phi_C(x) = x - c$ for some $c$, which implies that $C = cI$.

We now consider the case $\mathcal{F} = \mathcal{R}$ and obtain the following corollary of Theorem 7.14.

***Corollary.*** *Let $V$ be a finite-dimensional real vector space. If $\mathfrak{G}$ is a family of linear transformations which acts irreducibly on $V$, then a linear transformation $C$ which commutes with every $A \in \mathfrak{G}$ is one or the other of the following two types:*

**1.** *$C = cI$ with some $c \in \mathcal{R}$.*

**2.** *$C = aI + bJ$, where $a, b \in \mathcal{R}$, $b \neq 0$, and $J$ is a linear transformation of $V$ such that $J^2 = -I$. In this case,* dim *$V$ is even.*

*Proof.* The minimal polynomial $\phi_C(x)$ being irreducible over $\mathcal{R}$ by Theorem 7.14, we have either $\phi_C = x - c$ or $\phi_C = (x - a)^2 + b^2$, where $a, b, c \in \mathcal{R}$ and $b \neq 0$. In the first case, $C = cI$. In the second case, let

$$J = \frac{C - aI}{b}.$$

Since $(C - aI)^2 + b^2I = 0$, we have $J^2 = -I$; thus, we get $C = aI + bJ$, where $b \neq 0$ and $J^2 = -I$. Taking the determinants of both sides, we have

$$(\det J)^2 = (-1)^n, \qquad \text{where } n = \dim V.$$

Hence $n$ is even.

We shall now study in more detail linear transformations $J$ such that $J^2 = -I$. We begin with:

***Example 7.15.*** Let us consider $\mathbb{C}^n$ as a vector space $V$ over $\mathbb{R}$. The mapping $J_0$ defined by

$$J_0(\alpha) = i\alpha, \qquad \alpha \in V,$$

is a linear transformation of $V$ such that $J_0{}^2 = -I$. If $\{\epsilon_1, \ldots, \epsilon_n\}$ is the standard basis of the complex vector space $\mathbb{C}^n$, then $\epsilon_1, \ldots, \epsilon_n, i\epsilon_1, \ldots, i\epsilon_n$ form a basis of the real vector space $V$; every $\alpha = (c_1, \ldots, c_n)$, where $c_k \in \mathcal{C}$, can be expressed uniquely in the form

$$\alpha = a_1\epsilon_1 + \cdots + a_n\epsilon_n + b_1(i\epsilon_1) + \cdots + b_n(i\epsilon_n),$$

where $c_k = a_k + ib_k$, with $a_k, b_k \in \mathcal{R}$ for each $k$. Or, using $J_0$,

$$\{\epsilon_1, \ldots, \epsilon_n, J_0(\epsilon_1), \ldots, J_0(\epsilon_n)\}$$

is a basis of $V$. Thus $J_0$ can be represented by a matrix of the form

$$\begin{bmatrix} \mathbf{O}_n & -\mathbf{I}_n \\ \mathbf{I}_n & \mathbf{O}_n \end{bmatrix}$$

where $\mathbf{O}_n$ is the zero matrix and $\mathbf{I}_n$ is the identity matrix, both of degree $n$.

***Example 7.16.*** Let $W$ be a $2n$-dimensional real vector space. Given any linear isomorphism $\Phi$ of the real vector space $V$ in Example 7.15 onto $W$, the linear transformation $J = \Phi J_0 \Phi^{-1}$ of $W$ satisfies $J^2 = -I$.

**Definition 7.10**

A linear transformation $J$ of a finite-dimensional real vector space $V$ is called a *complex structure* if $J^2 = -I$. $J_0$ in Example 7.15 is called the *standard complex structure* on the real vector space $\mathcal{C}^n$ (namely, $\mathcal{C}^n$ regarded as a real vector space).

The terminology "complex structure" is justified in the following way. We shall show that a complex structure $J$ on a real vector space $V$ enables us to define the set $V$ as a complex vector space. First of all, we shall retain the same addition in $V$. Then we define scalar multiplication by complex numbers as follows: For any complex number $c = a + ib$, $a, b \in \mathcal{R}$, and for any $\alpha \in V$ we define

$$c\alpha = a\alpha + bJ(\alpha).$$

It is straightforward to verify that all the axioms for $V$ to be a vector space over $\mathcal{C}$ are satisfied. For example, if $c_1 = (a_1 + ib_1)$ and $c_2 = (a_2 + ib_2)$, then

$$\begin{aligned} c_1(c_2\alpha) &= c_1((a_2 + ib_2)\alpha) = c_1(a_2\alpha + b_2J(\alpha)) \\ &= (a_1 + ib_1)(a_2\alpha + b_2J(\alpha)) = a_1(a_2\alpha + b_2J(\alpha)) \\ &\qquad\qquad + b_1(a_2J(\alpha) + b_2J^2(\alpha)) \\ &= (a_1a_2 - b_1b_2)\alpha + (a_1b_2 + a_2b_1)J(\alpha), \end{aligned}$$

by virtue of $J^2 = -I$. On the other hand, we have

$$c_1c_2 = (a_1 + ib_1)(a_2 + ib_2) = a_1a_2 - b_1b_2 + i(a_1b_2 + a_2b_1)$$

and

$$(c_1c_2)(\alpha) = (a_1a_2 - b_1b_2)\alpha + (a_1b_2 + a_2b_1)J(\alpha).$$

Hence

$$c_1(c_2\alpha) = (c_1c_2)(\alpha).$$

When we regard $V$ as a complex vector space with respect to $J$, we shall write $V(\mathcal{C};J)$. If $\dim V(\mathcal{C};J) = n$ and if $\{\alpha_1, \ldots, \alpha_n\}$ is a basis of $V(\mathcal{C};J)$, then we can see, as in Example 7.15, that $\alpha_1, \ldots, \alpha_n, J(\alpha_1)\ (= i\alpha_1), \ldots, J(\alpha_n)\ (= i\alpha_n)$ form a basis of $V$. Thus $\dim V$ is even. (We may conclude this more directly from the existence of a complex structure $J$ by taking the determinants of both sides of $J^2 = -I$ as before.) If we take a linear mapping of the real vector space $\mathcal{C}^n$ into $V$ such that

$$\Phi(\epsilon_k) = \alpha_k \qquad \text{and} \qquad \Phi(J_0(\epsilon_k)) = J(\alpha_k), \qquad 1 \leq k \leq n,$$

it is clear that $\Phi$ is a linear isomorphism of $\mathcal{C}^n$ (as a vector space over $\mathcal{R}$) onto the real vector space $V$ such that $J = \Phi J_0 \Phi^{-1}$. In other words, the given complex structure $J$ on $V$ is obtained from the standard complex structure $J_0$ on $\mathcal{C}^n$ in the manner of Example 7.16.

Summarizing our discussions, we have:

***Proposition 7.15.*** *Let $V$ be an $m$-dimensional real vector space.*
**1.** *If there is a complex structure $J$ on $V$, we can make $V$ into a complex vector space in such a way that $i\alpha = J(\alpha)$ for every $\alpha \in V$;* dim *$V$ is even.*
**2.** *For any complex structure $J$ on $V$ there is a basis of $V$ of the form $\{\alpha_1, \ldots, \alpha_n, J(\alpha_1), \ldots, J(\alpha_n)\}$, where $2n = m$.*
**3.** *There is a one-to-one correspondence between the set of all complex structures $J$ on $V$ and the set of linear isomorphisms of the real vector space $\mathcal{C}^n$ onto $V$; the correspondence is given by $J = \Phi J_0 \Phi^{-1}$.*

We shall now discuss the relationship between complex structures on a real vector space $V$ and the complexification of $V$ defined in Sec. 5.7.

Let $V^c$ be the complexification of an $m$-dimensional real vector space $V$. If $J$ is a complex structure on $V$, let

$$W = \{\alpha - iJ(\alpha);\ \alpha \in V\}.$$

For any $\alpha, \beta \in V$, we have

$$(\alpha - iJ(\alpha)) + (\beta - iJ(\beta)) = (\alpha + \beta) - iJ(\alpha + \beta).$$

For any $c = a + ib$, $a, b \in \mathcal{R}$, we have

$$\begin{aligned} c(\alpha - iJ(\alpha)) &= (a + bi)\,(\alpha - iJ(\alpha)) \\ &= a\alpha + bJ(\alpha) - iJ(a\alpha + bJ(\alpha)). \end{aligned}$$

Thus $W$ is a (complex) subspace of $V^c$. Using the conjugate defined in Proposition 5.23, we have

$$\bar{W} = \{\bar{\gamma};\ \gamma \in W\} = \{\alpha + iJ(\alpha);\ \alpha \in V\},$$

which is also a (complex) subspace of $V^c$.

We shall now show that

$$V^c = W \oplus \bar{W}.$$

Let $\gamma \in W \cap \bar{W}$. Then there exist $\alpha, \beta \in V$ such that

$$\gamma = \alpha - iJ(\alpha) = \beta + iJ(\beta).$$

Hence $(\alpha - \beta) - iJ(\alpha + \beta) = 0$, which implies that $\alpha - \beta = 0$ and $\alpha + \beta = 0$. Thus $\alpha = \beta = 0$ and $\gamma = 0$, proving that $W \cap \bar{W} = (0)$. For any $\gamma \in V^c$, let $\gamma = \alpha + i\beta$, where $\alpha, \beta \in V$, and set

$$\alpha' = \frac{\alpha + J(\beta)}{2}, \qquad \beta' = \frac{\alpha - J(\beta)}{2}.$$

Then

$$\gamma = (\alpha' - iJ(\alpha')) + (\beta' + iJ(\beta')),$$

showing that $V^c = W + \bar{W}$. Thus $V^c = W \oplus \bar{W}$.

Conversely, suppose that $V^c$ is the direct sum of a subspace $W$ and its conjugate $\bar{W}$. We wish to show that $V$ admits a complex structure $J$ such that

$$W = \{\alpha - iJ(\alpha); \alpha \in V\}.$$

Let $\alpha \in V$ and let

$$\alpha = \gamma + \bar{\delta}, \qquad \text{where } \gamma, \delta \in W.$$

Since $\bar{\alpha} = \alpha$, we have $\bar{\gamma} + \delta = \gamma + \bar{\delta}$. Since $W \cap \bar{W} = (0)$, we have $\gamma = \delta$. Consider $\beta = i\gamma - i\bar{\gamma}$. Since

$$\bar{\beta} = -i\bar{\gamma} + i\gamma = \beta,$$

we see that $\beta$ is a real vector, that is, $\beta \in V$. We now define $J(\alpha) = \beta$. It is straightforward to verify that $J$ so defined is a complex structure of $V$. With this complex structure, we have

$$\alpha - iJ(\alpha) = \gamma + \bar{\gamma} - i(i\gamma - i\bar{\gamma}) = 2\gamma,$$

showing that $\alpha - iJ(\alpha) \in W$ for every $\alpha \in V$. On the other hand, for any $\gamma \in W$, let $\alpha = (\gamma + \bar{\gamma})/2$. Obviously, $\gamma = \alpha - iJ(\alpha)$, showing that $W = \{\alpha - iJ(\alpha); \alpha \in V\}$.

Summing up, we have:

***Proposition 7.16.*** *Let $V^c$ be the complexification of a finite-dimensional real vector space $V$. There is a one-to-one correspondence between the set of all complex structures $J$ on $V$ and the set of all subspaces $W$ of $V^c$ such that $V^c = W \oplus \bar{W}$. The correspondence is given by*

$$W = \{\alpha - iJ(\alpha);\ \alpha \in V\}.$$

***Remark.*** If $J$ is a complex structure on $V$, so is $-J$. If $W$ is the subspace of $V^c$ corresponding to $J$, then $\bar{W}$ is the subspace of $V^c$ corresponding to $-J$. $J$ and $-J$ are said to be *conjugate complex structures*.

Let $W$ be the subspace of $V^c$ corresponding to a complex structure $J$. We can define a mapping $\Phi: V \to W$ by

$$\Phi(\alpha) = \frac{\alpha - iJ(\alpha)}{2} \qquad \text{for every } \alpha \in V.$$

When we regard $W$ as a real vector space, $\Phi$ is a linear isomorphism of $V$ onto $W$ such that

$$\Phi(J(\alpha)) = i\Phi(\alpha) \qquad \text{for every } \alpha \in V.$$

This means that the complex vector space $W$ is isomorphic to the complex vector space $V(\mathcal{C};J)$ which we defined earlier. Likewise, the mapping $\bar{\Phi}: V \to \bar{W}$ defined by

$$\bar{\Phi}(\alpha) = \frac{\alpha + iJ(\alpha)}{2} \qquad \text{for every } \alpha \in V$$

is a linear isomorphism of $V$ onto the real vector space $\bar{W}$ such that

$$\bar{\Phi}(J(\alpha)) = -i\bar{\Phi}(\alpha) \qquad \text{for every } \alpha \in V.$$

Thus the complex vector space $W$ is isomorphic to the complex space $V(\mathcal{C};-J)$. We have, moreover,

$$\alpha = \Phi(\alpha) + \bar{\Phi}(\alpha).$$

We add one more observation. The linear transformation $J$ of $V$ can be extended to a linear transformation $J'$ of $V^c$ (Proposition 5.22). We have

$$J'(\alpha - iJ(\alpha)) = J(\alpha) - iJ^2(\alpha) = i(\alpha - iJ(\alpha))$$

and

$$J'(\alpha + iJ(\alpha)) = J(\alpha) + iJ^2(\alpha) = -i(\alpha + iJ(\alpha)),$$

in other words,

$$J' = iI \quad \text{on } W \qquad \text{and} \qquad J' = -iI \quad \text{on } \bar{W},$$

where $I$ is the identity mapping. Thus by taking any basis $\{\gamma_1, \ldots, \gamma_n\}$ in $W$, $J'$ can be represented by

$$\begin{bmatrix} i\mathbf{I}_n & \mathbf{O}_n \\ \mathbf{O}_n & -i\mathbf{I}_n \end{bmatrix}$$

with respect to the basis $\{\gamma_1, \ldots, \gamma_n, \bar{\gamma}_1, \ldots, \bar{\gamma}_n\}$.

We wish to point out that a complex structure on a real vector space is a special case of a more general concept. Let $V$ be a vector space over a field $\mathcal{F}$. For any extension $\mathcal{F}'$ of $\mathcal{F}$ we ask whether it is possible to make $V$ into a vector space over $\mathcal{F}'$ by extending scalar multiplication $\mathcal{F} \times V \to V$ to scalar multiplication $\mathcal{F}' \times V \to V$ while retaining the same addition.

The problem is equivalent to that of finding an isomorphism $f$ of $\mathcal{F}'$ into the ring $\mathrm{Hom}(V,V)$ of all linear transformations of $V$ such that $f(a) = aI$ for every $a \in \mathcal{F}$. Indeed, if there is such an isomorphism $f$, then we set

$$a'\alpha = f(a')(\alpha) \qquad \text{for } a' \in \mathcal{F}' \text{ and } \alpha \in V$$

and easily verify that this defines scalar multiplication $\mathcal{F}' \times V \to V$ to make $V$ into a vector space over $\mathcal{F}'$.

**Definition 7.11**

Let $V$ be a vector space over a field $\mathcal{F}$ and let $\mathcal{F}'$ be an extension of $\mathcal{F}$. A ring isomorphism $f\colon \mathcal{F}' \to \mathrm{Hom}(V,V)$ is called an *$\mathcal{F}'$-structure* on $V$.

Let us consider the case where $V$ is finite-dimensional and $\mathcal{F}' = \mathcal{F}(c')$ is a simple extension of $\mathcal{F}$. If $f$ is an $\mathcal{F}'$-structure on $V$, then $c'$ satisfies the same minimal polynomial as the linear transformation $f(c')$, so that $c'$ is algebraic. Suppose that $c'$ is algebraic and let $\phi(x)$ be the minimal polynomial of $c'$ (over $\mathcal{F}$). In order to find an isomorphism $f$ of $\mathcal{F}'$ into $\mathrm{Hom}(V,V)$, it is sufficient to define what $f(c')$ should be, since every element of $\mathcal{F}'$ is a polynomial of $c'$. Thus it is necessary and sufficient to find $A \in \mathrm{Hom}(V,V)$ whose minimal polynomial is equal to $\phi$ and define $f(c') = A$.

This question is solved by the following:

***Theorem 7.17.*** *Let $V$ be an $n$-dimensional vector space over a field $\mathcal{F}$. Let $\phi$ be a monic irreducible polynomial of degree $r$ in $\mathcal{F}[x]$. If $r$ divides $n$, then there exists a linear transformation $A$ of $V$ whose minimal polynomial is equal to $\phi$.*

*Proof.* We first consider the case where $r = n$. Let

$$\phi(x) = x^n + c_1x^{n-1} + \cdots + c_n.$$

Take any basis $\{\alpha_1, \ldots, \alpha_n\}$ of $V$ and define a linear transformation $A$ of $V$ by

$$A(\alpha_1) = \alpha_2,\ A(\alpha_2) = \alpha_3,\ \ldots,\ A(\alpha_{n-1}) = \alpha_n$$

and

$$A(\alpha_n) = -c_n\alpha_1 - c_{n-1}\alpha_2 - \cdots - c_1\alpha_n.$$

Then we have

$$\begin{aligned}\phi(A)\alpha_1 &= A^n(\alpha_1) + c_1A^{n-1}(\alpha_1) + \cdots + c_{n-1}A(\alpha_1) + c_n\alpha_1\\ &= -c_n\alpha_1 - c_{n-1}\alpha_2 - \cdots - c_1\alpha_n + c_1\alpha_n + \cdots + c_{n-1}\alpha_2 + c_n\alpha_1\\ &= 0.\end{aligned}$$

For any $k \geq 2$ we have

$$\phi(A)(\alpha_k) = \phi(A)A^{k-1}(\alpha_1) = A^{k-1}\phi(A)(\alpha_1) = 0.$$

Hence $\phi(A) = 0$. Since $\phi$ is irreducible, it is the minimal polynomial of $A$, proving our assertion.

In the general case, let $n = rp$. $V$ may be considered as the direct sum of $p$ isomorphic vector spaces $V_i$, $1 \leq i \leq p$, each of dimension $r$. In each $V_i$ we can define, as we have already proved, a linear transformation $A_i$ with minimal polynomial $\phi$. Let $A$ be the direct sum of $A_i$, $1 \leq i \leq p$. It is clear that the minimal polynomial of $A$ is $\phi$, proving Theorem 7.17.

When $\mathcal{F} = \mathcal{R}$ and $\mathcal{F}' = \mathcal{C} = \mathcal{R}(i)$, an $\mathcal{F}'$-structure is nothing but a complex structure; it is completely determined by the image $J = f(i)$ of the isomorphism $f\colon \mathcal{C} \to \operatorname{Hom}(V,V)$.

**EXERCISE 7.5**

›› **1.** Let $\Psi$ be a mapping of $\mathcal{C}^n$ onto $\mathcal{R}^{2n}$ defined by

$$\Psi(z^1,z^2, \ldots ,z^n) = (x^1, \ldots ,x^n,y^1, \ldots ,y^n),$$

where $z^k = x^k + iy^k$ with $x^k$, $y^k \in \mathcal{R}$. Find a complex structure $J_0$ on $\mathcal{R}^{2n}$ such that $\Psi(i\gamma) = J_0(\Psi(\gamma))$ for $\gamma \in \mathcal{C}^n$. (This $J_0$ is called the *standard complex structure* on $\mathcal{R}^{2n}$.)

**2.** How can we obtain all complex structures on $\mathcal{R}^{2n}$ from the standard one in number 1?

**3.** Keeping the same notation as in number 1, show that, for any linear transformation $\tilde{A}$ of $\mathcal{C}^n$, there exists a linear transformation $A$ of $\mathcal{R}^{2n}$ such that $\Psi\tilde{A} = A\Psi$ and $AJ_0 = J_0A$. Conversely, any linear transformation $A$ of $\mathcal{R}^{2n}$ which commutes with $J_0$ corresponds to a linear transformation $\tilde{A}$ of $\mathcal{C}^n$.

**4.** Knowing the matrix for $\tilde{A}$ with respect to the standard basis in $\mathcal{C}^n$, express the corresponding $A$ by a matrix with respect to the standard basis of $\mathcal{R}^{2n}$.

››☆ **5.** Let $\mathfrak{G}$ be a group of linear transformations which acts on an $n$-dimensional real vector space $V$ irreducibly. Let $S$ be the set of all linear transformations $A$ of $V$ which commute with every $B \in \mathfrak{G}$. Prove that $S$ is a skew field which is at the same time a vector space of finite dimension over $\mathcal{R}$. (We then say that $S$ is a *division algebra*. It is known that any division algebra over $\mathcal{R}$ is isomorphic with the real number field $\mathcal{R}$, the complex number field $\mathcal{C}$, or the skew field of quaternions.)

**6.** Let $c_1$, $c_2$ be two nonzero complex numbers whose ratio is not real. Show that there exists a $2 \times 2$ real matrix $\mathbf{J}$ such that

$$\mathbf{J}\begin{bmatrix} c_1 \\ c_2 \end{bmatrix} = i\begin{bmatrix} c_1 \\ c_2 \end{bmatrix} \quad \text{and} \quad \mathbf{J}^2 = -\mathbf{I}_2.$$

Moreover, show that $\mathbf{J}$ is uniquely determined by $c_1/c_2$.

›› **7.** Let $\mathcal{Q}$ be the rational number field. Define $\mathcal{Q}^2$ as a vector space over $\mathcal{Q}(\sqrt{2})$ by writing down scalar multiplication explicitly.

**8.** Let $\mathcal{Q}$ be the rational number field. Define $\mathcal{Q}^2$ as a vector space over $\mathcal{Q}(i)$, where $i = \sqrt{-1}$.

# 8 Inner Product

In this chapter we shall be concerned with real and complex vector spaces. After introducing the notion of inner product on such vector spaces we study the eigenvalue problem for several important classes of linear transformations. As is explained in *Suggestions for Class Use,* there are many ways of arriving at the main theorems in this chapter.

## 8.1 INNER PRODUCT

**Definition 8.1**

By an *inner product* on a *real* vector space $V$, we mean a symmetric bilinear function $V \times V \to \mathcal{R}$ which is positive-definite (Definition 4.15). In other words, it is a real-valued function $(\alpha,\beta)$, where $\alpha$, $\beta \in V$, which has the following properties:

**1.** $(\beta,\alpha) = (\alpha,\beta)$ (*symmetric*).
**2.** $(\alpha,\beta)$ is linear in $\alpha$ for any fixed $\beta \in V$.
$(\alpha,\beta)$ is linear in $\beta$ for any fixed $\alpha \in V$.
**3.** $(\alpha,\alpha) \geq 0$, and $(\alpha,\alpha) = 0$ if and only if $\alpha = 0$ (*positive-definite*).

We shall remark that the second statement in (2) is a consequence of the first and (1).

**Definition 8.2**

By an *inner product* on a *complex* vector space $V$ we mean a complex-valued function on $V \times V$, denoted by $(\alpha,\beta)$, which has the following properties:

**1′.** $(\beta,\alpha) = \overline{(\alpha,\beta)}$ (*hermitian*), where $\overline{\quad}$ denotes the complex conjugate.
**2′.** $(\alpha,\beta)$ is linear in $\alpha$ for any fixed $\beta \in V$;
$(\alpha,\beta)$ is conjugate-linear in $\beta$ for any fixed $\alpha \in V$, that is,

$$(\alpha, \beta_1 + \beta_2) = (\alpha,\beta_1) + (\alpha,\beta_2)$$

and
$$(\alpha,c\beta) = \bar{c}(\alpha,\beta), \qquad \text{where } c \in \mathcal{C}.$$

**3′.** $(\alpha,\alpha)$ is real and $\geq 0$, and $(\alpha,\alpha) = 0$ if and only if $\alpha = 0$ (*positive-definite*).

We shall give some examples.

***Example 8.1.*** On $\mathcal{R}^n$ $[\mathcal{C}^n]$, define for $\alpha = (a_i)$ and $\beta = (b_i)$

$$(\alpha,\beta) = \sum_{i=1}^{n} a_i b_i \qquad \left[(\alpha,\beta) = \sum_{i=1}^{n} a_i \bar{b}_i\right].$$

This is called the *standard inner product* on $\mathcal{R}^n$ $[\mathcal{C}^n]$.

***Example 8.2.*** Let $V$ be the set of all sequences of real [complex] numbers $\alpha = (a_n)_{n=1,2,\ldots}$ such that

$$\sum_{n=1}^{\infty} |a_n|^2 < \infty.$$

We say that the sequence $(a_n)$ is square-summable if it satisfies this condition. If $(a_n)$ and $(b_n)$ are square-summable, then both the sequences

$$(a_n + b_n) \qquad \text{and} \qquad (ca_n), \quad \text{where } c \in \mathcal{R}\ [\mathcal{C}],$$

are square-summable (we shall prove this in Example 8.5). Thus $V$ is a vector space over $\mathcal{R}$ $[\mathcal{C}]$. For $\alpha = (a_n)$ and $\beta = (b_n)$ in $V$, we see that

$$\sum_{n=1}^{\infty} a_n b_n \qquad \left[\sum_{n=1}^{\infty} a_n \bar{b}_n\right]$$

is convergent (as we shall prove later), and we define it to be $(\alpha,\beta)$. It can be seen that $(\alpha,\beta)$ is an inner product in $V$.

***Example 8.3.*** Let $V$ be the real [complex] vector space of all continuous functions on the interval $[0,1]$ that are real-valued [complex-valued]. For $\alpha = \alpha(t)$ and $\beta = \beta(t)$ in $V$, let

$$(\alpha,\beta) = \int_0^1 \alpha(t)\overline{\beta(t)}\, dt.$$

This is an inner product on $V$. Note that if $\beta(t)$ is real-valued, $\overline{\beta(t)} = \beta(t)$.

We observe that the properties 1′, 2′, and 3′ in Definition 8.2 reduce to properties 1, 2, and 3 in Definition 8.1 if $(\alpha,\beta)$ is assumed to be real and if $c$ is real. Many properties of the inner product we are going to study are common to real and complex vector spaces; the proofs for the complex case will be valid for the real case if we only note that $(\alpha,\beta)$ is real-valued and that scalars are real for the real vector space.

**Definition 8.3**

A real or complex vector space is called an *inner product space* when it is provided with a certain inner product.

**Definition 8.4**

In an inner product space $V$ the *norm* (or *length*) of a vector $\alpha$ is $\sqrt{(\alpha,\alpha)}$, where we note that $(\alpha,\alpha) \geq 0$. We denote the norm of $\alpha$ by $\|\alpha\|$.

The first important property we shall prove is:

***Theorem 8.1.*** *In an inner product space $V$ we have the following inequality of Schwarz:*

$$|(\alpha,\beta)| \leq \|\alpha\| \, \|\beta\| \qquad \text{for all } \alpha, \beta \in V;$$

*the equality holds if and only if $\alpha$ and $\beta$ are linearly dependent.*

*Proof.* For any $a, b \in \mathcal{C}$ we have $(a\alpha - b\beta, a\alpha - b\beta) \geq 0$. On the other hand, we have

$$\begin{aligned}(a\alpha - b\beta, a\alpha - b\beta) &= (a\alpha,a\alpha) - (a\alpha,b\beta) - (b\beta,a\alpha) + (b\beta,b\beta)\\ &= a\bar{a}(\alpha,\alpha) - a\bar{b}(\alpha,\beta) - b\bar{a}(\beta,\alpha) + b\bar{b}(\beta,\beta)\\ &= |a|^2 \|\alpha\|^2 - 2 \operatorname{Re} \{a\bar{b}(\alpha,\beta)\} + |b|^2 \|\beta\|^2,\end{aligned}$$

where Re { } denotes the real part of the complex number written inside the braces. Now let us take $a = \|\beta\|^2$ and $b = (\alpha,\beta)$. We obtain

$$\|\alpha\|^2 \|\beta\|^4 - 2|(\alpha,\beta)|^2 \|\beta\|^2 + |(\alpha,\beta)|^2 \|\beta\|^2 = \|\beta\|^2 (\|\alpha\|^2 \|\beta\|^2 - |(\alpha,\beta)|^2) \geq 0.$$

Hence

$$\|\alpha\|^2 \|\beta\|^2 \geq |(\alpha,\beta)|^2.$$

Taking the positive square roots, we obtain

$$|(\alpha,\beta)| \leq \|\alpha\| \, \|\beta\|,$$

proving the inequality.

Assume that $\alpha$ and $\beta$ are linearly dependent. If $\alpha = 0$, then obviously the equality holds. If $\alpha \neq 0$, let $\beta = c\alpha$ for some scalar $c$. We have

$$|(\alpha,\beta)| = |(\alpha,c\alpha)| = |\bar{c}(\alpha,\alpha)| = |c|(\alpha,\alpha)$$

and

$$\|\alpha\| \, \|\beta\| = \sqrt{(\alpha,\alpha)}\sqrt{(c\alpha,c\alpha)} = \sqrt{(\alpha,\alpha)}\sqrt{|c|^2(\alpha,\alpha)} = |c|(\alpha,\alpha),$$

so that again the equality holds. Conversely, suppose that

$$|(\alpha,\beta)| = \|\alpha\| \, \|\beta\|.$$

In the proof of the inequality, we then have

$$(a\alpha - b\beta, a\alpha - b\beta) = 0, \qquad \text{where } a = \|\beta\|^2 \text{ and } b = (\alpha,\beta).$$

If $\beta \neq 0$, then $a \neq 0$, and $\alpha$ and $\beta$ are linearly dependent. If $\beta = 0$, then, of course, $\alpha$ and $\beta$ are linearly dependent.

***Example 8.4.*** In $\mathcal{C}^n$ with the standard inner product, we have

$$\left|\sum_{i=1}^{n} x_i\bar{y}_i\right| \leq \sqrt{\sum_{i=1}^{n} |x_i|^2} \cdot \sqrt{\sum_{i=1}^{n} |y_i|^2}.$$

for any $\alpha = (x_i)$ and $\beta = (y_i)$. This is called the *Cauchy inequality.* In $\mathcal{R}^n$, we have

$$\left|\sum_{i=1}^{n} x_i y_i\right| \leq \sqrt{\sum_{i=1}^{n} x_i^2} \cdot \sqrt{\sum_{i=1}^{n} y_i^2}.$$

***Proposition 8.2.*** *The norm has the following properties:*

1. $\|\alpha\| \geq 0$; $\|\alpha\| = 0$ *if and only if* $\alpha = 0$.
2. $\|c\alpha\| = |c|\,\|\alpha\|$ *for* $\alpha \in V$ *and for any scalar* $c$.
3. $\|\alpha + \beta\| \leq \|\alpha\| + \|\beta\|$.

*Proof.* Since (1) and (2) are obvious, we shall prove only (3). Using the fact that $\text{Re}\,\{c\} \leq |c|$ for any complex number $c$, we have

$$\|\alpha + \beta\|^2 = (\alpha + \beta, \alpha + \beta) = (\alpha,\alpha) + (\beta,\beta) + 2\,\text{Re}\,\{(\alpha,\beta)\}$$
$$\leq (\alpha,\alpha) + (\beta,\beta) + 2\,|(\alpha,\beta)|$$

which is $\leq (\alpha,\alpha) + (\beta,\beta) + 2\,\|\alpha\|\,\|\beta\| = (\|\alpha\| + \|\beta\|)^2$

by the Schwarz inequality. Taking the positive square roots of both sides, we obtain

$$\|\alpha + \beta\| \leq \|\alpha\| + \|\beta\|.$$

***Example 8.5.*** Making use of Example 8.4 and Proposition 8.2, we shall prove the assertions stated in Example 8.2. Let

$$A = \sum_{n=1}^{\infty} |a_n|^2 \qquad \text{and} \qquad B = \sum_{n=1}^{\infty} |b_n|^2.$$

By (3) of Proposition 8.2 applied to the vector space $\mathcal{C}^n$ we have

$$\sqrt{\sum_{i=1}^{n} (a_i + b_i)\,\overline{(a_i + b_i)}} \leq \sqrt{\sum_{i=1}^{n} |a_i|^2} + \sqrt{\sum_{i=1}^{n} |b_i|^2} \leq \sqrt{A} + \sqrt{B}.$$

Since this holds for every $n$, we see that

$$\sum_{i=1}^{\infty} (a_i + b_i)\,\overline{(a_i + b_i)} \leq (\sqrt{A} + \sqrt{B})^2,$$

proving that the sum of two square-summable sequences $\alpha = (a_i)$ and $\beta = (b_i)$ is square-summable. The proof for $c\alpha = (ca_i)$ is easy. For any $\alpha = (a_i)$ and $\beta = (b_i)$ we have

$$\left|\sum_{i=m+1}^{n} a_i\bar{b}_i\right| \leq \sqrt{\sum_{i=m+1}^{n} |a_i|^2} \cdot \sqrt{\sum_{i=m+1}^{n} |b_i|^2}, \qquad \text{where } n > m,$$

by the Schwarz inequality applied to $\mathcal{C}^{n-m}$. Since each term of the right-hand side converges to 0 as $m, n \to \infty$, it follows that $\sum_{i=1}^{\infty} a_i\bar{b}_i$ is convergent.

**Definition 8.5**

A vector of norm 1 is called a *unit vector*. Two vectors $\alpha$ and $\beta$ are said to be *orthogonal* to each other if $(\alpha,\beta) = 0$. A set of nonzero vectors $S$ is said to be *orthogonal* if any two distinct vectors in $S$ are orthogonal to each other. If, furthermore, every vector in $S$ is a unit vector, we say that $S$ is *orthonormal*.

***Example 8.6.*** In $\mathbb{R}^2$ with standard inner product, the set of vectors $\{(\cos\theta, \sin\theta), (-\sin\theta, \cos\theta)\}$ is orthonormal for any value of $\theta$, $0 \leq \theta < 2\pi$.

We prove:

***Proposition 8.3.*** *An orthogonal set of vectors is linearly independent.*

*Proof.* We shall show that any finite number of elements $\alpha_1, \ldots, \alpha_k$ in the given set are linearly independent. Suppose that

$$\sum_{i=1}^{k} c_i\alpha_i = 0.$$

Taking the inner product with $\alpha_j$, $1 \leq j \leq k$, we have

$$\left(\sum_{i=1}^{n} c_i\alpha_i,\alpha_j\right) = \sum_{i=1}^{n} c_i(\alpha_i,\alpha_j) = c_j = 0.$$

***Theorem 8.4.*** *In an inner product space $V$ of finite dimension, let $\{\alpha_1, \ldots, \alpha_n\}$ be an arbitrary basis. Then there exists an orthonormal basis $\{\beta_1, \ldots, \beta_n\}$ such that each $\beta_i$ is a linear combination of $\alpha_1, \ldots, \alpha_i$.*

*Proof.* Let

$$\beta_1 = \frac{\alpha_1}{\|\alpha_1\|},$$

which is a unit vector. (We say that we *normalize* $\alpha_1$ to obtain $\beta_1$.) Assume that we can find an orthonormal set $\{\beta_1, \ldots, \beta_r\}$ such that each $\beta_i$, $1 \leq i \leq r$, is a linear combination of $\alpha_1, \ldots, \alpha_i$. We shall show that we can then find $\beta_{r+1}$ in such a way that $\{\beta_1, \ldots, \beta_{r+1}\}$ is orthonormal and $\beta_{r+1}$ is a linear combination of $\alpha_1, \ldots, \alpha_{r+1}$.

Consider

$$\gamma = \alpha_{r+1} - \sum_{i=1}^{r} x_i\beta_i,$$

where the coefficients $x_1, \ldots, x_r$ will soon be determined. Then we have

$$(\gamma,\beta_j) = (\alpha_{r+1},\beta_j) - \sum_{i=1}^{r} x_i(\beta_i,\beta_j) = (\alpha_{r+1},\beta_j) - x_j$$

for any $j$, $1 \leq j \leq r$. Let us now take

$$x_j = (\alpha_{r+1},\beta_j), \qquad 1 \leq j \leq r,$$

and
$$\gamma = \alpha_{r+1} - \sum_{i=1}^{r} x_i\beta_i,$$

which is not 0, since $\alpha_{r+1}$ is not linearly dependent on $\beta_1, \ldots, \beta_r$. Then we have $(\gamma,\beta_j) = 0$ for $1 \leq j \leq r$. Normalizing $\gamma$, we obtain $\beta_{r+1} = \gamma/\|\gamma\|$, which is orthogonal to $\beta_1, \ldots, \beta_r$. Since $\beta_{r+1}$ is a linear combination of $\beta_1, \ldots, \beta_r, \alpha_{r+1}$ and since each $\beta_i$, $1 \leq i \leq r$, is a linear combination of $\alpha_1, \ldots, \alpha_r$, we see that $\beta_{r+1}$ is a linear combination of $\alpha_1, \ldots, \alpha_{r+1}$.

By this induction process we have an orthonormal set $\{\beta_1, \ldots, \beta_n\}$ such that each $\beta_i$ is a linear combination of $\alpha_1, \ldots, \alpha_i$. By Proposition 8.3, $\beta_1, \ldots, \beta_n$ are linearly independent and hence form a basis of $V$, since $\dim V = n$.

The process above is called the *Gram-Schmidt orthogonalization process.*

***Corollary.*** *In an n-dimensional inner product space, there is an orthonormal basis.*

***Remark.*** For a real inner product space of finite dimension the existence of an orthonormal basis has already been proved in Theorem 4.19.

***Proposition 8.5.*** *Let $\{\alpha_1, \ldots, \alpha_n\}$ be an orthonormal basis in an inner product space V.*

**1.** *For any $\alpha \in V$, we have*
$$\alpha = \sum_{i=1}^{n} (\alpha,\alpha_i)\alpha_i.$$

**2.** *For any $\alpha = \sum_{i=1}^{n} x_i\alpha_i$ and $\beta = \sum_{i=1}^{n} y_i\alpha_i$, we have*
$$(\alpha,\beta) = \sum_{i=1}^{n} x_i\bar{y}_i.$$

*(When V is a real vector space, we omit the sign* $\bar{\ }$.)

*Proof.* 1. Let $\alpha = \sum_{i=1}^{n} x_i\alpha_i$ and take the inner product with each $\alpha_j$. We have
$$(\alpha,\alpha_j) = \left(\sum_{i=1}^{n} x_i\alpha_i,\alpha_j\right) = \sum_{i=1}^{n} x_i\delta_{ij} = x_j.$$

2. We have
$$(\alpha,\beta) = \sum_{i,j=1}^{n} x_i\bar{y}_j(\alpha_i,\alpha_j) = \sum_{i,j=1}^{n} x_i\,\bar{y}_j\delta_{ij} = \sum_{i=1}^{n} x_i\bar{y}_i.$$

***Example 8.7.*** In $V = \mathcal{R}^n$ or $\mathcal{C}^n$, the standard basis $\{\epsilon_1, \ldots, \epsilon_n\}$ is orthonormal with respect to the standard inner product. The expression in (2) of Proposition 8.5 is nothing but the definition of the standard inner product.

Corresponding to Proposition 3.7, we may interpret an orthonormal basis in the following way. First we give:

**Definition 8.6**

A linear mapping $A$ of an inner product space $V$ into another inner product space $W$ is said to *preserve inner product* (or called an *isometry*) if

$$(A(\alpha), A(\beta)) = (\alpha, \beta) \qquad \text{for all } \alpha, \beta \in V.$$

We have:

***Proposition 8.6.*** *For any n-dimensional real [complex] inner product space V, there is a one-to-one correspondence between the set of orthonormal bases in V and the set of all isometries of* $\mathcal{R}^n$ $[\mathcal{C}^n]$ *with standard inner product onto V in the following fashion. For any isometry* $\Phi$, $\{\Phi(\epsilon_1), \ldots, \Phi(\epsilon_n)\}$ *is an orthonormal basis of V, and for any orthonormal basis* $\{\alpha_1, \ldots, \alpha_n\}$ *of V there exists a unique isometry* $\Phi$ *such that* $\Phi(\epsilon_i) = \alpha_i$, $1 \leq i \leq n$.

*Proof.* Let $\Phi$ be an isometry of $\mathcal{R}^n$ $[\mathcal{C}^n]$ onto $V$ and let $\alpha_i = \Phi(\epsilon_i)$. Then

$$(\alpha_i, \alpha_j) = (\Phi(\epsilon_i), \Phi(\epsilon_j)) = (\epsilon_i, \epsilon_j) = \delta_{ij},$$

showing that $\{\alpha_1, \ldots, \alpha_n\}$ is an orthonormal basis. Conversely, for any orthonormal basis $\{\alpha_1, \ldots, \alpha_n\}$ let $\Phi$ be the unique linear isomorphism of $\mathcal{R}^n$ $[\mathcal{C}^n]$ onto $V$ such that $\Phi(\epsilon_i) = \alpha_i$ (cf. Proposition 3.7). For any $\alpha, \beta \in \mathcal{R}^n$ $[\mathcal{C}^n]$ let

$$\alpha = \sum_{i=1}^{n} x_i \epsilon_i \qquad \text{and} \qquad \beta = \sum_{i=1}^{n} y_i \epsilon_i.$$

Then

$$(\alpha, \beta) = \sum_{i=1}^{n} x_i \bar{y}_i = \left( \sum_{i=1}^{n} x_i \alpha_i, \sum_{i=1}^{n} y_i \alpha_i \right) = (\Phi(\alpha), \Phi(\beta)),$$

showing that $\Phi$ is an isometry.

We shall prove:

***Proposition 8.7.*** *Let A be a linear mapping of an inner product space V into an inner product space W.*

**1.** *A is an isometry if and only if* $\|A(\alpha)\| = \|\alpha\|$ *for every* $\alpha \in V$.

**2.** *If $A$ is an isometry, then it is one-to-one; in particular, if* $\dim V = \dim W = n$, *then it is a linear isomorphism of $V$ onto $W$.*

*Proof.* 1. If $A$ is an isometry, then we have

$$\|\alpha\|^2 = (\alpha,\alpha) = (A(\alpha),A(\alpha)) = \|A(\alpha)\|^2$$

so that $\|\alpha\| = \|A(\alpha)\|$ for every $\alpha \in V$. Conversely, assume that $A$ preserves norm: $\|A\alpha\| = \|\alpha\|$ for every $\alpha \in V$. For any $\alpha, \beta \in V$, we have

$$(A(\alpha+\beta),A(\alpha+\beta)) = (\alpha+\beta, \alpha+\beta).$$

Hence

$$(A(\alpha),A(\alpha)) + (A(\alpha),A(\beta)) + (A(\beta),A(\alpha)) + (A(\beta),A(\beta))$$
$$= (\alpha,\alpha) + (\alpha,\beta) + (\beta,\alpha) + (\beta,\beta).$$

Since $(A(\alpha),A(\alpha)) = (\alpha,\alpha)$ and $(A(\beta),A(\beta)) = (\beta,\beta)$, we have

$$(A(\alpha),A(\beta)) + (A(\beta),A(\alpha)) = (\alpha,\beta) + (\beta,\alpha),$$

that is,

$$\operatorname{Re}\{(A(\alpha),A(\beta))\} = \operatorname{Re}\{(\alpha,\beta)\}.$$

Replacing $\beta$ by $i\beta$, we obtain

$$\operatorname{Im}\{(A(\alpha),A(\beta))\} = \operatorname{Im}\{(\alpha,\beta)\},$$

where Im denotes the imaginary part of the complex number. Thus

$$(A(\alpha),A(\beta)) = (\alpha,\beta).$$

2. If $A(\alpha) = 0$, then

$$(\alpha,\alpha) = (A(\alpha),A(\alpha)) = 0,$$

which implies $\alpha = 0$.

**EXERCISE 8.1**

☆ **1.** In an inner product space $V$, define

$$d(\alpha,\beta) = \|\beta - \alpha\| \qquad \text{for } \alpha, \beta \in V.$$

Prove that $d$ is a *distance function* (or *metric*) on the set $V$, that is, it satisfies the following conditions:

(*a*) $d(\alpha,\beta) \geq 0$; $d(\alpha,\beta) = 0$ if and only if $\alpha = \beta$.

(*b*) $d(\beta,\alpha) = d(\alpha,\beta)$.

(*c*) $d(\alpha,\gamma) \leq d(\alpha,\beta) + d(\beta,\gamma)$.

**2.** In an inner product space $V$, show that

$$\left\|\frac{\alpha+\beta}{2}\right\|^2 + \left\|\frac{\alpha-\beta}{2}\right\|^2 = \frac{(\|\alpha\|^2 + \|\beta\|^2)}{2}.$$

›› **3.** In $\mathbb{R}^3$ with standard inner product, apply the Gram-Schmidt process to the following basis:

$$\alpha_1 = (1,0,1), \qquad \alpha_2 = (1,2,0), \qquad \alpha_3 = (0,2,3).$$

**4.** In $\mathcal{C}^2$ with standard inner product, apply the Gram-Schmidt process to the following basis: $\alpha_1 = (1,i)$, $\alpha_2 = (-i,-1)$.

**5.** In a complex inner product space, prove

$$2(\alpha,\beta) = \|\alpha + \beta\|^2 + i\|\alpha + i\beta\|^2 - (1 + i)(\|\alpha\|^2 + \|\beta\|^2).$$

In a real inner product space, prove

$$2(\alpha,\beta) = \|\alpha + \beta\|^2 - \|\alpha\|^2 - \|\beta\|^2.$$

»☆ **6.** Suppose that a real or complex vector space $V$ admits two inner products denoted by $(\alpha,\beta)_1$ and $(\alpha,\beta)_2$. Denote by $\| \ \|_1$ and $\| \ \|_2$ the corresponding norms. Prove that if $\|\alpha\|_1 = \|\alpha\|_2$ for every $\alpha \in V$, then the two inner products coincide.

**7.** Let $V$ be an inner product space of dimension $n$ and let $\{\alpha_1, \ldots, \alpha_n\}$ be an orthonormal basis of $V$. Prove:

(*a*) A linear transformation $A$ is an isometry of $V$ into itself if and only if $\{A(\alpha_1), \ldots, A(\alpha_n)\}$ is an orthonormal basis of $V$.

(*b*) For any orthonormal basis $\{\beta_1, \ldots, \beta_n\}$ of $V$ there exists a unique isometry $A$ of $V$ such that $A(\alpha_i) = \beta_i$ for $1 \leq i \leq n$.

»☆ **8.** Let $\mathfrak{G}$ be a finite group of linear transformations of an $n$-dimensional inner product space $V$. Prove that $V$ admits an inner product which is preserved by every $A \in \mathfrak{G}$.

☆ **9.** Let $V$ be a real vector space and assume that $f(\alpha,\beta)$ is a complex-valued function on $V \times V$ which has the following properties: (*a*) $f(\beta,\alpha) = \overline{f(\alpha,\beta)}$. (*b*) $f(\alpha,\beta)$ is linear in $\alpha$. (*c*) $f(\alpha,\alpha)$ is real and $\geq 0$; $f(\alpha,\alpha) = 0$ if and only if $\alpha = 0$. Prove that $(\alpha,\beta) = \operatorname{Re}\{f(\alpha,\beta)\}$ is an inner product on $V$. Conversely, any inner product on $V$ can be considered as the real part of a function $f$ satisfying (*a*), (*b*), and (*c*).

☆ **10.** A real or complex vector space $V$ is called a *normed space* if there is a function $\alpha \in V \to \|\alpha\| \in \mathcal{R}$ satisfying (1), (2), and (3) of Proposition 8.2. Such a function is called a *norm*, and $\|\alpha\|$ is the norm of $\alpha$. Show that $\mathcal{R}^n$ $[\mathcal{C}^n]$ is a normed space with respect to each of the following functions:

$$(a)\ \|\xi\| = \max_{1 \leq i \leq n} \{|x_i|\}, \qquad (b)\ \|\xi\| = \sum_{i=1}^{n} |x_i|,$$

$$(c)\ \|\xi\| = \sqrt{\sum_{i=1}^{n} |x_i|^2}, \qquad (d)\ \|\xi\| = \left(\sum_{i=1}^{n} |x_i|^p\right)^{1/p},$$

where $\xi = (x_i)$ and $p$ is a positive integer.

»☆ **11.** If $A$ is an $n \times n$ real [complex] matrix with row vectors $\alpha_1, \ldots, \alpha_n$, show that

$$|\det A| \leq \|\alpha_1\| \cdots \|\alpha_n\|,$$

where $\|\alpha_i\|$ is the norm of $\alpha_i$ in $\mathcal{R}^n$ $[\mathcal{C}^n]$ with standard inner product.

☆ **12.** Let $\alpha_1, \ldots, \alpha_p$ be elements in an inner product space $V$ and define the *Gramian* of $\alpha_1, \ldots, \alpha_p$ by

$$G(\alpha_1, \ldots, \alpha_p) = \det [(\alpha_i,\alpha_j)].$$

Prove that $\alpha_1, \ldots, \alpha_p$ are linearly dependent if and only if $G(\alpha_1, \ldots, \alpha_p) = 0$.

☆ **13.** Let $V$ be an $n$-dimensional inner product space. If $A$ is a mapping of $V$ into itself such that:

(*a*) $\|A(\alpha)\| = \|\alpha\|$ for all $\alpha \in V$,

(*b*) $A(c\alpha) = cA(\alpha)$ for all $\alpha \in V$ and for any scalar $c$,

then is $A$ linear (and hence an isometry)?

☆ **14.** Let $V$ be an $n$-dimensional real inner product space. A linear transformation $A$ of $V$ is called a *conformal transformation* if there exists a $c \in \mathcal{R}$, $c \neq 0$, such that $(A(\alpha),A(\beta)) = c(\alpha,\beta)$ for all $\alpha, \beta \in V$. Prove that:

(*a*) If $A$ is conformal, then $A$ is nonsingular.

(*b*) If $A$ is a nonsingular linear transformation such that $(\alpha,\beta) = 0$ implies $(A(\alpha),A(\beta)) = 0$, then $A$ is conformal.

(*c*) The product of two conformal transformations and the inverse of a conformal transformation are conformal.

☆**15.** Let $V$ be an $n$-dimensional inner product space (or normed space). Prove that a mapping $\alpha: t \to \alpha(t)$ of the unit interval $I$ into $V$ is continuous (cf. Exercise 3.3, number 9) if and only if the following condition is satisfied: Let $t_0 \in I$. For any $\epsilon > 0$, there exists $\delta > 0$ such that $|t - t_0| < \delta$ implies $\|\alpha(t) - \alpha(t_0)\| < \epsilon$.

☆**16.** In the vector space $\mathcal{C}^n{}_n$ $[\mathcal{R}^n{}_n]$, define

$$(A,B) = \sum_{i,j=1}^{n} a_{ij}\bar{b}_{ij} \qquad \text{for } A = [a_{ij}],\ B = [b_{ij}].$$

(*a*) Show that this is an inner product.

(*b*) Show that the norm for this inner product satisfies

$$\|AB\| \leq \|A\|\,\|B\| \qquad \text{for all } A,\ B.$$

☆**17.** In a normed space $V$ (cf. number 10), a sequence $\{\alpha_k\}$ is said to *converge* to $\alpha \in V$ if $\|\alpha - \alpha_k\| \to 0$ as $k \to \infty$. A sequence $\{\alpha_k\}$ is called a *Cauchy sequence* if $\|\alpha_m - \alpha_n\| \to 0$ as $m, n \to \infty$.

(*a*) Prove that a convergent sequence is a Cauchy sequence.

(*b*) Show that, in each example of number 10, every Cauchy sequence converges to some element. (We say that the normed space is *complete*. This and the following are based on completeness of $\mathcal{R}$ and $\mathcal{C}$.)

(*c*) Show that any inner product space of finite dimension is complete as a normed space. (In fact, any normed space of finite dimension is complete.)

☆**18.** (Cf. numbers 16 and 17) Let $A \in \mathcal{C}^n{}_n[\mathcal{R}^n{}_n]$. The sequence $\mathbf{A}_k = \sum_{i=0}^{k} (\mathbf{A}^i/i!)$ is a Cauchy sequence and hence converges to a matrix $\mathbf{B}$. We define $\mathbf{B} = \exp \mathbf{A}$, thus getting the *exponential function* of a matrix. (Cf. Exercise 7.3, number 3.)

## 8.2 SUBSPACES

Let $V$ be an inner product space. If $U$ is a subspace of $V$, the restriction of the inner product $(\alpha,\beta)$ to $\alpha, \beta \in U$ makes $U$ into an inner product space. On the other hand, let

$$U^\perp = \{\alpha \in V;\ (\beta,\alpha) = 0 \text{ for every } \beta \in U\}.$$

It is easily verified that $U^\perp$ is a subspace of $V$. We have $U \cap U^\perp = (0)$, since if $\alpha \in U \cap U^\perp$, then $(\alpha,\alpha) = 0$ and hence $\alpha = 0$.

We shall prove:

***Theorem 8.8.*** *Let $V$ be an inner product space of finite dimension. For any subspace $U$ of $V$, we have*

$$V = U \oplus U^\perp.$$

*Proof.* Let $\{\alpha_1, \ldots, \alpha_m, \alpha_{m+1}, \ldots, \alpha_n\}$ be a basis of $V$ such that $\{\alpha_1, \ldots, \alpha_m\}$ is a basis of $U$ (cf. Theorem 2.14). [We may assume that $U \neq (0)$, because for $U = (0)$ we have $U^\perp = V$ and our assertion is trivial.] If we apply the Gram-Schmidt process to $\{\alpha_1, \ldots, \alpha_n\}$ to obtain an orthonormal basis $\{\beta_1, \ldots, \beta_n\}$, then $\beta_1, \ldots, \beta_m$ form a basis of $U$. We show that $\beta_{m+1}, \ldots, \beta_n$ are in $U^\perp$. In fact, for each vector $\sum_{i=1}^{m} c_i\beta_i$ in $U$ we have

$$\left(\sum_{i=1}^{m} c_i\beta_i, \beta_j\right) = \sum_{i=1}^{m} c_i(\beta_i, \beta_j) = \sum_{i=1}^{m} c_i\delta_{ij} = 0$$

for each $j$, $m + 1 \leq j \leq n$. Since we know that $U \cap U^\perp = (0)$, we have $V = U \oplus U^\perp$. (As a matter of fact, $\{\beta_{m+1}, \ldots, \beta_n\}$ is a basis of $U^\perp$.)

**Definition 8.7**

$U^\perp$ is called the *orthogonal complement* of $U$.

It is evident that $(U^\perp)^\perp = U$, in view of Theorem 8.8.

In the decomposition $V = U \oplus U^\perp$, every $\alpha \in V$ can be written uniquely in the form

$$\alpha = \alpha_1 + \alpha_2, \qquad \text{where } \alpha_1 \in U \text{ and } \alpha_2 \in U^\perp.$$

We define linear transformations $P$ and $Q$ by

$$P(\alpha) = \alpha_1 \qquad \text{and} \qquad Q(\alpha) = \alpha_2.$$

We have

$$\alpha = P(\alpha) + Q(\alpha) \qquad \text{for every } \alpha \in V,$$

that is,

$$I = P + Q.$$

**Definition 8.8**

$P$ is called the *projection* (*orthogonal projection*) of $V$ onto $U$.

It follows that $Q = I - P$ is the projection of $V$ onto $U^\perp$.

We shall prove:

***Proposition 8.9.*** *Let $V$ be an inner product space of finite dimension. The projection $P$ of $V$ onto a subspace $U$ has the following properties:*

1. $P^2 = P$.
2. $(P(\alpha), \beta) = (\alpha, P(\beta))$ *for all* $\alpha, \beta \in V$.
3. $U = \{\alpha \in V;\ P(\alpha) = \alpha\} = \{P(\gamma);\ \gamma \in V\}$.

*Conversely, any linear transformation $P$ of $V$ which has properties* 1 *and* 2 *is the projection of $V$ onto the subspace $U = \{\alpha \in V;\ P(\alpha) = \alpha\}$.*

*Proof.* Let $P$ be the projection of $V$ onto $U$. To prove (1), it is sufficient to prove $P^2(\alpha) = P(\alpha)$ for $\alpha \in U$ and for $\alpha \in U^\perp$. If $\alpha \in U$, then $P(\alpha) = \alpha$ and $P^2(\alpha) = P(P(\alpha)) = P(\alpha) = \alpha$ by definition of $P$. If $\alpha \in U^\perp$, then $P(\alpha) = 0$ and $P^2(\alpha) = P(P(\alpha)) = P(0) = 0$. In both cases $P^2(\alpha) = P(\alpha)$, proving $P^2 = P$. Property 3 also is obvious.

In order to prove (2), it is sufficient to verify $(P(\alpha),\beta) = (\alpha,P(\beta))$ in the following four cases: (*a*) $\alpha, \beta \in U$; (*b*) $\alpha \in U, \beta \in U^\perp$; (*c*) $\alpha \in U^\perp$, $\beta \in U$; (*d*) $\alpha, \beta \in U^\perp$. In each case, the verification is straightforward.

Conversely, let $P$ satisfy (1) and (2), and let

$$U = \{\alpha \in V;\ P(\alpha) = \alpha\},$$

which is a subspace of $V$. For any $\alpha \in V$ we have

$$\alpha = P(\alpha) + (I - P)(\alpha),$$

where $P(P(\alpha)) = P^2(\alpha) = P(\alpha)$ by condition 1, that is, $P(\alpha) \in U$. We show that $(I - P)(\alpha) = \alpha - P(\alpha)$ is in $U^\perp$. For any $\beta \in U$ we have

$$\begin{aligned}(\beta, \alpha - P(\alpha)) &= (\beta,\alpha) - (\beta,P(\alpha)) \\ &= (\beta,\alpha) - (P(\beta),\alpha) = (\beta,\alpha) - (\beta,\alpha) = 0\end{aligned}$$

by using condition 2 and $P(\beta) = \beta$. This shows that $\alpha - P(\alpha) \in U^\perp$. Thus

$$\alpha = P(\alpha) + (\alpha - P(\alpha)), \qquad \text{where } P(\alpha) \in U \text{ and } \alpha - P(\alpha) \in U^\perp.$$

If $\alpha \in U \cap U^\perp$, then $(\alpha,\alpha) = 0$, which implies $\alpha = 0$. Thus we have $V = U \oplus U^\perp$. From the decomposition of $\alpha$ given above it follows that $P$ is the projection of $V$ onto $U$.

In Theorem 8.8 we have the direct-sum decomposition of a given inner product space. Reversing the situation, we may start with two inner product spaces $V_1$ and $V_2$ (both real or both complex) and define an inner product in the direct sum $V = V_1 \oplus V_2$ as follows:

For any $\alpha = (\alpha_1,\alpha_2) \in V$ and $\beta = (\beta_1,\beta_2) \in V$, we define

$$(\alpha,\beta) = (\alpha_1,\beta_1)_1 + (\alpha_2,\beta_2)_2,$$

where $(\ ,\ )_i$ denotes the given inner product in $V_i$, $i = 1, 2$. It is easily verified that $(\alpha,\beta)$ is an inner product on $V$. Moreover, if we identify $V_1$ and $V_2$ with the subspaces

$$\{(\alpha_1,0);\ \alpha_1 \in V_1\} \qquad \text{and} \qquad \{(0,\alpha_2);\ \alpha_2 \in V_2\},$$

respectively, as in Proposition 4.7, then $V_1$ and $V_2$ are orthogonal complements of each other in $V$.

### Definition 8.9

The inner product ( , ) defined above in the direct sum $V_1 \oplus V_2$ is called the *direct sum of the inner products* in $V_1$ and $V_2$. $V_1 \oplus V_2$ with this inner product is called the *direct sum of inner product spaces.*

It is easy to generalize the direct sum of two inner product spaces to the case of any finite number of inner product spaces $V_1, \ldots, V_k$. In $V = V_1 \oplus \cdots \oplus V_k$, we define

$$(\alpha,\beta) = \sum_{i=1}^{k} (\alpha_i,\beta_i)_i \qquad \text{for } \alpha = (\alpha_1, \ldots, \alpha_k),\ \beta = (\beta_1, \ldots, \beta_k),$$

where $( \ , \ )_i$ denotes the given inner product in $V_i$. Identifying each $V_i$ with a subspace of $V$ in the proper fashion, we see that $V_i$ and $V_j$ are orthogonal to each other whenever $i \neq j$.

#### EXERCISE 8.2

›› 1. Let $S$ be a subset of an inner product space $V$. Show that

$$S^{\perp} = \{\alpha \in V;\ (\alpha,\beta) = 0 \text{ for every } \beta \in S\}$$

is a subspace of $V$. Show that $(S^{\perp})^{\perp}$ contains the span of $S$ (cf. Definition 2.6).

2. If $V$ is finite-dimensional in number 1, then $(S^{\perp})^{\perp}$ is the span of $S$.

3. Let $V = \mathbb{R}^2$ with standard inner product and let $U = \mathrm{Sp}\{(1,1)\}$. Let $P$ be the projection of $V$ onto $U$ and find the images by $P$ of the following vectors:

$$(a)\ (1,0); \qquad (b)\ (0,1); \qquad (c)\ (3,4).$$

Illustrate $P$ geometrically.

4. Assume that an inner product space $V$ is the direct sum

$$V = V_1 \oplus V_2 \oplus \cdots \oplus V_k,$$

where the subspaces $V_i$ are orthogonal to each other. Show that $(V_i)^{\perp}$ is the direct sum of all $V_j$'s such that $j \neq i$. Let $P_i$ be the projection of $V$ onto $V_i$. Show then that:

(a) $I = P_1 + P_2 + \cdots + P_k$.
(b) $(P_i(\alpha),\beta) = (\alpha,P_i(\beta))$ for all $\alpha, \beta \in V$.
(c) $P_i^2 = P_i$ for each $i$, and $P_iP_j = 0$ for $i \neq j$.

☆ 5. Let $V$ be an inner product space of finite dimension and assume that $P_1, \ldots, P_k$ are a set of linear transformations satisfying conditions (a), (b), (c) of number 4. Prove that $V$ is the direct product of mutually orthogonal subspaces $V_1, \ldots, V_k$ and that $P_i$ is the projection of $V$ onto $V_i$ for each $i$.

6. Show that $\mathbb{C}^n$ with standard inner product is the direct sum of $\mathbb{C}^m$ and $\mathbb{C}^{n-m}$, both with standard inner product, where $0 < m < n$.

›› 7. Let $P$ be a projection of an $n$-dimensional inner product space $V$ onto an $m$-dimensional subspace $U$. What is the minimal polynomial of $P$? What is its characteristic polynomial?

8. Let $V = V_1 \oplus V_2$ be the direct sum of inner product spaces $V_1$ and $V_2$. If $\{\alpha_1, \ldots, \alpha_m\}$ and $\{\alpha_{m+1}, \ldots, \alpha_n\}$ are orthonormal bases of $V_1$ and $V_2$, respectively, then $\{\alpha_1, \ldots, \alpha_n\}$ is an orthonormal basis of $V$.

» **9.** Let $V$ be an $n$-dimensional inner product space. If $A$ is a projection of $V$ onto a subspace $U$, then

$$\|A(\alpha)\| \leq \|\alpha\| \qquad \text{for all } \alpha \in V.$$

If the equality holds for every $\alpha \in V$, then $A$ is the identity transformation.

**10.** Let $W$ be a subspace of an $n$-dimensional inner product space $V$. Show that any orthonormal basis of $W$ can be extended to an orthonormal basis of $V$.

☆ **11.** Let $A$ be a linear transformation of an $n$-dimensional complex inner product space $V$. Prove that there exists an orthonormal basis in $V$ with respect to which $A$ can be represented by a superdiagonal matrix (cf. Theorem 7.3).

## 8.3 DUAL SPACE

Let $V$ be an inner product space. For any $\alpha \in V$, the mapping

$$\xi \in V \rightarrow (\xi,\alpha)$$

is a linear function on $V$. We shall prove the converse as follows:

***Theorem 8.10.*** *Let $V$ be an inner product space of finite dimension. For any linear function $\alpha^*$ on $V$, there exists one and only one element $\alpha \in V$ such that*

$$\langle \xi, \alpha^* \rangle = (\xi,\alpha) \qquad \textit{for all } \xi \in V.$$

*Proof.* We shall first prove the uniqueness. Suppose that $\alpha_1, \alpha_2 \in V$ satisfy

$$\langle \xi,\alpha^* \rangle = (\xi,\alpha_1) = (\xi,\alpha_2) \qquad \text{for all } \xi \in V.$$

Then we have

$$(\xi, \alpha_1 - \alpha_2) = 0 \qquad \text{for all } \xi \in V,$$

and, in particular, taking $\xi = \alpha_1 - \alpha_2$,

$$(\alpha_1 - \alpha_2, \alpha_1 - \alpha_2) = 0.$$

This implies that

$$\alpha_1 - \alpha_2 = 0, \qquad \text{that is,} \qquad \alpha_1 = \alpha_2.$$

In order to prove the existence, let $\{\alpha_1, \ldots, \alpha_n\}$ be an orthonormal basis of $V$ and set

$$\alpha = \sum_{i=1}^{n} x_i\alpha_i,$$

where the coefficients are to be determined. Since

$$(\alpha_j,\alpha) = \sum_{i=1}^{n} \bar{x}_i(\alpha_j,\alpha_i) = \bar{x}_j, \qquad 1 \leq j \leq n,$$

we see that if we take

$$x_j = \overline{\langle \alpha_j,\alpha^* \rangle} \qquad \text{for every } j,$$

we have $\qquad (\alpha_j,\alpha) = \langle \alpha_j,\alpha^* \rangle \qquad$ for every $j$,

and hence $\qquad (\xi,\alpha) = \langle \xi,\alpha^* \rangle \qquad$ for every $\xi \in V$.

We shall denote by $\psi$ the mapping of $V$ into the dual space $V^*$ that maps $\alpha \in V$ into the linear function $\xi \in V \to (\xi,\alpha)$; this mapping $\psi$ is one-to-one and onto by virtue of Theorem 8.10. It is linear if $V$ is a real inner product space and conjugate-linear if $V$ is a complex inner product space:

$$\psi(\alpha_1 + \alpha_2) = \psi(\alpha_1) + \psi(\alpha_2),$$
$$\psi(c\alpha) = \bar{c}\psi(\alpha).$$

The first condition is nothing but

$$(\xi, \alpha_1 + \alpha_2) = (\xi,\alpha_1) + (\xi,\alpha_2);$$

the second follows from

$$(\xi,c\alpha) = \bar{c}(\xi,\alpha),$$

where $\bar{c} = c$ if $c$ is real in the case where $V$ is real.

The dual space $V^*$ can be made into an inner product space by defining

$$(\alpha^*,\beta^*) = (\psi^{-1}(\beta^*),\psi^{-1}(\alpha^*)) \qquad \text{for } \alpha^*, \beta^* \in V^*.$$

The verification that this is an inner product is straightforward.

**Definition 8.10**

The inner product in $V^*$ defined above is called the *dual inner product.* $V^*$ provided with this inner product is called the *dual space* of the inner product space $V$.

***Proposition 8.11.*** *Let $V^*$ be the dual space of an $n$-dimensional inner product space $V$. If $\{\alpha_1, \ldots, \alpha_n\}$ is an orthonormal basis of $V$, then the dual basis $\{\alpha_1^*, \ldots, \alpha_n^*\}$ is orthonormal in $V^*$.*

*Proof.* By using the mapping $\psi\colon V \to V^*$, we have

$$\langle \alpha_i,\psi(\alpha_j) \rangle = (\alpha_i,\alpha_j) = \delta_{ij},$$

which shows that $\psi(\alpha_j) = \alpha_j^*$, $1 \leq j \leq n$. By definition of the dual inner product we have

$$(\alpha_i^*,\alpha_j^*) = (\psi^{-1}(\alpha_j^*),\psi^{-1}(\alpha_i^*)) = (\alpha_j,\alpha_i) = \delta_{ji},$$

showing that $\{\alpha_1^*, \ldots, \alpha_n^*\}$ is orthonormal.

We shall now define the notion of *adjoint* of a linear transformation $A$ of an $n$-dimensional inner product space $V$. Let $A^* = \psi^{-1}({}^tA)\psi$, which is a mapping of $V$ into itself as illustrated by the following diagram:

$$\begin{array}{ccc} V & \underset{A^*}{\overset{A}{\rightleftarrows}} & V \\ \psi^{-1}\uparrow & & \downarrow\psi \\ V^* & \overset{{}^tA}{\longleftarrow} & V^*, \end{array}$$

where ${}^tA$ is the transpose of $A$ (Definition 4.5). We shall verify that $A^*$ is linear. First, $A^*(\alpha + \beta) = A^*(\alpha) + A^*(\beta)$ is obvious. Second, we have

$$\begin{aligned} A^*(c\alpha) &= \psi^{-1}({}^tA)\psi(c\alpha) = \psi^{-1}({}^tA\,(\bar{c}\psi(\alpha))) \\ &= \psi^{-1}(\bar{c}\,{}^tA\,(\psi(\alpha)) = c\psi^{-1}\,({}^tA\psi(\alpha)) = cA^*(\alpha), \end{aligned}$$

since $\psi$ and $\psi^{-1}$ are conjugate-linear (linear if $V$ is real) and ${}^tA$ is linear.

**Definition 8.11**

$A^*$ is called the *adjoint* of $A$.

***Proposition 8.12.*** *The adjoint $A^*$ satisfies*

$$(A(\alpha),\beta) = (\alpha,A^*(\beta)) \qquad \textit{for all } \alpha, \beta \in V.$$

*Proof.* From the definition of $A^*$, we have

$$\psi A^* = {}^tA\psi.$$

For any $\alpha, \beta \in V$, we have

$$\langle \alpha, {}^tA\,(\psi(\beta))\rangle = \langle A(\alpha),\psi(\beta)\rangle = (A(\alpha),\beta)$$

by recalling the definitions of ${}^tA$ and $\psi$. On the other hand, we have

$$\langle \alpha,\psi(A^*(\beta))\rangle = (\alpha,A^*(\beta)).$$

Hence

$$(A(\alpha),\beta) = (\alpha,A^*(\beta)).$$

***Remark.*** The property in Proposition 8.12 characterizes $A^*$. For any $\beta \in V$, let $A^*(\beta)$ be the unique element in $V$ such that the linear function $\alpha \to (A(\alpha),\beta)$ is equal to $\psi(A^*(\beta))$, that is,

$$(A(\alpha),\beta) = (\alpha,A^*(\beta)) \qquad \text{for all } \alpha \in V.$$

It is easy to verify that $A^*(\beta)$ depends linearly on $\beta$.

Let us find the matrix representation of the adjoint. Let $\{\alpha_1, \ldots, \alpha_n\}$ be an orthonormal basis of $V$. A given linear transformation $A$ of $V$ is represented by a matrix $\mathbf{A} = [a_{ij}]$, where

$$A(\alpha_i) = \sum_{j=1}^{n} a_{ji}\alpha_j.$$

Assume that $A^*$ is represented by a matrix $\mathbf{B} = [b_{ij}]$ with respect to $\{\alpha_1, \ldots, \alpha_n\}$:

$$A^*(\alpha_i) = \sum_{j=1}^{n} b_{ji}\alpha_j.$$

From Proposition 8.12 we obtain

$$(A(\alpha_i),\alpha_k) = (\alpha_i,A^*(\alpha_k)).$$

Since

$$(A(\alpha_i),\alpha_k) = a_{ki} \qquad \text{and} \qquad (\alpha_i,A^*(\alpha_k)) = \bar{b}_{ik},$$

we see that

$$b_{ik} = \bar{a}_{ki}, \qquad \text{that is,} \qquad B = {}^t\bar{A}.$$

Thus **B** is the transpose of the *conjugate* $\bar{\mathbf{A}} = [\bar{a}_{ij}]$ of $\mathbf{A} = [a_{ij}]$ (or the conjugate of the transpose of **A**).

### Definition 8.12

For any $m \times n$ complex matrix **A** we denote the $n \times m$ matrix ${}^t\bar{\mathbf{A}}$ by $\mathbf{A}^*$. A complex matrix **A** is called a *hermitian matrix* if $\mathbf{A}^* = \mathbf{A}$. (For a real matrix **A**, $\mathbf{A}^*$ is nothing but ${}^t\mathbf{A}$; a real hermitian matrix is nothing but a real symmetric matrix.) $\mathbf{A}^*$ is called the *adjoint* of **A**.

***Example 8.8***

$$\mathbf{A} = \begin{bmatrix} 2 & 1+i & -3i \\ 1 & 3 & 2+i \\ 0 & i & 4i \end{bmatrix}, \qquad \mathbf{A}^* = \begin{bmatrix} 2 & 1 & 0 \\ 1-i & 3 & -i \\ 3i & 2-i & -4i \end{bmatrix}.$$

***Example 8.9.*** Let $\{\alpha_1, \ldots, \alpha_n\}$ be an arbitrary basis of an inner product space $V$. Then the inner product is completely determined by the matrix $\mathbf{A} = [a_{ij}]$, where

$$a_{ij} = (\alpha_i,\alpha_j) \qquad \text{for } 1 \leq i, j \leq n.$$

Then **A** is a symmetric or hermitian matrix according as $V$ is real or complex. Moreover, **A** is *positive-definite*, that is,

$$\sum_{i,j=1}^{n} a_{ij}x_i\bar{x}_j \geq 0 \qquad \text{for any } (x_i) \in \mathcal{R}^n \text{ or } \mathcal{C}^n,$$

and the equality holds if and only if all $x_i$'s are 0.

***Example 8.10.*** Let **A** be an arbitrary $n \times n$ complex matrix and let $\mathbf{B} = \mathbf{A}^*\mathbf{A}$. Then **B** is hermitian and positive-semidefinite.

#### EXERCISE 8.3

**1.** Let $V$ be an inner product space of finite dimension and let $V^*$ be the dual space and $(V^*)^*$ the dual space of $V^*$. Show that the mapping $A: V \to (V^*)^*$ in Theorem 4.3 preserves inner product.

»☆ **2.** Let $V$ be a real vector space of finite dimension and $V^*$ the dual space of $V$. Assume that there is a linear isomorphism $\theta$ of $V$ onto $V^*$ which maps any basis $\{\alpha_1, \ldots, \alpha_n\}$ upon its dual basis. Prove that

$$(\alpha,\beta) = \langle \alpha,\theta(\beta)\rangle \qquad \text{for } \alpha, \beta \in V$$

defines an inner product in $V$.

☆ **3.** Find an analogue of number 2 for a complex vector space of finite dimension.

**4.** For each of the following matrices write down its adjoint $\mathbf{A}^*$:

››(*a*) $\mathbf{A} = \begin{bmatrix} 3 & 2+i \\ i & -1+i \end{bmatrix}$; (*b*) $\mathbf{A} = \begin{bmatrix} e^{i\omega} & 0 \\ 0 & e^{-i\omega} \end{bmatrix}$ ($\omega$: real);

››(*c*) $\mathbf{A} = \begin{bmatrix} 2 & i \\ -i & 1 \end{bmatrix}$; (*d*) $\mathbf{A} = \begin{bmatrix} 1 & 0 & 0 \\ i & -i & 1 \\ 2 & -1 & i \end{bmatrix}$.

Which matrices are hermitian?

☆ **5.** For linear transformations $A$, $B$ of an inner product space of finite dimension, prove the following: $(A + B)^* = A^* + B^*$; $(cA)^* = \bar{c}A^*$; $(A^*)^* = A$; $(AB)^* = B^*A^*$.

**6.** Prove the assertion in Example 8.10.

## 8.4 HERMITIAN AND UNITARY TRANSFORMATIONS

In this section, $V$ will denote an $n$-dimensional complex inner product space.

### Definition 8.13

A linear transformation $A$ of $V$ is called a *hermitian* transformation if $A = A^*$.

***Proposition 8.13.*** *For a linear transformation $A$ of $V$ the following conditions are mutually equivalent:*

**1.** *$A$ is hermitian.*

**2.** *$(A(\alpha),\beta) = (\alpha,A(\beta))$ for all $\alpha$, $\beta \in V$.*

**3.** *$A$ can be represented by a hermitian matrix with respect to an orthonormal basis of $V$.*

*Proof.* The equivalence of (1) and (2) follows from Proposition 8.12 and the subsequent remark. The equivalence of (1) and (3) follows from the fact we proved about the matrix representation of the adjoint.

***Proposition 8.14.*** *Every eigenvalue of a hermitian transformation is real.*

*Proof.* Let $c$ be an eigenvalue and let $A(\alpha) = c\alpha$, where $\alpha \neq 0$. Then

$$\begin{aligned}(A(\alpha),\alpha) &= (c\alpha,\alpha) = c(\alpha,\alpha) \\ &= (\alpha,A(\alpha)) = (\alpha,c\alpha) = \bar{c}(\alpha,\alpha).\end{aligned}$$

Since $(\alpha,\alpha) \neq 0$, we must have $c = \bar{c}$.

We shall consider another class of linear transformations.

### Definition 8.14

A linear transformation $A$ is called a *unitary transformation* if $A^*A = AA^* = I$, where $A^*$ is the adjoint of $A$ and $I$ is the identity transformation.

An $n \times n$ complex matrix $\mathbf{A}$ is called a *unitary matrix* if $\mathbf{A}^*\mathbf{A} = \mathbf{A}\mathbf{A}^* = \mathbf{I}_n$, where $\mathbf{I}_n$ is the identity matrix.

We know (corollary to Theorem 3.4) that condition $A^*A = I$ implies $AA^* = I$, and vice versa. The same is true for $\mathbf{A}^*\mathbf{A} = \mathbf{I}_n$ and $\mathbf{A}\mathbf{A}^* = \mathbf{I}_n$. A unitary transformation is, in fact, nothing but an isometry (Definition 8.6) of a complex inner product space, as we prove in the following:

***Proposition 8.15.*** *For a linear transformation $A$ of $V$ the following conditions are mutually equivalent:*
**1.** *$A$ is unitary.*
**2.** *$(A(\alpha),A(\beta)) = (\alpha,\beta)$ for all $\alpha, \beta \in V$.*
**3.** *$\|A(\alpha)\| = \|\alpha\|$ for all $\alpha \in V$.*
**4.** *$A$ can be represented by a unitary matrix with respect to an orthonormal basis of $V$.*

*Proof.* The equivalence of (2) and (3) has already been given in Proposition 8.7.

If $A$ is unitary: $A^*A = I$, then we have for any $\alpha, \beta \in V$

$$(\alpha,\beta) = (\alpha,A^*A(\beta)) = (A(\alpha),A(\beta)),$$

so that (1) implies (2). Suppose that $A$ satisfies (2). Then

$$(\alpha,A^*A(\beta)) = (A(\alpha),A(\beta)) = (\alpha,\beta)$$

and hence

$$(\alpha, A^*A(\beta) - \beta) = 0$$

for all $\alpha, \beta \in V$. Taking $\alpha = A^*A(\beta) - \beta$, we see that $A^*A(\beta) = \beta$ for all $\beta$, that is, $A^*A = I$. As we have already remarked, we then have $A^*A = I = AA^*$. Thus (2) implies (1). Finally, if $\Phi$ is an orthonormal basis of $V$, we have

$$(A^*A)_\Phi = (A_\Phi)^*A_\Phi$$

for any linear transformation $A$. Hence $A^*A = I$ if and only if $A_\Phi$ is unitary, proving the equivalence of (1) and (4).

***Example 8.11.*** For an $n \times n$ complex matrix $\mathbf{A} = [a_{ij}]$, condition $\mathbf{A}^*\mathbf{A} = \mathbf{I}_n$ is equivalent to

$$\sum_{k=1}^{n} a_{ki}\bar{a}_{kj} = \delta_{ij}, \qquad 1 \leq i, j \leq n.$$

Similarly, $\mathbf{A}\mathbf{A}^* = \mathbf{I}_n$ is equivalent to

$$\sum_{k=1}^{n} a_{ik}\bar{a}_{jk} = \delta_{ij}, \qquad 1 \leq i, j \leq n.$$

Thus $\mathbf{A}$ is unitary if and only if the $n$ column vectors (also, the $n$ row vectors) of $\mathbf{A}$ are orthonormal in $\mathfrak{C}^n$ with respect to the standard inner product.

***Example 8.12.*** Let $\{\alpha_1, \ldots ,\alpha_n\}$ and $\{\beta_1, \ldots ,\beta_n\}$ be orthonormal bases of an

$n$-dimensional complex inner product space. Then the matrix of transition $\mathbf{P} = [p_{ij}]$, where

$$\beta_i = \sum_{j=1}^{n} p_{ji}\alpha_j, \qquad 1 \leq i \leq n,$$

is unitary. In fact,

$$\delta_{ij} = (\beta_i,\beta_j) = \left(\sum_{k=1}^{n} p_{ki}\alpha_k, \sum_{k=1}^{n} p_{kj}\alpha_k\right) = \sum_{k=1}^{n} p_{ki}\bar{p}_{kj},$$

showing that the $n$ column vectors of $\mathbf{P}$ are orthonormal.

***Proposition 8.16.*** *Let $V$ be a complex inner product space.*

**1.** *Every eigenvalue of a unitary transformation $A$ has absolute value* 1.

**2.** *The determinant of a unitary transformation has absolute value* 1.

**3.** *The set of all unitary transformations of $V$ forms a group (under multiplication). The set of all unitary transformations with determinant* 1 *is a subgroup.*

*Proof.* 1. If $A(\alpha) = c\alpha$, where $\alpha \neq 0$, then

$$\|\alpha\| = \|A(\alpha)\| = \|c\alpha\| = |c|\ \|\alpha\|, \text{ and hence } |c| = 1.$$

2. It is sufficient to show that $|\det U| = 1$ for any unitary matrix $U$. We have $UU^* = I_n$, and hence

$$\begin{aligned} 1 = \det I_n = \det\,(UU^*) &= \det U \cdot \det U^* \\ &= \det U \cdot \det \bar{U} \\ &= \det U \cdot \overline{\det U} = |\det U|^2. \end{aligned}$$

3. If $A$ is unitary, then $AA^* = I$, so that $A^{-1}$ exists and is equal to $A^*$. $A^* = A^{-1}$ is unitary since

$$(A^{-1}(\alpha),A^{-1}(\beta)) = (A(A^{-1}\alpha),A(A^{-1}\beta)) = (\alpha,\beta)$$

for all $\alpha$, $\beta \in V$. If $A$ and $B$ are unitary, then $AB$ is unitary, because

$$(AB(\alpha),AB(\beta)) = (B(\alpha),B(\beta)) = (\alpha,\beta)$$

for all $\alpha$, $\beta \in V$. The last assertion is obvious.

***Example 8.13.*** The set of all unitary matrices of degree $n$ forms a group, called the *unitary group* $U(n)$. The subgroup consisting of all unitary matrices of determinant 1 is called the *special unitary group* $SU(n)$. For example,

$$\begin{bmatrix} e^{i\omega} & 0 \\ 0 & e^{-i\omega} \end{bmatrix} \in SU(2) \qquad \text{for any real number } \omega.$$

**EXERCISE 8.4**

» **1.** If $A$ and $B$ are hermitian transformations of a complex inner product space $V$, then prove that $AB$ is hermitian if and only if $AB = BA$.

**2.** Prove that for any transformation $A$ of a complex inner product space $V$, $A^*A$ is hermitian. If $A$ is hermitian, then $B^*AB$ is hermitian for any transformation $B$.

›› **3.** Prove that any $n \times n$ complex matrix $\mathbf{A}$ can be expressed uniquely in the form $\mathbf{B} + i\mathbf{C}$, where $\mathbf{B}$ and $\mathbf{C}$ are hermitian matrices. State the analogue for linear transformations.

**4.** Prove that if $A$ is a hermitian transformation of an $n$-dimensional complex vector space $V$, then the set $\{(A(\alpha),\alpha);\ \alpha \in V\}$ is a convex set in the complex plane which contains all eigenvalues of $A$. (A subset $S$ of the plane is *convex* if for any two points in $S$ the segment joining them is contained in $S$.)

›› **5.** Prove that if $A$ is a hermitian or unitary transformation of $V$ and if $U$ is a subspace invariant by $A$, then $U^\perp$ is invariant by $A$.

**6.** Show that a diagonal matrix

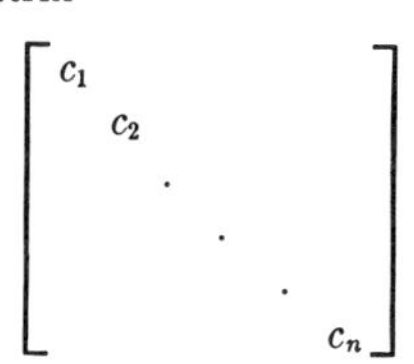

is unitary if and only if $|c_i| = 1$ for every $i$.

›› **7.** Show that the set of all $n \times n$ hermitian matrices forms a vector space over $\mathcal{R}$ and find its dimension.

**8.** An $n \times n$ matrix $\mathbf{A}$ is called a *skew-hermitian* matrix if $\mathbf{A}^* = -\mathbf{A}$. Show that the set of skew-hermitian $n \times n$ matrices forms a vector space over $\mathcal{R}$ and find its dimension.

››☆ **9.** In the vector space of all $n \times n$ skew-hermitian matrices, show that

$$(\mathbf{A},\mathbf{B}) = -\text{trace}\,(\mathbf{A}\mathbf{B}^*)$$

is an inner product. (Cf. Exercise 8.1, number 16.)

››☆ **10.** Prove that if $\mathbf{A}$ is an $n \times n$ hermitian matrix, then $\mathbf{A} + i\mathbf{I}_n$ is invertible and

$$\mathbf{U} = (\mathbf{A} + i\mathbf{I}_n)^{-1}(\mathbf{A} - i\mathbf{I}_n)$$

is unitary. ($\mathbf{U}$ is called the *Cayley transform* of $\mathbf{A}$.)

☆ **11.** Prove that every eigenvalue of a skew-hermitian transformation (that is, transformation $A$ such that $A^* = -A$) is 0 or purely imaginary.

››☆ **12.** Find the center of the unitary group $U(n)$. (The *center* of a group $\mathfrak{G}$ is the set of all elements which commute with every element of $\mathfrak{G}$.)

☆ **13.** Let $V$ be an $n$-dimensional complex vector space. A mapping $f\colon V \times V \to \mathcal{C}$ is called a *hermitian form* if

1. $f(\alpha,\beta)$ is linear in $\alpha$, whenever $\beta$ is fixed.
2. $f(\beta,\alpha) = \overline{f(\alpha,\beta)}$ for all $\alpha$, $\beta \in V$.

[As a result, $f(\alpha,\beta)$ is conjugate-linear in $\beta$ whenever $\alpha$ is fixed.] Prove that when $V$ has an inner product, any hermitian form $f$ is of the form

$$f(\alpha,\beta) = (A(\alpha),\beta)$$

where $A$ is a hermitian transformation.

**14.** Prove that if a hermitian matrix $\mathbf{A}$ is nilpotent, then $\mathbf{A} = \mathbf{0}$.

☆ **15.** Let $A$ be a hermitian transformation of an $n$-dimensional complex inner product space $V$. Prove Theorem 8.21 in the next section in the following steps:

(*a*) Use induction on dim $V$.

(*b*) For an eigenvalue $c_1$ of $A$, let

$$W = \{\alpha \in V;\ A(\alpha) = c_1\alpha\}.$$

$W$ is a subspace invariant by $A$.

(*c*) The orthogonal complement $W^{\perp}$ is invariant by $A$ (cf. number 5), and the restriction of $A$ to $W^{\perp}$ is hermitian.

(*d*) Apply the inductive assumption to $A$ on $W^{\perp}$ and observe $V = W \oplus W^{\perp}$.

☆**16.** Let $A$ be a unitary transformation of an $n$-dimensional complex inner product space $V$. Prove Theorem 8.22 in the next section by the same method as in number 15.

☆**17.** For any $n \times n$ complex matrix $\mathbf{A}$, prove that there exists a unitary matrix $\mathbf{U}$ such that $\mathbf{UAU}^{-1}$ is a superdiagonal matrix. (Cf. corollary to Proposition 7.4 and Exercise 8.2, number 11.)

## 8.5 NORMAL TRANSFORMATIONS

Hermitian and unitary transformations, which we studied in the preceding section, have a common property that $AA^* = A^*A$. This property is also shared by symmetric and orthogonal transformations of a real inner product space, which we shall study later. Since it is possible to treat transformations having this formal property in a unified way, we shall make the following definition.

### Definition 8.15

A linear transformation $A$ of a real or complex inner product space $V$ is called a *normal transformation* if $A^*A = AA^*$. An $n \times n$ real or complex matrix $\mathbf{A}$ is said to be *normal* if $\mathbf{AA}^* = \mathbf{A}^*\mathbf{A}$.

***Example 8.14.*** Assume that a linear transformation $A$ of $V$ can be represented by a diagonal matrix (with diagonal elements $c_1, \ldots, c_n$) with respect to a certain orthonormal basis $\Phi = \{\alpha_1, \ldots, \alpha_n\}$ of $V$. This means that

$$A(\alpha_i) = c_i\alpha_i, \qquad 1 \leq i \leq n,$$

and, consequently, $\alpha_1, \ldots, \alpha_n$ are eigenvectors corresponding to the eigenvalues $c_1, \ldots, c_n$, respectively. Since

$$\begin{bmatrix} c_1 & & & \\ & \cdot & & \\ & & \cdot & \\ & & & \cdot \\ & & & & c_n \end{bmatrix}^* = \begin{bmatrix} \bar{c}_1 & & & \\ & \cdot & & \\ & & \cdot & \\ & & & \cdot \\ & & & & \bar{c}_n \end{bmatrix}$$

and since these two diagonal matrices commute, it follows that $AA^* = A^*A$, that is, $A$ is normal. As a matter of fact, we have

$$A^*(\alpha_i) = \bar{c}_i\alpha_i, \qquad 1 \leq i \leq n.$$

The main problem of this section is concerned with the converse of the situation in Example 8.14, namely, the problem of representing any given normal transformation by a diagonal matrix (with respect to a suitable orthonormal basis). We solve this problem and apply the result to the case of hermitian and unitary transformations (as well as symmetric and orthogonal transformations, in the next two sections).

Let us recall Theorem 7.8. It says that a linear transformation $A$ of an $n$-dimensional vector space $V$ over a field $\mathcal{F}$ can be represented by a diagonal matrix (with respect to a suitable basis) if and only if the minimal polynomial $\phi_A$ decomposes into distinct linear factors in $\mathcal{F}$. When $V$ is an inner product space, the basis which we use is required to be orthonormal; on the other hand, we assume that $A$ is normal. In view of this comparison of the present problem to that which is answered by Theorem 7.8, it is natural to see what we can say about the minimal polynomial of a normal transformation.

We first prepare some facts which we shall need.

***Lemma.*** *For linear transformations $A, B$ of an inner product space $V$, we have*

$$(A + B)^* = A^* + B^*, \quad (A^k)^* = (A^*)^k \quad \textit{for any positive integer } k.$$

*Proof.* By the remark following Proposition 8.12, it is sufficient to show $((A + B)(\alpha), \beta) = (\alpha, (A^* + B^*)(\beta))$ for all $\alpha, \beta \in V$. This follows from

$$\begin{aligned}((A + B)(\alpha), \beta) &= (A(\alpha) + B(\alpha), \beta) = (A(\alpha),\beta) + (B(\alpha),\beta) \\ &= (\alpha,A^*(\beta)) + (\alpha,B^*(\beta)) = (\alpha, (A^* + B^*)(\beta)).\end{aligned}$$

The second assertion can be proved in a similar fashion by using induction on $k$.

***Proposition 8.17.*** *Let $A$ be a normal transformation of an inner product space $V$.*

**1.** *For any $\alpha \in V$, $\|A(\alpha)\| = \|A^*(\alpha)\|$.*

**2.** *If $A(\alpha) = c\alpha$, then $A^*(\alpha) = \bar{c}\alpha$.*

**3.** *If $A(\alpha) = c_1\alpha$ and $A(\beta) = c_2\beta$ and if $c_1 \neq c_2$, then $(\alpha,\beta) = 0$.*

**4.** *If $A^p(\alpha) = 0$ for some $\alpha \in V$ and for some integer $p \geq 1$, then $A(\alpha) = 0$.*

**5.** *For any polynomial $g(x)$, $g(A)$ is normal.*

*Proof.* 1. This follows from

$$(A(\alpha),A(\alpha)) = (\alpha,A^*A(\alpha)) = (\alpha,AA^*(\alpha)) = (A^*(\alpha),A^*(\alpha)).$$

2. By assumption, we have $(A - cI)\alpha = 0$, where $I$ is the identity transformation. Since $(A - cI)^* = A^* - \bar{c}I$, it follows that $A - cI$ is normal together with $A$. Applying (1), we see that

$$0 = \|(A - cI)(\alpha)\| = \|(A^* - \bar{c}I)(\alpha)\|,$$

that is,

$$A^*(\alpha) = \bar{c}\alpha.$$

3. We have by (2)

$$c_1(\alpha,\beta) = (c_1\alpha,\beta) = (A(\alpha),\beta)$$
$$= (\alpha,A^*(\beta)) = (\alpha,\bar{c}_2\beta) = c_2(\alpha,\beta).$$

Since $c_1 \neq c_2$, we obtain $(\alpha,\beta) = 0$.

4. Let $p$ be the smallest integer such that $A^p(\alpha) = 0$ and assume that $p > 1$. Let $\beta = A^{p-1}(\alpha)$. Since $A(\beta) = A^p(\alpha) = 0$, we have $\|A^*(\beta)\| = \|A(\beta)\| = 0$ by (1), and hence $A^*(\beta) = 0$. Therefore

$$0 = (A^*(\beta),A^{p-2}(\alpha)) = (\beta,A(A^{p-2}(\alpha))) = (\beta,\beta),$$

and hence $\beta = 0$, that is, $A^{p-1}(\alpha) = 0$, contrary to the assumption. Thus $A(\alpha) = 0$.

5. For $g(x) = \sum_{i=0}^{r} c_i x^i$ we have by the lemma

$$g(A)^* = \left(\sum_{i=0}^{r} c_i A^i\right)^* = \sum_{i=0}^{r} \bar{c}_i A^{*i},$$

which commutes with $A$ and hence with $g(A)$.

We now prove:

***Proposition 8.18.*** *Let $A$ be a normal transformation of an inner product space $V$. If $\phi_1$ is an irreducible factor of the minimal polynomial $\phi_A$, then*

$$\phi_A = \phi_1 \cdot \psi,$$

*where $\phi_1$ and $\psi$ are relatively prime.*

*Proof.* Assume that

$$\phi_A = \phi_1^p \cdot \psi,$$

where $\phi_1$ and $\psi$ are relatively prime. We shall show that $p = 1$. Let $\alpha$ be an arbitrary element of $V$. We have

$$\phi_1(A)^p(\psi(A)(\alpha)) = 0.$$

Since $\phi_1(A)$ is normal together with $A$ [cf. (5) of Proposition 8.17], we apply (4) of Proposition 8.17 to $\phi_1(A)$ and the vector $\psi(A)(\alpha)$. We obtain

$$\phi_1(A)\ (\psi(A)(\alpha)) = 0.$$

This being the case for any $\alpha \in V$, we see that $\phi_1(A)\psi(A) = 0$, in other words, $p = 1$.

***Corollary.*** *If $A$ is a normal transformation of a complex inner product space, then $\phi_A$ decomposes into distinct linear factors.*

*Proof.* We know that

$$\phi_A(x) = (x - c_1)^{p_1} \cdots (x - c_k)^{p_k},$$

where $c_1, \ldots, c_k$ are distinct complex numbers. Applying Proposition 8.18 to each factor $x - c_i$, we see that $p_i = 1$ for $1 \leq i \leq n$.

We are now in a position to prove the following main result (called the *spectral decomposition theorem for a normal transformation*).

***Theorem 8.19.*** *Let $A$ be a normal transformation of an $n$-dimensional complex inner product space $V$. Then there exists an orthonormal basis of $V$ with respect to which $A$ can be represented by a diagonal matrix; the diagonal elements exhaust all eigenvalues of $A$ counting their multiplicities.*

*Proof.* By the corollary to Proposition 8.18 and by Theorem 7.8, we see that

$$V = V_1 \oplus V_2 \oplus \cdots \oplus V_k,$$

where

$$V_i = \{\alpha \in V; A(\alpha) = c_i\alpha\}, \qquad 1 \leq i \leq k$$

and $c_1, \ldots, c_k$ are all distinct roots of $\phi_A$ (so they are all distinct eigenvalues of $A$). By (3) of Proposition 8.17 we see that $V_i$ and $V_j$ are orthogonal to each other whenever $i \neq j$. We may choose an orthonormal basis in each $V_i$ and make up an orthonormal basis of $V$ which has the required property.

***Corollary.*** *For any normal matrix $\mathbf{A}$ of degree $n$, there exists a unitary matrix $\mathbf{U}$ such that $\mathbf{UAU}^{-1}$ is a diagonal matrix.*

*Proof.* Consider $\mathbf{A}$ as a normal transformation of the vector space $\mathcal{C}^n$ with standard inner product. By Theorem 8.19 there exists an orthonormal basis $\{\alpha_1, \ldots, \alpha_n\}$ with respect to which $\mathbf{A}$ can be represented by a diagonal matrix $\mathbf{D}$. By the corollary to Proposition 3.12 we have $\mathbf{D} = \mathbf{UAU}^{-1}$, where $\mathbf{U}$ is the matrix of transition from the standard basis to $\{\alpha_1, \ldots, \alpha_n\}$ and is a unitary matrix as we have seen in Example 8.12.

***Theorem 8.20.*** *Let $A$ be a normal transformation of an $n$-dimensional real vector space $V$. If all the characteristic roots of $A$ are real, then $V$ has an orthonormal basis with respect to which $A$ can be represented by a diagonal matrix.*

*Proof.* If all the characteristic roots are real, we know that $\phi_A$ decomposes into distinct linear factors, since $\phi_A$ and the characteristic polynomial $f_A$ have the same roots. The rest of the proof is similar to that of the complex case.

In order to state the analogue of the corollary to Theorem 8.19, we make the following definition.

### Definition 8.16

An $n \times n$ real matrix $\mathbf{A}$ is called an *orthogonal matrix* if $\mathbf{A}{}^t\mathbf{A} = {}^t\mathbf{A}\mathbf{A} = \mathbf{I}_n$. In other words, an orthogonal matrix is a real unitary matrix (for a real matrix $\mathbf{A}$ we have $\mathbf{A}^* = {}^t\mathbf{A}$). (Of course, $\mathbf{A}{}^t\mathbf{A} = \mathbf{I}_n$ implies ${}^t\mathbf{A}\mathbf{A} = \mathbf{I}_n$, and vice versa.)

***Remark.*** The condition $\mathbf{A}{}^t\mathbf{A} = \mathbf{I}_n$ is equivalent to

$$\sum_{k=1}^{n} a_{ik}a_{jk} = \delta_{ij},$$

that is, the $n$ row vectors of $\mathbf{A}$ are orthonormal in $\mathbb{R}^n$ with standard inner product. Similarly, ${}^t\mathbf{A}\mathbf{A} = \mathbf{I}_n$ is equivalent to

$$\sum_{k=1}^{n} a_{ki}a_{kj} = \delta_{ij},$$

that is, the $n$ column vectors of $\mathbf{A}$ are orthonormal in $\mathbb{R}^n$ with standard inner product. (Compare with Example 8.11.)

As in Example 8.12, it is easy to see that the matrix of transition from one orthonormal basis to another in an $n$-dimensional real inner product space is an orthogonal matrix. Using this last fact, we obtain the following:

**Corollary.** *If* $\mathbf{A}$ *is a real normal matrix of degree n such that all the characteristic roots are real, then there exists an orthogonal matrix* $\mathbf{S}$ *such that* $\mathbf{S}\mathbf{A}\mathbf{S}^{-1}$ *is a diagonal matrix.*

Applying Theorem 8.19, we can now obtain the spectral decomposition theorems for hermitian and unitary transformations.

***Theorem 8.21.*** *Let $A$ be a hermitian transformation of an n-dimensional complex inner product space $V$. Then there exists an orthonormal basis $\{\alpha_1, \ldots, \alpha_n\}$ of $V$ with respect to which $A$ can be represented by a diagonal matrix; the diagonal elements are all the eigenvalues counting their multiplicities, and they are real.*

**Corollary.** *For any $n \times n$ hermitian matrix* $\mathbf{A}$ *there exists a unitary matrix* $\mathbf{U}$ *such that* $\mathbf{U}\mathbf{A}\mathbf{U}^{-1}$ *is a diagonal matrix; the diagonal elements are real.*

***Theorem 8.22.*** *Let $A$ be a unitary transformation of an n-dimensional complex inner product space $V$. Then there exists an orthonormal basis $\{\alpha_1, \ldots, \alpha_n\}$ of $V$ with respect to which $A$ can be represented by a diagonal matrix; the diagonal elements are all the eigenvalues counting their multiplicities, and they are of absolute value* 1.

**Corollary.** *For any $n \times n$ unitary matrix* $\mathbf{U}$, *there exists a unitary matrix* $\mathbf{S}$ *such that* $\mathbf{SUS}^{-1}$ *is a diagonal matrix; the diagonal elements are of absolute value* 1.

EXERCISE 8.5

1. If $\mathbf{A}$ is a normal matrix and $\mathbf{U}$ is a unitary matrix, both of degree $n$, show that $\mathbf{U}^*\mathbf{AU}$ is normal.

»☆ 2. Prove that a linear transformation $A$ of an inner product space $V$ is normal if and only if $\|A(\alpha)\| = \|A^*(\alpha)\|$ for every $\alpha \in V$.

3. Let $\mathbf{A}$ and $\mathbf{B}$ be $n \times n$ hermitian matrices. Show that $\mathbf{A} + i\mathbf{B}$ is normal if and only if $\mathbf{AB} = \mathbf{BA}$.

» 4. Prove that a normal transformation $A$ which is nilpotent is 0.

☆ 5. Prove that a normal transformation $A$ of an inner product space is semi-simple, that is, if $U$ is an $A$-invariant subspace, then $U^\perp$ is $A$-invariant (cf. Exercise 7.2, number 10).

☆ 6. Let $A$ be a normal transformation of a complex inner product space $V$. Prove:

(*a*) $A$ is hermitian if and only if all the eigenvalues of $A$ are real.

(*b*) $A$ is skew-hermitian (cf. Exercise 8.4, number 11) if and only if all the eigenvalues are 0 or purely imaginary.

(*c*) $A$ is unitary if and only if all the eigenvalues of $A$ have absolute value 1.

(*d*) $A$ is 0 if and only if all the eigenvalues are 0.

☆ 7. If $A$ is a normal transformation of an $n$-dimensional real inner product space $V$, then show that $V$ is the direct sum of one- or two-dimensional subspaces which are invariant by $A$. Prove that *there exists an orthonormal basis of* $V$ *with respect to which* $A$ *can be represented by a matrix of the form*

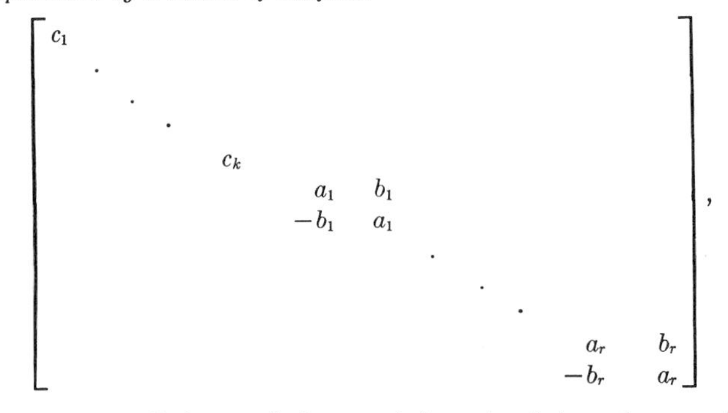

where $c_1, \ldots, c_k$ are distinct real characteristic roots of $A$, each appearing the same number of times as its multiplicity, and $a_j$, $b_j$, $1 \le j \le r$, are real numbers such that $(x - a_j)^2 + b_j^2$ are all the distinct factors of degree 2 of the characteristic polynomial $f_A$ of $A$, each block

$$\begin{bmatrix} a_j & b_j \\ -b_j & a_j \end{bmatrix}$$

appearing the same number of times as the multiplicity of $(x - a_j)^2 + b_j^2$ as a factor of $f_A$.

☆ 8. Give an alternative proof of Theorem 8.19 in the following steps:

(*a*) Use induction on dim $V$.

(*b*) Let $\alpha$ be an eigenvector of $A$ and let $W = \text{Sp}(\alpha)$.

(*c*) Show that $W^{\perp}$ is invariant by $A$ as follows: If $\beta \in W$, then $(\alpha,A(\beta)) = (A^*(\alpha),\beta) = (\bar{c}\alpha,\beta) = \bar{c}(\alpha,\beta) = 0$ by virtue of (2) of Proposition 8.17. Thus $A(\beta) \in W^{\perp}$.

(*d*) Similarly, $W^{\perp}$ is invariant by $A^*$. Show that the restriction of $A$ to $W^{\perp}$ is normal, and use $V = W \oplus W^{\perp}$ and the inductive assumption on $W$.

☆ **9.** Give an alternative proof of Theorem 8.20 along the same lines as number 8.

☆ **10.** A hermitian transformation of an $n$-dimensional complex inner product space $V$ is said to be *positive-semidefinite* [*negative-semidefinite*] if $(A(\alpha),\alpha) \geq 0$ [$\leq 0$] for all $\alpha \in V$. It is said to be *positive-definite* [*negative-definite*] if, furthermore, $(A(\alpha),\alpha) = 0$ implies $\alpha = 0$. Prove that:

(*a*) $A$ is positive-definite if and only if all eigenvalues of $A$ are positive,

(*b*) $A$ is negative-definite if and only if all eigenvalues of $A$ are negative.

State analogous results for the semidefinite cases.

»☆ **11.** Prove that if an $n \times n$ complex matrix **A** is hermitian and positive-definite, then there exists an $n \times n$ matrix **B** such that $\mathbf{A} = \mathbf{B}^*\mathbf{B}$.

»☆ **12.** Prove that a nonsingular $n \times n$ complex matrix **A** can be written uniquely in the form $\mathbf{A} = \mathbf{BC}$, where $\mathbf{B} \in U(n)$ and **C** is a positive-definite hermitian matrix.

☆ **13.** Prove that $U(n)$ is arcwise-connected (cf. Exercise 6.3, number 15). How about $SU(n)$?

**14.** Prove that a complex superdiagonal matrix **A** is diagonal if and only if it is normal.

☆ **15.** Prove the corollary to Theorem 8.19 by using number 17 of Exercise 8.4 and numbers 1 and 14 above.

## 8.6 SYMMETRIC TRANSFORMATIONS

In this section, $V$ will denote an $n$-dimensional real inner product space.

**Definition 8.17**

A linear transformation $A$ of $V$ is called a *symmetric transformation* if $A = A^*$.

In other words, $A$ is symmetric if and only if

$$(A(\alpha),\beta) = (\alpha,A(\beta)) \qquad \text{for all } \alpha, \beta \in V.$$

***Example 8.15.*** When we regard an $n \times n$ real matrix **A** as a linear transformation of $\mathcal{R}^n$ with standard inner product, the transformation is symmetric if and only if ${}^t\mathbf{A} = \mathbf{A}$ (that is, **A** is a symmetric matrix).

More generally,

***Proposition 8.23.*** *A transformation is symmetric if and only if it is represented by a symmetric matrix with respect to an orthonormal basis of $V$.*

*Proof.* This is obvious from the remark preceding Definition 8.12.

We shall now establish a theorem similar to Theorem 8.21 for symmetric transformations. Since a symmetric transformation $A$ is obviously normal, we may appeal to Theorem 8.20 provided we can prove that all the characteristic roots of a symmetric transformation are real. We thus prove:

***Theorem 8.24.*** *Let $A$ be a symmetric transformation of an $n$-dimensional real inner product space $V$. Then all the characteristic roots of $A$ are real.*

We shall present two proofs.

*Proof* 1. By choosing any orthonormal basis of $V$, $A$ can be represented by a symmetric matrix $\mathbf{A}$ (thus $\mathbf{A}$ is a real matrix such that ${}^t\mathbf{A} = \mathbf{A}$) and the characteristic polynomial $f_A$ of $A$ is equal to the characteristic polynomial $f_{\mathbf{A}}$ of $\mathbf{A}$. Now $\mathbf{A}$ may be considered as a hermitian matrix (and also a hermitian transformation of $\mathcal{C}^n$ with standard inner product), so that all the roots of $f_{\mathbf{A}}$ are real. Hence all the roots of $f_A$ are real. (A more intrinsic argument which explains the meaning of this proof will be given in Exercise 8.6, number 15.)

*Proof* 2. This proof does not depend on the theorem for hermitian transformations. The minimal polynomial $\phi_A$ can be decomposed over $\mathcal{R}$ into a product of linear factors and factors of degree 2, namely, quadratic polynomials of the form $(x - a)^2 + b^2$ with $b \neq 0$. We shall show that a factor of this form does not exist. If it does exist, then we may write

$$\phi_A(x) = ((x - a)^2 + b^2) \cdot \psi(x),$$

where $\psi(x)$ is a polynomial over $\mathcal{R}$. For any $\alpha \in V$, let $\beta = \psi(A)(\alpha)$. Then

$$\phi_A(A)(\alpha) = ((A - aI)^2 + b^2I)(\beta) = 0.$$

Since $A - aI$ is symmetric, we have

$$\begin{aligned} 0 &= ((A - aI)^2(\beta) + b^2\beta, \beta) = ((A - aI)^2(\beta), \beta) + (b^2\beta, \beta) \\ &= ((A - aI)(\beta), (A - aI)(\beta)) + b^2(\beta,\beta). \end{aligned}$$

Since each term is nonnegative, this is possible only if $(A - aI)(\beta) = 0$ and $b^2(\beta,\beta) = 0$. Since $b \neq 0$, we have $\beta = 0$. This means that $\psi(A)(\alpha) = 0$ for all $\alpha \in V$, contrary to the assumption that $\psi$ is a proper factor of $\phi_A$.

Thus $\phi_A$ is a product of linear factors. Since $f_A$ and $\phi_A$ have the same roots, we see that all the characteristic roots of $A$ are real.

From Theorems 8.20 and 8.24 we may now obtain:

***Theorem 8.25.*** *Let $A$ be a symmetric transformation of an $n$-dimensional real inner product space $V$. Then there exists an orthonormal basis of $V$ with respect to which $A$ can be represented by a diagonal matrix; the diagonal elements are all the eigenvalues of $A$ counting their multiplicities.*

***Corollary.*** *For any (real) symmetric matrix $\mathbf{A}$ of degree $n$, there exists an orthogonal matrix $\mathbf{S}$ of degree $n$ such that $\mathbf{SAS}^{-1}$ is a diagonal matrix.*

We shall now discuss the relationship between symmetric transformations and bilinear forms. Recall that a bilinear form is a symmetric bilinear function on $V \times V$, that is, a bilinear function $f(\alpha,\beta)$ satisfying

$$f(\alpha,\beta) = f(\beta,\alpha) \qquad \text{for all } \alpha,\ \beta \in V.$$

(Note that a bilinear form can be defined on any vector space without reference to inner product, whereas symmetric transformations are defined only with reference to a given inner product. An inner product itself is a bilinear form which is positive-definite.)

***Proposition 8.26.*** *Let $V$ be an $n$-dimensional real inner product space. For any symmetric transformation $A$ of $V$, $f(\alpha,\beta) = (A(\alpha),\beta)$ is a bilinear form. Conversely, for any bilinear form $f$ there exists a unique symmetric transformation $A$ of $V$ such that $f(\alpha,\beta) = (A(\alpha),\beta)$.*

*Proof.* The first assertion is easy to verify. We prove the second. Given a bilinear form $f$, consider the linear function

$$\beta \rightarrow f(\alpha,\beta),$$

where $\alpha$ is a fixed element of $V$. By Theorem 8.10 there exists a unique element, say, $\gamma \in V$, such that

$$(\gamma,\beta) = f(\alpha,\beta) \qquad \text{for all } \beta \in V.$$

Since $\gamma$ is uniquely determined by $\alpha$, we define $A(\alpha) = \gamma$. It is now easy to verify that $A$ is a symmetric linear transformation. Thus $f(\alpha,\beta) = (A(\alpha),\beta)$. The uniqueness of such $A$ is also clear.

Let us remark that it is possible to obtain Theorem 4.19 by using Theorem 8.25 and Proposition 8.26. Let $f$ be a bilinear form on any $n$-dimensional real vector space $V$. Choose any inner product on $V$ (for example, choose any basis and consider it orthonormal). Then by Proposition 8.26 there is a symmetric transformation $A$ of $V$ such that $f(\alpha,\beta) = (A(\alpha),\beta)$. By Theorem 8.25 there is an orthonormal basis $\alpha_1, \ldots, \alpha_n$ of $V$ such that

$$A(\alpha_i) = c_i\alpha_i, \qquad 1 \leq i \leq n,$$

where $c_1, \ldots, c_n$ are the eigenvalues of $A$ (not necessarily all distinct) and $c_i > 0$ for $1 \leq i \leq p$, $c_i < 0$ for $p + 1 \leq i \leq r$, and $c_i = 0$ for $r + 1 \leq i \leq n$. Then

$$f(\alpha_i,\alpha_i) = (A(\alpha_i),\alpha_i) = (c_i\alpha_i,\alpha_i) = c_i, \qquad 1 \leq i \leq n,$$

and

$$f(\alpha_i,\alpha_j) = (A(\alpha_i),\alpha_j) = (c_i\alpha_i,\alpha_j) = 0, \qquad i \neq j.$$

We may now change $\alpha_i$, $1 \leq i \leq p$, into $\alpha_i/\sqrt{c_i}$, and $\alpha_i$, $p + 1 \leq i \leq r$, into $\alpha_i/\sqrt{-c_i}$ in order to obtain a basis required for Theorem 4.19.

In connection with this, we make the following definition.

**Definition 8.18**

A symmetric transformation $A$ of a real inner product space $V$ is said to be *positive-definite* [*negative-definite*] if

$$(A(\alpha),\alpha) \geq 0; \quad (A(\alpha),\alpha) = 0 \quad \text{if and only if } \alpha = 0$$

$$[(A(\alpha),\alpha) \leq 0; \quad (A(\alpha),\alpha) = 0 \quad \text{if and only if } \alpha = 0].$$

When the additional condition concerning $(A(\alpha),\alpha) = 0$ is not required, $A$ is said to be *positive-semidefinite* [*negative-semidefinite*]. (See Definition 4.15.)

Similar definitions can be made for real symmetric matrices. For example, an $n \times n$ real symmetric matrix $\mathbf{A} = [a_{ij}]$ is positive-definite if the transformation $A: \mathcal{R}^n \to \mathcal{R}^n$ (which is symmetric with respect to the standard inner product) is positive-definite. Thus $\mathbf{A}$ is positive-definite if and only if $\sum_{i,j=1}^{n} a_{ij}x_ix_j > 0$ for all $\xi = (x_i) \in \mathcal{R}^n$ other than $\mathbf{0}$.

EXERCISE 8.6

**1.** State and prove the analogue of Exercise 8.4, number 1, for symmetric transformations of a real inner product space.

**2.** Do the same as in number 1 for Exercise 8.4, number 2.

**3.** State and prove the analogue of Exercise 8.4, number 3, for real matrices.

**4.** Let $A_1, \ldots, A_k$ be symmetric transformations of a real inner product space $V$. Prove that if $A_1^2 + \cdots + A_k^2 = 0$, then $A_1 = A_2 = \cdots = A_k = 0$.

**5.** If a real matrix $\mathbf{A}$ of degree 2 is normal, show that either $\mathbf{A}$ is symmetric or $\mathbf{A}$ is of the form $\begin{bmatrix} a & b \\ -b & a \end{bmatrix}$.

**6.** Find an orthogonal matrix $\mathbf{S}$ such that $\mathbf{S}\begin{bmatrix} 1 & 0 & 2 \\ 0 & 1 & 2 \\ 2 & 2 & 0 \end{bmatrix}\mathbf{S}^{-1}$ is diagonal.

☆ **7.** A linear transformation $A$ of a real inner product space is said to be *skew-symmetric* if $A^* + A = 0$. A real matrix $\mathbf{A}$ of degree $n$ is called a *skew-symmetric* matrix if $\mathbf{A} + {}^t\mathbf{A} = \mathbf{0}$. How are they related?

›› **8.** The set of all real symmetric matrices of degree $n$ forms a real vector space. Find its dimension.

☆ **9.** The set of all skew-symmetric matrices of degree $n$ forms a real vector space. Find its dimension. Show also that

$$(\mathbf{A},\mathbf{B}) = -\text{trace}\,(\mathbf{AB})$$

is an inner product in this vector space. (Cf. Exercise 8.1, number 16.)

››☆ **10.** Prove that a symmetric transformation $A$ is positive-definite if and only if all the eigenvalues of $A$ are positive. Also state and prove the analogue for the negative-definite case as well as for the semidefinite cases.

☆**11.** Let $A$ be a symmetric transformation of an $n$-dimensional real inner product space $V$ and $f$ the corresponding bilinear form (as in Proposition 8.26). Prove that a linear transformation $B$ of $V$ leaves the form $f$ invariant [that is, $f(B(\alpha),B(\beta)) = f(\alpha,\beta)$ for all $\alpha,\beta \in V$] if and only if $AB = BA$.

›› **12.** On the vector space $\mathcal{R}^n$ with standard inner product, find the symmetric matrix $\mathbf{A}$ corresponding to the following bilinear form of signature $(p, r - p)$ and rank $r$:

$$f(\xi,\eta) = x_1y_1 + \cdots + x_py_p - x_{p+1}y_{p+1} - \cdots - x_ry_r,$$

where $\xi = (x_1, \ldots, x_n)$ and $\eta = (y_1, \ldots, y_n)$.

☆**13.** Let $V$ be an $n$-dimensional real vector space and $V^c$ the complexification of $V$ (Proposition 5.23). Let $f$ be a bilinear form on $V$. Prove:

(*a*) If we define a function $f^c$ on $V^c \times V^c$ by

$$f^c(\alpha + i\beta,\ \alpha' + i\beta') = f(\alpha,\alpha') - f(\beta,\beta') + if(\alpha,\beta') + if(\beta,\alpha'),$$

then $f^c$ is a bilinear form on $V^c$ (called the *linear extension* of $f$ to $V^c$).

(*b*) If we define a function $\tilde{f}$ on $V^c \times V^c$ by

$$\tilde{f}(\alpha + i\beta,\ \alpha' + i\beta') = f(\alpha,\alpha') + f(\beta,\beta') + if(\beta,\alpha') - if(\alpha,\beta'),$$

then $\tilde{f}$ is a hermitian form on $V^c$ (called the *hermitian extension* of $f$ to $V^c$).

››☆**14.** Let $V = \mathcal{R}^n$ and $V^c = \mathcal{C}^n$. If $f$ is a bilinear form on $\mathcal{R}^n$ defined by a symmetric matrix $\mathbf{A}$ with respect to the standard basis, find the matrices representing $f^c$ and $\tilde{f}$ on $V^c$ (defined in number 13) with respect to the standard basis of $\mathcal{C}^n$.

☆**15.** (*a*) In number 13, suppose $f$ is an inner product in $V$. Show that $\tilde{f}$ is an inner product in $V^c$, called the *extension of inner product.*

(*b*) If $\{\alpha_1, \ldots, \alpha_n\}$ is an orthonormal basis in $V$ with respect to a given inner product, then the same elements, regarded as elements of $V^c$, are orthonormal with respect to the extended inner product.

(*c*) If $A$ is a symmetric transformation of $V$, then the extension $A^c$ (cf. Proposition 5.22) is a hermitian transformation of $V^c$.

(*d*) If $A$ is an orthogonal transformation (that is, a transformation which preserves inner product) of $V$, then $A^c$ is a unitary transformation.

(*e*) Interpret the statements in (*c*) and (*d*) in terms of matrices.

☆**16.** Let $A$ be a symmetric transformation of an $n$-dimensional real inner product space $V$. Prove that $a_1 = \max\{(A(\alpha),\alpha);\ \|\alpha\| = 1\}$ is equal to the largest eigenvalue of $A$ and that $a_1 = (A(\alpha_1),\alpha_1)$ if and only if $\alpha_1$ is an eigenvector for the eigenvalue $a_1$. Find an analogous result for $a_2 = \min\{(A(\alpha),\alpha);\ \|\alpha\| = 1\}$. [One may use Theorem 8.25. On the other hand, without using it, one can prove that the function $(A(\alpha),\alpha)$, which is a continuous function in $\alpha \in V$, takes the maximum value $a_1$ and the minimum value $a_2$ on the unit sphere $\{\alpha;\ \|\alpha\| = 1\}$ (by a basic theorem in analysis) and that they are eigenvalues of $A$ (by using the so-called Lagrange multiplier rule). This is one way of proving that $A$ has *real* eigenvalues.]

☆**17.** Prove that the characteristic roots of a symmetric matrix $\mathbf{A} = [a_{ij}]$ are real along the following lines: Let $c$ be a characteristic root (possibly complex) of $\mathbf{A}$. Then there exists a nonzero vector $(z_i) \in \mathcal{C}^n$ such that $\sum_{j=1}^{n} a_{ij}z_j = cz_i,\ 1 \leq i \leq n.$ Hence

$$\sum_{i,j=1}^{n} a_{ij}z_j\bar{z}_i = c\sum_{i=1}^{n} |z_i|^2.$$

Show that the left-hand side is a real number and conclude that $c$ is real. (We have also shown that the characteristic roots of a symmetric transformation are real; in particular, a symmetric transformation has at least one eigenvector.)

☆**18.** Give an alternative proof of Theorem 8.25 as follows:

($a$) Since we know that $A$ has at least one eigenvalue (cf. number 16 or 17), let $\alpha$ be an eigenvector of $A$ and let $W = \mathrm{Sp}(\alpha)$.

($b$) Show that $W^{\perp}$ is invariant by $A$ and that the restriction of $A$ to $W^{\perp}$ is symmetric.

($c$) Note that $V = W \oplus W^{\perp}$ and use the inductive assumption on $W^{\perp}$.

☆**19.** Derive Theorem 8.25 from the result in Exercise 8.5, number 7.

»☆**20.** Let $\mathbf{A}$ be a positive-definite symmetric matrix of degree $n$. Prove that for any positive integer $k$ there is a positive-definite symmetric matrix $\mathbf{B}$ such that $\mathbf{A} = \mathbf{B}^k$.

## 8.7 ORTHOGONAL TRANSFORMATIONS

In this section let $V$ be an $n$-dimensional real inner product space. The following definition and proposition are analogous to those for unitary transformations.

### Definition 8.19

A linear transformation $A$ of $V$ is called an *orthogonal transformation* if $AA^* = A^*A = I$, where $A^*$ is the adjoint of $A$ and $I$ is the identity transformation.

***Proposition 8.27.*** *For a linear transformation $A$ of $V$ the following conditions are equivalent:*

1. *$A$ is orthogonal.*
2. *$(A(\alpha),A(\beta)) = (\alpha,\beta)$ for all $\alpha,\ \beta \in V$.*
3. *$\|A(\alpha)\| = \|\alpha\|$ for all $\alpha \in V$.*
4. *$A$ can be represented by an orthogonal matrix with respect to an orthonormal basis of $V$.*

The definition of orthogonal matrix was given in Definition 8.16. An orthogonal matrix $\mathbf{A}$ can be considered as an orthogonal transformation of $\mathcal{R}^n$ with standard inner product. Let us also recall that the matrix representing the transition of one orthonormal basis of $V$ to another is orthogonal (cf. the remarks following Definition 8.16).

***Example 8.16.*** The set of all $n \times n$ orthogonal matrices forms a group, called the *orthogonal group* $O(n)$. The set of all elements in $O(n)$ with determinant 1 is a subgroup of $O(n)$; we denote this group by $SO(n)$ and call it the *special orthogonal group* or the *rotation group*. Observe that if $\mathbf{A} \in O(n)$, then $\det \mathbf{A} = 1$ or $-1$, because $\mathbf{A}{}^t\mathbf{A} = \mathbf{I}_n$ implies $\det(\mathbf{A}{}^t\mathbf{A}) = (\det \mathbf{A})(\det {}^t\mathbf{A}) = (\det \mathbf{A})^2 = 1$.

***Example 8.17.*** An $n \times n$ complex matrix $\mathbf{A} = [a_{ij}]$ is also said to be *orthogonal* if $\mathbf{A}{}^t\mathbf{A} = {}^t\mathbf{A}\mathbf{A} = \mathbf{I}_n$. Such a matrix will be called a *complex orthogonal matrix*, whereas an orthogonal matrix refers to a real matrix. For example, $\begin{bmatrix} i & \sqrt{2} \\ \sqrt{2} & -i \end{bmatrix}$ is complex orthogonal but not unitary, while for $\omega \neq n\pi$, $\begin{bmatrix} e^{i\omega} & 0 \\ 0 & e^{i\omega} \end{bmatrix}$ is unitary but not complex orthogonal. On the other hand, a (real) orthogonal matrix is a unitary matrix.

We shall now study how we can represent an orthogonal transformation by choosing a suitable orthonormal basis of $V$. We cannot expect to get a diagonal matrix, in contrast to the case of unitary transformations. For example, the matrix

$$\mathbf{A} = \begin{bmatrix} \cos\theta & -\sin\theta \\ \sin\theta & \cos\theta \end{bmatrix}$$

is orthogonal and defines an orthogonal transformation of $\mathcal{R}^2$ with standard inner product. If we could represent this transformation by a diagonal matrix (over $\mathcal{R}$) with respect to a suitable basis of $\mathcal{R}^2$, then the characteristic roots of $A$ would be real, but this is not the case unless $\theta = n\pi$ for some integer $n$.

We first prove:

***Proposition 8.28.*** *If $A$ is an orthogonal transformation of a two-dimensional real inner product space $V$, then there exists an orthonormal basis $\{\alpha_1,\alpha_2\}$ of $V$ with respect to which $A$ can be represented by either one of the following two matrices:*

**1.** $\begin{bmatrix} \cos\theta & -\sin\theta \\ \sin\theta & \cos\theta \end{bmatrix}$, *where* $0 \leq \theta \leq 2\pi$.

**2.** $\begin{bmatrix} 1 & 0 \\ 0 & -1 \end{bmatrix}$.

*Proof.* Let $\{\beta_1,\beta_2\}$ be an arbitrary orthonormal basis of $V$ and represent $A$ by an orthogonal matrix $\begin{bmatrix} a & c \\ b & d \end{bmatrix}$. Since $a^2 + b^2 = 1$, there is a unique $\theta$, $0 \leq \theta < 2\pi$, such that $a = \cos\theta$ and $b = \sin\theta$. Using $a^2 + c^2 = b^2 + d^2 = 1$ and $ac + bd = 0$, we have two cases. In the first case, $c = -\sin\theta$ and $d = \cos\theta$; then we already have a matrix of type 1. In the second case, $c = \sin\theta$ and $d = -\cos\theta$. In this case we may find a unit vector $\alpha_1$ such that $A(\alpha_1) = \alpha_1$:

$$\begin{bmatrix} \cos\theta & \sin\theta \\ \sin\theta & -\cos\theta \end{bmatrix} \begin{bmatrix} x_1 \\ x_2 \end{bmatrix} = \begin{bmatrix} x_1 \\ x_2 \end{bmatrix}, \quad \text{where } \alpha_1 = (x_1,x_2);$$

indeed, this is equivalent to the system of linear equations

$$\begin{aligned} (\cos\theta - 1)x_1 + (\sin\theta)x_2 &= 0, \\ (\sin\theta)x_1 - (1 + \cos\theta)x_2 &= 0, \end{aligned}$$

where the determinant of the coefficients is 0. We may normalize any nonzero solution $(x_1,x_2)$ and call it $\alpha_1$. If we choose a unit vector $\alpha_2$ orthogonal to $\alpha_1$, then the transformation $A$ can be represented, with respect to the orthonormal basis $\{\alpha_1,\alpha_2\}$, by an orthogonal matrix of the form $\begin{bmatrix} 1 & p \\ 0 & q \end{bmatrix}$.

Here $q = -1$, because

$$q = \det \begin{bmatrix} 1 & p \\ 0 & q \end{bmatrix} = \det \begin{bmatrix} \cos\theta & \sin\theta \\ \sin\theta & -\cos\theta \end{bmatrix} = -1.$$

But, then, $p = 0$. Thus $A$ can be represented by the matrix (2) with respect to $\{\alpha_1,\alpha_2\}$.

***Remark.*** The matrices of type 1 include the following as special cases:

$$\begin{bmatrix} 1 & 0 \\ 0 & 1 \end{bmatrix} (\theta = 0); \quad \begin{bmatrix} 0 & -1 \\ 1 & 0 \end{bmatrix} (\theta = \pi/2);$$

$$\begin{bmatrix} -1 & 0 \\ 0 & -1 \end{bmatrix} (\theta = \pi); \quad \begin{bmatrix} 0 & 1 \\ -1 & 0 \end{bmatrix} (\theta = 3\pi/2).$$

Also observe that in Proposition 8.28 we have type 1 if and only if $\det A = 1$, and type 2 if and only if $\det A = -1$.

***Proposition 8.29.*** *Let $A$ be an orthogonal transformation of $V$.*
**1.** *If $A$ has an eigenvalue $c$ (that is, a real characteristic root), then $c = 1$ or $-1$. Every characteristic root of $A$ has absolute value 1.*
**2.** *If $U$ is a subspace invariant by $A$, then $U^\perp$ is invariant by $A$.*
*Proof.* 1. Let $A(\alpha) = c\alpha$, where $\alpha \neq 0$. Then

$$(\alpha,\alpha) = (A(\alpha),A(\alpha)) = (c\alpha,c\alpha) = c^2(\alpha,\alpha)$$

so that $c = \pm 1$. For the second assertion, represent $A$ by an orthogonal matrix $\mathbf{A}$. Since $\mathbf{A}$ is unitary as well, its characteristic roots have absolute value 1 by Proposition 8.16.

2. Let $\beta \in U^\perp$. For any $\alpha \in U$, we have

$$(A(\beta),\alpha) = (\beta,A^{-1}(\alpha)) = 0,$$

since $A^{-1}(\alpha) \in U$. [The restriction of $A$ to $U$ is isometric and hence one-to-one; thus $A(U) = U$, so that $A^{-1}(U) = U$.]

We are now able to prove the following main result.

***Theorem 8.30.*** *Let $A$ be an orthogonal transformation of an n-dimensional real inner product space $V$. Then there exists an orthonormal basis $\{\alpha_1, \ldots ,\alpha_n\}$ of $V$ with respect to which $A$ can be represented by a matrix of the form*

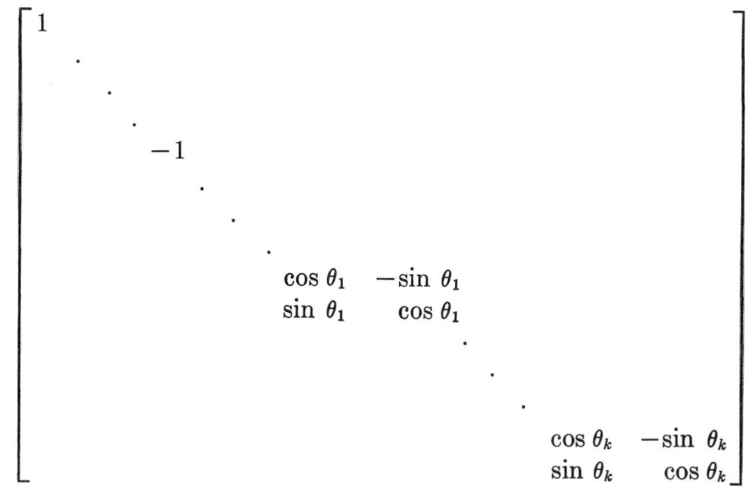

*where* 1 *and* $-1$ *appear the same number of times as their multiplicities as characteristic roots and* $\theta_1, \ldots, \theta_k$, $0 \leq \theta_j < 2\pi$, *are such that* $\cos\theta_j \pm i\sin\theta_j$, $1 \leq j \leq k$, *are the distinct characteristic roots of* $A$ *other than* $\pm 1$, *each block appearing the same number of times as the multiplicity of* $\cos\theta_j + i\sin\theta_j$ *as a characteristic root.*

*Proof.* We shall use induction on $\dim V$. If $\dim V = 1$, any orthogonal transformation is either $I$ or $-I$ and the theorem is evident. Assume that the theorem is valid for $\dim V < n$ and now let $\dim V = n$.

The minimal polynomial $\phi_A$ is a product of factors of the form $x - c$ or $(x - a)^2 + b^2$, where $b \neq 0$ and $a, b, c$ are real numbers. (Since $A$ is normal, we know by Proposition 8.18 that all the factors are distinct.) We consider the following two cases.

1. $\phi_A$ has a factor of the form $x - c$. In this case, $c$ is a real characteristic root, so that $c = 1$ or $-1$ by Proposition 8.29. Let $\alpha$ be an eigenvector corresponding to $c$. Then $U = \mathrm{Sp}\{\alpha\}$ is invariant by $A$. By (2) of Proposition 8.29, $U^\perp$ is invariant by $A$. We have $V = U \oplus U^\perp$. Since $\dim U = n - 1$, we may apply the inductive assumption to the restriction of $A$ to $U^\perp$ to obtain an orthonormal basis $\{\alpha_2, \ldots, \alpha_n\}$ with respect to which $A$ on $U^\perp$ can be represented in the desired form. It is obvious, then, that $\{\alpha_1, \ldots, \alpha_n\}$ is an orthonormal basis of $V$ with respect to which $A$ has the desired matrix representation.

2. $\phi_A$ has only factors of the form $(x - a)^2 + b^2$, $b \neq 0$. In this case, let

$$\phi_A = ((x - a)^2 + b^2)\psi(x),$$

where $\psi(x)$ and $(x - a)^2 + b^2$ are relatively prime. By Theorem 7.5 there exists an $A$-invariant subspace $U$ on which $A$ has the minimal polynomial $(x - a)^2 + b^2$. Thus we may express $A$ on $U$ in the form

$$A = aI + bJ,$$

where $I$ is the identity transformation and $J = (A - aI)/b$ so that $J^2 = -I$. Let $\beta_1$ be an arbitrary nonzero vector in $U$ and set $\beta_2 = J\beta_1$. Then the subspace $W$ of $U$ spanned by $\beta_1$, $\beta_2$ is obviously invariant by $A$. Since $\beta_1$ and $\beta_2$ are linearly independent, we have dim $W = 2$. By Proposition 8.28 there exists an orthonormal basis $\{\alpha_1,\alpha_2\}$ of $W$ with respect to which $A$ on $W$ can be represented by

$$\begin{bmatrix} \cos\theta & -\sin\theta \\ \sin\theta & \cos\theta \end{bmatrix}, \qquad \text{where } a = \cos\theta \text{ and } b = \sin\theta.$$

(Note that the other case in Proposition 8.28 does not occur, since $A$ does not have $\pm 1$ as eigenvalues.) Now again we have $V = W \oplus W^{\perp}$ and we can apply the inductive assumption to the restriction of $A$ to $W^{\perp}$. It is clear that we can thus get an orthonormal basis of $V$ with the desired property.

***Corollary.*** *For any orthogonal matrix* $\mathbf{A}$ *of degree* $n$, *there is an orthogonal matrix* $\mathbf{S}$ *such that* $\mathbf{SAS}^{-1}$ *is of the form stated in Theorem* 8.30 (called the *canonical form* of $\mathbf{A}$).

***Example 8.18.*** Let $\mathbf{A}$ be an orthogonal matrix of degree 3. Then the canonical form is either

$$\begin{bmatrix} 1 & 0 & 0 \\ 0 & \cos\theta & -\sin\theta \\ 0 & \sin\theta & \cos\theta \end{bmatrix} \qquad \text{with } 0 \leq \theta < 2\pi$$

or

$$\begin{bmatrix} -1 & 0 & 0 \\ 0 & \cos\theta & -\sin\theta \\ 0 & \sin\theta & \cos\theta \end{bmatrix},$$

depending on whether det $\mathbf{A} = 1$ or det $\mathbf{A} = -1$. It follows that every orthogonal transformation $A$ of a three-dimensional real inner product space $V$ leaves a one-dimensional subspace invariant.

***Example 8.19.*** Let $V$ be a $2m$-dimensional real vector space with a complex structure $J$ (Definition 7.10). For any inner product $(\ ,\ )$ on $V$, we may define a new inner product:

$$((\alpha,\beta)) = (\alpha,\beta) + (J(\alpha),J(\beta)) \qquad \text{for all } \alpha, \beta \in V,$$

which is invariant by $J$: $((J(\alpha),J(\beta))) = ((\alpha,\beta))$.

Thus, changing the notation, we assume that $V$ has an inner product $(\alpha,\beta)$ for which $J$ is orthogonal. Since the minimal polynomial of $J$ is $x^2 + 1$, the canonical form of $J$ consists of blocks along the diagonal of the forms

$$\begin{bmatrix} 0 & -1 \\ 1 & 0 \end{bmatrix} \text{ for } \theta = \pi/2 \qquad \text{and} \qquad \begin{bmatrix} 0 & 1 \\ -1 & 0 \end{bmatrix} \text{ for } \theta = 3\pi/2.$$

In the second form we may switch the two vectors of the orthonormal basis to bring the block into the first form. Thus there is an orthonormal basis of $V$ with respect to which $J$ is represented by a matrix consisting of blocks along the diagonal of the form $\begin{bmatrix} 0 & -1 \\ 1 & 0 \end{bmatrix}$. Or, rearranging the order of the elements, we have a basis of the form

$\{\alpha_1, \ldots, \alpha_m, J(\alpha_1), \ldots, J(\alpha_m)\}$. (We have already proved this fact in Proposition 7.15.)

We shall now prove:

***Theorem 8.31.*** *Let $\mathfrak{G}$ be a group of orthogonal transformations of an n-dimensional real inner product space V. Assume that $\mathfrak{G}$ acts irreducibly on V. Then every symmetric bilinear form $f(\alpha,\beta)$ on $V \times V$ which is invariant by $\mathfrak{G}$ is a scalar multiple of the inner product.*

*Proof.* Let $A$ be the symmetric transformation such that

$$f(\alpha,\beta) = (A(\alpha),\beta) \qquad \text{for all } \alpha, \beta \in V,$$

as in Proposition 8.26. For any $B \in \mathfrak{G}$ we have

$$(AB(\alpha),\beta) = f(B(\alpha),\beta) = f(\alpha,B^{-1}(\beta)) = (A(\alpha),B^{-1}(\beta)) = (BA(\alpha),\beta)$$

since both $f$ and the inner product are invariant by $B, B^{-1} \in \mathfrak{G}$. Thus we have $AB = BA$. By the corollary to Theorem 7.14, we have either $A = aI$ or $A = aI + bJ$, where $J^2 = -I$ and $b \neq 0$. In the first case we have

$$f(\alpha,\beta) = (A(\alpha),\beta) = (a\alpha,\beta) = a(\alpha,\beta)$$

for all $\alpha, \beta \in V$, as we wanted to prove. We now show that the second case is impossible. Since $A$ is symmetric, so is $A - aI$. Thus $J = (A - aI)/b$ is symmetric, and

$$(J(\alpha),J(\alpha)) = (\alpha,J^2(\alpha)) = (\alpha,-\alpha) = -(\alpha,\alpha).$$

Thus we have a contradiction by taking any nonzero $\alpha \in V$.

**EXERCISE 8.7**

**1.** Show that the matrix

$$\begin{bmatrix} \sin\theta\cos\varphi & \cos\theta\cos\varphi & -\sin\varphi \\ \sin\theta\sin\varphi & \cos\theta\sin\varphi & \cos\varphi \\ \cos\theta & -\sin\theta & 0 \end{bmatrix}$$

is orthogonal for any values of $\theta$ and $\varphi$.

»☆ **2.** Let **A** and **B** be two $n \times n$ real matrices. Prove that the complex matrix $\mathbf{A} + i\mathbf{B}$ is unitary if and only if $\begin{bmatrix} \mathbf{A} & -\mathbf{B} \\ \mathbf{B} & \mathbf{A} \end{bmatrix}$ is orthogonal. In this case, find the determinant of $\begin{bmatrix} \mathbf{A} & -\mathbf{B} \\ \mathbf{B} & \mathbf{A} \end{bmatrix}$. (See Exercise 6.3, number 10.)

☆ **3.** (*a*) Let $V$ be a complex inner product space. Regarding $V$ as a real vector space $V'$, show that

$$(\alpha,\beta)' = \operatorname{Re}\{(\alpha,\beta)\}, \qquad \alpha, \beta \in V'$$

is an inner product in $V'$.

(*b*) If $\{\alpha_1, \ldots, \alpha_n\}$ is an orthonormal basis in $V$, show that

$$\{\alpha_1, \ldots, \alpha_n, i\alpha_1, \ldots, i\alpha_n\}$$

is an orthonormal basis in $V'$.

(*c*) Let $C$ be a linear transformation of $V$ defined by a complex matrix $\mathbf{C} = \mathbf{A} + i\mathbf{B}$ with respect to the orthonormal basis $\{\alpha_1, \ldots, \alpha_n\}$ of $V$, where $\mathbf{A}$ and $\mathbf{B}$ are real matrices. Show that the transformation of $V'$ which is defined by a matrix $\begin{bmatrix} \mathbf{A} & -\mathbf{B} \\ \mathbf{B} & \mathbf{A} \end{bmatrix}$ with respect to the basis $\{\alpha_1, \ldots, \alpha_n, i\alpha_1, \ldots, i\alpha_n\}$ is orthogonal if and only if $C$ is unitary (cf. number 2).

›› **4.** Find the canonical form of $\begin{bmatrix} \cos\theta & \sin\theta \\ \sin\theta & -\cos\theta \end{bmatrix}$.

☆ **5.** Prove:

››(*a*) If $\mathbf{A}$ is an orthogonal matrix for which $-1$ is not an eigenvalue, then $\mathbf{I}_n + \mathbf{A}$ is nonsingular and

$$\mathbf{S} = (\mathbf{I}_n - \mathbf{A})(\mathbf{I}_n + \mathbf{A})^{-1}$$

is skew-symmetric. ($\mathbf{S}$ is called the *Cayley transform of* $\mathbf{A}$.)

(*b*) For any skew-symmetric matrix $\mathbf{S}$, $\mathbf{I}_n + \mathbf{S}$ is nonsingular and

$$\mathbf{A} = (\mathbf{I}_n - \mathbf{S})(\mathbf{I}_n + \mathbf{S})^{-1}$$

is an orthogonal matrix for which $-1$ is not an eigenvalue.

(*c*) The correspondence $\mathbf{A} \leftrightarrow \mathbf{S}$ is one-to-one (cf. Exercise 8.4, number 10).

**6.** In part *b* of number 5, let $n = 2$ and $\mathbf{S} = \begin{bmatrix} 0 & c \\ -c & 0 \end{bmatrix}$, $c$ real. Find the corresponding $\mathbf{A}$.

››☆ **7.** Let $V$ be an $n$-dimensional real vector space and let $V^c$ be its complexification. Given an inner product in $V$, extend it to an inner product of $V^c$ as in Exercise 8.6, number 15. Let $A$ be an orthogonal transformation of $V$ and let $A^c$ be its extension (which is unitary as in Exercise 8.6, number 15). Let $c = \cos\theta + i\sin\theta$ be an eigenvalue of $A^c$ which is not real and let $\gamma \in V^c$ be an eigenvector corresponding to $c$. Writing $\gamma = \alpha + i\beta$, where $\alpha, \beta \in V$, prove that $\mathrm{Sp}\{\alpha,\beta\}$ is a two-dimensional subspace of $V$ invariant by $A$.

☆ **8.** Give an alternative proof of Theorem 8.30 in the following steps:

(*a*) Use induction on dim $V$. It is sufficient to prove that $A$ has a one- or two-dimensional invariant subspace.

(*b*) If $A$ has a real eigenvalue (namely, 1 or $-1$), then there is a one-dimensional invariant subspace. So assume that $A$ has no real eigenvalue.

(*c*) The extension $A^c$ of $A$ to the complexification $V^c$ of $V$ is unitary. Take an eigenvalue $\cos\theta + i\sin\theta\ (\neq \pm 1)$ and a corresponding eigenvector $\alpha + i\beta$. By number 7, $\mathrm{Sp}\{\alpha,\beta\}$ is a two-dimensional invariant subspace.

☆ **9.** Still another proof of Theorem 8.30 can be given as follows: We follow part *a* of Exercise 8. Let $B = A + A^* = A + A^{-1}$. $B$ is symmetric. By Theorem 8.16, $B$ has a real eigenvalue, say, $c$. Let $\alpha$ be an eigenvector corresponding to $c$. Then the subspace $\mathrm{Sp}\{\alpha,A(\alpha)\}$ is invariant by $A$ [observe that $(A + A^{-1})(\alpha) = c\alpha$ implies $A^2(\alpha) + \alpha = cA(\alpha)$, that is, $A(A(\alpha)) = cA(\alpha) - \alpha$]. Thus there is a one- or two-dimensional subspace invariant by $A$.

☆ **10.** Let $V$ be a $2m$-dimensional real inner product space with a complex structure and assume that $J$ is orthogonal (cf. Example 8.19). The complexification $V^c$ admits an extended inner product (Exercise 8.6, number 15). On the other hand, we have

$$V^c = W \oplus \bar{W},$$

where (Proposition 7.16)

$$W = \{\alpha - iJ(\alpha);\ \alpha \in V\} \quad \text{and} \quad \bar{W} = \{\alpha + iJ(\alpha);\ \alpha \in V\}$$

(*a*) Prove that $W$ and $\bar{W}$ are orthogonal to each other.

(*b*) Prove that the mapping $\phi\colon \alpha \in V \to \phi(\alpha) = (\alpha - iJ(\alpha))/2 \in W$, which is linear if $W$ is considered as a real vector space, preserves inner product: $(\alpha,\beta) = (\phi(\alpha),\phi(\beta))$.

(*c*) If $A$ is an orthogonal transformation of $V$ which commutes with $J$, then its extension $A^c$ is a unitary transformation which leaves $W$ and $\bar{W}$ invariant.

☆**11.** Recall that (Exercise 8.6, number 7) a linear transformation $A$ of an $n$-dimensional real inner product space $V$ is called a *skew-symmetric* transformation if $A + A^* = 0$. Prove the following:

(*a*) A linear transformation $A$ is skew-symmetric if and only if it can be represented by a skew-symmetric matrix with respect to an orthonormal basis.

(*b*) Any eigenvalue of a skew-symmetric transformation is 0.

(*c*) The characteristic roots of a skew-symmetric transformation are 0 or purely imaginary.

(*d*) The extension $A^c$ of a linear transformation $A$ to the complexification $V^c$ is skew-hermitian (with respect to the extended inner product in $V^c$) if and only if $A$ is skew-symmetric.

(*e*) If $A$ is skew-symmetric and if $\alpha + i\beta \in V^c$ with $\alpha, \beta \in V$ is an eigenvector for $A^c$ corresponding to $ic$, where $c$ is real and $\neq 0$, then $\mathrm{Sp}\{\alpha,\beta\}$ is a two-dimensional subspace of $V$ invariant by $A$.

☆**12.** Continuing number 11, prove the following theorem. *If $A$ is a skew-symmetric transformation of an $n$-dimensional real inner product space $V$, then there exists an orthonormal basis of $V$ with respect to which $A$ can be represented by a matrix of the form*

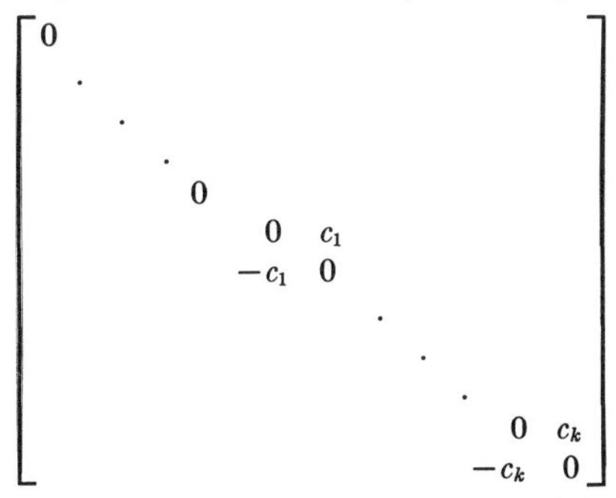

*where $c_1, \ldots, c_k$ are distinct nonzero real numbers, each block* $\begin{bmatrix} 0 & c_j \\ -c_j & 0 \end{bmatrix}$ *appears the same number of times as the multiplicity of the characteristic root $ic_j$ of $A$, and, before the first block* $\begin{bmatrix} 0 & c_1 \\ -c_1 & 0 \end{bmatrix}$, *0 appears the same number of times as its multiplicity as a characteristic root of $A$.* (Use induction on dim $V$. Show, by using number 11, that $V$ has a one- or two-dimensional subspace invariant by $A$.)

☆**13.** Derive Theorem 8.30 as well as the result in number 12 above from the result in number 7 of Exercise 8.5.

» **14.** Prove that the rank of a skew-symmetric transformation is even.

☆**15.** Let $V$ be an $n$-dimensional real inner product space. Prove that there is a one-to-one correspondence between the set of skew-symmetric transformations $A$ and the set of skew-symmetric bilinear functions $f$ on $V \times V$, the correspondence being given by $f(\alpha,\beta) = (A(\alpha),\beta)$, where $\alpha, \beta \in V$.

›› **16.** If $\mathbf{A} = [a_{ij}]$ is an $n \times n$ orthogonal matrix and $\Delta_{ij}$ is the $(i,j)$ cofactor, prove that

$$\Delta_{ij} = (\det \mathbf{A})\, a_{ij} \qquad \text{for all } i, j.$$

**17.** Prove that if $\mathbf{A}$ is an $n \times n$ real matrix, then there exist orthogonal matrices $\mathbf{S}$ and $\mathbf{T}$ such that

$$\mathbf{SAT} = \begin{bmatrix} c_1 & & & 0 \\ & \cdot & & \\ & & \cdot & \\ 0 & & & c_n \end{bmatrix},$$

where $c_1^2, \ldots, c_n^2$ are the eigenvalues of the symmetric matrix ${}^t\mathbf{AA}$.

››☆**18.** Prove that a nonsingular $n \times n$ real matrix $\mathbf{A}$ can be expressed in the form $\mathbf{A} = \mathbf{BC}$, where $\mathbf{B} \in O(n)$ and $\mathbf{C}$ is a real superdiagonal matrix (cf. Theorem 7.3).

››☆**19.** Prove that a nonsingular $n \times n$ real matrix $\mathbf{A}$ can be expressed in the form $\mathbf{A} = \mathbf{BC}$, where $\mathbf{B} \in O(n)$ and $\mathbf{C}$ is a positive-definite symmetric matrix.

**20.** Prove that $|\text{trace } \mathbf{A}| \leq n$ for every $\mathbf{A} \in O(n)$.

☆**21.** Prove that $SO(n)$ is arcwise-connected (cf. Exercise 6.3, number 15). How about $O(n)$?

# 9 Affine Spaces

Starting from the notion of vector space, we define an affine space by a set of axioms given by H. Weyl and study basic concepts such as affine coordinate systems, affine subspaces, and affine transformations. After treating various ways of representing affine subspaces (including barycentric coordinates), we discuss convex sets in a real affine space.

## 9.1 AFFINE SPACES

Let $V$ be a vector space of finite dimension over an arbitrary field $\mathfrak{F}$.

**Definition 9.1**

A nonempty set $X$ is called an *affine space* associated to $V$ if there is a mapping of $X \times X$ into $V$, denoted by

$$(P,Q) \in X \times X \to \overrightarrow{PQ} \in V,$$

which has the following properties:

**1.** For any three points $P$, $Q$, and $R$ in $X$, we have $\overrightarrow{PR} = \overrightarrow{PQ} + \overrightarrow{QR}$.

**2.** For any $P \in X$ and for any $\alpha \in V$, there is one and only one point $Q \in X$ such that $\overrightarrow{PQ} = \alpha$.

We shall say that $\overrightarrow{PQ}$ is the vector determined by the *initial point* $P$ and the *end point* $Q$. We define dim $X$ to be dim $V$.

Condition 2 may be stated as follows. For any point $P \in X$ the mapping $Q \in X \to \overrightarrow{PQ} \in V$ is a one-to-one mapping of $X$ onto $V$. In this sense the affine space $X$ may be identified as a set with the vector space $V$. The vector space $V$ has a special element, namely, the zero vector 0, which has the property that $\alpha + 0 = \alpha$ for every $\alpha \in V$, whereas the affine space $X$ has no particular point which is distinguished from other points.

***Example 9.1.*** Let $V = \mathfrak{F}^n$ be the standard $n$-dimensional vector space over a field $\mathfrak{F}$. In the same set $X = \mathfrak{F}^n$, we define a mapping $X \times X \to V$ by

$$\overrightarrow{PQ} = \beta - \alpha \qquad \text{for } P = \alpha \text{ and } Q = \beta.$$

Then the set $X = \mathfrak{F}^n$ is an affine space associated with the vector space $V = \mathfrak{F}^n$. This affine space is called the *standard n-dimensional affine space* $\mathfrak{F}^n$ (*standard n-dimensional real* or *complex affine space* according as $\mathfrak{F} = \mathfrak{R}$ or $\mathfrak{C}$).

We shall derive some consequences of Definition 9.1.

***Proposition 9.1.*** *In an affine space X we have:*

1. $\overrightarrow{PP} = 0$ (*zero vector*) *for every* $P \in X$.
2. $\overrightarrow{QP} = -\overrightarrow{PQ}$ *for any* $P, Q \in X$.
3. *If* $\overrightarrow{PQ} = \overrightarrow{P'Q'}$, *then* $\overrightarrow{PP'} = \overrightarrow{QQ'}$.

*Proof.* 1. In condition 1 of Definition 9.1, assume $P = Q = R$. Then $\overrightarrow{PP} = \overrightarrow{PP} + \overrightarrow{PP}$, which implies $\overrightarrow{PP} = 0$.

2. In condition 1 assume $P = R$. Then $\overrightarrow{PP} = \overrightarrow{PQ} + \overrightarrow{QP}$. Since we already know that $\overrightarrow{PP} = 0$, it follows that $\overrightarrow{QP} = -\overrightarrow{PQ}$.

3. We have

$$\overrightarrow{PQ'} = \overrightarrow{PQ} + \overrightarrow{QQ'} = \overrightarrow{PP'} + \overrightarrow{P'Q'}.$$

Since $\overrightarrow{PQ} = \overrightarrow{P'Q'}$ by assumption, we obtain $\overrightarrow{QQ'} = \overrightarrow{PP'}$.

Statement 3 may be interpreted as a law of a parallelogram. If $\overrightarrow{PQ} = \overrightarrow{P'Q'}$ in the plane, then $P, Q, Q', P'$ are the vertices of a parallelogram in that order and $\overrightarrow{PP'} = \overrightarrow{QQ'}$.

We now make the following definition.

**Definition 9.2**

An ordered $(n + 1)$-tuple of points $\{P_0,P_1, \ldots ,P_n\}$ in an affine space $X$ is called an *affine frame* if the vectors $\overrightarrow{P_0P_i}$, $1 \leq i \leq n$, form a basis of $V$ The point $P_0$ is called the *origin*, and $P_i$ the *i*th *unit point*, of the affine frame.

Given any point $P_0$, there exists an affine frame with origin $P_0$. In fact, let $\{\alpha_1, \ldots ,\alpha_n\}$ be an arbitrary basis of $V$ and let $P_i$ be the unique point of $X$ such that $\overrightarrow{P_0P_i} = \alpha_i$ for each $i$. Then $\{P_0,P_1, \ldots ,P_n\}$ is an affine frame. This argument shows that, once a point is chosen as the origin, affine frames with this origin and bases in $V$ correspond to each other.

An affine frame $\{P_0,P_1, \ldots ,P_n\}$ will determine a certain coordinate system as follows. For each point $P \in X$ we may write the vector $\overrightarrow{P_0P}$

in the form

$$\overrightarrow{P_0P} = \sum_{i=1}^{n} a^i \overrightarrow{P_0P_i},$$

where the coefficients $a^1, \ldots, a^n$ are uniquely determined. We define a set of $\mathcal{F}$-valued functions $x^i$, $1 \leq i \leq n$, on the affine space $X$ by

$$x^i(P) = a^i, \qquad 1 \leq i \leq n.$$

We associate with $P$ the element

$$(x^1(P), \ldots, x^n(P))$$

of the standard affine space $\mathcal{F}^n$ and call this ordered $n$-tuple of scalars the *coordinates* of $P$.

Conversely, given any ordered set of $n$ scalars $a^1, \ldots, a^n$, there exists one and only one point $P \in X$ whose coordinates are precisely $(a^1, \ldots, a^n)$. In fact, consider the vector $\sum_{i=1}^{n} a^i \overrightarrow{P_0P_i} \in V$ and let $P$ be the unique point such that $\overrightarrow{P_0P}$ is equal to this vector. (The existence of $P$ is assured by condition 2 of Definition 9.1.)

We have thus obtained a system of $\mathcal{F}$-valued functions $\{x^1, \ldots, x^n\}$, which gives rise to a one-to-one mapping of $X$ onto the standard affine space $\mathcal{F}^n$. This system of functions is called the *affine coordinate system* determined by the affine frame $\{P_0, P_1, \ldots, P_n\}$.

***Example 9.2.*** For the affine coordinate system determined by an affine frame $\{P_0, P_1, \ldots, P_n\}$ in an affine space, the coordinates of $P_0, P_1, \ldots, P_n$ are respectively

$$(0,0, \ldots, 0), (1,0, \ldots, 0), \ldots, (0, \ldots, 0,1).$$

This fact justifies the names of origin and $i$th unit points for $P_0$ and $P_i$'s.

***Example 9.3.*** In the standard affine space $\mathfrak{F}^n$, let

$$E_0 = (0,0, \ldots, 0), E_1 = (1,0, \ldots, 0), \ldots, E_n = (0, \ldots, 0,1).$$

The frame $\{E_0, E_1, \ldots, E_n\}$ is called the *standard affine frame.* The corresponding coordinate system, called the *standard coordinate system,* assigns to any point $P = (a^1, \ldots, a^n)$ the components $a^1, \ldots, a^n$ themselves.

***Remark.*** In this chapter we shall represent vectors of the standard vector space $\mathfrak{F}^n$ or points of the standard affine space $\mathfrak{F}^n$ by $(a^i) = (a^1, \ldots, a^n)$ or, more frequently, by

$$[a^i] = \begin{bmatrix} a^1 \\ a^2 \\ \cdot \\ \cdot \\ \cdot \\ a^n \end{bmatrix},$$

where in both cases we use the *superscripts* $1, 2, \ldots, n$. We shall also denote a matrix $A$ by $[a^i{}_j]$, where $a^i{}_j$ is the $(i,j)$ component.

We now study how two different affine coordinate systems are related.

Let $\{P_i\}$ and $\{Q_i\}$ be two affine frames in an affine space $X$. The "relative position" of the frame $\{P_i\}$ with respect to the frame $\{Q_i\}$ can be expressed as follows:

$$\overrightarrow{Q_0P_0} = \sum_{i=1}^{n} a^i \overrightarrow{Q_0Q_i}, \tag{9.1}$$

$$\overrightarrow{P_0P_i} = \sum_{j=1}^{n} a^j{}_i \overrightarrow{Q_0Q_j}, \qquad 1 \leq i \leq n. \tag{9.2}$$

Equation (9.1) means that the origin $P_0$ of the frame $\{P_i\}$ has coordinates $[a^i]$ with respect to the frame $\{Q_i\}$. In (9.2), the matrix $\mathbf{A} = [a^i{}_j]$ is nonsingular, since both $\{\overrightarrow{P_0P_i}; 1 \leq i \leq n\}$ and $\{\overrightarrow{Q_0Q_i}; 1 \leq i \leq n\}$ form a basis of $V$. We may reverse the roles of $\{P_i\}$ and $\{Q_i\}$ and write

$$\overrightarrow{P_0Q_0} = \sum_{i=1}^{n} b^i \overrightarrow{P_0P_i}, \tag{9.3}$$

$$\overrightarrow{Q_0Q_i} = \sum_{j=1}^{n} b^j{}_i \overrightarrow{P_0P_j}, \qquad 1 \leq i \leq n, \tag{9.4}$$

where $\mathbf{B} = [b^j{}_i]$ is the inverse of $\mathbf{A} = [a^j{}_i]$, that is,

$$\sum_{k=1}^{n} b^i{}_k a^k{}_j = \delta^i{}_j = \begin{cases} 1 & \text{if } i = j, \\ 0 & \text{if } i \neq j. \end{cases}$$

Now for an arbitrary point $P$ in $X$ we have

$$\overrightarrow{Q_0P} = \overrightarrow{Q_0P_0} + \overrightarrow{P_0P}.$$

The left-hand side is equal to $\sum_{i=1}^{n} y^i \overrightarrow{Q_0Q_i}$, where $[y^i]$ are the coordinates of $P$ with respect to the frame $\{Q_i\}$. On the other hand, we have

$$\overrightarrow{P_0P} = \sum_{i=1}^{n} x^i \overrightarrow{P_0P_i},$$

where $[x^i]$ are the coordinates of $P$ with respect to the frame $\{P_i\}$. Using (9.1) and (9.2), we obtain

$$\begin{aligned} \overrightarrow{Q_0P_0} + \overrightarrow{P_0P} &= \sum_{i=1}^{n} a^i \overrightarrow{Q_0Q_i} + \sum_{j=1}^{n} x^j \overrightarrow{P_0P_j} \\ &= \sum_{i=1}^{n} a^i \overrightarrow{Q_0Q_i} + \sum_{j=1}^{n} \left( \sum_{i=1}^{n} x^j a^i{}_j \overrightarrow{Q_0Q_i} \right) \\ &= \sum_{i=1}^{n} \left( \sum_{j=1}^{n} a^i{}_j x^j + a^i \right) \overrightarrow{Q_0Q_i}. \end{aligned}$$

Hence we obtain

$$y^i = \sum_{j=1}^{n} a^i{}_j x^j + a^i, \qquad 1 \leq i \leq n. \tag{9.5}$$

Since this is the case for an arbitrary point $P$, the relationship between the two coordinate systems determined by the frames $\{P_i\}$ and $\{Q_i\}$ is given by (9.5). It can be expressed very conveniently by the following matrix notation:

$$\begin{bmatrix} y^1 \\ y^2 \\ \cdot \\ \cdot \\ \cdot \\ y^n \\ 1 \end{bmatrix} = \begin{bmatrix} a^1{}_1 & \cdots & a^1{}_n & a^1 \\ a^2{}_1 & \cdots & a^2{}_n & a^2 \\ \cdots & \cdots & \cdots & \cdots \\ a^n{}_1 & \cdots & a^n{}_n & a^n \\ 0 & \cdots & 0 & 1 \end{bmatrix} \begin{bmatrix} x^1 \\ x^2 \\ \cdot \\ \cdot \\ \cdot \\ x^n \\ 1 \end{bmatrix} \tag{9.6}$$

or

$$\eta = \mathbf{A}\xi + \alpha, \tag{9.7}$$

where $\xi$, $\eta$, and $\alpha$ are the column vectors $[x^i]$, $[y^i]$, and $[a^i]$, respectively, and $\mathbf{A} = [a^i{}_j]$.

Interchanging the roles of the two coordinate systems, we have

$$\begin{bmatrix} x^1 \\ x^2 \\ \cdot \\ \cdot \\ \cdot \\ x^n \\ 1 \end{bmatrix} = \begin{bmatrix} b^1{}_1 & \cdots & b^1{}_n & b^1 \\ b^2{}_1 & \cdots & b^2{}_n & b^2 \\ \cdots & \cdots & \cdots & \cdots \\ b^n{}_1 & \cdots & b^n{}_n & b^n \\ 0 & \cdots & 0 & 1 \end{bmatrix} \begin{bmatrix} y^1 \\ y^2 \\ \cdot \\ \cdot \\ \cdot \\ y^n \\ 1 \end{bmatrix} \tag{9.8}$$

or

$$\xi = \mathbf{B}\eta + \beta, \qquad \mathbf{B} = [b^i{}_j], \tag{9.9}$$

where $\beta = [b^i]$ can be found by solving (9.7) as follows:

$$\xi = \mathbf{A}^{-1}(\eta - \alpha) = \mathbf{B}\eta - \mathbf{B}\alpha,$$

so that

$$\beta = -\mathbf{B}\alpha, \qquad \text{where } \mathbf{B} = \mathbf{A}^{-1}. \tag{9.10}$$

We sum up these discussions in:

***Proposition 9.2.*** *If two affine frames $\{P_i\}$ and $\{Q_i\}$ are related by* (9.1) *and* (9.2), *then the corresponding coordinate systems $\{x^i\}$ and $\{y^i\}$ are related by* (9.6) *and* (9.8), *where $\mathbf{B} = \mathbf{A}^{-1}$ and $\beta = [b^i]$ is given by* (9.10).

There are two special cases of interest.

***Corollary 1.*** *In particular, if $P_0 = Q_0$, then we have*

$$(9.11) \qquad \begin{bmatrix} y^1 \\ \cdot \\ \cdot \\ \cdot \\ y^n \end{bmatrix} = \mathbf{A} \begin{bmatrix} x^1 \\ \cdot \\ \cdot \\ \cdot \\ x^n \end{bmatrix}.$$

(Note that this is the same form as a linear transformation of the vector space $\mathfrak{F}^n$.)

***Corollary 2.*** *In particular, if $\overrightarrow{P_0P_i} = \overrightarrow{Q_0Q_i}$ for every $i$, $1 \le i \le n$, then we have*

$$(9.12) \qquad y^i = x^i + a^i, \qquad 1 \le i \le n.$$

(This is called the *translation of coordinate axes.*)

***Example 9.4.*** In the standard affine space $\mathfrak{F}^n$, let

$$P_i = \begin{bmatrix} a^1{}_i \\ \cdot \\ \cdot \\ \cdot \\ a^n{}_i \end{bmatrix}, \qquad 0 \le i \le n.$$

Then

$$(9.13) \qquad \overrightarrow{P_0P_i} = \begin{bmatrix} p^1{}_i - p^1{}_0 \\ \cdot \\ \cdot \\ \cdot \\ p^n{}_i - p^n{}_0 \end{bmatrix}, \qquad 1 \le i \le n.$$

The points $P_0, P_1, \ldots, P_n$ form an affine frame if and only if the determinant of the matrix with column vectors (9.13) is not 0. But this determinant is equal to

$$(9.14) \qquad \begin{vmatrix} p^1{}_0 & p^1{}_1 & \cdots & p^1{}_n \\ \cdot & \cdot & \cdots & \cdot \\ p^n{}_0 & p^n{}_1 & \cdots & p^n{}_n \\ 1 & 1 & \cdots & 1 \end{vmatrix}$$

up to the sign $(-1)^n$. Thus $\{P_i\}$ is an affine frame if and only if (9.14) is not 0.

## EXERCISE 9.1

**1.** For any points $P_1, P_2, \ldots, P_k$ in an affine space, show that

$$\overrightarrow{P_1P_2} + \overrightarrow{P_2P_3} + \cdots + \overrightarrow{P_{k-1}P_k} + \overrightarrow{P_kP_1} = 0.$$

›› **2.** If $\{P_0,P_1, \ldots ,P_n\}$ is an affine frame in an affine space, show that, for any permutation $\pi$ of $\{0, 1, \ldots ,n\}$, $\{P_{\pi(0)},P_{\pi(1)}, \ldots ,P_{\pi(n)}\}$ is an affine frame.

**3.** In the standard real affine space $\mathfrak{R}^3$, determine whether each of the following sets of points is an affine frame:

(a) $P_0 = \begin{bmatrix} 1 \\ 0 \\ 1 \end{bmatrix}, \quad P_1 = \begin{bmatrix} 1 \\ 1 \\ 0 \end{bmatrix}, \quad P_2 = \begin{bmatrix} 0 \\ 1 \\ 1 \end{bmatrix}, \quad P_3 = \begin{bmatrix} 0 \\ 0 \\ 1 \end{bmatrix}.$

(b) $P_0 = \begin{bmatrix} -1 \\ 0 \\ 2 \end{bmatrix}, \quad P_1 = \begin{bmatrix} 0 \\ 1 \\ -2 \end{bmatrix}, \quad P_2 = \begin{bmatrix} 0 \\ 0 \\ 1 \end{bmatrix}, \quad P_3 = \begin{bmatrix} 1 \\ 0 \\ -3 \end{bmatrix}.$

» **4.** In the standard complex affine space $\mathcal{C}^2$, let

$$P_0 = \begin{bmatrix} 1 \\ i \end{bmatrix}, \quad P_1 = \begin{bmatrix} 0 \\ 1 \end{bmatrix}, \quad P_2 = \begin{bmatrix} 1+i \\ 1 \end{bmatrix}.$$

Show that they form an affine frame. Find the relationship between the standard coordinate system and the coordinate system determined by $\{P_0,P_1,P_2\}$.

## 9.2 AFFINE TRANSFORMATIONS

Let $V_1$ and $V_2$ be vector spaces over the same field $\mathcal{F}$ and let $X_1$ and $X_2$ be affine spaces associated with $V_1$ and $V_2$, respectively.

Consider a mapping $f$ of $X_1$ into $X_2$. For any point $P$ in $X_1$, we define a mapping $\phi_P$ of $V_1$ into $V_2$ as follows. For $\alpha \in V_1$, let $Q$ be the unique point in $X_1$ such that $\alpha = \overrightarrow{PQ}$, and set

$$\phi_P(\alpha) = \overrightarrow{f(P)f(Q)}. \tag{9.15}$$

We prove:

***Proposition 9.3.*** *If $\phi_P$ is a linear mapping of $V_1$ into $V_2$, then for any point $P_1$, the mapping $\phi_{P_1}$ coincides with $\phi_P$.*

*Proof.* For an arbitrary point $Q$ in $X_1$, we have

$$\overrightarrow{P_1Q} = \overrightarrow{PQ} - \overrightarrow{PP_1} \quad \text{and} \quad \overrightarrow{f(P_1)f(Q)} = \overrightarrow{f(P)f(Q)} - \overrightarrow{f(P)f(P_1)}.$$

Since the mapping $\phi_P$ is linear by assumption, we obtain

$$\begin{aligned} \phi_P(\overrightarrow{P_1Q}) &= \phi_P(\overrightarrow{PQ}) - \phi_P(\overrightarrow{PP_1}) \\ &= \overrightarrow{f(P)f(Q)} - \overrightarrow{f(P)f(P_1)} \\ &= \overrightarrow{f(P_1)f(Q)} = \phi_{P_1}(\overrightarrow{P_1Q}). \end{aligned}$$

Given any $\alpha \in V_1$, we may choose $Q$ in $X_1$ so that $\overrightarrow{P_1Q} = \alpha$. Then $\phi_P(\alpha) = \phi_{P_1}(\alpha)$ as above, showing that $\phi_P = \phi_{P_1}$.

Proposition 9.3 shows that, for a given mapping $f$ of $X_1$ into $X_2$, the condition that the mapping $\phi_P$ is linear for some point $P$ implies that this linear mapping is determined independently of the choice of $P$.

**Definition 9.3**

A mapping $f$ of $X_1$ into $X_2$ is called an *affine mapping* if $\phi_P$ is linear for some point $P$ in $X_1$. We shall denote this linear mapping simply by $\phi$ and call it the *associated linear mapping.*

**Definition 9.4**

A one-to-one affine mapping of $X_1$ onto $X_2$ is called an *affine isomorphism* of $X_1$ onto $X_2$. In particular, an affine isomorphism of an affine space $X$ onto itself is called an *affine transformation* of $X$.

If $f$ is an affine isomorphism of $X_1$ onto $X_2$, then the inverse mapping $f^{-1}$ is also affine and its associated linear mapping is $\phi^{-1}$.

We shall now prove:

***Proposition 9.4.*** *Let $X_1$ and $X_2$ be affine spaces associated with vector spaces $V_1$ and $V_2$, respectively, where* $\dim V_1 = \dim V_2 = n$. *Then*

1. *If $\{P_0,P_1, \ldots ,P_n\}$ is an affine frame of $X_1$ and if $f$ is an affine isomorphism of $X_1$ onto $X_2$, then $\{f(P_0),f(P_1), \ldots ,f(P_n)\}$ is an affine frame of $X_2$.*
2. *Given affine frames $\{P_i\}$ and $\{Q_i\}$ in $X_1$ and $X_2$, respectively, there is one and only one affine isomorphism $f$ of $X_1$ onto $X_2$ such that $f(P_i) = Q_i$ for $0 \le i \le n$.*

*Proof.* 1. We have $\overrightarrow{f(P_0)f(P_i)} = \phi(\overrightarrow{P_0P_i})$. Since $\phi$ is a linear isomorphism of $V_1$ onto $V_2$, it follows that $\overrightarrow{f(P_0)f(P_i)}$, $1 \le i \le n$, form a basis of $V_2$.

2. For any point $P$ in $X_1$, let $\overrightarrow{P_0P} = \sum_{i=1}^{n} a^i\overrightarrow{P_0P_i}$ and define $f(P)$ as the unique point of $X_2$ such that

$$\overrightarrow{Q_0f(P)} = \sum_{i=1}^{n} a^i\overrightarrow{Q_0Q_i}.$$

(In other words, a point with coordinates $[a^i]$ with respect to the frame $\{P_i\}$ is mapped upon the point of $X_2$ with the same coordinates with respect to the frame $\{Q_i\}$.) Obviously, $f(P_i) = Q_i$ for $0 \le i \le n$. The mapping $\phi_{P_0}$ maps a vector $\sum_{i=1}^{n} a^i\overrightarrow{P_0P_i}$ in $V_1$ upon the vector $\sum_{i=1}^{n} a^i\overrightarrow{Q_0Q_i}$ and hence is linear. This shows that our mapping $f$ is affine. Finally, it is obvious that it is an affine isomorphism.

***Example 9.5.*** Let $X$ be an affine space associated with an $n$-dimensional vector space $V$ over a field $\mathfrak{F}$. Let $\mathfrak{F}^n$ be the standard affine space with standard affine frame $\{E_0,E_1, \ldots ,E_n\}$. For any affine frame $\{P_i\}$ in $X$, there is an affine isomorphism $f$ of

$\mathfrak{F}^n$ onto $X$ such that

$$f(E_0) = P_0, \qquad f(E_i) = P_i, \qquad 1 \le i \le n.$$

Conversely, an affine isomorphism $f$ of $\mathfrak{F}^n$ onto $X$ determines an affine frame $\{P_i\}$, where $P_0 = f(E_0)$ and $P_i = f(E_i)$, $1 \le i \le n$. In terms of affine coordinate systems, we may say that the affine coordinate system $\{y^i\}$ on $X$ determined by $\{P_i\}$ is given by $y^i(P) = x^i(f^{-1}(P))$, $P \in X$, where $\{x^i\}$ is the standard coordinate system in $\mathfrak{F}^n$.

***Example 9.6.*** Let $\mathbf{A} = [a^i{}_j]$ be an $m \times n$ matrix over $\mathfrak{F}$ and $\alpha = [a^i]$ an $m$-dimensional column vector (in $\mathfrak{F}^m$). Then

$$f\begin{bmatrix} x^1 \\ \cdot \\ \cdot \\ \cdot \\ x^n \end{bmatrix} = \mathbf{A}\begin{bmatrix} x^1 \\ \cdot \\ \cdot \\ \cdot \\ x^n \end{bmatrix} + \begin{bmatrix} a^1 \\ \cdot \\ \cdot \\ \cdot \\ a^m \end{bmatrix}$$

defines an affine mapping of the standard affine space $\mathfrak{F}^n$ into the standard affine space $\mathfrak{F}^m$. The associated linear mapping is given by

$$\phi\begin{bmatrix} x^1 \\ \cdot \\ \cdot \\ \cdot \\ x^n \end{bmatrix} = \mathbf{A}\begin{bmatrix} x^1 \\ \cdot \\ \cdot \\ \cdot \\ x^n \end{bmatrix}.$$

We shall now express an affine transformation $f$ of an affine space $X$ in terms of an arbitrarily fixed affine coordinate system. Let $\{P_i\}$ be an arbitrary affine frame in $X$ and let $\{x^i\}$ be the corresponding coordinate system. Then $f(P_i)$ is an affine frame and the corresponding coordinate system $\{y^i\}$ is related to $\{x^i\}$ by

$$x^i = \sum_{j=1}^{n} b^i{}_j y^j + b^i, \qquad 1 \le i \le n,$$

as in Sec. 9.1, where $\mathbf{B} = [b^i{}_j] \in GL(n,\mathfrak{F})$ and $\beta = [b^i] \in \mathfrak{F}^n$.

Now a point $P$ such that $x^i(P) = p^i$ is mapped upon the point $f(P)$ with $y^i(f(P)) = p^i$. Therefore, we have

$$x^i(f(P)) = \sum_{j=1}^{n} b^i{}_j p^j + b^i, \qquad 1 \le i \le n.$$

By using only the coordinate system $\{x^i\}$, the mapping $f$ is expressed by

$$\bar{x}^i = \sum_{j=1}^{n} b^i{}_j x^j + b^i, \qquad 1 \le i \le n, \tag{9.16}$$

or equivalently,

$$\begin{bmatrix} \bar{x}^1 \\ \cdot \\ \cdot \\ \cdot \\ \bar{x}^n \\ 1 \end{bmatrix} = \begin{bmatrix} \mathbf{B} & \beta \\ 0 & 1 \end{bmatrix}\begin{bmatrix} x^1 \\ \cdot \\ \cdot \\ \cdot \\ x^n \\ 1 \end{bmatrix}, \tag{9.17}$$

where $\bar{x}^i$'s are the functions such that $\bar{x}^i(P) = x^i(f(P))$. In (9.17) the vector $\beta$ and the matrix $\mathbf{B}$ have the following significance. The image $f(P_0)$ of the origin $P_0$ has coordinates $\beta = [b^i]$, since

$$\begin{bmatrix} \mathbf{B} & \beta \\ 0 & 1 \end{bmatrix} \begin{bmatrix} 0 \\ 1 \end{bmatrix} = \begin{bmatrix} \beta \\ 1 \end{bmatrix}.$$

Using the standard basis $\{\epsilon_i\}$ of the vector space $\mathcal{F}^n$, we have

$$\begin{bmatrix} \mathbf{B} & \beta \\ 0 & 1 \end{bmatrix} \begin{bmatrix} \epsilon_i \\ 1 \end{bmatrix} = (i\text{th column vector of } \mathbf{B}) + \beta,$$

which implies that the $j$th coordinate of $f(P_i)$ is $b^j{}_i + b^j$. The matrix $\mathbf{B}$ represents the associated linear mapping with respect to the basis $\{\overrightarrow{P_0P_i}\}$ of the vector space $\mathcal{F}^n$.

We summarize the result in:

***Proposition 9.5.*** *With respect to a fixed affine coordinate system* $\{x^i\}$, *an affine transformation* $f$ *is expressed by* (9.17), *where* $\mathbf{B} \in GL(n,\mathcal{F})$ *and* $\beta \in \mathcal{F}^n$ *are determined by the images of the origin and the unit points of the affine frame.*

We shall denote by $A(X)$ the set of all affine transformations of an affine space $X$. It can be easily verified that $A(X)$ forms a group under the composition of mappings. If we fix an affine coordinate system, then the group $A(X)$ can be represented as the group of all $(n + 1) \times (n + 1)$ matrices of the form

$$\begin{bmatrix} \mathbf{A} & \alpha \\ 0 & 1 \end{bmatrix}, \qquad \text{where } \mathbf{A} \in GL(n,\mathcal{F}) \text{ and } \alpha \in \mathcal{F}^n. \tag{9.18}$$

**Definition 9.5**

The group consisting of all matrices of the form (9.18) is called the *affine group* and is denoted by $A(n,\mathcal{F})$.

We shall consider some special affine transformations as examples.

***Example 9.7.*** An affine transformation is called a *centroaffine transformation* if it fixes a certain point. By taking this point as the origin, the centroaffine transformation can be expressed by

$$\bar{x}^i = \sum_{j=1}^{n} a^i{}_j x^j, \qquad 1 \leq i \leq n. \tag{9.19}$$

Note that this formula is the same as that for a linear transformation of the vector space $\mathfrak{F}^n$.

***Example 9.8.*** A *dilation* is a centroaffine transformation whose associated linear mapping is $cI$, where $I$ is the identity transformation and $c$ is a certain nonzero element of $\mathfrak{F}$. In (9.19) we have

$$\bar{x}^i = cx^i, \qquad 1 \leq i \leq n. \tag{9.20}$$

***Example 9.9.*** An affine transformation $f$ is called a *translation* if the associated linear mapping is the identity transformation. For a translation $f$, the vector $\overrightarrow{Pf(P)}$ is a constant vector in $V$ for all points $P$ in $X$. In fact, if $P, Q \in X$, then we have

$$\phi(\overrightarrow{PQ}) = \overrightarrow{f(P)f(Q)}$$

by definition of $\phi$. Since $\phi$ is the identity, we have

$$\overrightarrow{f(P)f(Q)} = \overrightarrow{PQ},$$

which implies $\overrightarrow{Pf(P)} = \overrightarrow{Qf(Q)}$ by (3) of Proposition 9.1. This constant vector is called the *vector of translation*, since it determines $f$ completely.

We shall prove:

***Proposition 9.6.*** *Let $O$ be an arbitrary point of an affine space $X$. Any affine transformation $f$ of $X$ can be written uniquely as a product $f = hg$, where $g$ is a centroaffine transformation fixing $O$ and $h$ is a translation.*

*Proof.* Let $h$ be the translation given by the vector $\overrightarrow{Of(O)}$; then $g = h^{-1}f$ leaves $O$ invariant and $f = hg$. This representation is unique, because $f(O) = hg(O) = h(O)$, so that $h$ must be the translation given by $\overrightarrow{Of(O)}$.

Proposition 9.6 is equivalent to the following proposition on matrices.

***Proposition 9.7.*** *Every matrix in $A(n,\mathfrak{F})$ can be written uniquely in the form*

$$\begin{bmatrix} \mathbf{A} & \alpha \\ 0 & 1 \end{bmatrix} = \begin{bmatrix} \mathbf{I}_n & \alpha \\ 0 & 1 \end{bmatrix} \begin{bmatrix} \mathbf{A} & 0 \\ 0 & 1 \end{bmatrix}.$$

***Example 9.10.*** An affine transformation $f$ is called a *homothety* if the associated linear mapping is $cI$, where $c \neq 0$ is in $\mathfrak{F}$. It is a product of a dilation and a translation. In terms of an affine coordinate system, it can be represented by a matrix of the form

$$\begin{bmatrix} c\mathbf{I}_n & \alpha \\ 0 & 1 \end{bmatrix}.$$

Finally, we prove:

***Proposition 9.8.*** *Let $X$ be an affine space associated with a vector space $V$ over a field $\mathcal{F}$ of characteristic $\neq 2$. A transformation $f$ of $X$ (that is, a one-to-one mapping of $X$ onto itself) is an affine transformation if and only if it has the following property:*

$$\text{(9.21)} \qquad \text{If } \overrightarrow{PR} = c\overrightarrow{PQ}, \text{ then } \overrightarrow{f(P)f(R)} = c\overrightarrow{f(P)f(Q)}.$$

*Proof.* Let $f$ be an affine transformation and let $\phi$ be the associated linear mapping. Then we have

$$\phi(\overrightarrow{PR}) = \phi(c\overrightarrow{PQ}) = c\phi(\overrightarrow{PQ}),$$

that is, $\overrightarrow{f(P)f(R)} = c\overrightarrow{f(P)f(Q)}$. Conversely, let $f$ be a transformation with property (9.21). Let $O$ be a point of $X$. For any two vectors $\alpha, \beta \in V$, let $P$, $Q$, and $R$ be points such that $\overrightarrow{OP} = \alpha$, $\overrightarrow{OQ} = \beta$, and $\overrightarrow{OR} = \alpha + \beta$. We prove:

***Lemma.*** *If $T$ is a point such that $\overrightarrow{PT} = (\beta - \alpha)/2$, then*

$$\overrightarrow{OT} = \frac{\alpha + \beta}{2}.$$

In fact, this follows from

$$\overrightarrow{OT} = \overrightarrow{OP} + \overrightarrow{PT} = \alpha + \frac{\beta - \alpha}{2} = \frac{\alpha + \beta}{2}.$$

Now, applying condition (9.21) to $\overrightarrow{PQ} = 2\overrightarrow{PT}$, we have

$$\overrightarrow{f(P)f(Q)} = 2\overrightarrow{f(P)f(T)}.$$

Applying (9.21) to $\overrightarrow{OR} = 2\overrightarrow{OT}$, we get

$$\overrightarrow{f(O)f(R)} = 2\overrightarrow{f(O)f(T)}.$$

Therefore we have

$$\begin{aligned} \overrightarrow{f(O)f(R)} &= 2\left(\overrightarrow{f(O)f(P)} + \overrightarrow{f(P)f(T)}\right) \\ &= \overrightarrow{f(O)f(P)} + \overrightarrow{f(O)f(P)} + \overrightarrow{f(P)f(Q)} \\ &= \overrightarrow{f(O)f(P)} + \overrightarrow{f(O)f(Q)}, \end{aligned}$$

which shows $\phi_O(\alpha + \beta) = \phi_O(\alpha) + \phi_O(\beta)$. On the other hand, (9.21) implies immediately that $\phi_O(c\alpha) = c\phi_O(\alpha)$ for any vector $\alpha$ and $c \in \mathcal{F}$. Thus

$\phi_O$ is a linear mapping and $f$ is an affine mapping. Since $f$ is a transformation, it is an affine transformation.

***Remark.*** The lemma above has the following geometric meaning. If $OPRQ$ is a parallelogram, then the segments $OR$ and $PQ$ bisect each other.

**EXERCISE 9.2**

» **1.** In the standard affine space $\mathcal{R}^2$, find an affine transformation $f$ which maps

$$E_0 = \begin{bmatrix} 0 \\ 0 \end{bmatrix}, \quad E_1 = \begin{bmatrix} 1 \\ 0 \end{bmatrix}, \quad E_2 = \begin{bmatrix} 0 \\ 1 \end{bmatrix}$$

upon

$$P_0 = \begin{bmatrix} 1 \\ 2 \end{bmatrix}, \quad P_1 = \begin{bmatrix} 1 \\ -1 \end{bmatrix}, \quad P_2 = \begin{bmatrix} 0 \\ 1 \end{bmatrix}.$$

[Express $f$ in the form (9.17).]

**2.** In the standard affine space $\mathcal{R}^3$, find an affine transformation $f$ which maps

$$P_0 = \begin{bmatrix} 1 \\ 0 \\ 1 \end{bmatrix}, \quad P_1 = \begin{bmatrix} 0 \\ 1 \\ 1 \end{bmatrix}, \quad P_2 = \begin{bmatrix} 1 \\ 0 \\ 0 \end{bmatrix}, \quad P_3 = \begin{bmatrix} -1 \\ 1 \\ 2 \end{bmatrix}$$

upon

$$Q_0 = \begin{bmatrix} 1 \\ 2 \\ 1 \end{bmatrix}, \quad Q_1 = \begin{bmatrix} -1 \\ 2 \\ 0 \end{bmatrix}, \quad Q_2 = \begin{bmatrix} 0 \\ 3 \\ 4 \end{bmatrix}, \quad Q_3 = \begin{bmatrix} 1 \\ -2 \\ 5 \end{bmatrix}.$$

[Express $f$ in the form (9.17).]

» **3.** In the standard affine space $\mathcal{C}^2$, find an affine mapping which maps

$$P_0 = \begin{bmatrix} 0 \\ 0 \end{bmatrix}, \quad P_1 = \begin{bmatrix} 1 \\ i \end{bmatrix}, \quad P_2 = \begin{bmatrix} -1 \\ 1 \end{bmatrix}$$

upon

$$Q_0 = \begin{bmatrix} 1 \\ 2i \end{bmatrix}, \quad Q_1 = \begin{bmatrix} 2 \\ i \end{bmatrix}, \quad Q_2 = \begin{bmatrix} 3 \\ 0 \end{bmatrix}.$$

Is this an affine transformation?

☆ **4.** About the group of affine transformations $A(X)$ of an affine space $X$, prove the following:

(*a*) The set of centroaffine transformations leaving a point $O$ invariant is a subgroup.

(*b*) The set of all translations is a subgroup, say, $T$.

(*c*) Show that $fhf^{-1} \in T$ for any $h \in T$ and $f \in A(X)$. [$T$ is a normal subgroup of $A(X)$.]

☆ **5.** If $V$ is an $n$-dimensional vector space over the field $\mathfrak{F} = \{0, 1\}$, how many affine transformations does an affine space $X$ associated to $V$ admit?

☆ **6.** Let $X$ be an $n$-dimensional real affine space. By an *orientation* of $X$ we mean an orientation of the associated vector space $V$ (cf. Exercise 6.3, number 16). An affine transformation $f$ of $X$ is said to be *orientation-preserving* if the associated linear mapping is orientation-preserving (cf. Exercise 6.5, number 5). Prove that an affine transformation $f$ is orientation-preserving if and only if the matrix $\mathbf{B}$ in (9.17) has positive determinant.

## 9.3 AFFINE SUBSPACES

Let $X$ be an affine space associated with a vector space $V$.

**Definition 9.6**

A nonempty subset $Y$ of $X$ is called an *affine subspace* of $X$ if, for some $P \in Y$, the set of vectors

$$W_P(Y) = \{\overrightarrow{PQ}; Q \in Y\}$$

is a subspace of $V$.

This definition is actually independent of the choice of $P$. Namely, we have:

***Proposition 9.9.*** *If for some $P \in Y$, the set $W_P(Y)$ is a subspace of $V$, then for any $P' \in Y$, the set*

$$W_{P'}(Y) = \{\overrightarrow{P'Q}; Q \in Y\}$$

*is a subspace of $V$ and, in fact, $W_{P'}(Y) = W_P(Y)$.*

*Proof.* We shall show that $W_{P'}(Y) = W_P(Y)$. For any $Q \in Y$ we have

$$\overrightarrow{P'Q} = \overrightarrow{P'P} + \overrightarrow{PQ} = -\overrightarrow{PP'} + \overrightarrow{PQ}, \qquad \text{where } \overrightarrow{PP'}, \overrightarrow{PQ} \in W_P(Y).$$

Since $W_P(Y)$ is a subspace of $V$, we see that $\overrightarrow{P'Q} \in W_P(Y)$. This proves $W_{P'}(Y) \subseteq W_P(Y)$. In order to prove $W_P(Y) \subseteq W_{P'}(Y)$, let $\alpha \in W_P(Y)$. If we denote $\overrightarrow{PP'} \in W_P(Y)$ by $\beta$, then $\alpha + \beta$ is in the subspace $W_P(Y)$. Let $Q'$ be a point of $Y$ such that $\overrightarrow{PQ'} = \alpha + \beta$. Then $\overrightarrow{PQ'} = \overrightarrow{PP'} + \overrightarrow{P'Q'}$ implies $\alpha + \beta = \beta + \overrightarrow{P'Q'}$, that is, $\overrightarrow{P'Q'} = \alpha$, showing that $\alpha$ is in $W_{P'}(Y)$. Since $\alpha$ is an arbitrary element of $W_P(Y)$, we have $W_P(Y) \subseteq W_{P'}(Y)$. Hence $W_P(Y) = W_{P'}(Y)$.

**Definition 9.7**

For an affine subspace $Y$ of $X$, the vector space $W_P(Y)$, which is independent of $P \in Y$, is called the *vector space associated with* $Y$. We shall denote it simply by $W(Y)$. We define the *dimension* of $Y$ to be $\dim W(Y)$.

If $Y$ is an affine subspace of $X$, then it is an affine space associated with $W(Y)$ in an obvious way. The following is also obvious:

***Proposition 9.10.*** *Given a point $P \in X$ and a subspace $W$ of $V$, the set*

$$Y = \{Q \in X; \overrightarrow{PQ} \in W\}$$

*is an affine subspace such that $W(Y) = W$.*

**Definition 9.8**

An affine subspace of dimension 0 is a subset consisting of one single point. An affine subspace is called a *line, plane, hyperplane* according as $\dim Y = 1$, $2, n - 1$, where $n = \dim X$. (For $n = 2$, a line and a hyperplane mean the same thing, and for $n = 3$, a plane and a hyperplane mean the same thing.)

***Proposition 9.11.*** *Let $Y_\lambda$ be a family of affine subspaces of an affine space $X$. If $\bigcap Y_\lambda$ is not empty, it is an affine subspace associated with $\bigcap W_\lambda$, where $W_\lambda$ is the vector space associated with $Y_\lambda$.*

*Proof.* Let $P \in \bigcap Y_\lambda$. For any $Q \in \bigcap Y_\lambda$ we have $\overrightarrow{PQ} \in W_\lambda$ for every $\lambda$, and hence $\overrightarrow{PQ} \in \bigcap W_\lambda$. Conversely, given any $\alpha \in \bigcap W_\lambda$, there exists, for each $\lambda$, a point $Q_\lambda \in W_\lambda$ such that $\overrightarrow{PQ_\lambda} = \alpha$. Since there is one and only one point $Q \in X$ such that $\overrightarrow{PQ} = \alpha$, we must have $Q_\lambda = Q$ for every $\lambda$. Thus $Q \in \bigcap Y_\lambda$ and $\overrightarrow{PQ} = \alpha$, showing that

$$\bigcap Y_\lambda = \{Q \in X; \overrightarrow{PQ} \in \bigcap W_\lambda\},$$

that is, $\bigcap Y_\lambda$ is an affine subspace associated with $\bigcap W_\lambda$.

***Corollary.*** *Given a nonempty subset $S$ of $X$, there is a smallest affine subspace $Y$ containing $S$.*

*Proof.* Let $Y_\lambda$ be the family of all affine subspaces of $X$ which contain $S$ (for example, $X$ itself is a member of this family). Then $\bigcap Y_\lambda$ is an affine subspace containing $S$. If $Y$ is any affine subspace containing $S$, it is a member of the family $Y_\lambda$ and hence $Y \supset \bigcap Y_\lambda$.

The smallest affine subspace containing a given subset $S$ is called the *affine subspace spanned* by $S$. In particular, given two affine subspaces $Y_1$ and $Y_2$, the affine subspace spanned by the union $Y_1 \cup Y_2$ is denoted by $Y_1 \vee Y_2$.

***Example 9.11.*** In the standard affine space $\mathbb{R}^3$, let $S$ be the set of three points

$$(0,0,0),\ (1,0,0),\ (0,1,0).$$

The affine subspace spanned by $S$ is equal to $\{(x^1,x^2,0);\ x^1,\ x^2 \text{ arbitrary real numbers}\}$.

We shall prove:

***Proposition 9.12.*** *If $Y_1$ and $Y_2$ are affine subspaces such that $Y_1 \cap Y_2$ is not empty, then*

$$\dim (Y_1 \vee Y_2) + \dim (Y_1 \cap Y_2) = \dim Y_1 + \dim Y_2.$$

*Proof.* Let $P \in Y_1 \cap Y_2$ and let $W_1$, $W_2$, and $W$ be the vector spaces associated with $Y_1$, $Y_2$, and $Y_1 \vee Y_2$, respectively. We already know that $W_1 \cap W_2$ is the vector space associated with $Y_1 \cap Y_2$. We shall show that $W = W_1 + W_2$.

In fact, if we let

$$Z = \{Q; \overrightarrow{PQ} \in W_1 + W_2\},$$

$Z$ is an affine subspace containing $Y_1$ and $Y_2$. Thus $Z \supseteq Y_1 \vee Y_2$, and hence $W_1 + W_2 \supseteq W$. On the other hand, if $\overrightarrow{PQ} \in W_1 + W_2$, then there exist $Q_1 \in Y_1$ and $Q_2 \in Y_2$ such that $\overrightarrow{PQ} = \overrightarrow{PQ_1} + \overrightarrow{PQ_2}$. Since $Q_1, Q_2 \in Y_1 \vee Y_2$, we have $\overrightarrow{PQ_1}, \overrightarrow{PQ_2} \in W$ and hence $\overrightarrow{PQ} \in W$. This shows that $W_1 + W_2 \subseteq W$, proving $W = W_1 + W_2$. Now our assertion follows from the corresponding theorem on dimensions for vector subspaces (Theorem 2.15).

We now introduce the notion of parallelism for affine subspaces. Let $Y_1$ and $Y_2$ be affine subspaces of the same dimension, say, $r$, in an affine space $X$ of dimension $n$.

**Definition 9.9**

$Y_1$ and $Y_2$ are said to be *parallel* if either $Y_1 = Y_2$ or

$$\dim (Y_1 \vee Y_2) = r + 1 \qquad \text{and} \qquad Y_1 \cap Y_2 \text{ is empty.}$$

***Example 9.12.*** Two distinct lines $Y_1$ and $Y_2$ in $X$ are parallel if they are contained in one plane and $Y_1 \cap Y_2$ is empty. (In $\mathcal{R}^3$, two lines are *skew* if they are not contained in one plane.)

***Proposition 9.13***

**1.** *Two subspaces $Y_1$ and $Y_2$ of the same dimension are parallel if and only if $W(Y_1) = W(Y_2)$.*

**2.** *For subspaces of dimension $r$, parallelism is an equivalence relation.*

**3.** *For any affine subspace $Y_1$ and for any point $P$ in $X$, there is one and only one affine subspace $Y_2$ which contains $P$ and which is parallel to $Y_1$.*

*Proof.* 1. Suppose that $W = W(Y_1) = W(Y_2)$ and $\dim W = r$. If $Y_1 \cap Y_2$ is not empty, let $P \in Y_1 \cap Y_2$. Then

$$Y_1 = \{Q; \overrightarrow{PQ} \in W\} = Y_2.$$

If $Y_1 \cap Y_2$ is empty, let $P_1 \in Y_1$ and $P_2 \in Y_2$. Then the vector $\alpha = \overrightarrow{P_1P_2}$ is not in $W$, so that $\dim (W + \text{Sp}\{\alpha\}) = r + 1$. Let

$$Z = \{Q; \overrightarrow{P_1Q} \in W + \text{Sp}\ \{\alpha\}\}.$$

Then we see that $Z = Y_1 \vee Y_2$ and $\dim Z = r + 1$. Hence $Y_1$ and $Y_2$ are parallel.

Conversely, suppose that $Y_1 \neq Y_2$ and that they are parallel. In order to prove $W(Y_1) \subseteq W(Y_2)$, assume that there is a $\beta \in W(Y_1)$ which is not in $W(Y_2)$. Since $Y_1$ and $Y_2$ are parallel, $Y_1 \vee Y_2$ has dimension $r + 1$, where $r = \dim Y_1 = \dim Y_2$ and $Y_1 \cap Y_2$ is empty. If $W$ is the vector space associated with $Y_1 \vee Y_2$, then $W(Y_1) \subseteq W$, $W(Y_2) \subseteq W$, and $\dim W = r + 1$. Since $\beta$ is not in $W(Y_2)$, we have $W = \text{Sp}\ \{\beta\} + W(Y_2)$. Now for $P_1 \in Y_1$, $P_2 \in Y_2$, the vector $\alpha = \overrightarrow{P_1P_2}$ is in $W$; we have

$$\alpha = c\beta + \gamma, \qquad \text{where } \gamma \in W(Y_2).$$

Let $P$ be the point of $Y_1$ such that $\overrightarrow{P_1P} = c\beta$ and let $Q$ be the point of $Y_2$ such that $\overrightarrow{P_2Q} = -\gamma$. Then we have $\overrightarrow{P_1P_2} = \overrightarrow{P_1P} + \overrightarrow{QP_2}$, that is, $\overrightarrow{P_2Q} = \overrightarrow{P_2P_1} + \overrightarrow{P_1P}$. Combined with $\overrightarrow{P_2P_1} + \overrightarrow{P_1P} = \overrightarrow{P_2P}$, this gives $\overrightarrow{P_2Q} = \overrightarrow{P_2P}$, and hence $Q = P$. Thus this point lies in $Y_1 \cap Y_2$, contrary to the assumption that $Y_1 \cap Y_2$ is empty. We have thus proved $W(Y_1) \subseteq W(Y_2)$. Similarly, we have $W(Y_2) \subseteq W(Y_1)$, and hence $W(Y_1) = W(Y_2)$.

2. This follows immediately from (1).

3. Let $Y_2 = \{Q \in X; \overrightarrow{PQ} \in W(Y_1)\}$. By Proposition 9.10, $Y_2$ is an affine subspace containing $P$ whose associated vector space is $W(Y_1)$. By (1), $Y_2$ is parallel to $Y_1$. If $Y$ is an affine subspace containing $P$ and is parallel to $Y_1$, then we must have $Y = \{Q; \overrightarrow{PQ} \in W(Y_1)\}$ and hence $Y = Y_2$, proving the uniqueness.

## EXERCISE 9.3

» **1.** Given two distinct points $P$ and $Q$ in an affine space, show that there is one and only one line containing $P$ and $Q$. (This line is nothing but $P \vee Q$.)

**2.** If $Y_1$ and $Y_2$ are two lines which have a common point, then there is a unique plane which contains $Y_1$ and $Y_2$.

**3.** If $Y$ is a line and $P$ is a point outside $Y$ (that is, $P \notin Y$), then there is a unique plane which contains $P$ and $Y$.

**4.** If $Y_1$ and $Y_2$ are two distinct lines contained in a plane, then either $Y_1$ and $Y_2$ are parallel or $Y_1$ and $Y_2$ have a single common point.

» **5.** If $Y_1$ and $Y_2$ are two planes in a three-dimensional affine space $X$, then either $Y_1$ and $Y_2$ are parallel or $Y_1 \cap Y_2$ is a line.

**6.** Let $S$ be a nonempty subset in an $n$-dimensional affine space $X$. Let $r$ be the maximum number of linearly independent vectors in the set $\{\overrightarrow{PQ}; P, Q \in X\}$. Prove that $S$ is contained in a unique $r$-dimensional subspace of $X$.

☆ **7.** In a three-dimensional affine space $X$, let $L_1$ and $L_2$ be two lines which are skew (cf. Example 9.12) and let $O$ be a point which is neither on $L_1$ nor on $L_2$. Prove that there exists at most one line going through $O$ and intersecting $L_1$ and $L_2$.

## 9.4 REPRESENTATIONS OF AFFINE SUBSPACES

We shall now consider various ways of representing a given affine subspace in an affine space $X$. Let us first make the following general remark. Suppose that $\{x^1, \ldots, x^n\}$ is an affine coordinate system and that $f_i(x^1, \ldots, x^n)$, $1 \le i \le m$, are functions of $x^1, \ldots, x^n$. If $Y$ is the set of all points $P$ whose coordinates satisfy

$$f_i(x^1(P), \ldots, x^n(P)) = 0, \qquad 1 \le i \le m,$$

then we say that the set of equations

$$(*) \qquad f_i(x^1, \ldots, x^n) = 0, \qquad 1 \le i \le m,$$

*defines* $Y$ (or *is the set of equations for* $Y$). We shall also say that $Y$ is *defined* (or *represented*) *by* (*).

Now let $Y$ be an affine subspace of dimension $r$ in an affine space $X$ of dimension $n$. Let $\{P_0, P_1, \ldots, P_r\}$ be an affine frame in $Y$ and choose $P_{r+1}, \ldots, P_n$ so that $\{P_0, P_1, \ldots, P_n\}$ is an affine frame in $X$. Then a point $P$ of $X$ is in $Y$ if and only if

$$\overrightarrow{P_0P} = \sum_{i=1}^{r} a^i \, \overrightarrow{P_0P_i}.$$

If $\{x^1, \ldots, x^n\}$ is the affine coordinate system in $X$ determined by $\{P_0, P_1, \ldots, P_n\}$, then a point $P$ is in $Y$ if and only if

$$x^{r+1}(P) = \cdots = x^n(P) = 0.$$

Thus we have:

***Proposition 9.14.*** *For an $r$-dimensional affine subspace $Y$ of an $n$-dimensional affine space $X$, there exists an affine coordinate system $\{x^1, \ldots, x^n\}$ such that $Y$ is represented by*

$$(9.22) \qquad x^{r+1} = \cdots = x^n = 0.$$

Let $\{y^1, \ldots, y^n\}$ be an arbitrary affine coordinate system. The relationship between $\{y^1, \ldots, y^n\}$ and $\{x^1, \ldots, x^n\}$ is given by

$$x^i = \sum_{j=1}^{n} a^i{}_j y^j + a^i, \qquad 1 \le i \le n,$$

where $\mathbf{A} = [a^i{}_j]$ is a nonsingular $n \times n$ matrix and $\alpha = [a^i]$ is an element in $\mathcal{F}^n$ (Proposition 9.2). In view of Proposition 9.14, we see that a point $P$ is in $Y$ if and only if

$$\sum_{j=1}^{n} a^i{}_j y^j(P) + a^i = 0, \qquad r + 1 \le i \le n,$$

where the matrix $[a^i{}_j]$, $r + 1 \le i \le n$, $1 \le j \le n$, has rank $n - r$ (since the row vectors of $\mathbf{A}$ and hence the $n - r$ row vectors of this matrix are linearly independent). Changing the notation, we have:

***Proposition 9.15.*** *Let $\{x^1, \ldots, x^n\}$ be an arbitrary affine coordinate system in an affine space $X$. For any $r$-dimensional affine subspace $Y$ in $X$, there exist a matrix $\mathbf{B} = [b^i{}_j]$, $1 \le i \le n - r$, $1 \le j \le n$, of rank $n - r$ and $\beta = [b^i]$, $1 \le i \le n - r$, such that $Y$ is represented by*

$$\sum_{j=1}^{n} b^i{}_j x^j + b^i = 0, \qquad 1 \le i \le n - r. \tag{9.23}$$

Conversely, any set of equations (9.23), where the rank of $[b^i{}_j]$ is $n - r$, defines an $r$-dimensional subspace. Let $P = (p^1, \ldots, p^n)$ be a point satisfying (9.23) whose existence is guaranteed by the results in Sec. 3.5. If $Q = (q^1, \ldots, q^n)$ is another point satisfying (9.23), then we have

$$\sum_{j=1}^{n} b^i{}_j (q^j - p^j) = 0, \qquad 1 \le i \le n - r,$$

and

$$\overrightarrow{PQ} = \sum_{i=1}^{n} (q^i - p^i)\alpha_i,$$

where $\{\alpha_1, \ldots, \alpha_n\}$ is the basis of $V$ corresponding to the affine frame we are using. Now in the vector space $V$ the set of all vectors $\sum_{i=1}^{n} a^i\alpha_i$, where $a^1, \ldots, a^n$ satisfy

$$\sum_{j=1}^{n} b^i{}_j x^j = 0, \qquad 1 \le i \le n - r, \tag{9.24}$$

forms an $r$-dimensional subspace $W$. Let $Y = \{R \in X; \overrightarrow{PR} \in W\}$, which is an affine subspace containing $P$ with $W(Y) = W$ by Proposition 9.10. Clearly, $Y$ is defined by (9.23).

***Remark.*** As we see from the discussions above, an affine subspace $Y$ is the set of all solutions of an inhomogeneous system (9.23), and the corresponding vector subspace

$W(Y)$ is the set of all vectors whose coordinates are solutions of the associated homogeneous system (9.24).

***Example 9.13.*** A hyperplane $Y$ in $X$ can be represented by one equation

$$\sum_{i=1}^{n} a_i x^i + b = 0, \qquad \text{where not all } a_i\text{'s are } 0. \tag{9.25}$$

***Example 9.14.*** An $r$-dimensional affine subspace $Y$ is the intersection of $n - r$ hyperplanes defined by each of the equations (9.23). In particular, in a three-dimensional affine space $X$, a line $Y$ is the intersection of two nonparallel planes defined by

$$\sum_{i=1}^{3} a^1{}_i x^i + b^1 = 0 \qquad \text{and} \qquad \sum_{i=1}^{3} a^2{}_i x^i + b^2 = 0.$$

***Example 9.15.*** Let

$$A = \begin{bmatrix} a^1 \\ a^2 \\ a^3 \end{bmatrix}, \qquad B = \begin{bmatrix} b^1 \\ b^2 \\ b^3 \end{bmatrix}, \qquad C = \begin{bmatrix} c^1 \\ c^2 \\ c^3 \end{bmatrix}$$

be three noncollinear points in the three-dimensional standard real affine space $\mathcal{R}^3$. Then the equation for the plane determined by $A$, $B$, and $C$ is

$$\begin{vmatrix} x^1 & a^1 & b^1 & c^1 \\ x^2 & a^2 & b^2 & c^2 \\ x^3 & a^3 & b^3 & c^3 \\ 1 & 1 & 1 & 1 \end{vmatrix} = 0. \tag{9.26}$$

In fact, this is an equation of the form (9.25) whose coefficients are not all 0 (verify this). The coordinates of $A$, $B$, and $C$ satisfy the equation, since, for example, the determinant in (9.26) is 0 for $x^i = a^i$, $1 \leq i \leq 3$.

***Example 9.16.*** In an $n$-dimensional real affine space $X$, let $Y$ be a hyperplane defined by

$$\sum_{i=1}^{n} a_i x^i + b = 0$$

with respect to a certain affine coordinate system. The set of all points which satisfy

$$\sum_{i=1}^{n} a_i x^i + b > 0 \; [\geq 0]$$

and the set of all points which satisfy

$$\sum_{i=1}^{n} a_i x^i + b < 0 \; [\leq 0]$$

are called [closed] *half-spaces* determined by $Y$. The space $X$ is the disjoint union of $Y$ and the two half-spaces determined by $Y$.

We shall now discuss two other ways of representing affine subspaces.

Let $Y$ be an $r$-dimensional affine subspace and let $\{P_0, P_1, \ldots, P_r\}$ be an affine frame in $Y$. When we fix an arbitrary point $O$ in $X$, a point $P \in X$ is in $Y$ if and only if

$$\overrightarrow{OP} = \overrightarrow{OP_0} + \sum_{j=1}^{r} t^j \overrightarrow{P_0 P_j} \tag{9.27}$$

for some coefficients $t^1, \ldots, t^r$. (9.27) is called a *parametric representation* of $Y$, and $(t^1, \ldots, t^r)$ are called the *parameters* for the point $P$ (with respect to the point $O$ and the affine frame $\{P_0,P_1, \ldots ,P_r\}$). Note that $(t^1, \ldots ,t^r)$ are nothing but the coordinates of $P$ with respect to the affine coordinate system in $Y$ determined by the affine frame $\{P_i\}$.

We shall now modify (9.27) as follows. Since $\overrightarrow{P_0P_j} = \overrightarrow{OP_j} - \overrightarrow{OP_0}$, we may rewrite (9.27) as

$$\overrightarrow{OP} = \overrightarrow{OP_0} + \sum_{j=1}^{r} t^j(\overrightarrow{OP_j} - \overrightarrow{OP_0})$$

$$= \left(1 - \sum_{j=1}^{r} t^j\right)\overrightarrow{OP_0} + \sum_{j=1}^{r} t^j\overrightarrow{OP_j}.$$

Setting

$$s^0 = 1 - \sum_{j=1}^{r} t^j \qquad \text{and} \qquad s^j = t^j \qquad \text{for } 1 \leq j \leq r,$$

we have

$$\overrightarrow{OP} = \sum_{j=0}^{r} s^j\overrightarrow{OP_j}, \qquad \text{where } \sum_{j=0}^{r} s^j = 1. \tag{9.28}$$

We shall show that the set of numbers $(s^0,s^1, \ldots ,s^r)$ with $\sum_{j=0}^{r} s^j = 1$ in (9.28) is independent of the choice of the point $O$. In fact, if $O'$ is any other point, then

$$\overrightarrow{O'P} = \overrightarrow{O'O} + \overrightarrow{OP} = \overrightarrow{O'O} + \sum_{j=0}^{r} s^j\overrightarrow{OP_j}.$$

Since $\overrightarrow{OP_j} = \overrightarrow{OO'} + \overrightarrow{O'P_j}$, we have

$$\overrightarrow{O'P} = \overrightarrow{O'O} + \sum_{j=0}^{r} s^j(\overrightarrow{OO'} + \overrightarrow{O'P_j})$$

$$= \overrightarrow{O'O} + \left(\sum_{j=0}^{r} s^j\right)\overrightarrow{OO'} + \sum_{j=0}^{r} s^j\overrightarrow{O'P_j} = \sum_{j=0}^{r} s^j\overrightarrow{O'P_j},$$

because $\sum_{j=0}^{r} s^j = 1$ and $\overrightarrow{O'O} + \overrightarrow{OO'} = 0$.

This shows that, when an affine frame $\{P_0,P_1, \ldots ,P_r\}$ in $Y$ is fixed, there is a unique way of associating with every point $P$ of $Y$ a set of $r + 1$ numbers $(s^0,s^1, \ldots ,s^r)$ with $\sum_{j=0}^{r} s^j = 1$, so that (9.28) is valid for an arbitrary point $O$ in $X$. We call $(s^0,s^1, \ldots ,s^r)$ the *barycentric coordinates* of $P$. By taking $O = P_0$, we see that $(s^1, \ldots ,s^r)$ are the coordinates of $P$ with respect to the affine frame $\{P_0,P_1, \ldots ,P_r\}$.

***Example 9.17.*** In an affine space $X$ with an arbitrary coordinate system, a line can be represented by

$$x^i = a^it + b^i, \qquad 1 \leq i \leq n \quad (n = \dim X),$$

where $t$ is a parameter and $(a^1, \ldots ,a^n) \neq (0, \ldots ,0)$.

***Example 9.18.*** Let $A$ and $B$ be two distinct points in a real affine space $X$. The *segment* $AB$ is defined to be the set of all points $C$ such that $\overrightarrow{AC} = c\overrightarrow{AB}$, where $0 \leq c \leq 1$. If $C$ is a point of the segment $AB$, then the barycentric coordinates of $C$ with respect to the affine frame $\{A,B\}$ (in the affine subspace, namely, the line determined by $A$ and $B$) are $(s,t)$, where $s + t = 1$, $0 \leqq s$, and $0 \leqq t$. In fact, if $\overrightarrow{AC} = c\overrightarrow{AB}$, then for any point $O$ we have

$$\overrightarrow{OC} = \overrightarrow{OA} + c\overrightarrow{AB} = \overrightarrow{OA} + c(\overrightarrow{OB} - \overrightarrow{OA})$$

$$= (1 - c)\overrightarrow{OA} + c\,\overrightarrow{OB},$$

so that $s = 1 - c$ and $t = c$ are the barycentric coordinates of $C$, where $0 \leq s$, $0 \leq t$. In particular, the midpoint $M$ of $AB$ (that is, the point $M$ such that $\overrightarrow{AM} = \frac{1}{2}\overrightarrow{AB}$) has $(\frac{1}{2},\frac{1}{2})$ as barycentric coordinates.

***Example 9.19.*** Let $\{P_0,P_1, \ldots ,P_n\}$ be an affine frame in a real affine space $X$. Let $A$ and $B$ be two distinct points with barycentric coordinates $(a^0,a^1, \ldots ,a^n)$ and $(b^0,b^1, \ldots ,b^n)$, respectively. For any point $C$ such that

$$\overrightarrow{OC} = s\,\overrightarrow{OA} + t\,\overrightarrow{OB}, \qquad s + t = 1,$$

we have

$$\overrightarrow{OC} = s\left(\sum_{i=0}^{n} a^i\,\overrightarrow{OP_i}\right) + t\left(\sum_{i=0}^{n} b^i\,\overrightarrow{OP_i}\right)$$

$$= \sum_{i=0}^{n} (sa^i + tb^i)\overrightarrow{OP_i},$$

where

$$\sum_{i=0}^{n} (sa^i + tb^i) = s\left(\sum_{i=0}^{n} a^i\right) + t\left(\sum_{i=0}^{n} b^i\right) = s + t = 1.$$

Thus the barycentric coordinates of $C$ are

$$(sa^0 + tb^0,\ sa^1 + tb^1, \ldots ,\ sa^n + tb^n),$$

which we may write as $s(a^0,a^1, \ldots ,a^n) + t(b^0,b^1, \ldots ,b^n)$.

***Example 9.20.*** Using Examples 9.18 and 9.19, we shall prove the theorem in elementary geometry that the three medians of a triangle meet at one point (cf. Examples 1.5 and 1.8). Let $A$, $B$, $C$ be the vertices of a triangle and let $L$, $M$, $N$ be the midpoints of the sides $BC$, $CA$, and $AB$, respectively. By using the affine frame $\{A,B,C\}$, the barycentric coordinates of $L$, $M$, and $N$ are

$$(0,\tfrac{1}{2},\tfrac{1}{2}),\ (\tfrac{1}{2},0,\tfrac{1}{2}),\ (\tfrac{1}{2},\tfrac{1}{2},0),$$

respectively. Let $G$ be the intersection of $AL$ and $CN$. Then

$$\overrightarrow{OG} = s\,\overrightarrow{OL} + t\,\overrightarrow{OA} = u\overrightarrow{OC} + v\overrightarrow{ON},$$

where $s + t = u + v = 1$ and $0 \leq s, t, u, v$. Thus the barycentric coordinates of $G$ are

$$s(0,\tfrac{1}{2},\tfrac{1}{2}) + t(1,0,0) = u(0,0,1) + v(\tfrac{1}{2},\tfrac{1}{2},0),$$

from which we obtain

$$t = \frac{v}{2}, \qquad \frac{s}{2} = \frac{v}{2}, \qquad \text{and} \qquad \frac{s}{2} = u,$$

so that $t = s/2$, that is, $s = \frac{2}{3}$ and $t = \frac{1}{3}$. Thus the barycentric coordinates of $G$ are $(\frac{1}{3},\frac{1}{3},\frac{1}{3})$. Similarly, the intersection of $AL$ and $BM$ has $(\frac{1}{3},\frac{1}{3},\frac{1}{3})$ as barycentric coordinates and coincides with $G$. This proves our theorem. ($G$ is called the *center of gravity,* or *barycenter,* of the triangle.)

Finally, we shall introduce the notion of convex set in a real affine space $X$.

**Definition 9.10**

A nonempty subset $K$ of $X$ is said to be *convex* if, for any two points $P, Q \in K$, the segment $PQ$ is contained in $K$.

The following sets are convex:

1. The entire space $X$.
2. Any affine subspace of $X$.
3. A half-space. In fact, let $H$ be the half-space defined by

$$\sum_{i=1}^{n} a_i x^i + b > 0,$$

determined by the hyperplane $Y$: $\sum_{i=1}^{n} a_i x^i + b = 0$ (we are using any affine coordinate system in $X$). If

$$P = (p^1, \ . \ . \ . \ ,p^n) \qquad \text{and} \qquad Q = (q^1, \ . \ . \ . \ ,q^n)$$

are in $H$, then for any point $R$ of the segment $PQ$, we have $R = (tp^1 + sq^1, \ . \ . \ . \ ,tp^n + sq^n)$, where $t + s = 1$ and $0 \leq t, s$. Hence

$$\sum_{i=1}^{n} a_i(tp^i + sq^i) + b = t\left(\sum_{i=1}^{n} a_i p^i + b\right) + s\left(\sum_{i=1}^{n} a_i q^i + b\right) > 0.$$

4. An $r$-dimensional *parallelepiped,* which is defined as follows. Let $P_0, P_1, \ . \ . \ . \ , P_r$ be $r + 1$ independent points (namely, $\overrightarrow{P_0P_i}$, $1 \leq i \leq r$, are linearly independent). Then

$$\left\{P \in X; \overrightarrow{P_0P} = \sum_{i=1}^{r} t^i \overrightarrow{P_0P_i}, \ 0 \leq t^i \leq 1\right\}$$

is called an $r$-dimensional parallelepiped with vertices $P_0, P_1, \ . \ . \ . \ , P_r$.

5. An $r$-dimensional *cone,* which is defined as follows. For $r + 1$ points $P_0, P_1, \ . \ . \ . \ , P_r$ which are independent the set

$$\{P; \overrightarrow{P_0P} = \sum_{i=1}^{r} t^i \overrightarrow{P_0P_i}, \ 0 \leq t^i\}$$

is called an $r$-dimensional cone with vertex $P_0$.

6. An $r$-dimensional *simplex*, which is defined as follows. For $r + 1$ points $P_0, P_1, \ldots, P_r$ which are independent, the set

$$\{P; \overrightarrow{OP} = \sum_{i=0}^{r} s^i \overrightarrow{OP_i}; \sum_{i=0}^{r} s^i = 1, s^i \geq 0\}$$

is called the $r$-dimensional simplex determined by $P_0, \ldots, P_r$ and is denoted by $\Delta(P_0,P_1, \ldots ,P_r)$. Observe that this set is determined independently of the choice of a point $O$; in fact, it is the set of all points $P$ whose barycentric coordinates with respect to $P_0, P_1, \ldots, P_r$ are $(s^0,s^1, \ldots ,s^r)$, where $\sum_{i=0}^{r} s^i = 1, s^i \geq 0$.

The verification that (4), (5), and (6) are convex is left to the reader. In (6) the point $G$ with barycentric coordinates

$$\left(\frac{1}{r+1}, \frac{1}{r+1}, \ldots, \frac{1}{r+1}\right)$$

is called the *barycenter* (or *center of gravity*) of the simplex

$$\Delta(P_0,P_1, \ldots ,P_r).$$

## EXERCISE 9.4

**1.** In an affine space $X$ with an affine coordinate system $\{x^i\}$, show that two linear equations

$$\sum_{i=1}^{n} a_i x^i + b = 0, \qquad (a^1, \ldots ,a^n) \neq (0, \ldots ,0),$$

$$\sum_{i=1}^{n} c_i x^i + d = 0, \qquad (c^1, \ldots ,c^n) \neq (0, \ldots ,0)$$

define the same hyperplane if and only if there is a nonzero scalar $k$ such that

$$c_i = ka_i, \quad 1 \leq i \leq n, \qquad \text{and} \qquad d = kb.$$

**2.** Show that the two equations in number 1 define two parallel and nonidentical hyperplanes if and only if there is a nonzero scalar $k$ such that

$$c_i = ka_i, \quad 1 \leq i \leq n, \qquad \text{and} \qquad d \neq kb.$$

›› ☆**3.** In a two-dimensional affine space $\mathfrak{F}^2$, where $\mathfrak{F}$ is the field $\{0,1\}$, how many distinct points and how many distinct lines are there?

**4.** Show that, for an affine transformation $f$ of an $n$-dimensional affine space $X$, the image $f(Y)$ of an $r$-dimensional affine subspace $Y$ is an $r$-dimensional affine subspace.

**5.** Consider an affine transformation of $\mathfrak{R}^2$ defined by

$$\begin{bmatrix} y^1 \\ y^2 \\ 1 \end{bmatrix} = \begin{bmatrix} 3 & 2 & -1 \\ 1 & 1 & 2 \\ 0 & 0 & 1 \end{bmatrix} \begin{bmatrix} x^1 \\ x^2 \\ 1 \end{bmatrix}.$$

Find the equation of $f(Y)$ when $Y$ is the line defined by ››(*a*) $x^1 = x^2$; (*b*) $2x^1 - x^2 = 3$.

**6.** Let $L_i$, $i = 1, 2, 3$, be lines in the affine space $\mathcal{R}^2$ defined by $a_ix^1 + b_ix^2 + c_i = 0$. If they have a common point, then

$$\det \begin{bmatrix} a_1 & b_1 & c_1 \\ a_2 & b_2 & c_2 \\ a_3 & b_3 & c_3 \end{bmatrix} = 0.$$

How about the converse?

» **7.** Let $\pi_i$, $1 \leq i \leq 4$, be the planes in $\mathcal{R}^3$ defined by $a_ix^1 + b_ix^2 + c_ix^3 + d_i = 0$. If they have a common point, then

$$\det \begin{bmatrix} a_1 & b_1 & c_1 & d_1 \\ a_2 & b_2 & c_2 & d_2 \\ a_3 & b_3 & c_3 & d_3 \\ a_4 & b_4 & c_4 & d_4 \end{bmatrix} = 0.$$

**8.** Prove that, for an affine transformation $f$ of a real affine space, the image of a half-space is a half-space.

» **9.** Let $A$, $B$, $C$, and $D$ be four points in a real affine space which do not lie on one plane. Find the position of the point $G$ with barycentric coordinates $(\frac{1}{4},\frac{1}{4},\frac{1}{4},\frac{1}{4})$ with respect to the affine frame $\{A,B,C,D\}$.

**10.** Prove that if $K_\lambda$ is an indexed family of convex sets in a real affine space and if $\bigcap_\lambda K_\lambda$ is nonempty, then $\bigcap_\lambda K_\lambda$ is convex.

»☆ **11.** Given any nonempty subset $S$ of a real affine space, prove that there is a smallest convex set containing $S$. (It is called the *convex hull* of $S$.)

☆ **12.** (See number 11) Prove that the convex hull of the set of $r + 1$ independent points $P_0, P_1, \ldots, P_r$ in a real affine space is the simplex $\Delta(P_0,P_1, \ldots ,P_r)$.

» **13.** Express the $n$-dimensional simplex $\Delta(E_0,E_1, \ldots ,E_n)$ in the standard $n$-dimensional real affine space $\mathcal{R}^n$, where $\{E_0,E_1, \ldots ,E_n\}$ is the standard affine frame, as the intersection of $n + 1$ closed half-spaces.

☆ **14.** Let $X$ be an $n$-dimensional affine space over a field $\mathfrak{F}$ of characteristic $\neq 2$. A subset $Q$ of $X$ is called a *quadratic hypersurface* in $X$ if there exists an affine coordinate system $x^i$ in $X$ with respect to which $Q$ *is defined by a quadratic equation:* $\sum_{i,j=1}^{n} a_{ij}x^ix^j + 2\sum_{i=1}^{n} a_ix^i + a_0 = 0$, where $a_{ji} = a_{ij}$.

(*a*) Show that this condition is independent of the choice of an affine coordinate system in $X$.

(*b*) For a quadratic hypersurface $Q$, the rank of $\mathbf{A} = [a_{ij}]$ is independent of the choice of an affine coordinate system; we define it as the *rank* of $Q$.

(*c*) Given a quadratic hypersurface $Q$, show that there is an affine coordinate system in $X$ such that $Q$ is defined by $\sum_{i=1}^{n} c_i(x^i)^2 + 2\sum_{i=1}^{n} a_ix^i + a_0 = 0$ (cf. Theorem 4.17). If the rank of $Q$ is $n$, there is an affine coordinate system such that $Q$ is defined by

$$\sum_{i=1}^{n} c_i(x^i)^2 = 0 \qquad \text{or} \qquad \sum_{i=1}^{n} c_i(x^i)^2 + 1 = 0.$$

(*d*) For the cases where $\mathfrak{F} = \mathcal{R}$ or $\mathcal{C}$, study how we can further simplify the equation for $Q$ in (*c*). (Cf. Theorems 4.18, 4.19.)

(*e*) Study the case where the rank of $Q$ is less than $n$.

# 10 Euclidean Spaces

A real affine space $X$ is called a euclidean space when the associated vector space $V$ is given a certain inner product. By means of this inner product, the metric properties—length, angle, orthogonality, distance—are introduced in $X$. Rectangular coordinate systems and isometries are discussed. In particular, we classify all isometries of *n-dimensional* euclidean spaces.

## 10.1 EUCLIDEAN SPACES

### Definition 10.1

An affine space $X$ associated with an $n$-dimensional real inner product space $V$ is called an *n-dimensional euclidean space.*

***Example 10.1.*** Let $\mathcal{R}^n$ be the standard affine space. When we consider the standard inner product in the vector space $\mathcal{R}^n$ (Example 8.1), the affine space $\mathcal{R}^n$ is called the *standard euclidean space* of dimension $n$.

Although we considered in Chap. 9 affine spaces over arbitrary fields, we are now considering only real affine spaces. Since any $n$-dimensional real vector space $V$ always admits an inner product, any $n$-dimensional real affine space may be considered as a euclidean space by fixing any inner product in $V$. All concepts that we have defined for a real affine space make sense for a euclidean space. But it is important to distinguish real affine spaces and euclidean spaces. A real affine space $X$ associated with $V$ can have different structures as a euclidean space, depending on the inner product we choose in $V$.

In the following let $X$ be a euclidean space associated with an $n$-dimensional real inner product space $V$.

### Definition 10.2

For any two points $P, Q$ in $X$, the *distance* between $P$ and $Q$, denoted by $d(P,Q)$, is defined to be $\|\overrightarrow{PQ}\|$, where $\|\ \ \|$ denotes the norm in $V$ (Definition 8.4).

***Proposition 10.1.*** *$d(P,Q)$ satisfies the following conditions:*
**1.** *$d(P,Q) \geq 0$; $d(P,Q) = 0$ if and only if $P = Q$.*
**2.** *$d(Q,P) = d(P,Q)$.*
**3.** *$d(P,R) \leq d(P,Q) + d(Q,R)$, triangle inequality.*

*Proof.* 1. $d(P,Q) \geq 0$, since $\|\alpha\| \geq 0$ for all $\alpha \in V$. If $d(P,Q) = 0$, then $\|\overrightarrow{PQ}\| = 0$ and hence $\overrightarrow{PQ} = 0$, that is, $P = Q$. The converse is obvious.

2. $\overrightarrow{QP} = -\overrightarrow{PQ}$ implies $\|\overrightarrow{QP}\| = \|\overrightarrow{PQ}\|$ and $d(Q,P) = d(P,Q)$.
3. By (3) of Proposition 8.2, we have

$$\|\overrightarrow{PR}\| = \|\overrightarrow{PQ} + \overrightarrow{QR}\| \leq \|\overrightarrow{PQ}\| + \|\overrightarrow{QR}\|,$$

which implies the triangle inequality.

For an arbitrary set $X$, a real-valued function $d(P,Q)$ on $X \times X$ having the three properties stated in Proposition 10.1 is called a *distance function* (or *metric*) on $X$. The metric defined above in the euclidean space $X$ is often called the *euclidean metric* in order to distinguish it from any other possible metric on $X$.

**Definition 10.3**

Given three distinct points $P$, $Q$, and $R$ in $X$, the *angle* $\angle QPR$ is defined to be the real number $\theta$, $0 \leq \theta \leq \pi$, such that

$$\cos\theta = \frac{(\overrightarrow{PQ}, \overrightarrow{PR})}{\|\overrightarrow{PQ}\|\,\|\overrightarrow{PR}\|}.$$

Observe that the right-hand side is a number between $-1$ and $1$ by virtue of Schwarz's inequality in Theorem 8.1, so that $\theta$ is uniquely determined by the equality above. If $Q = R$ (different from $P$), then the angle $\angle QPQ$ is defined to be 0 (and this is the case from the equality above as well).

***Example 10.2.*** Given three distinct points $P$, $Q$, and $R$ in the two-dimensional euclidean plane, let $\theta = \angle QPR$. Then

$$d(Q,R)^2 = d(P,Q)^2 + d(P,R)^2 - 2d(P,Q)d(P,R)\cos\theta.$$

In order to prove this, let $\alpha = \overrightarrow{PQ}$, $\beta = \overrightarrow{PR}$, $\gamma = \overrightarrow{QR}$, so that $\gamma = \beta - \alpha$. From

$$(\gamma,\gamma) = (\beta - \alpha, \beta - \alpha) = (\beta,\beta) - 2(\alpha,\beta) + (\alpha,\alpha)$$

we have

$$\|\gamma\|^2 = \|\beta\|^2 - 2\,\|\alpha\|\,\|\beta\|\cos\theta + \|\alpha\|^2,$$

and hence the formula.

**Definition 10.4**

An ordered $n$-tuple of points $\{P_0, P_1, \ldots, P_n\}$ is called a *rectangular frame* (or *euclidean frame*) if the vectors $\overrightarrow{P_0P_i}$, $1 \leq i \leq n$, form an orthonormal basis in $V$ (cf. Definition 8.5). The affine coordinate system determined by such a frame is called a *rectangular* (or *euclidean*) *coordinate system.*

***Example 10.3.*** In the standard $n$-dimensional euclidean space $\mathcal{R}^n$, the points

$$E_0 = (0, \ldots, 0),\ E_1 = (1, 0, \ldots, 0),\ \ldots,\ E_n = (0, \ldots, 0, 1)$$

form a rectangular frame. In other words, the standard affine coordinate system is a rectangular coordinate system.

***Example 10.4.*** In the standard two-dimensional euclidean space $\mathcal{R}^2$, the points

$$P_0 = (a, b),\ P_1 = (a + \cos\theta,\ b + \sin\theta),\ P_2 = (a - \sin\theta,\ b + \cos\theta)$$

form a rectangular frame for any fixed values of $a$, $b$, $\theta$.

***Proposition 10.2.*** *An affine coordinate system $\{x^i\}$ in a euclidean space is rectangular if and only if the distance between two points $A = (a^i)$ and $B = (b_i)$ can be expressed by*

$$d(A,B) = \sqrt{\sum_{i=1}^{n} (b^i - a^i)^2}. \tag{10.1}$$

*Proof.* Let $\{P_i\}$ be the affine frame which determines the given coordinate system $\{x^i\}$. If $\{P_i\}$ is rectangular, then for

$$\overrightarrow{P_0A} = \sum_{i=1}^{n} a^i \overrightarrow{P_0P_i} \quad \text{and} \quad \overrightarrow{P_0B} = \sum_{i=1}^{n} b^i \overrightarrow{P_0P_i}$$

we have

$$\overrightarrow{AB} = \sum_{i=1}^{n} (b^i - a^i)\overrightarrow{P_0P_i}$$

and

$$d(A,B) = \|\overrightarrow{AB}\| = \sqrt{\sum_{i=1}^{n} (b^i - a^i)^2}$$

by (2) of Proposition 8.5.

Conversely, assume that (10.1) is valid for any two points $A$ and $B$. Then for

$$P_0 = (0, \ldots, 0),\ P_1 = (1, 0, \ldots, 0),\ \ldots,\ P_n = (0, 0, \ldots, 1)$$

we have

$$d(P_0, P_i) = 1 \quad \text{and} \quad d(P_i, P_j) = \sqrt{2} \quad \text{for } i \neq j,\ 1 \leq i, j \leq n.$$

From the formula in Example 10.2, we have

$$\cos\theta = 0 \quad \text{for the angle } \theta = \angle P_iP_0P_j, \qquad i \neq j,\ 1 \leq i, j \leq n.$$

This means that $\overrightarrow{P_0P_i}$, $1 \leq i \leq n$, form an orthonormal basis in $V$, that is, $\{P_i\}$ is a rectangular frame.

***Remark.*** For an arbitrary affine coordinate system $\{x^i\}$, there exists a positive-definite symmetric matrix $\mathbf{G} = [g_{ij}]$, $1 \leq i, j \leq n$, such that

$$d(A,B) = \sqrt{\sum_{i,j=1}^{n} g_{ij}(b^i - a^i)(b^j - a^j)}$$

for any two points $A = (a^i)$ and $B = (b^i)$. In fact, for the corresponding affine frame $\{P_i\}$, we set

$$g_{ij} = (\overrightarrow{P_0P_i}, \overrightarrow{P_0P_j}), \qquad 1 \leq i, j \leq n.$$

$\{P_i\}$ is rectangular if and only if $\mathbf{G}$ is the identity matrix.

It is almost obvious that, given any point $P_0$ in $X$, there exists a rectangular frame with origin $P_0$. Indeed, for any orthonormal basis $\{\alpha_i\}$ in $V$ (whose existence was proved in Theorem 8.4), let $P_i$ be the unique point such that $\overrightarrow{P_0P_i} = \alpha_i$, $1 \leq i \leq n$. Then $\{P_i\}$ is a rectangular frame with origin $P_0$. We now prove:

***Proposition 10.3.*** *If $\{x^1, \ldots, x^n\}$ and $\{y^1, \ldots, y^n\}$ are two rectangular coordinate systems in the euclidean space $X$, then*

$$\begin{bmatrix} y^1 \\ \cdot \\ \cdot \\ \cdot \\ y^n \\ 1 \end{bmatrix} = \begin{bmatrix} a^1{}_1 & \cdots & a^1{}_n & a^1 \\ \cdot & \cdots & \cdot & \cdot \\ a^n{}_1 & \cdots & a^n{}_n & a^n \\ 0 & \cdots & 0 & 1 \end{bmatrix} \begin{bmatrix} x^1 \\ \cdot \\ \cdot \\ \cdot \\ x^n \\ 1 \end{bmatrix}, \qquad (10.2)$$

*where $\mathbf{A} = [a^i{}_j]$ is an orthogonal matrix.*

*Proof.* By Proposition 9.2, $\{x^i\}$ and $\{y^i\}$ are related by (9.6), where $\mathbf{A} = [a^i{}_j]$ is a nonsingular matrix which represents the change of bases corresponding to the two affine frames for $\{x^i\}$ and $\{y^i\}$. In our case, since the two bases are orthonormal, $\mathbf{A}$ is an orthogonal matrix according to the remark following Definition 8.16.

**EXERCISE 10.1**

**1.** In the standard euclidean space $\mathbb{R}^4$, find $d(A,B)$ for $A = (1,0,-1,1)$ and $B = (-1,2,1,-2)$.

›› **2.** In the standard euclidean space $\mathbb{R}^2$, let

$$P_0 = (1,1), \qquad P_1 = (2,2), \qquad P_2 = (-3,5).$$

Find the distance formula with respect to the affine coordinate system determined by the frame $\{P_0,P_1,P_2\}$.

**3.** Find the relationship [in the form of (10.2)] between the standard coordinate system and the rectangular coordinate system determined by $\{P_0,P_1,P_2\}$ in Example 10.4.

**4.** Consider the standard affine space $\mathcal{R}^2$ as a euclidean space associated with the inner product $(\alpha,\beta) = 2a^1b^1 + a^2b^2$ for $\alpha = (a^1,a^2)$ and $\beta = (b^1,b^2)$ in the vector space $\mathcal{R}^2$.

(*a*) Show that $P_0 = (0,0)$, $P_1 = (1/\sqrt{2},0)$, $P_2 = (0,1)$ form a rectangular frame.

(*b*) Show that a point $(x^1,x^2) \in \mathcal{R}^2$ is at distance 1 from $P_0$ if and only if $2(x^1)^2 + (x^2)^2 = 1$. (Thus this is the equation in the standard coordinate system for a *circle* in the given euclidean metric.)

»☆ **5.** On the standard affine space $\mathcal{R}^n$, define

$$d'(A,B) = \max_{1 \le i \le n} \{|b^i - a^i|\} \qquad \text{for } A = (a^i) \text{ and } B = (b^i).$$

Show that $d'$ is a metric on $\mathcal{R}^n$. Prove that it is not the euclidean metric for any structure of euclidean space on $\mathcal{R}^n$ $(n > 1)$.

☆ **6.** In the standard euclidean space $\mathcal{R}^n$, let

$$A_k = (a^1{}_k, \ldots, a^n{}_k), \qquad 1 \le k \le m.$$

Find the point $G$ in $\mathcal{R}^n$ such that $\sum_{j=1}^{m} d(G,A_k)^2$ has the smallest possible value.

☆ **7.** For an $r$-simplex $\Delta(P_0,P_1, \ldots ,P_r)$ in a euclidean space $X$, prove that $\sum_{i=0}^{r} d(P,P_i)^2$ has the smallest value if and only if $P$ coincides with the barycenter of the simplex.

## 10.2 SUBSPACES

Let $X$ be a euclidean space associated with an inner product space $V$ of dimension $n$. If $Y$ is an affine subspace of $X$ associated with a subspace $W$ of $V$, then we may regard $Y$ as a euclidean space with respect to the inner product in $W$ that is induced by the inner product in $V$. In this sense, we call $Y$ a *euclidean subspace* of $X$. For $P, Q \in Y$, the euclidean distance between $P$ and $Q$ in $Y$ is equal to the euclidean distance between $P$ and $Q$ in $X$.

Results in Secs. 9.3 and 9.4 for affine subspaces remain valid for euclidean subspaces. We shall discuss a new feature pertinent only to euclidean geometry.

Let $Y$ be an $r$-dimensional euclidean subspace in $X$, where $1 \le r < n$. A vector $\alpha \in V$ is said to be *normal* to $Y$ if

$$(\alpha,\beta) = 0 \qquad \text{for all } \beta \in W,$$

where $W$ is the vector subspace of $V$ associated with $Y$. By Theorem 8.8, we have

$$V = W \oplus W^{\perp},$$

where $\dim W = r$ and $\dim W^{\perp} = n - r$. Thus any vector in $W^{\perp}$ is normal to $Y$. In particular, a unit vector normal to $Y$ is called a *unit normal* to $Y$. Conversely, let $N$ be any subspace of $V$ with $1 \leq \dim N < n$. For any point $P_0$ of $X$,

$$Y = \{P \in X;\ (\overrightarrow{P_0P}, \alpha) = 0 \text{ for } \alpha \in N\}$$

is a euclidean subspace associated with $W = N^{\perp}$.

***Example 10.5.*** Let $Y$ be a hyperplane in $X$. If $\alpha$ is a unit normal, so is $-\alpha$. A unit normal to $Y$ is determined up to a sign. Let $\{x^i\}$ be a rectangular coordinate system determined by a rectangular frame $\{P_i\}$, where we set $\alpha_i = \overrightarrow{P_0P_i}$, $1 \leq i \leq n$. Let $\alpha = \sum_{i=1}^{n} a^i\alpha_i$ be a unit normal and let $P = (p^1, \ldots, p^n)$ be an arbitrary point of $Y$. Then a point $Q = (q^1, \ldots, q^n)$ in $X$ is in $Y$ if and only if

$$(\overrightarrow{PQ},\alpha) = \sum_{i=1}^{n} (q^i - p^i)a^i = 0,$$

that is,

$$\sum_{i=1}^{n} a^iq^i = b, \qquad \text{where } b = \sum_{i=1}^{n} a^ip^i.$$

Thus the equation for $Y$ is obtained as follows:

$$\sum_{i=1}^{n} a^ix^i = b, \qquad \text{where } \sum_{i=1}^{n} (a^i)^2 = 1. \tag{10.3}$$

[Observe that $b$ is independent of the choice of a point $P = (p^i)$ in $Y$.]

***Example 10.6.*** Let $Y$ be an $r$-dimensional euclidean subspace of $X$. Let

$$\{\alpha_1, \ldots, \alpha_{n-r}\}$$

be an orthonormal basis of $W^{\perp}$, where $W$ is the vector subspace associated with $Y$. For an arbitrary point $P$ of $Y$, we have

$$Y = \{Q \in X;\ (\overrightarrow{PQ},\alpha_i) = 0 \text{ for } 1 \leq i \leq n - r\}.$$

We also see that $Y$ is the intersection of $n - r$ hyperplanes:

$$Y_i = \{Q \in X;\ (\overrightarrow{PQ},\alpha_i) = 0\}, \qquad 1 \leq i \leq n - r,$$

as we saw in Example 9.14.

We shall now prove:

***Theorem 10.4.*** *Let $Y$ be an $r$-dimensional euclidean subspace of an $n$-dimensional euclidean space $X$, where $1 \leq r < n$. For any point $A \notin Y$, there is a unique point $B \in Y$ such that $\overrightarrow{AB}$ is normal to $Y$. We have*

$$d(A,B) < d(A,P) \qquad \textit{for all points } P \neq B \textit{ in } Y.$$

$d(A,B)$ is called the *shortest distance* from $A$ to the subspace $Y$ and is denoted by $d(A,Y)$.

*Proof.* Let $V$ be the inner product space associated with $X$, and $W$ the subspace of $V$ associated with $Y$. For any point $P_0 \in Y$, let

$$\overrightarrow{P_0A} = \beta + \alpha, \qquad \text{where } \beta \in W \text{ and } \alpha \in W^{\perp}.$$

If we take a point $B \in Y$ such that $\overrightarrow{P_0B} = \beta$, then

$$\overrightarrow{BA} = \overrightarrow{P_0A} - \overrightarrow{P_0B} = \alpha$$

and hence $\overrightarrow{AB} = -\overrightarrow{BA}$ is normal to $Y$. To prove the uniqueness, let $B'$ be any point of $Y$ such that $\overrightarrow{AB'}$ is normal to $Y$. Then

$$\overrightarrow{BB'} = \overrightarrow{AB'} - \overrightarrow{AB}$$

is normal to $Y$ and it is in $W$. Thus $(\overrightarrow{BB'},\overrightarrow{BB'}) = 0$, which implies $\overrightarrow{BB'} = 0$, that is, $B = B'$.

For any point $P \in Y$, we have

$$\overrightarrow{AP} = \overrightarrow{AB} + \overrightarrow{BP}, \qquad \text{where } (\overrightarrow{AB},\overrightarrow{BP}) = 0.$$

Thus

$$\|\overrightarrow{AP}\|^2 = \|\overrightarrow{AB}\|^2 + \|\overrightarrow{BP}\|^2 > \|\overrightarrow{AB}\|^2 \qquad \text{unless } B = P.$$

Hence

$$d(A,P) > d(A,B) \qquad \text{for any } P \neq B \text{ in } Y.$$

***Example 10.7.*** Consider the equation (10.3) of a hyperplane $Y$. If $d(A,Y) = d(A,B)$, then $\overrightarrow{AB} = c\alpha$, where $\alpha = \sum_{i=1}^{n} a^i\alpha_i$ is a unit normal. If $A = (p^1, \ldots, p^n)$, then $B = (p^1 + ca^1, \ldots, p^n + ca^n)$. Substituting in (10.3), we have

$$\sum_{i=1}^{n} a^i(p^i + ca^i) = b, \qquad \text{that is,} \qquad c = b - \sum_{i=1}^{n} a^ip^i.$$

Hence

$$d(A,Y) = |c| = \left| \sum_{i=1}^{n} a^ip^i - b \right|. \tag{10.4}$$

In particular, for the origin $P_0$ of the coordinate system we have

$$d(P_0,Y) = |b|.$$

***Remark.*** If $b < 0$ in (10.3), we may change it into $-b > 0$ by changing $(a^i)$ into $(-a^i)$. Thus we may assume that $b > 0$ and hence $b = d(P_0,Y)$. The equation (10.3) in which $b > 0$ is called the *normal form* of the hyperplane $Y$. In this case we have a unit normal given by

$$\alpha = \sum_{i=1}^{n} a^i\alpha_i,$$

and

$$d(P_0,B) = d(P_0,Y) = b \qquad \text{and} \qquad B \in Y.$$

***Example 10.8.*** In the standard euclidean space $\mathcal{R}^3$, let $Y$ be a plane defined by

$$x^1 + 2x^2 - 2x^3 + 4 = 0.$$

The normal form of $Y$ is

$$-\frac{x^1}{3} - 2\frac{x^2}{3} + 2\frac{x^3}{3} = \frac{4}{3},$$

and

$$d(E_0,Y) = \frac{4}{3},$$

where $E_0$ is the origin of the standard coordinate system.

Although a full development of the theory of volume in euclidean space is outside the scope of this book, we shall define here the $k$-dimensional volume of a $k$-dimensional parallelepiped in a euclidean space $X$.

Let $(P_0,P_1, \ldots ,P_k)$ be a parallelepiped (Example 4 following Definition 9.10) determined by $P_0, P_1, \ldots , P_k$, where the vectors $\alpha_i = \overrightarrow{P_0P_i}$, $1 \leq i \leq k$, are linearly independent. We shall define the *$k$-dimensional volume* $V_k(P_0,P_1, \ldots ,P_k)$ by induction on $k$.

For $k = 1$, we define $V_1(P_0,P_1) = \|\overrightarrow{P_0P_1}\|$. Suppose that

$$V_{k-1}(P_0,P_1, \ldots ,P_{k-1})$$

is already defined. Let $h$ be the shortest distance (more intuitively, the height) of $P_k$ from the $(k - 1)$-dimensional subspace $Y$ spanned by $P_0$, $P_1, \ldots , P_{k-1}$, and define

$$V_k(P_0,P_1, \ldots ,P_k) = hV_{k-1}(P_0,P_1, \ldots ,P_{k-1}). \tag{10.5}$$

That $V_k(P_0,P_1, \ldots ,P_k)$ is independent of the order of the points $P_i$, $1 \leq i \leq n$, will follow from the following results.

***Proposition 10.5***

$$V_k(P_0,P_1, \ldots ,P_k)^2 = \det \begin{bmatrix} (\alpha_1,\alpha_1) & \cdots & (\alpha_1,\alpha_k) \\ \cdots & \cdots & \cdots \\ (\alpha_k,\alpha_1) & \cdots & (\alpha_k,\alpha_k) \end{bmatrix}, \tag{10.6}$$

*where* $\alpha_i = \overrightarrow{P_0P_i}$, $1 \leq i \leq k$. (The determinant of the right-hand side is called the *Gramian* of $\alpha_1, \ldots , \alpha_k$; cf. Exercise 8.1, number 12.)

*Proof.* Formula (10.6) is valid for $k = 1$. Assume that it is valid for a $(k - 1)$-dimensional parallelepiped. If $B$ is the unique point of the subspace $Y$ spanned by $P_0, P_1, \ldots , P_{k-1}$ such that $d(P_k,B) = d(P_k,Y) = h$, then $\overrightarrow{BP_k} = h\alpha$, where $\alpha$ is a unit normal to $Y$. Then $\overrightarrow{P_0P_k} = \overrightarrow{P_0I} + \overrightarrow{BP_k}$ implies

$$\alpha_k = h\alpha + \sum_{i=1}^{k-1} c^i\alpha_i,$$

where $c^1, \ldots, c^{k-1}$ are suitable coefficients. We have

$$(\alpha_j,\alpha_k) = \sum_{i=1}^{k-1} c^i(\alpha_j,\alpha_i) \qquad \text{for } 1 \leq j \leq k-1$$

and

$$(\alpha_k,\alpha_k) = h^2 + \sum_{i,j=1}^{k-1} c^i c^j (\alpha_i,\alpha_j).$$

Thus the Gramian of $\alpha_1, \ldots, \alpha_k$ is equal to

$$\det \begin{bmatrix} (\alpha_1,\alpha_1) & \cdots & (\alpha_1,\alpha_{k-1}) & \sum_{i=1}^{k-1} c^i(\alpha_1,\alpha_i) \\ \cdots & \cdots & \cdots & \cdots \\ (\alpha_{k-1},\alpha_1) & \cdots & (\alpha_{k-1},\alpha_{k-1}) & \sum_{i=1}^{k-1} c^i(\alpha_{k-1},\alpha_i) \\ \sum_{i=1}^{k-1} c^i(\alpha_i,\alpha_1) & \cdots & \sum_{i=1}^{k-1} c^i(\alpha_i,\alpha_{k-1}) & h^2 + \sum_{i,j=1}^{k-1} c^i c^j(\alpha_i,\alpha_j) \end{bmatrix}.$$

Subtracting $c^i$ times the $i$th column from the last column for each $i$, $1 \leq i \leq k-1$, we see that the determinant above is equal to

$$\det \begin{bmatrix} (\alpha_1,\alpha_1) & \cdots & (\alpha_1,\alpha_{k-1}) & 0 \\ \cdots & \cdots & \cdots & \cdots \\ (\alpha_{k-1},\alpha_1) & \cdots & (\alpha_{k-1},\alpha_{k-1}) & 0 \\ \sum_{i=1}^{k-1} c^i(\alpha_i,\alpha_1) & \cdots & \sum_{i=1}^{k-1} c^i(\alpha_i,\alpha_{k-1}) & h^2 \end{bmatrix},$$

which is equal to the Gramian of $\alpha_1, \ldots, \alpha_{k-1}$ times $h^2$. By the formula (10.6) for $k-1$, we see that

$$\det \begin{bmatrix} (\alpha_1,\alpha_1) & \cdots & (\alpha_1,\alpha_k) \\ \cdots & \cdots & \cdots \\ (\alpha_k,\alpha_1) & \cdots & (\alpha_k,\alpha_k) \end{bmatrix} = h^2 V_{k-1}(P_0,P_1, \ldots ,P_{k-1})^2$$

$$= V_k(P_0,P_1, \ldots ,P_k)^2,$$

proving the formula (10.6).

***Remark 1.*** $V_k(P_0,P_1, \ldots ,P_k)$ is thus the positive square root of the Gramian of $\alpha_1, \ldots, \alpha_k$. Although (10.6) shows that the Gramian of $\alpha_1, \ldots, \alpha_k$ (when they are linearly independent) is always positive, this fact can also be verified as follows. Consider the $n \times n$ matrix $\mathbf{A} = [a_{ij}]$, where $a_{ij} = (\alpha_i,\alpha_j)$. This matrix is symmetric and positive-definite; for any $(x^1, \ldots ,x^n) \in \mathcal{R}^n$ we have

$$\sum_{i,j=1}^{n} a_{ij}x^i x^j = \left( \sum_{i=1}^{n} x^i\alpha_i, \sum_{i=1}^{n} x^i\alpha_i \right) > 0,$$

unless all $x^i$'s are equal to 0. Since all the eigenvalues of $\mathbf{A}$ are positive (cf. Exercise 8.6, number 10), their product, which is equal to $\det \mathbf{A}$, is positive.

***Remark 2.*** If we interchange $P_0$ and $P_i$ for some $i$, $1 \leq i \leq k$, then $\alpha_i$ is changed into $-\alpha_i$. By the changes of sign for the $i$th row and the $i$th column, the Gramian of $\alpha_1, \ldots, -\alpha_i, \ldots, \alpha_k$ is equal to that of $\alpha_1, \ldots, \alpha_i, \ldots, \alpha_k$. If we interchange $P_i$ and $P_j$, where $1 \leq i, j \leq k$, then the Gramian of $\alpha_1, \ldots, \alpha_j, \ldots, \alpha_i, \ldots, \alpha_k$ is equal to the Gramian of $\alpha_1, \ldots, \alpha_i, \ldots, \alpha_j, \ldots, \alpha_k$ (because the $i$th column and the $j$th column as well as the $i$th row and the $j$th row are interchanged in the determinant). Now any change of the order of points $P_i$ in the parallelepiped $(P_0,P_1, \ldots, P_k)$ is obtained by a number of interchanges of two points. Hence the volume $V_k(P_0,P_1, \ldots, P_k)$ is independent of the order of the points.

We can give another expression for the volume of a parallelepiped $(P_0,P_1, \ldots, P_k)$, where $\alpha_i = \overrightarrow{P_0P_i}$, $1 \leq i \leq k$.

***Proposition 10.6.*** *Let* $\{\beta_1, \ldots, \beta_k\}$ *be an arbitrary orthonormal basis of* $\mathrm{Sp}\{\alpha_1, \ldots, \alpha_k\}$, *and let*

$$\alpha_i = \sum_{j=1}^{k} a_{ji}\beta_j, \qquad 1 \leq i \leq k.$$

*Then* $V_k(P_0,P_1, \ldots, P_k)$ *is equal to the absolute value of* $\det \mathbf{A}$, *where* $\mathbf{A} = [a_{ji}]$.

*Proof.* We have

$$(\alpha_i,\alpha_j) = \left(\sum_{p=1}^{k} a_{pi}\beta_p, \sum_{q=1}^{k} a_{qj}\beta_q\right) = \sum_{p=1}^{k} a_{pi}a_{pj},$$

since $(\beta_p,\beta_q) = \delta_{pq}$. Thus $(\alpha_i,\alpha_j)$ is equal to the $(i,j)$ component of the matrix ${}^t\mathbf{A}\mathbf{A}$. Hence the Gramian of $\alpha_1, \ldots, \alpha_k$ is equal to

$$\det({}^t\mathbf{A}\mathbf{A}) = \det({}^t\mathbf{A}) \det \mathbf{A} = (\det \mathbf{A})^2.$$

Since $V_k(P_0,P_1, \ldots, P_k)^2$ is equal to this value by Proposition 10.5, we have

$$V_k(P_0,P_1, \ldots, P_k) = |\det \mathbf{A}|.$$

**EXERCISE 10.2**

1. In the euclidean space $\mathcal{R}^3$, find the normal form of each of the following planes:

$$\pi_1\colon x^1 + x^2 - x^3 + 3 = 0; \qquad \pi_2\colon 2x^1 - x^2 - 1 = 0.$$

Also find the distance $d(A,\pi_1)$ and $d(A,\pi_2)$, where $A = (-1,1,1)$.

» 2. Find the distance from the origin to the intersection $\pi_1 \cap \pi_2$, where $\pi_1$ and $\pi_2$ are the planes defined in number 1.

3. Given two distinct points $A$ and $B$ in a euclidean space, show that the set $Y$ of all points $P$ such that $d(P,A) = d(P,B)$ is a hyperplane. In particular, when $A = (a^1, \ldots, a^n)$ and $B = (b^1, \ldots, b^n)$ in the standard euclidean space $\mathcal{R}^n$, find the equation of $Y$.

» 4. Given three points $A$, $B$, and $C$ which are not collinear in a euclidean space $X$, find the set of all points $P$ such that $d(P,A) = d(P,B) = d(P,C)$.

☆ **5.** In a three-dimensional euclidean space, let $L_1$ and $L_2$ be two lines which are skew. Prove that there exists one and only one line $L$ which intersects $L_1$ and $L_2$ orthogonally. If $A$ and $B$ are the points of intersection, then

$$d(A,B) \leqq d(P,Q) \qquad \text{for all } P \in L_1 \text{ and } Q \in L_2.$$

» **6.** For the subspace $Y = \{Q;\ (\overrightarrow{PQ},\alpha_i) = 0 \text{ for } 1 \leq i \leq n - r\}$ in Example 10.6 and for a point $O$ outside $Y$, find $d(O,Y)$ in terms of $\alpha = \overrightarrow{OP}$ and $\alpha_1, \ldots, \alpha_{n-r}$.

**7.** Let $P_i = (a^1{}_i, \ldots, a^n{}_i)$, $0 \leq i \leq n$, be independent in the standard euclidean space $\mathcal{R}^n$. Prove that $V_n(P_0,P_1, \ldots, P_n)$ is equal to the absolute value of

$$\det \begin{bmatrix} a^1{}_0 & a^1{}_1 & \cdots & a^1{}_n \\ \cdots & \cdots & \cdots & \cdots \\ a^n{}_0 & a^n{}_1 & \cdots & a^n{}_n \\ 1 & 1 & \cdots & 1 \end{bmatrix}.$$

» **8.** Let $f$ be an affine transformation of a euclidean space $X$. Show that for the volume of a parallelepiped we have

$$V_n(f(P_0),f(P_1), \ldots, f(P_n)) = |c|\, V_n(P_0,P_1, \ldots, P_n),$$

where $c$ is the determinant of the linear transformation associated with $f$.

**9.** (See Exercise 9.4, number 14) Let $Q$ be a quadratic hypersurface of an $n$-dimensional euclidean space $X$. Prove that there is a rectangular coordinate system in $X$ such that $Q$ is defined by the equation

$$\sum_{i=1}^{n} c_i(x^i)^2 + 2\sum_{i=1}^{n} a_i x^i + a_0 = 0.$$

Show that if the rank of $Q$ is $n$ and if $Q$ is not empty, then there is a rectangular coordinate system such that $Q$ is defined by

$$\sum_{i=1}^{p} \frac{(x^i)^2}{(a_i)^2} - \sum_{i=p+1}^{n} \frac{(x^i)^2}{(a_i)^2} = 0 \text{ (or 1)}, \qquad 1 \leq p \leq n.$$

Study the case where the rank of $Q$ is less than $n$.

## 10.3 ISOMETRIES

Let $X_1$ and $X_2$ be euclidean spaces associated with inner product spaces $V_1$ and $V_2$, both of dimension $n$, respectively. By regarding $X_1$ and $X_2$ as affine spaces, we know what is meant by an affine mapping of $X_1$ into $X_2$.

### Definition 10.5

An affine mapping $f$ of $X_1$ into $X_2$ is called an *isometry* if the associated linear mapping $\phi: V_1 \to V_2$ preserves inner product, that is,

$$(\phi(\alpha),\phi(\beta)) = (\alpha,\beta) \qquad \text{for all } \alpha, \beta \in V_1.$$

***Proposition 10.7.*** *An isometry $f$ has the following properties:*

**1.** *$f$ is one-to-one and onto.*

**2.** *$f$ preserves distance, that is,*

$$d(f(P),f(Q)) = d(P,Q) \qquad \textit{for all } P, Q \in X_1.$$

**3.** *If $\{x^i\}$ and $\{y^i\}$ are rectangular coordinate systems in $X_1$ and $X_2$, respectively, then $f$ can be represented by*

$$\begin{bmatrix} x^1 \\ \cdot \\ \cdot \\ \cdot \\ x^n \\ 1 \end{bmatrix} \to \begin{bmatrix} y^1 \\ \cdot \\ \cdot \\ \cdot \\ y^n \\ 1 \end{bmatrix} = \begin{bmatrix} a^1{}_1 & \cdots & a^1{}_n & a^1 \\ \cdots & \cdots & \cdots & \cdots \\ a^n{}_1 & \cdots & a^n{}_n & a^n \\ 0 & \cdots & 0 & 1 \end{bmatrix} \begin{bmatrix} x^1 \\ \cdot \\ \cdot \\ \cdot \\ x^n \\ 1 \end{bmatrix}, \tag{10.7}$$

*where $\mathbf{A} = [a^i{}_j]$ is an orthogonal matrix.*

*Proof.* 1, 2. Since $\phi$ preserves inner product, we have

$$(\phi(\alpha),\phi(\alpha)) = (\alpha,\alpha), \qquad \text{that is,} \qquad \|\phi(\alpha)\| = \|\alpha\|$$

for every $\alpha \in V_1$. For any points $P$, $Q$ in $X_1$, we have

$$d(P,Q) = \|\overrightarrow{PQ}\| = \|\phi(\overrightarrow{PQ})\| = \|\overrightarrow{f(P)f(Q)}\| = d(f(P),f(Q)),$$

that is, $f$ preserves distance. It is also clear that $\phi$ is one-to-one; in fact, if $\phi(\alpha) = 0$, then $\|\phi(\alpha)\| = \|\alpha\| = 0$ and $\alpha = 0$. Since $\dim V_1 = \dim V_2$, it also follows that $\phi$ is onto. We now show that $f$ is onto. Let $P \in X_1$. For any $Q' \in X_2$, there exists an $\alpha \in V_1$ such that $\phi(\alpha) = \overrightarrow{f(P)Q'}$. This means that if we take $Q \in X_1$ such that $\overrightarrow{PQ} = \alpha$, then $\overrightarrow{f(P)f(Q)} = \overrightarrow{f(P)Q'}$, and hence $f(Q) = Q'$, proving our assertion.

3. Let $\{P_i\}$ be the rectangular frame which corresponds to the coordinate system $\{x^i\}$. Since

$$\begin{aligned} (\overrightarrow{f(P_0)f(P_i)},\overrightarrow{f(P_0)f(P_j)}) &= (\phi(\overrightarrow{P_0P_i}),\phi(\overrightarrow{P_0P_j})) \\ &= (\overrightarrow{P_0P_i},\overrightarrow{P_0P_j}) = \delta_{ij}, \end{aligned}$$

it follows that $\{f(P_i)\}$ is a rectangular frame in $X_2$. Let $\{\bar{x}^i\}$ be the corresponding coordinate system in $X_2$. For any point $P$ with coordinates $[p^i]$ in $X_1$, we have

$$\overrightarrow{P_0P} = \sum_{i=1}^{n} p^i\overrightarrow{P_0P_i}$$

and hence

$$\overrightarrow{f(P_0)f(P)} = \sum_{i=1}^{n} p^i\overrightarrow{f(P_0)f(P_i)}.$$

This means that $f(P)$ has coordinates $[p^i]$ with respect to the coordinate system $\{\bar{x}^i\}$. Now the two coordinate systems $\{y^i\}$ and $\{\bar{x}^i\}$ are related by

$$y^i = \sum_{j=1}^{n} a^i{}_j \bar{x}^j + a^i, \qquad 1 \leq i \leq n,$$

where $\mathbf{A} = [a^i{}_j]$ is a certain orthogonal matrix and $\alpha = [a^i]$ is a certain vector. Thus $f(P)$ has coordinates $[q^i]$, where

$$q^i = \sum_{j=1}^{n} a^i{}_j p^j + a^i, \qquad 1 \leq i \leq n.$$

Since $P$ is arbitrary in $X_1$, we see that $f$ can be represented by (10.7).

**Corollary.** *If $f$ is an isometry of $X_1$ onto $X_2$, then $f^{-1}$ is an isometry of $X_2$ onto $X_1$.*

We shall now prove the following result.

**Theorem 10.8.** *A mapping $f$ of a euclidean space $X_1$ into a euclidean space $X_2$, where $n = \dim X_1 = \dim X_2$, is an isometry if it preserves distance.*

*Proof.* Let $P$ be a point of $X_1$ and consider the mapping $\phi_P$ of $V_1$ into $V_2$ such that

$$\phi_P(\overrightarrow{PQ}) = \overrightarrow{f(P)f(Q)} \qquad \text{for every } Q \in X_1$$

(namely, the mapping which we considered prior to Proposition 9.3). We shall show that $\phi_P$ is linear and preserves inner product; then it follows that $f$ is an isometry. Although our mapping $\phi_P$ is not known to be independent of the choice of $P$ before we prove that it is linear (cf. Proposition 9.3), we shall denote it by $\phi$ for the sake of brevity.

By the identity in Example 10.2, we have

$$\begin{aligned} 2(\overrightarrow{PQ},\overrightarrow{PR}) &= d(P,Q)^2 + d(P,R)^2 - d(Q,R)^2, \\ &= d(f(P),f(Q))^2 + d(f(P),f(R))^2 - d(f(Q),f(R))^2, \end{aligned}$$

since $f$ preserves distance. But the right-hand side is equal to

$$2(\overrightarrow{f(P)f(Q)}, \overrightarrow{f(P)f(R)}),$$

again by the identity in Example 10.2 applied to the three points $f(P)$, $f(Q)$, and $f(R)$. Thus

$$(\overrightarrow{f(P)f(Q)},\overrightarrow{f(P)f(R)}) = (\overrightarrow{PQ},\overrightarrow{PR}) \qquad \text{for all } Q, R \in X_1,$$

that is,

$$(\phi(\alpha),\phi(\beta)) = (\alpha,\beta) \qquad \text{for all } \alpha, \beta \in V_1.$$

In order to prove that $\phi$ is linear, it is sufficient to prove:

***Lemma.*** *Let $\phi$ be a mapping of an inner product space $V_1$ into an inner product space $V_2$, where* $\dim V_1 = \dim V_2$. *If $(\phi(\alpha),\phi(\beta)) = (\alpha,\beta)$ for all $\alpha, \beta \in V_1$, then $\phi$ is linear.*

*Proof.* Let $\{\alpha_1, \ldots, \alpha^n\}$ be an orthonormal basis of $V_1$. Then

$$\{\phi(\alpha_1), \ldots, \phi(\alpha_n)\}$$

is an orthonormal basis of $V_2$, since

$$(\phi(\alpha_i),\phi(\alpha_j)) = (\alpha_i,\alpha_j) = \delta_{ij}.$$

For any $\alpha \in V_1$, we have by (1) of Proposition 8.5

$$\alpha = \sum_{i=1}^{n} (\alpha,\alpha_i)\alpha_i, \tag{10.8}$$

and similarly,

$$\phi(\alpha) = \sum_{i=1}^{n} (\phi(\alpha),\phi(\alpha_i))\phi(\alpha_i).$$

Since $(\phi(\alpha),\phi(\alpha_i)) = (\alpha,\alpha_i)$, we have

$$\phi(\alpha) = \sum_{i=1}^{n} (\alpha,\alpha_i)\phi(\alpha_i). \tag{10.9}$$

Equations (10.8) and (10.9) show that $\phi$ maps a vector $\sum_{i=1}^{n} c^i\alpha_i \in V_1$ upon $\sum_{i=1}^{n} c^i\phi(\alpha_i) \in V_2$ and hence is linear. This proves the lemma and completes the proof of Theorem 10.8.

For a euclidean space $X$ of dimension $n$, let us denote by $I(X)$ the set of all isometries of $X$ into itself. It is easy to see that $I(X)$ is a group under the composition of mappings, which we call the *group of isometries* of $X$. If we fix a rectangular coordinate system $\{x^i\}$, then $I(X)$ can be represented by the group of all $(n+1) \times (n+1)$ matrices of the form

$$\begin{bmatrix} \mathbf{A} & \alpha \\ 0 & 1 \end{bmatrix}, \qquad \text{where } \mathbf{A} \in O(n) \text{ and } \alpha \in \mathcal{R}^n. \tag{10.10}$$

**Definition 10.6**

An isometry $f$ is called a *motion* if, for some rectangular coordinate system $\{x^i\}$, the matrix representation (10.10) for $f$ is such that $\det \mathbf{A} = 1$, namely, $\mathbf{A} \in SO(n)$.

Let us observe that the value of det $\mathbf{A}$ is in fact independent of the choice of a rectangular coordinate system. If $\{y^i\}$ is any other rectangular coordinate system, then we have by Proposition 10.3

$$\begin{bmatrix} y^1 \\ \cdot \\ \cdot \\ \cdot \\ y^n \\ 1 \end{bmatrix} = \begin{bmatrix} \mathbf{C} & \gamma \\ 0 & 1 \end{bmatrix} \begin{bmatrix} x^1 \\ \cdot \\ \cdot \\ \cdot \\ x^n \\ 1 \end{bmatrix}, \tag{10.11}$$

where $\mathbf{C} \in O(n)$ and $\gamma \in \mathcal{R}^n$. In terms of $\{y^i\}$, $f$ can be expressed by

$$\begin{bmatrix} \bar{y}^1 \\ \cdot \\ \cdot \\ \cdot \\ \bar{y}^n \\ 1 \end{bmatrix} = \begin{bmatrix} \mathbf{B} & \beta \\ 0 & 1 \end{bmatrix} \begin{bmatrix} y^1 \\ \cdot \\ \cdot \\ \cdot \\ y^n \\ 1 \end{bmatrix}, \tag{10.12}$$

where $\mathbf{B} \in O(n)$ and $\beta \in \mathcal{R}^n$. It follows that

$$\begin{bmatrix} \mathbf{B} & \beta \\ 0 & 1 \end{bmatrix} = \begin{bmatrix} \mathbf{C} & \gamma \\ 0 & 1 \end{bmatrix} \begin{bmatrix} \mathbf{A} & \alpha \\ 0 & 1 \end{bmatrix} \begin{bmatrix} \mathbf{C} & \gamma \\ 0 & 1 \end{bmatrix}^{-1},$$

in particular,

$$\mathbf{B} = \mathbf{CAC}^{-1} \qquad \text{and} \qquad \det \mathbf{B} = \det \mathbf{A}.$$

**Definition 10.7**

An isometry $f$ is called a *rotation* around a point $O$ if $f(O) = O$. A rotation is called a *proper rotation* if it is a motion and an *improper rotation* if it is not a motion.

If $f$ is a rotation around $O$, then, for any rectangular coordinate system $\{x^i\}$ with origin $O, f$ can be expressed by an orthogonal matrix $\mathbf{A}$:

$$\begin{bmatrix} \bar{x}^1 \\ \cdot \\ \cdot \\ \cdot \\ \bar{x}^n \end{bmatrix} = \mathbf{A} \begin{bmatrix} x^1 \\ \cdot \\ \cdot \\ \cdot \\ x^n \end{bmatrix}.$$

$f$ is a proper rotation if and only if $\mathbf{A} \in SO(n)$.

***Example 10.9.*** A translation is a motion. The set of all translations is a subgroup of $I(X)$.

***Example 10.10.*** The set of all rotations [proper rotations] around a point $O$ is a subgroup of $I(X)$. The set of all motions is a subgroup of $I(X)$.

***Example 10.11.*** For any point $O$, define the following transformation $f$ of $X$. For any point $P$, let $f(P)$ be the point such that $\overrightarrow{Of(P)} = -\overrightarrow{OP}$. With respect to any rectangular coordinate system with origin $O$, $f$ can be expressed by

$$y^i = -x^i, \qquad 1 \leq i \leq n,$$

and is obviously an isometry. It is a rotation (a proper rotation if $\dim X$ is even), called the *symmetry with respect to the point O*.

***Example 10.12.*** Let $Y$ be an arbitrary euclidean subspace of $X$. We define a transformation $f$ as follows. For any point $P \in Y$, we set $f(P) = P$. For any point $P \notin Y$, let $Q$ be the unique point of $Y$ such that $\overrightarrow{PQ}$ is orthogonal to $Y$ (cf. Theorem 10.4) and define $f(P)$ to be the point such that $\overrightarrow{Qf(P)} = -\overrightarrow{QP}$. If we choose a rectangular coordinate system $\{x^i\}$ such that $Y$ is given by $x^i = 0$ for $r + 1 \leq i \leq n$, where $r = \dim Y$, then $f$ is expressed by

$$\begin{aligned} y^i &= x^i, & 1 &\leq i \leq r, \\ y^i &= -x^i, & r+1 &\leq i \leq n, \end{aligned}$$

which implies that $f$ is an isometry. This isometry is called the *reflection with respect to the subspace Y*.

***Proposition 10.9.*** *Let $O$ be an arbitrary point of $X$. Then any isometry $f$ of $X$ can be written uniquely in the form $f = gh$, where $h$ is a rotation around $O$ and $g$ is a translation. $f$ is a motion if and only if $h$ is a proper rotation.*

*Proof.* Let $\{x^i\}$ be a rectangular coordinate system with origin $O$. Then $f$ can be represented by a matrix $\begin{bmatrix} \mathbf{A} & \alpha \\ 0 & 1 \end{bmatrix}$, where $\mathbf{A} \in O(n)$ and $\alpha \in \mathcal{R}^n$. Let $h$ be the rotation around $O$ defined by $\mathbf{A}$ and let $g$ be the translation whose vector is $\alpha$. Then we have $f = gh$. If $f$ is a motion, we have $\mathbf{A} \in SO(n)$, that is, $h$ is a proper rotation, and vice versa. To prove the uniqueness, let $f = g'h'$ be another representation of $f$, where $g'$ is a translation and $h'$ a rotation around $O$. Then we have

$$g^{-1}g' = hh'^{-1}.$$

Since $g^{-1}g'(O) = O$, we have $hh'^{-1}(O) = O$. Since $hh'^{-1}$ is a translation, it follows that it is the identity transformation, that is, $h' = h$. Consequently, we have $g' = g$.

## EXERCISE 10.3

**1.** Let $\{P_i\}$, $0 \leq i \leq n$, and $\{Q_i\}$, $0 \leq i \leq n$, be rectangular frames in euclidean spaces $X_1$ and $X_2$, respectively. Prove that there is one and only one isometry $f$ of $X_1$ onto $X_2$ such that $f(P_i) = Q_i$ for $0 \leq i \leq n$ (cf. Proposition 9.4).

**2.** Let $X$ be a euclidean space of dimension $n$. Prove that there is a one-to-one correspondence between the set of rectangular frames $\{P_i\}$ in $X$ and the set of isometries $f$ of the standard euclidean space $\mathcal{R}^n$ onto $X$ in the manner analogous to that in Example 9.5.

›› **3.** Let $f$ be an affine transformation of a euclidean space $X$ into itself. If there exists a rectangular frame $\{P_i\}$ such that $\{f(P_i)\}$ is also a rectangular frame, then $f$ is an isometry.

**4.** State and prove the results analogous to Exercise 9.2, number 4, for the group of isometries $I(X)$ of a euclidean space.

☆ **5.** For the standard euclidean space $\mathcal{R}^n$, let $L(\mathcal{R}^n)$ denote the set of all rectangular frames in $X$. For any $\{P_i\} \in L(\mathcal{R}^n)$, let

$$\omega(\{P_i\}) = (P_0,\mathbf{A}) \in \mathcal{R}^n \times O(n),$$

where $\mathbf{A}$ is the orthogonal matrix with column vectors $\overrightarrow{P_0P_i}$, $1 \leq i \leq n$. Prove that $\omega$ is a one-to-one mapping of $L(\mathcal{R}^n)$ onto $\mathcal{R}^n \times O(n)$.

☆ **6.** Let $L(X)$ denote the set of all rectangular frames of a euclidean space $X$. Choose any $\{P_i\} \in L(X)$. For any isometry $f$ of $X$, let $\iota(f)$ be the rectangular frame $\{f(P_i)\}$. Show that $\iota$ is a one-to-one mapping of the group $I(X)$ onto $L(X)$. How does this mapping change when we choose another frame $\{Q_i\} \in L(X)$ instead of $\{P_i\}$?

☆ **7.** Let $X$ be a euclidean space associated with an inner product space $V$ of dimension $n$. An affine transformation $f$ of $X$ is called a *conformal transformation* if for the associated linear mapping $\phi$ there is a $c > 0$ such that

$$(\phi(\alpha),\phi(\beta)) = c^2(\alpha,\beta) \qquad \text{for all } \alpha,\ \beta \in V.$$

(*a*) Prove that an isometry is a conformal transformation.

(*b*) Prove that a conformal transformation preserves angle, that is, $\angle QPR = \angle f(Q)f(P)f(R)$.

(*c*) If $\{x^i\}$ is a rectangular coordinate system in $X$, then a conformal transformation $f$ can be represented by

$$y^i = \sum_{j=1}^{n} a^i{}_j\, x^j + a^i, \qquad 1 \leq i \leq n,$$

where $\alpha = [a^i]$ is a vector and $\mathbf{A} = [a^i{}_j]$ is of the form $c\mathbf{B}$ with an orthogonal matrix $\mathbf{B}$.

(*d*) The set of all conformal transformations $C(X)$ is a group under the composition of mappings.

☆ **8.** An isometry $f$ of a euclidean space is orientation-preserving (cf. Exercise 9.2, number 6) if and only if it is a motion.

## 10.4 CLASSIFICATION OF ISOMETRIES

We shall now classify all isometries of a euclidean space $X$. We say that two isometries $f_1$ and $f_2$ of $X$ are *equivalent* if there is an isometry $g$ of $X$ such that $f_2 = gf_1g^{-1}$. If $f_1$ and $f_2$ are equivalent, then they have the same geometric properties. For example, if $f_1$ is a motion, so is $f_2$. If $f_1$ is a rotation, say, around $O$, then $f_2$ is a rotation around $g(O)$. If $f_1$ has a fixed line $L$ [that is, $f_1(L) = L$], then $f_2$ has a fixed line, namely, the line $g(L)$. We shall be interested in finding all equivalence classes of isometries with respect to this equivalence relation.

We start with a two-dimensional case.

***Theorem 10.10.*** *Let $X$ be a euclidean plane (that is, a two-dimensional euclidean space). An isometry $f$ is one of the following types:*
**1.** *A translation (including the identity transformation), which can be expressed by*

$$x' = x + a, \qquad y' = y. \tag{10.13}$$

*(Here and in the following $\{x,y\}$ is a suitable rectangular coordinate system depending on the given isometry $f$.)*
**2.** *A proper rotation by angle $\theta$, $0 < \theta < 2\pi$, around a certain point $O$, which can be expressed by*

$$x' = x \cos\theta - y \sin\theta, \qquad y' = x \sin\theta + y \cos\theta, \tag{10.14}$$

*where $O$ is the origin of the coordinate system.*
**3.** *A reflection with respect to a certain line, which can be expressed by*

$$x' = x, \qquad y' = -y. \tag{10.15}$$

**4.** *A reflection with respect to a line $L$ followed by a nontrivial translation parallel to $L$, which can be expressed by*

$$x' = x + a, \quad (a \neq 0) \qquad y' = -y. \tag{10.16}$$

*Isometries belonging to two different types are not equivalent; we have the following equivalence within each type:*

**1.** *Two translations are equivalent if and only if the vectors of translation have the same length.*

**2.** *Two proper rotations are equivalent if and only if the angles of rotation $\theta_1$ and $\theta_2$ satisfy either $\theta_1 = \theta_2$ or $\theta_1 + \theta_2 = 2\pi$.*

**3.** *Any two reflections with respect to lines are equivalent.*

**4.** *Two isometries of this type are equivalent if and only if the vectors of translation have the same length.*

*Proof.* Let $\phi$ be the associated linear mapping of the associated inner product space $V$. By Proposition 8.28, we see that there is an orthonormal basis $\{\alpha_1,\alpha_2\}$ of $V$ such that $\phi$ is represented by

1. $$\begin{bmatrix} \cos\theta & -\sin\theta \\ \sin\theta & \cos\theta \end{bmatrix}, \qquad 0 \leq \theta < 2\pi,$$

or

2. $$\begin{bmatrix} 1 & 0 \\ 0 & -1 \end{bmatrix}.$$

Let $\{x,y\}$ be a rectangular coordinate system with respect to a rectangular frame $\{P_0,P_1,P_2\}$ such that $\overrightarrow{P_0P_1} = \alpha_1$ and $\overrightarrow{P_0P_2} = \alpha_2$. Then $f$ can be expressed by

$$(a)\quad \begin{bmatrix} \cos\theta & -\sin\theta & a \\ \sin\theta & \cos\theta & b \\ 0 & 0 & 1 \end{bmatrix}$$

or

$$(b)\quad \begin{bmatrix} 1 & 0 & a \\ 0 & -1 & b \\ 0 & 0 & 1 \end{bmatrix}.$$

In case $(a)$, we examine if $f$ has a fixed point, namely, a point $P_0$ with coordinates $(x_0,y_0)$ such that

$$(\cos\theta - 1)x_0 - (\sin\theta)y_0 + a = 0$$
$$(\sin\theta)x_0 + (\cos\theta - 1)y_0 + b = 0.$$

If

$$\begin{vmatrix} \cos\theta - 1 & -\sin\theta \\ \sin\theta & \cos\theta - 1 \end{vmatrix} = (\cos\theta - 1)^2 + \sin^2\theta = 2(1 - \cos\theta)$$

is not 0, then there is a unique solution $(x_0,y_0)$, that is, a unique fixed point $P_0$. It follows that $f$ is a proper rotation around $P_0$ and that it can be expressed by (10.14) with respect to any rectangular coordinate system with $P_0$ as the origin. If $\cos\theta = 1$, then $(a)$ becomes

$$\begin{bmatrix} 1 & 0 & a \\ 0 & 1 & b \\ 0 & 0 & 1 \end{bmatrix},$$

and $f$ is a translation. If the translation is not trivial, we take a rectangular frame $\{P_0,P_1,P_2\}$ such that the vector of translation is a scalar multiple of $\overrightarrow{P_0P_1}$. Then $f$ can be expressed by (10.13).

In case $(b)$, $f$ is the transformation defined by $x' = x + a$ and $y' = -y + b$. By the change of coordinates, $\bar{x} = x$ and $\bar{y} = y - b/2$, we see that $f$ can be expressed by

$$\bar{x}' = \bar{x} + a \qquad \text{and} \qquad \bar{y}' = -\bar{y}.$$

If $a = 0$, we have type 3. If $a \neq 0$, we have type 4.

We show that isometries of different types are not equivalent. Isometries of types 1 and 2 are motions, and those of types 3 and 4 are not. An isometry of type 1 has no fixed point (unless it is the identity transformation), and an isometry of type 2 has one and only one fixed point. An isometry of type 3 has a line as its set of fixed points, and an isometry of type 4 has no fixed point.

Finally, the equivalence condition for isometries of the same type can be verified in each case.

***Example 10.13.*** An isometry $f$ such that $f^2$ is the identity transformation is called an *involution*. An involution of the euclidean plane is

$$\text{either} \quad \begin{aligned} x' &= -x \\ y' &= -y \end{aligned} \quad \text{or} \quad \begin{aligned} x' &= x \\ y' &= -y \end{aligned}$$

(with respect to a suitable rectangular coordinate system). This follows easily from Theorem 10.10.

We shall now consider the case of an $n$-dimensional euclidean space $X$ associated with an inner product space $V$. Let $f$ be an isometry of $X$ and let $\phi$ be the associated linear mapping of $V$, which is orthogonal. By Theorem 8.30, there is an orthonormal basis of $V$ with respect to which $\phi$ can be represented by a matrix of the canonical form stated in Theorem 8.30. If $-1$ appears an even number of times, then we may consider each block $\begin{bmatrix} -1 & 0 \\ 0 & -1 \end{bmatrix}$ as $\begin{bmatrix} \cos\omega & -\sin\omega \\ \sin\omega & \cos\omega \end{bmatrix}$ with $\omega = \pi$. Hence the matrix is of the form

$$(a) \quad \begin{bmatrix} 1 & & & & & & \\ & \ddots & & & & & \\ & & 1 & & & & \\ & & & \begin{matrix} \cos\theta_1 & -\sin\theta_1 \\ \sin\theta_1 & \cos\theta_1 \end{matrix} & & & \\ & & & & \ddots & & \\ & & & & & \begin{matrix} \cos\theta_p & -\sin\theta_p \\ \sin\theta_p & \cos\theta_p \end{matrix} \end{bmatrix},$$

where 1 appears $r$ times, $0 \leq r \leq n$, and $0 < \theta_i < 2\pi$, for $1 \leq i \leq p$, where $r + 2p = n$.

On the other hand, if $-1$ appears an odd number of times, the same convention will put the matrix in the form

$$(b) \quad \begin{bmatrix} 1 & & & & & & & \\ & \ddots & & & & & & \\ & & 1 & & & & & \\ & & & -1 & & & & \\ & & & & \begin{matrix} \cos\theta_1 & -\sin\theta_1 \\ \sin\theta_1 & \cos\theta_1 \end{matrix} & & & \\ & & & & & \ddots & & \\ & & & & & & \begin{matrix} \cos\theta_p & -\sin\theta_p \\ \sin\theta_p & \cos\theta_p \end{matrix} \end{bmatrix},$$

where 1 appears $r - 1$ times, $1 \leq r \leq n$, and $0 < \theta_i < 2\pi$ for $1 \leq i \leq p$, where $r + 2p = n$.

In case (*a*), $\det \phi = 1$ and $f$ is a motion. By taking a rectangular coordinate system (with an arbitrary origin) corresponding to the basis of $V$, we can represent $f$ by

$$y^1 = x^1 + a^1, \quad \ldots, \quad y^r = x^r + a^r, \tag{10.17}$$

$$\begin{aligned} y^{r+2k-1} &= x^{r+2k-1} \cos \theta_k - x^{r+2k} \sin \theta_k + a^{r+2k-1}, \\ y^{r+2k} &= x^{r+2k-1} \sin \theta_k + x^{r+2k} \cos \theta_k + a^{r+2k}, \end{aligned} \qquad 1 \leq k \leq p. \tag{10.18}$$

We may consider (10.17) as a translation in the euclidean subspace $x^{r+1} = \cdots = x^n = 0$. By modifying the coordinate system $\{x^1, \ldots, x^r\}$ in this subspace, we may assume that (10.17) takes the form

$$y^1 = x^1 + a^1, \; y^2 = x^2, \; \ldots, \; y^r = x^r. \tag{10.19}$$

On the other hand, the pair of equations for each $k$ in (10.18) may be considered as a motion, $g_k$, in the two-dimensional euclidean subspace $Y_k$ (for which all coordinates other than $x^{r+2k-1}$ and $x^{r+2k}$ are 0). Since $1 - \cos \theta_k \neq 0$ $(0 < \theta_k < 2\pi)$, the argument in the proof of Theorem 10.10 shows that $g_k$ is a proper rotation of the subspace $Y_k$. Thus by changing the coordinate system $\{x^{r+2k-1}, x^{r+2k}\}$ of $Y_k$ suitably, we may assume that (10.18) now takes the form

$$\begin{aligned} y^{r+2k-1} &= x^{r+2k-1} \cos \theta_k - x^{r+2k} \sin \theta_k, \\ y^{r+2k} &= x^{r+2k-1} \sin \theta_k + x^{r+2k} \cos \theta_k. \end{aligned} \tag{10.20}$$

Thus we conclude that, in case (*a*), there is a rectangular coordinate system in $X$ with respect to which $f$ can be represented by (10.19) and (10.20).

In case (*b*), Eq. (10.17) is changed into

$$y^1 = x^1 + a^1, \; \ldots, \; y^{r-1} = x^{r-1} + a^{r-1}, \; y^r = -x^r + a^r, \tag{10.17$'$}$$

while (10.18) remains the same. By modifying the $r$th coordinate so that the last equation of (10.17′) becomes

$$y^r = -x^r,$$

and by modifying $x^1, \ldots, x^{r-1}$ in the same manner as in case (*a*), we see that (10.17′) takes the form

$$y^1 = x^1 + a^1, \; y^2 = x^2, \; \ldots, \; y^{r-1} = x^{r-1}, \; y^r = -x^r. \tag{10.19$'$}$$

The argument for (10.18) remains the same as case (*a*). Thus we conclude that, in case (*b*), there is a rectangular coordinate system in $X$ with respect to which $f$ can be represented by (10.19′) and (10.20).

Summarizing our discussions, we may state:

***Theorem 10.11.*** *An isometry $f$ of an $n$-dimensional euclidean space $X$ is one of the following types:*

**1.** *A translation (including the identity transformation), which can be expressed by*

$$(10.21) \qquad y^1 = x^1 + a^1, \qquad y^i = x^i \quad (2 \leq i \leq n),$$

*(Here and in the following* $\{x^i\}$ *is a suitable rectangular coordinate system depending on* $f$.)

**2.** *A nontrivial proper rotation followed by a translation (possibly trivial) by a vector fixed by the rotation, which may be expressed by* (10.19) *and* (10.20), *where* $0 \leq r < n$, $r + 2p = n$, *and* $0 < \theta_k < 2\pi$, $1 \leq k \leq p$.

**3.** *A proper rotation (possibly trivial) followed by a reflection with respect to a hyperplane whose points are fixed by the rotation, which may be expressed by*

$$(10.22) \qquad y^i = x^i \quad (1 \leq i \leq r-1), \qquad y^r = -x^r,$$

*and* (10.20), *where* $1 \leq r \leq n$, $r + 2p = n$, *and* $0 < \theta_k < 2\pi$, $1 \leq k \leq p$ *(the rotation involved is trivial for* $r = n$).

**4.** *An isometry of type* 3 *followed by a nontrivial translation, which may be expressed by* (10.19′) *and* (10.20), *where* $2 \leq r < n$, $r + 2p = n$, *and* $0 < \theta_k < 2\pi$, $1 \leq k \leq p$, $a^1 \neq 0$.

*Isometries belonging to two different types are not equivalent; we have the following equivalence within each type:*

**1.** *Equivalent if and only if the vectors of translation have the same length.*

**2.** *Equivalent if and only if the integers* $r$ *are equal and the corresponding* $\theta_i$*'s are equal or add up to* $2\pi$ *for each* $i$, $1 \leq i \leq p$.

**3.** *Equivalent if and only if the conditions stated for type* 2 *are satisfied.*

**4.** *Equivalent if and only if the conditions stated for type* 2 *are satisfied and the vectors of translation have the same length.*

*Proof.* The classification into the four types is almost immediate from the preceding discussions. An isometry expressed by (10.19) and (10.20) is of type 1 or 2 according as $r = n$ or $r < n$. An isometry expressed by (10.19′) and (10.20) is of type 3 or 4 according as $r = n$ or $r < n$.

Isometries of types 1 and 2 are motions, and those of types 3 and 4 are not. An isometry $f$ of type 1 has the property that the vector $\overrightarrow{Pf(P)}$ is a constant vector for all $P$, but this property is not valid for an isometry of type 2. An isometry of type 3 has a fixed point, namely, the origin of the coordinate system used for (10.22) and (10.20); an isometry of type 4 has no fixed point. Thus, isometries belonging to different types are not equivalent. Finally, equivalence within each type can be obtained, under the conditions stated, by using the following principles:

1. Suppose that two isometries $f$ and $g$ are expressed by the same formula with respect to two rectangular coordinate systems $\{x^i\}$ and $\{u^i\}$, respec-

tively. Then $f$ and $g$ are equivalent. Indeed, if $h$ is an isometry which maps a point with coordinates $(a^1, \ldots, a^n)$ with respect to $\{x^i\}$ upon the point with coordinates $(a^1, \ldots, a^n)$ with respect to $\{u^i\}$, then $f = h^{-1}gh$.

2. Two translations, $y^1 = x^1 + a$, $y^i = x^i$, $2 \leq i \leq n$, and $y^1 = x^1 - a$, $y^i = x^i$, $2 \leq i \leq n$, are equivalent. In fact, if we change the coordinate system $\{x^1, \ldots, x^n\}$ into $\{-x^1, x^2, \ldots, x^n\}$, then the second isometry is expressed by the same formula as the first expression. Hence (1) can be applied.

3. Two proper rotations, say,

$$f: \begin{array}{l} y^1 = x^1 \cos\theta - x^2 \sin\theta, \\ y^2 = x^1 \sin\theta + x^2 \cos\theta, \end{array} \qquad y^i = x^i \quad (3 \leq i \leq n),$$

and

$$g: \begin{array}{l} y^1 = x^1 \cos\theta + x^2 \sin\theta, \\ y^2 = -x^1 \sin\theta + x^2 \cos\theta, \end{array} \qquad y^i = x^i \ (3 \leq i \leq n)$$

are equivalent. In fact, if we change

$$\{x^1, x^2, x^3, \ldots, x^n\} \qquad \text{into} \qquad \{x^2, x^1, x^3, \ldots, x^n\},$$

then the first pair of equations for $g$ will take the same form as the first pair of equations for $f$. Hence (1) can be applied.

**Corollary.** *For a three-dimensional euclidean space $X$, isometries can be classified as follows:*

1. *A translation (including the identity transformation), which may be expressed by*

$$y^1 = x^1 + a^1, \qquad y^2 = x^2, \qquad y^3 = x^3.$$

2. *A proper rotation around a straight line followed by a translation (possibly trivial) parallel to the line, which may be expressed by*

$$y^1 = x^1 + a^1, \qquad \begin{array}{l} y^2 = x^2 \cos\theta - x^3 \sin\theta \\ y^3 = x^2 \sin\theta + x^3 \cos\theta \end{array} \qquad (0 < \theta < 2\pi).$$

3. *A rotation (possibly trivial) followed by the reflection relative to the plane fixed by the rotation, which may be expressed by*

$$\begin{array}{l} y^1 = x^1 \cos\theta - x^2 \sin\theta \\ y^2 = x^1 \sin\theta + x^2 \cos\theta \\ y^3 = -x^3. \end{array}$$

4. *A reflection with respect to a plane followed by a nontrivial translation parallel to the plane, which may be expressed by*

$$y^1 = x^1 + a^1, \qquad y^2 = x^2, \qquad y^3 = -x^3.$$

*Proof.* This follows from Theorem 10.11 if we observe that the integer $r$ for types 2, 3, and 4 is either 1 or 3, since $n = 3$.

**EXERCISE 10.4**

**1.** In the euclidean plane $\mathcal{R}^2$ with standard coordinates $\{x,y\}$, find the formula for the reflection with respect to the line $2x - y = 1$.

» **2.** In the euclidean plane $\mathcal{R}^2$ with standard coordinates $\{x,y\}$, determine the type of each of the isometries defined by

$$(a)\ \begin{aligned} x' &= -x + a \\ y' &= -y + b \end{aligned}; \qquad (b)\ \begin{bmatrix} x' \\ y' \end{bmatrix} = \begin{bmatrix} \cos\theta & \sin\theta \\ \sin\theta & -\cos\theta \end{bmatrix} \begin{bmatrix} x \\ y \end{bmatrix};$$

$$(c)\ \begin{bmatrix} x' \\ y' \\ 1 \end{bmatrix} = \begin{bmatrix} 1/\sqrt{2} & -1/\sqrt{2} & 2 \\ 1/\sqrt{2} & 1/\sqrt{2} & -1 \\ 0 & 0 & 1 \end{bmatrix} \begin{bmatrix} x \\ y \\ 1 \end{bmatrix}.$$

Find the set of all fixed points for each isometry.

**3.** Let $f_1$ and $f_2$ be the reflections with respect to the lines $L_1$ and $L_2$, respectively. Find the type of the product $f_1f_2$.

» **4.** In the euclidean space $\mathcal{R}^3$ with standard coordinates $\{x^1,x^2,x^3\}$, find the formula for the reflection with respect to the plane $x^1 + x^2 - x^3 = 0$.

**5.** In the euclidean space $\mathcal{R}^3$ with standard coordinates $\{x^1,x^2,x^3\}$, determine the type of each of the isometries defined by

$$(a)\ \begin{aligned} y^1 &= x^1 + a^1 \\ y^2 &= -x^2 + a^2 \\ y^3 &= x^3 \end{aligned}; \qquad (b)\ \begin{bmatrix} y^1 \\ y^2 \\ y^3 \\ 1 \end{bmatrix} = \begin{bmatrix} \cos\theta & -\sin\theta & 0 & a \\ \sin\theta & \cos\theta & 0 & b \\ 0 & 0 & -1 & 0 \\ 0 & 0 & 0 & 1 \end{bmatrix} \begin{bmatrix} x^1 \\ x^2 \\ x^3 \\ 1 \end{bmatrix};$$

$$(c)\ \begin{bmatrix} y^1 \\ y^2 \\ y^3 \\ 1 \end{bmatrix} = \begin{bmatrix} 1/\sqrt{2} & 0 & -1/\sqrt{2} & 1 \\ 0 & 1 & 0 & -1 \\ 1/\sqrt{2} & 0 & 1/\sqrt{2} & 1 \\ 0 & 0 & 0 & 1 \end{bmatrix} \begin{bmatrix} x^1 \\ x^2 \\ x^3 \\ 1 \end{bmatrix}.$$

»☆ **6.** Find all involutions of a three-dimensional euclidean space.

**7.** Show that in a four-dimensional euclidean space there is a proper rotation which has no fixed line.

# *Appendix*

## A.1 SETS

In the text we shall freely use the notations in elementary set theory. $x \in M$ means that the object $x$ is an *element* of the *set* $M$ (or $x$ *belongs to* $M$). $x \notin M$ means that $x$ is not an element of $M$. A set $M$ is described by listing all its elements as

$$M = \{a_1, \ . \ . \ . \ ,a_n\},$$

or by writing the property which characterizes all the elements and only the elements of the set as

$$M = \{x; P\},$$

meaning that $M$ is the set of all objects $x$ that have the property $P$. The only exception is the *empty set*, denoted by $\varnothing$, that has no element.

By $N \subseteq M$ we mean that $N$ is a *subset* of the set $M$ (that is, every element of $N$ is an element of $M$). By $N \subset M$ we mean that $N$ is a *proper subset* of $M$, that is, $N \subseteq M$ and $N \neq M$.

For two subsets $N_1$ and $N_2$ of a set $M$, $N_1 \cup N_2$ denotes the *union* of $N_1$ and $N_2$, namely,

$$N_1 \cup N_2 = \{x \in M; x \in N_1 \text{ or } x \in N_2\}.$$

$N_1 \cap N_2$ denotes the *intersection* of $N_1$ and $N_2$, namely,

$$N_1 \cap N_2 = \{x \in M; x \in N_1 \text{ and } x \in N_2\}.$$

$N_1$ and $N_2$ are said to be *disjoint* if $N_1 \cap N_2 = \varnothing$. The notions of union and intersection are easily extended to the case of an arbitrary family of subsets, or more precisely, an *indexed family* of subsets $N_\lambda$, $\lambda \in \Lambda$ (that is, for each element $\lambda$ of a certain set $\Lambda$ there is associated a subset $N_\lambda$ of $M$):

$$\bigcup_{\lambda \in \Lambda} N_\lambda = \{x \in M; x \in N_\lambda \text{ for some } \lambda \in \Lambda\},$$

$$\bigcap_{\lambda \in \Lambda} N_\lambda = \{x \in M; x \in N_\lambda \text{ for every } \lambda \in \Lambda\}.$$

Given two sets $M$ and $N$, the *cartesian product* $M \times N$ is the set of all pairs $(x,y)$, where $x \in M$ and $y \in N$. This notion can be extended to the case of more than two sets.

## A.2 MAPPINGS

By a *mapping* $f$ of a set $M$ into a set $N$ we mean a rule which associates to each $x \in M$ a certain well-determined element $y$ of $N$; this element $y$ is denoted by $f(x)$. We can give a more rigorous definition of this notion as follows. Given two sets $M$ and $N$, a subset $A$ of $M \times N$ is called a *relation*. [Indeed, this is a mathematical definition of the ordinary sense of the word "relation"; if $(x,y) \in A$, then $x$ is related to $y$, and if $(x,y) \notin A$, then $x$ is not related to $y$.] Given a relation $A \subseteq M \times N$, the set

$$M_0 = \{x \in M; \text{ there exists a } y \in N \text{ such that } (x,y) \in A\}$$

is called the *domain*, and the set

$$N_0 = \{y \in N; \text{ there exists an } x \in M \text{ such that } (x,y) \in A\}$$

is called the *range*. If a relation $A$ satisfies the condition:

(*) If $(x,y_1) \in A$ and $(x,y_2) \in A$, then $y_1 = y_2$,

then it is called a *mapping*. In this case, for any $x \in M_0$, there exists at least one $y \in N$ such that $(x,y) \in A$ (by definition of $M_0$) and such $y$ is unique by condition (*). Thus the element $y$ is determined by $x$. Therefore, we have a rule which associates to each $x \in M_0$ a certain $y \in N$ which is uniquely determined by $x$. In particular, when $M_0 = M$, we have a mapping $f$ of $M$ into $N$ in the sense we explained in the beginning. To be more precise, $(x,y) \in A$ if and only if $y = f(x)$.

Once this rigorous definition of mapping is established, we find it more convenient to use the notation $f$ for a mapping of $M$ into $N$ (rather than the notation $A \subseteq M \times N$). The mapping $f$ is said to be *onto* $N$ if for any $y \in N$ there exists (at least one) $x \in M$ such that $y = f(x)$. $f$ is said to be *one-to-one* if $f(x_1) = f(x_2)$ implies $x_1 = x_2$.

## A.3 EQUIVALENCE RELATION

An *equivalence relation* in a set $M$ is a relation $A \subseteq M \times M$ which has the following properties:

1. $(x,x) \in A$ for every $x \in M$.
2. $(x,y) \in A$ implies $(y,x) \in A$.
3. $(x,y) \in A$ and $(y,z) \in A$ imply $(x,z) \in A$.

If we write $x \approx y$ for $(x,y) \in A$, then the preceding conditions are equivalent to

1′. $x \approx x$ for every $x \in M$.
2′. $x \approx y$ implies $y \approx x$.
3′. $x \approx y$ and $y \approx z$ imply $x \approx z$.

Thus a relation $x \approx y$ is an equivalence relation if it satisfies (1′), (2′), and (3′).

Given an equivalence relation in $M$, expressed by $x \approx y$, the *equivalence class* of $x \in M$ is the subset $\pi(x)$ of $M$ defined by

$$\pi(x) = \{y \in M; x \approx y\}.$$

The set $M$ is then the union of the family of disjoint subsets each of which is the equivalence class of a certain element of $M$. In fact, each $x \in M$ belongs to $\pi(x)$ by condition 1′. If $\pi(x) \cap \pi(y)$ has an element $z$, then $x \approx z$ and $y \approx z$ (so that $z \approx y$ by condition 2′). It follows by (3′) that $x \approx y$. This implies that $\pi(x) = \pi(y)$ by a similar argument; $x \approx y$ if and only if $\pi(x) = \pi(y)$.

If we denote by $M^*$ the set of all equivalence classes, then $x \in M \to \pi(x) \in M^*$ may be considered as a mapping of $M$ onto $M^*$ which associates to each $x \in M$ the equivalence class $\pi(x)$ of $x$. This mapping $\pi$ is called the *natural projection* of $M$ onto $M^*$.

## A.4 INDUCTION

Suppose that for each positive integer $n$ there is given a proposition $P(n)$. In order to prove that $P(n)$ is valid for every $n$, we may proceed as follows:

1. We prove that $P(1)$ is valid,
2. Let $n$ be an arbitrary positive integer $n$. Assuming that $P(n-1)$ is valid [or $P(k)$ is valid for all $k < n$], we prove that $P(n)$ is valid.

This method of proof is called *mathematical induction.* In step 2, the assumption that $P(n-1)$ is valid [or $P(k)$ is valid for all $k < n$] is called the *inductive assumption.*

A similar process can be used to define a sequence of objects $a_n$, $n = 1, 2, 3, \ldots,$ as follows:

1. We define what $a_1$ is.
2. Let $n$ be an arbitrary positive integer. Assuming that we know what $a_{n-1}$ is (or what $a_k$'s are for all $k < n$), we define what $a_n$ is.

This process is called *inductive definition.*

# Notation

$\mathcal{R}$ real number system *32*

$\mathcal{Q}$ rational number field *35*

$\mathcal{C}$ complex number system *32*

$\mathcal{Z}$ set of integers *35*

$\mathcal{F}$ field *32, 112*

$\mathcal{R}^n$ $n$-dimensional real standard vector (affine or euclidean) space *32, 260, 285*

$\mathcal{C}^n$ $n$-dimensional complex standard vector (or affine) space *32, 260*

$\mathcal{F}^n$ $n$-dimensional standard vector (or affine) space over a field $\mathcal{F}$ *32, 260*

$\mathcal{F}^m{}_n$ set of all $m \times n$ matrices over $\mathcal{F}$ *39*

**A**, **B**, **C**, etc. matrices; $\mathbf{A} = [a_{ij}]$ or $[a^i{}_j]$ *3, 38, 261*

**x**, **y**, **z**, etc. column or row vectors; $\mathbf{x} = [x_i]$, $(x_i)$ *3, 39*

$\mathbf{I}_n$ identity matrix of degree $n$ *26*

$\mathbf{O}_n$ zero matrix of degree $n$ *193*

$\mathbf{O}_{m,n}$ $m \times n$ zero matrix *39*

$\{\epsilon_1, \ldots, \epsilon_n\}$ standard basis of $\mathcal{F}^n$ *34*

$\alpha, \beta, \gamma$, etc. elements of a vector space *32, 36*

$\Phi = \{\alpha_1, \ldots, \alpha_n\}$ basis of a vector space $V$ or the corresponding linear isomorphism of $\mathcal{F}^n \to V$ *44, 61*

$A_\Phi$ matrix representing a linear transformation $A: V \to V$ with respect to a basis $\Phi$ in $V$ *64*

$A_{\Phi,\Psi}$ matrix representing a linear mapping $A: V \to W$ with respect to bases $\Phi$ and $\Psi$ in $V$ and $W$, respectively *62*

$\delta_{ij}$ Kronecker's delta *27*

det **A**, $|\mathbf{A}|$ determinant of a matrix **A** *161–162*

det $A$ determinant of a linear transformation $A$ *176*

${}^t\mathbf{A}$ transpose of a matrix **A** *93*

$\bar{\mathbf{A}}$ conjugate of a matrix **A** *234*

$\mathbf{A}^* = {}^t\bar{\mathbf{A}} = \overline{{}^t\mathbf{A}}$ adjoint of **A** *234*

† Page number where notation is explained.

$f_{\mathbf{A}}$ characteristic polynomial of a matrix **A** *183*
$f_A$ characteristic polynomial of a linear transformation $A$ *183*
$\phi_{\mathbf{A}}$ minimal polynomial of a matrix **A** *188*
$\phi_A$ minimal polynomial of a linear transformation $A$ *188*

$\mathrm{Hom}(V,W)$ set of all linear mappings $V \to W$ *57*
$\mathfrak{gl}(V) = \mathrm{Hom}(V,V)$ set of all linear transformations of $A$ *57*
$GL(V)$ set of all nonsingular linear transformations of $V$ *149*
$GL(n,\mathcal{F})$ set of all nonsingular matrices of degree $n$ over $\mathcal{F}$ ($\mathcal{F}$ is often $\mathcal{R}$ or $\mathcal{C}$) *149*
deg degree of a polynomial *119*
Re real part *220*
Im imaginary part *225*
$^\perp$ orthogonal complement *227*

$U \subseteq V$ $U$ is a subset of $V$ *309*
$x \in X$ $x$ belongs to $X$ ($x$ is an element of $X$) *309*
$\cap$ intersection *309*
$\bigcap$ intersection of an indexed family *41*
$\cup$ union *309*
$\vee$ affine subspace spanned by the union *273*
Sp span of (linear subspace spanned by) *41*

# *Index*

# Index

# SOLUTIONS

# TO

# SELECTED EXERCISES

Page 7

Exercise 1.1

1. (a) $\begin{bmatrix} -1 & 1 \\ 19 & -5 \end{bmatrix}$ (b) $\begin{bmatrix} 4 \\ -2 \end{bmatrix}$ (c) $\begin{bmatrix} 8 & 5 & 13 \\ 9 & 2 & 11 \\ 19 & 6 & 25 \end{bmatrix}$

(d) $\begin{bmatrix} 5 & 22 \\ 15 & 14 \\ 4 & 6 \end{bmatrix}$ (e) 4

2. (a) No (b) Yes; $\begin{bmatrix} 2 & -2 & 2 \\ 1 & -1 & 1 \\ 3 & -3 & 3 \end{bmatrix}$

3. (a) $x = 0$, $y = -1/4$, $z = 5/4$

Page 10

Exercise 1.2

1. $c_1 \cos t + c_2 \sin t + 1$

Page 20

Exercise 1.3

4. If N is the midpoint of AB and K is the point such that $\overrightarrow{GK} = 2\,\overrightarrow{GN}$, then $\overrightarrow{GA} + \overrightarrow{GB} = 2\,\overrightarrow{GN}$. Also $\overrightarrow{GC} = -2\,\overrightarrow{GN}$, by Example 1.5 or 1.8.

7. $((a_1 + b_1 + c_1)/3,\ (a_2 + b_2 + c_2)/3)$

11. $\overrightarrow{PR} = 2\,\overrightarrow{AB}$

15. (a) $(3/5)\,x_1 + (4/5)\,x_2 = 2$

(b) $(-3/5)\,x_1 - (4/5)\,x_2 = 2$

16. $|a_1 y_1 + a_2 y_2 - d|$

Page 31

Exercise 1,4

1. (a) $\begin{bmatrix} 1/2 & 0 \\ 0 & 1 \end{bmatrix}$; (c) $\begin{bmatrix} -1 & 0 \\ 0 & 1 \end{bmatrix}$

3. Solve $\begin{matrix} ax + bu = 1 \\ cx + du = 0 \end{matrix}$ and $\begin{matrix} ay + bv = 0 \\ cy + dv = 1 \end{matrix}$

5. $\begin{bmatrix} 1/2 & -3/2 \\ 3/2 & 1/2 \end{bmatrix}$; $\begin{bmatrix} 0 & -1 \\ 1 & 0 \end{bmatrix}$; $\begin{bmatrix} -1 & 0 \\ 0 & -1 \end{bmatrix}$

7. $x_1^2 + x_2^2 = 1$

8. $(x_1')^2 + (x_2')^2 = 25$; circle of radius 5

Page 34

Exercise 2.1

1. $(-6, -4, -12)$; $(0, 0, 0)$

3. $-(1/2)(0, 1, 1) + (1/2)(1, 0, 1) + (1/2)(1, 1, 0)$

Page 39

Exercise 2,3

4, No. Property 4 in Definition 2.3 is violated.

7. $(c_1, \ldots, c_n) = \sum_{k=1}^{n} a_k \epsilon_k + \sum_{k=1}^{n} b_k \epsilon_{n+k}$,

where $c_k = a_k + i b_k$, $1 \le k \le n$,

Page 42

Exercise 2.3

1. (a), (d) are subspaces.

3. No (unless one contains the other).

4. No, $\mathrm{Sp}(\mathcal{R}^n)$ in $\mathcal{C}^n$ is $\mathcal{C}^n$.

Page 47
Exercise 2.4

1. Yes; No

3. The dimension is $2n$.

4. $n(n+1)/2$.

6. (b) implies (a) by Theorem 2.13. (c) implies (a) by Theorem 2.12 and Theorem 2.11.

Page 51
Exercise 2.5

1. Choose a basis $\{\alpha_1, \ldots, \alpha_m, \alpha_{m+1}, \ldots, \alpha_n\}$ as in Theorem 2.14, and let $U = \mathrm{Sp}\{\alpha_{m+1}, \ldots, \alpha_n\}$. But a subspace $U$ such that $V = U \oplus W$ is not unique. For example, let $V = R^2$ and $W = \mathrm{Sp}\{(1, 0)\}$. Then $U$ can be $\mathrm{Sp}\{(0, 1)\}$ or $\mathrm{Sp}\{(1, 1)\}$ or $\mathrm{Sp}\{(1, 2)\}$, etc.

4. (a) $W_1 \cap W_2$ consists of all diagonal matrices.

(b) $\dim W_1 = \dim W_2 = n(n+1)/2$, $\dim(W_1 \cap W_2) = n$.

5. $\mathcal{R}^3$ is not the direct sum of $W_1$, $W_2$ and $W_3$.

Page 58
Exercise 3,1

4, Extend $\{\alpha_1, \ldots, \alpha_k\}$ to a basis by adjoining, say, $n-k$ vectors $\alpha_{k+1}, \ldots, \alpha_n$, where $n = \dim V$, Define $A$ by

$$A\left(\sum_{i=1}^{n} c_i \alpha_i\right) = \sum_{i=1}^{k} c_i \beta_i,$$

5, If $\{\alpha_1, \ldots, \alpha_m\}$ is a basis of $V_1$, then $A(\alpha_1), \ldots, A(\alpha_m)$ span $A(V_1)$,

6, Let $\{\beta_1, \ldots, \beta_m\}$ be a basis of the range $A(V)$ and choose $\alpha_1, \ldots, \alpha_r \in V$ such that $A(\alpha_i) = \beta_i$, $1 \leq i \leq r$, Next, choose a basis $\{\alpha_{r+1}, \ldots, \alpha_{r+s}\}$ of the null space, Then $\{\alpha_1, \ldots, \alpha_{r+s}\}$ is a basis of $V$,

Exercise 3.1 (cont'd)

8. $(A+B)\alpha = A\alpha + B\alpha$ is in the subspace $A(V)+B(V)$, namely, $(A+B)(V) \subset A(V)+B(V)$.

9. If W is finite-Dimensional, so are $A(V)$ and $B(W)$.

10. Consider A as a linear mapping $\mathfrak{F}^n \to \mathfrak{F}^m$ and B as a linear mapping $\mathfrak{F}^m \to \mathfrak{F}^n$. The assumption $AB = I_m$ implies $m = \text{rank } AB \leq \text{rank } B \leq n$, by Proposition 3.5. Thus $m \leq n$. Similarly, $n \leq m$.

12. For any $\alpha \in V$, let $\beta = A(\alpha)$ and $\gamma = \alpha - A(\alpha)$. Then $A(\beta) = A^2(\alpha) = A(\alpha) = \beta$, $A(\gamma) = A(\alpha) - A^2(\alpha) = 0$, and $\alpha = \beta + \gamma$.

13. For any $\alpha \in V$, let $\beta = (\alpha + A(\alpha))/2$, $\gamma = (\alpha - A(\alpha))/2$. Then $A(\beta) = \beta$, $A(\gamma) = -\gamma$, and $\alpha = \beta + \gamma$.

17. $B(V)$ is contained in the null space of A, hence rank $B \leq$ nullity A. But nullity A = n - rank A.

20. Assume that $\sum_{i=1}^{m} c_i A^{i-1}(\alpha) = 0$ and that $c_k$ is the first nonzero coefficient. Thus $\sum_{i=k}^{m} c_k A^{i-1}(\alpha) = 0$.

Applying $A^{m-k}$, we get $c_k A^{m-1}(\alpha) = 0$ and hence $c_k = 0$.

21. If $\alpha_1 \neq 0$, then $\alpha_1$ and $A\alpha_1$ are linearly independent, since $a\alpha_1 + bA\alpha_1 = 0$ implies

$$A(a\alpha_1 + bA\alpha_1) = aA\alpha_1 - b\alpha_1 = 0,$$

which leads to $a = b = 0$. Assuming that we have chosen $\alpha_1, \ldots, \alpha_k$ such that $\alpha_1, \ldots, \alpha_k, A\alpha_1, \ldots, A\alpha_k$ are linearly independent, pick $\alpha_{k+1}$ which is not in $Sp\{\alpha_1, \ldots, \alpha_k, A\alpha_1, \ldots, A\alpha_k\}$. Show that $\alpha_1, \ldots, \alpha_{k+1}$, $A\alpha_1, \ldots, A\alpha_{k+1}$ are linearly independent.

Page 65

Exercise 3.2

1. (a) $\begin{bmatrix} 6 \\ -3 \end{bmatrix}$, $\begin{bmatrix} 16 \\ -7 \end{bmatrix}$

(b) 2

(c) 1

3. (a) $\begin{bmatrix} a_{11}b_{11} & & & \\ 0 & a_{22}b_{22} & & * \\ & & & \\ 0 & 0 & & a_{nn}b_{nn} \end{bmatrix}$,

(c) $\begin{bmatrix} 0 & 0 & 0 \\ 0 & 0 & 0 \\ 0 & 0 & 0 \end{bmatrix}$

(e) n even: $\begin{bmatrix} 1 & 0 & 0 \\ 0 & 1 & 0 \\ 0 & 0 & 1 \end{bmatrix}$; n odd: $\begin{bmatrix} 0 & 0 & 1 \\ 0 & 1 & 0 \\ 1 & 0 & 0 \end{bmatrix}$

6. Fix a basis $\Phi = \{\alpha_1, \ldots, \alpha_n\}$ of V. Take $\beta \neq 0$ in A(V) (assuming rank A = 1). Then $A(\alpha_j) = c_j\beta$, $1 \leq j \leq n$. If $\beta = \Sigma b_i \alpha_i$, then $A_\Phi = [a_{ij}]$ with $a_{ij} = b_i c_j$.

Exercise 3.2 (cont'd)

17. Let $\mathbf{E}_{i,j}$ be the standard basis as in number 9 of this exercise. Then

$$\mathbf{A}\,\mathbf{E}_{i,j} = \left(\sum_{p,q} \mathbf{E}_{p,q}\right)\mathbf{E}_{i,j} = \sum_p a_{pi}\,\mathbf{E}_{p,j}$$

Relative to the basis

$$(\mathbf{E}_{1,1}, \mathbf{E}_{2,1}, \ldots, \mathbf{E}_{n,1}, \mathbf{E}_{1,2}, \ldots, \mathbf{E}_{n,2}, \ldots, \mathbf{E}_{1,n}, \mathbf{E}_{2,n}, \ldots, \mathbf{E}_{n,n})$$

the transformation $f_{\mathbf{A}}$ is expressed by the $n^2 \times n^2$ matrix

$$\begin{bmatrix} \mathbf{A} & 0 & \cdots & 0 \\ 0 & \mathbf{A} & \cdots & 0 \\ \vdots & \vdots & \ddots & \vdots \\ 0 & 0 & \cdots & \mathbf{A} \end{bmatrix}$$

Thus rank $f_{\mathbf{A}} = n$ rank $\mathbf{A}$.

Page 73

Exercise 3,3

1. (a) $\begin{bmatrix} 1/4 & 1/4 \\ 1/4 & -3/4 \end{bmatrix}$ (b) $\begin{bmatrix} 2 & 1 \\ -4 & -1 \end{bmatrix}$

2. $\begin{bmatrix} 1 & 0 & a \\ 0 & 1 & b \end{bmatrix}$; $\begin{bmatrix} 1 & a & 0 \\ 0 & 0 & 1 \end{bmatrix}$; $\begin{bmatrix} 0 & 1 & 0 \\ 0 & 0 & 1 \end{bmatrix}$;

$\begin{bmatrix} 0 & 1 & a \\ 0 & 0 & 0 \end{bmatrix}$; $\begin{bmatrix} 0 & 0 & 1 \\ 0 & 0 & 0 \end{bmatrix}$; $\begin{bmatrix} 0 & 0 & 0 \\ 0 & 0 & 0 \end{bmatrix}$.

4. Considering $\mathbf{A}$ as a linear transformation A of $V = \mathcal{F}^n$, take U and W as in Exercise 3.1, number 12. Take a basis $\{\alpha_1, \ldots \alpha_r, \alpha_{r+1}, \ldots \alpha_r\}$ of $\mathcal{F}^n$ such that $\{\alpha_1, \ldots, \alpha_r\}$ is a basis of U and $\{\alpha_{r+1}, \ldots, \alpha_b\}$ is a basis of W. The matrix representing A is the given one.

5. Follow the same idea s in 4., using Exercise 3.1, number 13.

6. Use the same idea as in 4., using Exercise 3.1, number 21.

7. Use the fact that each column vector of **AP** is a linear combination of column vectors of **A.** (See Exercise 3.2, number 11, (b).)

Page 81
Exercise 3.4

1. 2; 3; 3

4. $$\begin{bmatrix} 2 & 1 \\ 1 & 4 \end{bmatrix} = \begin{bmatrix} 0 & 1 \\ 1 & 0 \end{bmatrix}\begin{bmatrix} 1 & 0 \\ 2 & 1 \end{bmatrix}\begin{bmatrix} 1 & 0 \\ 0 & -7 \end{bmatrix}\begin{bmatrix} 1 & 4 \\ 0 & 1 \end{bmatrix}$$

Page 87
Exercise 3.5

1. $x_1 = c, \quad x_2 = c, \quad x_3 = c.$ (c: arbitrary)

3. $2y_1 + 3y_2 - 7y_3 = 0$

6. Suppose the augmented matrix $\widetilde{\mathbf{A}} = [\mathbf{A}\,\mathbf{y}]$ changes to $\widetilde{\mathbf{R}} = [\mathbf{R}\,\mathbf{z}]$ as **A** is row-reduced to **R** (cf. page 85). The system $\mathbf{A}\mathbf{x} = \mathbf{y}$ has a solution if and only if $\widetilde{\mathbf{R}}$ = rank **R.** But rank **R** = rank **A** and rank $\widetilde{\mathbf{R}}$ = rank **R.**

Page 93
Exercise 4.1

1. (1/3, 2/3). (1/3, -1/3)

4. sp$\{(1, 1, -1, 0)\}$ in $\mathcal{R}^4$.

6. For $\alpha \in V$, $\gamma^* \in U^*$, one has $\langle \alpha, {}^t A\, {}^t B\gamma^* \rangle = \langle A\alpha, {}^t B\gamma^* \rangle =$ $\langle A\alpha, {}^t B\gamma^* \rangle = \langle BA\alpha, \gamma^* \rangle$ as well as $\langle \alpha, {}^t(BA)\gamma^* \rangle = \langle BA\alpha, \gamma^* \rangle$.

8. For fixed $\beta \in W$, $\alpha \in V \to f(\alpha, \beta) \in \mathcal{F}$ is a linear function on V by (a). Call it $C(\beta) \in V^*$. $C: W \to V^*$ is linear, by (b). If $C(\beta) = 0$, then $F(\alpha, \beta) = 0$ for all $\alpha \in V$, so that $\beta = 0$ by (d), and hence C is one to one. To show that C is onto, it is sufficient (why?) to show that if $\langle \alpha, C(\beta) \rangle = 0$ for all $\beta \in W$, then $\alpha = 0$. This follows from (c).

Exercise 4.1 (cont'd)

10. (Converse.) Let $\{\mathbf{E}_{i,j}\}$ be the standard basis in $\mathcal{F}^n_n$ (cf. Exercise 2.2, number 5 and Exercise 3.2, number 9). For a given linear function f on $\mathcal{F}^n_n$, let

$$a_i^j = f(\mathbf{E}_{i,j}).$$

Then

$$f(\mathbf{X}) = f\Big(\sum_{i,j} x_j^i \mathbf{E}_{i,j}\Big) = \sum_{i,j} x_j^i a_i^j$$
$$= \text{trace}\,(\mathbf{AX}).$$

15. Let $\{\alpha_1, \ldots, \alpha_n\}$ be a basis of V and $\{\alpha_1^*, \ldots, \alpha_n^*\}$ the dual basis in V*. Write

$$\alpha = \sum a^i \alpha_i.$$
$$\alpha^* = \sum a_j (\alpha_j)^*.$$

Then the linear transformation $\xi \to \langle \xi, \alpha^* \rangle$ is given by

$$\alpha_j \to \langle \alpha_j, \alpha^* \rangle \alpha = \sum a_j a^i \alpha_i,$$

and hence is represented by the matrix $[a^i a_j]$, whose trace is

$$\sum a^i a_i = \langle \alpha, \alpha^* \rangle.$$

Page 99

Exercise 4.2

4. Suppose $W$ is a proper subspace (and hence of dimension 1) invariant by $A$. If

$$\boldsymbol{a} = \begin{bmatrix} a \\ b \end{bmatrix}$$

is a basis of $W$, one must have

$$\begin{bmatrix} 1 & -5 \\ 3 & 2 \end{bmatrix} \begin{bmatrix} a \\ b \end{bmatrix} = t \begin{bmatrix} a \\ b \end{bmatrix}$$

for some $t$, i.e.,

$$(1 - t)a - 5b = 0,$$

$$3a + (2 - t)b = 0.$$

But there is no real number $t$ for which these equations have a solution $\boldsymbol{a} \neq 0$.

5. Yes.

Exercise 4.3

2. Let $\epsilon_1, \epsilon_2$ be the standard basis in $\mathcal{R}^2$. Then

$$\begin{aligned} f(x_1\epsilon_1 + x_2\epsilon_2, y_1\epsilon_1 + y_2\epsilon_2) &= x_1y_1 f(\epsilon_1, \epsilon_1) + x_1y_2 f(\epsilon_1, \epsilon_2) \\ &\quad + x_2y_1 f(\epsilon_1, \epsilon_1) + x_2y_2 f(\epsilon_2, \epsilon_2) \\ &= (x_1y_2 - x_2y_1) f(\epsilon_1, \epsilon_2), \end{aligned}$$

because

$$f(\epsilon_1, \epsilon_1) = f(\epsilon_2, \epsilon_2) = 0,$$

$$f(\epsilon_2, \epsilon_1) = -f(\epsilon_1, \epsilon_2).$$

4. f is nondegenerate because $\mathbf{A}$ is invertible. The signature is $(2, 1)$. $\mathbf{P}$ is, for example,

$$\begin{bmatrix} 1 & 2 & -2 \\ 0 & -1 & 1 \\ 0 & 4 & 0 \end{bmatrix}$$

and

$${}^t\mathbf{PAP} = \begin{bmatrix} 1 & & \\ & 4 & \\ & & -4 \end{bmatrix}$$

7. $\dim W = n^2, \quad \dim W_1 = \frac{n(n+1)}{2}, \quad \dim W_2 = \frac{n(n-1)}{2}$

9. (a) Let $\alpha_1, \ldots, \alpha_k$ be a basis of W such that

$$f(\alpha_i, \alpha_i) = c_i \neq 0, \qquad f(\alpha_i, \alpha_j) = 0 \quad \text{for} \quad i \neq j.$$

For any given $\alpha \in V$, let

$$\beta = \sum_{i=1}^{k} \frac{f(\alpha, \alpha_i)}{c_i} \alpha_i \in W$$

$$\gamma = \alpha - \beta \qquad (\text{so that } \gamma = \beta + \alpha).$$

Then

$$f(\gamma, \alpha_j) = f(\alpha, \alpha_j) - \sum_{i=1}^{k} \frac{f(\alpha, \alpha_i)}{c_i} f(\alpha_i, \alpha_j)$$

$$= f(\alpha, \alpha_j) - \frac{f(\alpha, \alpha_j)}{c_j} c_j = 0,$$

and thus $\gamma \in W'$. This shows that $V = W + W'$. $W \cap W' = (0)$ because $f$ on $W$ is nondegenerate.

Page 115

Exercise 5.1

2. If $a + b\sqrt{2} \neq 0$, then $a^2 - b^2 \neq 0$. Now

$$(a + b\sqrt{2}) \left( \frac{a}{a^2 - 2b^2} - \frac{b}{a^2 - 2b^2}\sqrt{2} \right) = 1.$$

5. No. In the field $\mathcal{F}$ in Example 5.1, $1 + 1 = 0$ but $1 \neq 0$.

7. (c) If $\mathbf{A} = \begin{bmatrix} a & -b \\ b & a \end{bmatrix} \neq 0$, then $a^2 + b^2 \neq 0$. Now

$$\mathbf{A}^{-1} = \begin{bmatrix} \frac{a}{a^2+b^2} & \frac{b}{a^2+b^2} \\ \frac{-b}{a^2+b^2} & \frac{a}{a^2+b^2} \end{bmatrix}$$

8. (c) If $\mathbf{A} \neq 0$,

$$\mathbf{A}^{-1} = \begin{bmatrix} \frac{a}{e} & \frac{b}{e} & \frac{c}{e} & \frac{d}{e} \\ \frac{-b}{e} & \frac{a}{e} & \frac{d}{e} & \frac{-c}{e} \\ \frac{-c}{e} & \frac{-d}{e} & \frac{a}{e} & \frac{b}{e} \\ \frac{-d}{e} & \frac{c}{e} & \frac{-b}{e} & \frac{a}{e} \end{bmatrix}, \quad \text{with } e = a^2 + b^2 + c^2 + d^2.$$

Page 121
Exercise 5.2

1. $2^n$ polynomials.

4. Quotient $5x^3 - 8x^2 + 3x + 8$;
remainder $-12x - 2$.

7. (Second question.) It is a subring but not a commutative subring.

Page 127
Exercise 5.3

3. $x^2 + 1$

5. $x^3 + x^2 + 1$ and $x^3 + x + 1$.

6. Let $\mathbf{E}_{i,j}$ be the standard basis in $\mathcal{F}_n^n$ (cf. Exercise 3.2, number 9). Suppose an ideal $I$ contains $\mathbf{A} \neq 0$, say, $a_{js} \neq 0$. Then for any pair $(i, t)$, one has $\mathbf{E}_{i,j}\mathbf{A}\mathbf{E}_{s,t} = a_{js}\mathbf{E}_{i,t} \in I$ and hence $\mathbf{E}_{i,t} \in I$. Thus $I = \mathcal{F}_n^n$

Page 133
Exercise 5.4

1. The polynomial function is 0.

$$x^3 + 2x = x(x+1)(x+2).$$

3. $x^4 - 10x^2 + 1$.

4. If $\{c_1, \ldots, c_n\}$ is a basis of $\mathcal{F}''$ over $\mathcal{F}'$ and if $\{a_1, \ldots, a_m\}$ is a basis of $\mathcal{F}'$ over $\mathcal{F}$, then the mn elements $c_i a_j$ form a basis of $\mathcal{F}''$ over $\mathcal{F}$.

6. For $c \in \mathcal{F}'$, let $f(x) \in \mathcal{F}'[x]$ be its minimal polynomial. If $a_0, a_1, \ldots, a_n$ are the coefficients (in $\mathcal{F}'$) of $f(x)$, consider $\mathcal{F}^* = \mathcal{F}(a_0, a_1, \ldots, a_n)$. Since each $a_i$ is algebraic over $\mathcal{F}$, $\mathcal{F}(a_0, a_1, \ldots, a_n)$ is a finite extension of $\mathcal{F}$ (by Theorem 5.16 and number 4 of this exercise.) Then $\mathcal{F}^*(c)$ is a finite extension of $\mathcal{F}$. Hence $c$ is algebraic over $\mathcal{F}$.

Page 137

Exercise 5.5

3. $2^* \circ 3^*$

4. 4 elements.

Page 140

Exercise 5.6

1. No.

Page 143

Exercise 5.7

3. If **A** is real and $\xi \in \mathcal{R}^n$, then $A\xi \in \mathcal{R}^n$.
(Converse.) Let $\epsilon_1, \ldots, \epsilon_n$ be the standard basis of $\mathcal{R}^n$. Then $\mathbf{A}\epsilon_i$ is the i-th column vector of **A**. By assumption, this is real. Hence **A** is real.

Page 154

Exercise 6.1

1. (a) (16)(13)(27)(24)(25)

sign -1

11. If a, b, c, d are distinct, then

$$(ab)(cd) = (acb)(cda).$$

If $a = c$, then

$$(ab)(ad) = (abd).$$

If $a = c$, $b = d$, then

$$(ab)(ab) = \text{Identity} = (abe)(eba)$$

for any $e \neq a, b$.

Page 160

Exercise 6.2

1. 1 - (a); 2 - (a); 4 - (a); 5 - (b).

Page 166

Exercise 6.3

1. (a) $a^2 + b^2$ (c) 0
   (e) -1 (g) 0

2. (a) -14 (c) 1

3. Make use of Theorem 6.11.

5. $(x - 1)^3 (x + 3)$.

7. Reverse the order of columns (thus changing the sign of the determinant $n(n - 1)/2$ times), then reverse the order of rows (with the same change of the sign), and finally take the transpose.

10.
$$\det \begin{bmatrix} \mathbf{A} & -\mathbf{B} \\ \mathbf{B} & \mathbf{A} \end{bmatrix} = \det \begin{bmatrix} \mathbf{A} + i\mathbf{B} & -\mathbf{B} + i\mathbf{A} \\ \mathbf{B} & \mathbf{A} \end{bmatrix}$$

$$= \det \begin{bmatrix} \mathbf{A} + i\mathbf{B} & (-\mathbf{B} + i\mathbf{A}) - i(\mathbf{A} + i\mathbf{B}) \\ \mathbf{B} & \mathbf{A} - i\mathbf{B} \end{bmatrix}$$

$$= \det \begin{bmatrix} \mathbf{A} + i\mathbf{B} & 0 \\ \mathbf{B} & \mathbf{A} - i\mathbf{B} \end{bmatrix}$$

$$= \det(\mathbf{A} + i\mathbf{B}) \cdot \det(\mathbf{A} - i\mathbf{B})$$

$$= \left|\det(\mathbf{A} + i\mathbf{B})\right|^2.$$

Page 166

Exercise 6.3 (cont'd)

13. $a_{\sigma(i)i} = 1 \qquad 1 \leq i \leq n$

$a_{ji} = 0 \qquad j \neq \sigma(i)$

determinant = sign of $\sigma$.

Page 174

Exercise 6.4

1. -11; -11.

4. (a) invertible; $\begin{bmatrix} 2 & 1 \\ 5 & 3 \end{bmatrix}$

(b) invertible; $\begin{bmatrix} 1 & -2 \\ 1 & -3 \end{bmatrix}$

(c) not invertible

4. (d) invertible; $\begin{bmatrix} 1-x & x \\ x & -x-1 \end{bmatrix}$

(e) not invertible.

7. For $x = 1$, the rank is 1.
For $x = -(1/3)$, the rank is 3.

9. The determinant in question is equal to $\det \tilde{\mathbf{A}}$. From Theorem 6.13 one gets

$$\det \mathbf{A} \;\; \det \tilde{\mathbf{A}} = (\det \mathbf{A})^n.$$

If $\det \mathbf{A} \neq 0$, then $\det \tilde{\mathbf{A}} = (\det \mathbf{A})^{n-1}$. If $\det \mathbf{A} = 0$, then (1) and (2) in the proof of Theorem 6.13 show that the column vectors of $[\Delta_{ij}]$ are linearly dependent. Thus the matrix has determinant 0.

Exercise 6.5

2. See the matrix for $f_A$ in the solution of Exercise 3.2, number 17.

4. (a) is obvious.

(b) If $\alpha_1, \ldots, \alpha_n$ is a basis of V and

$$A(\alpha_j) = \sum_{i=1}^{n} a_j^i \alpha_i,$$

then

$$\begin{aligned} & f(A(\alpha_1), \alpha_2, \ldots, \alpha_n) \\ &= f\left(\sum_{i=1}^{n} a_1^i \alpha_i, \alpha_2, \ldots, \alpha_n\right) \\ &= a_1^1 f(\alpha_1, \alpha_2, \ldots, \alpha_n), \qquad \text{etc.} \end{aligned}$$

Thus

$$\begin{aligned} (Df)(\alpha_1, \ldots, \alpha_n) &= \left(\sum_{i=1}^{n} a_i^i\right) f(\alpha_1, \ldots, \alpha_n) \\ &= (\text{trace } A)\, f(\alpha_1, \ldots, \alpha_n). \end{aligned}$$

Page 180

Exercise 6.6

1. (a)

$$x_1 = \frac{\begin{vmatrix} 1 & -2 & 1 \\ 3 & 4 & -5 \\ -4 & 3 & 2 \end{vmatrix}}{\begin{vmatrix} 3 & -2 & 1 \\ 1 & 4 & -5 \\ -1 & 3 & 2 \end{vmatrix}}; \quad x_2 = \frac{\begin{vmatrix} 3 & 1 & 1 \\ 1 & 3 & -5 \\ -1 & -4 & 2 \end{vmatrix}}{\begin{vmatrix} 3 & -2 & 1 \\ 1 & 4 & -5 \\ -1 & 3 & 2 \end{vmatrix}}$$

$$x_3 = \frac{\begin{vmatrix} 3 & -2 & 1 \\ 1 & 4 & 3 \\ -1 & 3 & -4 \end{vmatrix}}{\begin{vmatrix} 3 & -2 & 1 \\ 1 & 4 & -5 \\ -1 & 3 & 2 \end{vmatrix}}$$

Page 186

Exercise 7.1

1. (a) $x^2 - 5x + 7$; (c) $x^2(x+3)$; (e) $(x-1)(x^2-1)^2$.

3. The proof of Theorem 7.3 works.

5. $(x-1)^m x^{n-m}$.

6. Observe that $x\mathbf{I}_n - {}^t\mathbf{A} = {}^t(x\mathbf{I}_n - \mathbf{A})$.

9. (a) $n = 2$ and $\mathbf{A} = \begin{bmatrix} 0 & -1 \\ 1 & 0 \end{bmatrix}$.

Exercise 7.1 (cont'd)

(b) From number 7,

$$f_A(0) = (-1)^n \det A < 0$$

by assumption on $n$ and $\det A \circ f_A(x)$ being of even degree, $f_A(x) > 0$ for large $|x|$. Hence $f_A(x) = 0$ must have at least two real roots.

11. If $A$ is nonsingular, then $AB = A(BA)A^{-1}$. By proposition 7.1,

$$f_{AB} = f_{BA}.$$

Page 195

Exercise 7.2

1. (a) $(x-1)^2$ (c) $(x-1)^3$

2. (a) None (c) $P = \begin{bmatrix} \frac{-i}{2} & \frac{1}{2} \\ \frac{i}{2} & \frac{1}{2} \end{bmatrix}$

3. For any polynomial $g(x)$, one has

$${}^t(g(A)) = g({}^tA).$$

4. If $A$ is non-singular, $AB = A(BA)A^{-1}$ and $AB$ and $BA$ have the same minimal polynomial. The same for the case where $B$ is non-singular. But the result is generally false.

Counter-example: $A = \begin{bmatrix} 0 & 0 \\ 1 & 0 \end{bmatrix}$, $B = \begin{bmatrix} 0 & 0 \\ 0 & 1 \end{bmatrix}$. $AB = \begin{bmatrix} 0 & 0 \\ 0 & 0 \end{bmatrix}$ has minimal polynomial $x$. $BA = \begin{bmatrix} 0 & 0 \\ 1 & 0 \end{bmatrix}$ has minimal polynomial $x^2$.

7. Since $\Phi_A$ has simple roots, say, $a_1, \ldots, a_n$, in $\mathcal{F}$, Theorem 7.8 says that there exists a basis $\{\alpha_1, \ldots, \alpha_n\}$ such that $A\alpha_i = a_i\alpha_i$, $1 \le i \le n$. From $AB = BA$, we have

$$A(B\alpha_i) = B(A\alpha_i) = B(a_i\alpha_i) = a_iB\alpha_i.$$

7. (cont'd)

Since $a_1, \ldots, a_n$ are distinct, a vector $\beta$ such that $A\beta = a_i\beta$ must be a scalar multiple of $\alpha_i$. Hence $B\alpha_i = b_i\alpha_i$ for some $b_i$, $1 \le i \le n$.

Page 204
Exercise 7.3

1. If $PAP^{-1}$ is a superdiagonal matrix as in Theorem 7.3, the diagonal elements $c_1, \ldots, c_n$ must be 0, since they are eigenvalues of $A$. Then $A^n = 0$, as can be easily checked.

4. Write down the canonical forms as in the Corollary to Theorem 7.11 according to all the possible invariant systems

$$\{6\},\quad \{5, 1\},\quad \{4, 2\},\quad \{4, 1, 1\},$$
$$\{3, 3\},\quad \{3, 2, 1\},\quad \{3, 1, 1, 1\},\quad \{2, 2, 2\},$$
$$\{2, 2, 1, 1\},\quad \{2, 1, 1, 1, 1\},\quad \{1, 1, 1, 1, 1, 1\},$$

for example,

$$\begin{bmatrix} 0 & 0 & 0 & & & \\ 1 & 0 & 0 & & & \\ 0 & 1 & 0 & & & \\ & & & 0 & 0 & 0 \\ & & & 1 & 0 & 0 \\ & & & 0 & 1 & 0 \end{bmatrix} \text{ for } \{3, 3\}; \quad \begin{bmatrix} 0 & 0 & & & & \\ 1 & 0 & & & & \\ & & 0 & 0 & & \\ & & 1 & 0 & & \\ & & & & 0 & 0 \\ & & & & 1 & 0 \end{bmatrix} \text{ for } \{2, 2, 2\};$$

$$\begin{bmatrix} 0 & 0 & & & & \\ 1 & 0 & & & & \\ & & 0 & & & \\ & & & 0 & & \\ & & & & 0 & \\ & & & & & 0 \end{bmatrix} \text{ for } \{2, 1, 1, 1, 1\}.$$

5. Use the Corollary to Theorem 7.11.

Page 208

Exercise 7.4

1. (a) $\begin{bmatrix} 2 & 0 & 0 \\ 1 & 2 & 0 \\ 0 & 1 & 2 \end{bmatrix}$ (c) $\begin{bmatrix} i & 0 & 0 & 0 & 0 \\ 1 & i & 0 & 0 & 0 \\ 0 & 1 & i & 0 & 0 \\ 0 & 0 & 0 & i & 0 \\ 0 & 0 & 0 & 1 & i \end{bmatrix}$

(e) $\begin{bmatrix} 2 & 0 & 0 & & & & & & & \\ 1 & 2 & 0 & & & & & & & \\ 0 & 1 & 2 & & & & & & & \\ & & & 1 & 0 & 0 & 0 & & & \\ & & & 1 & 1 & 0 & 0 & & & \\ & & & 0 & 1 & 1 & 0 & & & \\ & & & 0 & 0 & 1 & 1 & & & \\ & & & & & & & 1 & 0 & 0 \\ & & & & & & & 1 & 1 & 0 \\ & & & & & & & 0 & 1 & 1 \end{bmatrix}$

3. (a) $\begin{bmatrix} \frac{1+\sqrt{17}}{2} & 0 \\ 0 & \frac{1-\sqrt{17}}{2} \end{bmatrix}$ (b) $\begin{bmatrix} 1 & 0 \\ 1 & 1 \end{bmatrix}$

(d) $\begin{bmatrix} 1 & & \\ & 2 & \\ & & -3 \end{bmatrix}$ (e) $\begin{bmatrix} 1 & 0 & 0 \\ 1 & 1 & 0 \\ 0 & 0 & 1 \end{bmatrix}$

6. $\mathbf{A}$ and ${}^t\mathbf{A}$ have the same characteristic poynomial and the same minimal polynomial. Also, use the following lemma: If $\mathbf{B}$ is nilpotent of index $k$, then so is ${}^t\mathbf{B}$, and $\mathbf{B}$ and ${}^t\mathbf{B}$ have the same invariant system.

Page 217

Exercise 7.5

1. $\mathbf{J}_0 = \begin{bmatrix} 0 & -\mathbf{I}_n \\ \mathbf{I}_n & 0 \end{bmatrix}$

Exercise 7.5 (cont'd)

5. If $A \in S$, $A \neq 0$, then $A$ has an inverse, because the null space $\{\alpha \in V;\ A\alpha = 0\}$ is invariant by $G$ and is hence $(0)$, that is, $A$ is one-to-one.

7. Define scalar multiplication by $\sqrt{2}$ as follows:

$$\sqrt{2}(1, 0) = (0, 1),$$

$$\sqrt{2}(0, 1) = (2, 0)$$

Page 225

Exercise 8.1

3 $\beta_1 = (1/\sqrt{2}, 0, 1/\sqrt{2})$, $\beta_2 = (1/3\sqrt{2}, 4/3\sqrt{2}, -1/3\sqrt{2})$, $\beta_3 = (-2/3, 1/3, 2/3)$

6. Use the identities in number 5.

8. Let $G = \{A_1, A_2, \ldots, A_r\}$. For any inner product $(\ ,\ )$, define a new inner product $\langle\ ,\ \rangle$ by

$$\langle \alpha, \beta \rangle = \sum_{i=1}^{r} (A_i\alpha, A_i\beta)$$

11. We may assume that $\alpha_1, \ldots, \alpha_n$ are linearly independent. By the Gram-Schmidt process, we can find an orthonormal basis $\beta_1, \ldots, \beta_n$ such that

$$(*) \qquad \alpha_i = \sum_{j=1}^{i} c_{ji}\beta_j.$$

Then

$$\det[\alpha_1 \alpha_2 \cdots \alpha_n] = \pm c_{11}c_{22} \cdots c_{nn},$$

since $\det[\beta_1, \beta_2 \cdots \beta_n] = \pm 1$. From (*)

$$|c_{ii}| \leq \|\alpha_i\| \text{ for each } i.$$

Hence

$$\det[\alpha_1 \alpha_2 \cdots \alpha_n] \leq \|\alpha_1\| \, \|\alpha_2\| \cdots \|\alpha_n\|$$

Page 230

Exercise 8.2

1. Clearly, $S \subset (S^\perp)^\perp$. Since $(S^\perp)^\perp$ is a subspace, it contains the span of S.

7. The minimal polynomial is $(x - 1)x$. The characteristic polynomial is $(x - 1)^m x^{n-m}$.

9. If B is the projection of V onto $U^\perp$, then for any $\alpha \in V$,

$$\alpha = A\alpha + B\alpha$$

and

$$\|\alpha\|^2 = \|A\alpha\|^2 + \|B\alpha\|^2.$$

Page 234

Exercise 8.3

2. $(\alpha, \beta)$ is bilinear.

$$(\alpha_i, \alpha_j) = \langle \alpha_i, \theta(\alpha_j) \rangle = \delta_{ij} \quad \text{so} \quad (\alpha_i, \alpha_j) = \delta_{ij}.$$

Thus $(\alpha, \beta)$ is a positive-definite inner product such that $\{\alpha_1, \dots, \alpha_n\}$ is orthonormal.

4. (a) $\mathbf{A}^* = \begin{bmatrix} 3 & -i \\ 2 - 1 & -1 - i \end{bmatrix}$

(c) $\mathbf{A}^* = \begin{bmatrix} 2 & i \\ -i & 1 \end{bmatrix}$ (c) is hermitian.

Page 237

Exercise 8.4

1. $(AB)^* = B^*A^* = BA$ when A, B are hermitian. Thus $(AB)^* = AB$ if and only if $AB = BA$.

3. If $\mathbf{A} = \mathbf{B} + i\mathbf{C}$ with hermitian $\mathbf{B}, \mathbf{C}$, then

$$\mathbf{A}^* = \mathbf{B}^* - i\mathbf{C}^* = \mathbf{B} - i\mathbf{C}.$$

Thus $\mathbf{B} = \frac{1}{2}(\mathbf{A} + \mathbf{A}^*)$, $\mathbf{C} = \frac{1}{2i}(\mathbf{A} - \mathbf{A}^*)$.

Exercise 8.4 (cont'd)

5. Suppose A is hermitian. If $\beta \in U^{\perp}$, then

$$(A\beta, \alpha) = (\beta, A\alpha) = 0 \text{ for every } \alpha \in U,$$

because $A\alpha \in U$.

7. If $A^* = A$, then $(A + B)^* = A + B$ and $(cA)^* = cA$ for real c. The dimension of this real vector space is $n^2$.

9. If $A = [a_{ij}]$, $B = [b_{ij}]$, then

$$- \text{trace}(AB^*) = \sum a_{ik}\bar{b}_{ik}.$$

10. If $(A + iI_n)\alpha = 0$, then $A\alpha = -i$. From $(A\alpha, \alpha) = (\alpha, A\alpha)$, one gets

$$(-i\alpha, \alpha) = (\alpha, -i\alpha)$$

and thus

$$-i(\alpha, \alpha) = i(\alpha, \alpha) \text{ and } (\alpha, \alpha) = 0, \text{ i.e., } \alpha = 0.$$

Hence $A + iI_n$ is nonsingular. From $(A + iI_n)U = A - iI_n$, one gets

$$U^*(A - iI_n) = A + iI_n.$$

Since U commutes with $A + iI_n$, one has

$$U^*(A - iI_n) = U^*U(A + iI_n).$$

Hence $U^*U = I_n$.

12. The center consists of all matrices of the form $e^{i\theta}I_n$.

Page 244

Exercise 8.5

2. Using the identity in Exercise 8.1, number 5, page 226, we see that $(A\alpha, A\beta) = (A^*\alpha, A^*\beta)$ for all $\alpha, \beta \in V$ if and only if $\|A\alpha\|^2 = \|A^*\alpha\|^2$ for all $\alpha \in V$. Now

$$(A\alpha, A\beta) = (\alpha, A^*A\beta)$$

$$(A^*\alpha, A^*\beta) = (\alpha, AA^*\beta).$$

Thus

$$(A\alpha, A\beta) = (A^*\alpha, A^*\beta) \text{ for all } \alpha, \beta \in V$$

if and only if $A^*A = AA^*$.

Exercise 8.5 (cont'd)

4. (i) Since $AA^* = A^*A$, $B = AA^*$ is hermitian and nilpotent, say $B^r = 0$. For any vector $\alpha$, $(B^{r-1}\alpha, B^{r-1}\alpha) = (B^{r-2}\alpha, B^r\alpha) = 0$ and hence $B^{r-1}\alpha = 0$. Thus we find $B = 0$. Then $(A\alpha, A\alpha) = (\alpha, A^*A\alpha) = 0$, i.e., $A\alpha = 0$, and hence $A = 0$.

or (ii) Use Theorem 8.19. If $A^r = 0$, then $c_i^r = 0$ and hence $c_i = 0$.

11. Take $\mathbf{U}$ such that $\mathbf{D} = \mathbf{UAU}^{-1}$ is a diagonal matrix as in Corollary to Theorem 8.21. The diagonal elements $d_1, \ldots, d_n$ are positive. Take the diagonal matrix $\mathbf{D}_1$ with diagonal elements $\sqrt{d_1}, \ldots, \sqrt{d_n}$, and let $\mathbf{B} = \mathbf{U}^{-1}\mathbf{D}_1\mathbf{U}$.

12. Since $\mathbf{A}^*\mathbf{A}$ is hermitian and positive-definite there is a hermitian matrix $\mathbf{C}$ (nonsingular) such that $\mathbf{A}^*\mathbf{A} = \mathbf{C}^2$ (similar to number 11). Let $\mathbf{B} = \mathbf{AC}^{-1}$.

Page 248

Exercise 8.6

8. $n(n+1)/2$.

10. Let $\{\alpha_1, \ldots, \alpha_n\}$ be an orthonormal basis such that $A\alpha_i = c_i\alpha_i$, $1 \le i \le n$, where $c_i, \ldots, c_n$ are the eigenvalues of A. If A is positive-definite, then $(A\alpha_i, \alpha_i) = c_i > 0$. Conversely, assume $c_i > 0$ for all i. For any

$\alpha = \sum_{i=1}^{n} a_i\alpha_i$, we get

$$(A\alpha, \alpha) = \sum_{i=1}^{n} c_i a_i^2 > 0$$

unless $a_1 = \ldots = a_n = 0$ (i.e., $\alpha = 0$).

12. The diagonal matrix $a_1 = \ldots = a_p = 1$, $a_{p+1} = \ldots = a_r = -1$ and $a_{r+1} = \ldots = a_n = 0$ on the diagonal.

14. A for $f^c$ and $\tilde{f}$.

Exercise 8.6 (cont'd)

20. If $\mathbf{S}$ is an orthonormal matrix such that $\mathbf{SAS}^{-1}$ is diagonal with positive eigenvalues $c_1, \ldots, c_n$ on the diagonal, take $\mathbf{B} = \mathbf{S}^{-1}\mathbf{D}_1\mathbf{S}$, where $\mathbf{D}_1$ is the diagonal matrix with $\sqrt[k]{c_1}, \ldots, \sqrt[k]{c_n}$ on the diagonal.

Page 255

Exercise 8.7

2. Use $(\mathbf{A} + i\mathbf{B})^* = {}^t\mathbf{A} - i\,{}^t\mathbf{B}$

and
$$ {}^t\begin{bmatrix} A & -B \\ B & A \end{bmatrix} = \begin{bmatrix} {}^t\mathbf{A} & {}^t\mathrm{B} \\ -{}^t\mathrm{B} & {}^t\mathrm{A} \end{bmatrix} $$

4. $\begin{bmatrix} 1 & 0 \\ 0 & -1 \end{bmatrix}$

5. (a) If $(\mathbf{I} + \mathbf{A})\alpha = 0$ for $\alpha \in \mathcal{R}^n$, then $\mathbf{A}\alpha = -\alpha$. If $-1$ is not an eigenvalue, then $\alpha = 0$. Thus $\mathbf{I}_n + \mathbf{A}$ is nonsingular. From $\mathbf{I}_n - \mathbf{A} = \mathbf{S}(\mathbf{I}_n + \mathbf{A})$, we get
$$ \mathbf{I}_n - {}^t\mathbf{A} = (\mathbf{I}_n + {}^t\mathbf{A})\,{}^t\mathbf{S}. $$
Multiplying by $\mathbf{A}$ on the left, we have
$$ \mathbf{A} - \mathbf{I}_n = (\mathrm{A} + \mathbf{I}_n)\,{}^t\mathbf{S}. $$
Also,
$$ \mathbf{S}(\mathbf{I}_n + \mathrm{A}) = (\mathbf{I}_n + \mathrm{A})\mathbf{S}. $$
Hence
$$ (\mathbf{A} + \mathbf{I}_n)\,{}^t\mathbf{S} = -(\mathbf{A} + \mathbf{I}_n)\mathbf{S} \quad \text{and} \quad {}^t\mathbf{S} = -\mathbf{S}. $$

7. $\mathrm{A}^c(\alpha + i\beta) = \mathrm{A}\alpha + i\mathrm{A}\beta$
$$ (\cos\theta + i\sin\theta)(\alpha + i\beta) = \cos\theta\,\alpha - \sin\theta\,\beta + i(\sin\theta\,\alpha + \cos\theta\,\beta). $$
Hence
$$ \mathrm{A}\alpha = \cos\theta\,\alpha - \sin\theta\,\beta $$
$$ \mathrm{A}\beta = \sin\theta\,\alpha + \cos\theta\,\beta. $$

Exercise 8.7 (cont'd)

14. Obvious from number 12.

16. Use the Corollary to Theorem 6.14, page 172, and the assumption $\mathbf{A}^{-1} = {}^t\mathbf{A}$.

18. The column vectors $\alpha_1, \ldots, \alpha_n$ of $\mathbf{A}$ form a basis. Let $\beta_1, \ldots \beta_n$ be an orthonormal basis obtained from $\alpha_1, \ldots, \alpha_n$, as in Theorem. 8.4. Then

$$\alpha_i = \sum_{j=1}^{i} C_{ji}\beta_j, \qquad 1 \le i \le n.$$

Let $\mathbf{B}$ be the orthogonal matrix with column vectors $\beta_1, \ldots, \beta_n$. Let

$$\mathbf{C} = \begin{bmatrix} C_{11} & C_{12} & C_{13} & \cdots & C_{1n} \\ 0 & C_{22} & C_{23} & \cdots & C_{2n} \\ \cdot & 0 & C_{33} & & \\ \cdot & \cdot & & & \cdot \\ \cdot & \cdot & \cdot & & \cdot \\ 0 & 0 & 0 & & C_{nn} \end{bmatrix}$$

Then

$$\mathbf{A} = \mathbf{BC}.$$

19. See the answer to Exercise 8.5, number 12, page 245.

Page 264

Exercise 9.1

2. Let $\alpha_i = \overrightarrow{P_0P_i}$, $1 \le i \le n$. If $\pi(0) = 0$, $\overrightarrow{P_{\pi(0)}P_{\pi(i)}}$, $1 \le i \le n$, are obtained from $\alpha_1, \ldots, \alpha_n$ by a permutation and remain a basis. Suppose $\pi(0) = 1$, say. Then

$$\overrightarrow{P_1P_0} = -\alpha_1, \quad \overrightarrow{P_1P_2} = \alpha_2 - \alpha_1, \quad \ldots, \quad \overrightarrow{P_1P_n} = \alpha_n - \alpha_1$$

are linearly independent.

Exercise 9.1 (cont'd)

4. $\begin{cases} y^1 = -x^1 + ix^2 + 1 \\ y^2 = (1-i)x^1 + (1-i)x^2 + i, \end{cases}$

where $\{x^1, x^2\}$ is based on the given frame and $\{y^1, y^2\}$ is the standard coordinate system.

Page 271

Exercise 9.2

1. $$\begin{bmatrix} 0 & -1 & 1 \\ -3 & -1 & 2 \\ 0 & 0 & 1 \end{bmatrix}$$

3. $$\begin{bmatrix} \frac{-1-3i}{2} & \frac{3-3i}{2} & 1 \\ \frac{-3+i}{2} & \frac{-3-3i}{2} & 2i \\ 0 & 0 & 1 \end{bmatrix}$$

Page 275

Exercise 9.3

1. $P \vee Q = \{Q_1;\ \overrightarrow{PQ_1} \in Sp\{\overrightarrow{PQ}\}\}$

4. If $Y_1 \cap Y_2$ is empty, or if $Y_1$ and $Y_2$ coincide, then $Y_1$ and $Y_2$ are parallel (Definition 9.9). Otherwise, $Y_1 \cap Y_2$ is a line.

Page 282

Exercise 9.4

3. 4 points; 6 lines.

5. (a) $2x^1 - 5x^2 + 12 = 0$

Exercise 9.4 (cont'd)

7. If the determinant is not 0, then the system of linear equations

$$a_i x^1 + b_i x^2 + c_i x^3 + d_i x^4 = 0, \quad 1 \le i \le 4$$

has only the trivial solution $x^1 = x^2 = x^3 = x^4 = 0$.

9. Let $A^1$, $B^1$, $C^1$, $D^1$ be the barycenters of the triangles BCD, CDA, BDA, ABC, respectively. The lines $AA^1$, $BB^1$, $CC^1$, $DD^1$ meet at one point, whose barycentric coordinates are (1/4, 1/4, 1/4, 1/4).

11. Take the intersection of all convex sets containing S.

13. $x^1 \ge 0$, $x^2 \ge 0$, ..., $x^n \ge 0$, and $x^1 + x^2 + \dots + x^n \le 1$, where $\{x^1, x^2, \dots, x^n\}$ is the standard coordinate system.

Page 287

Exercise 10.1

2. $\sqrt{2(b^1 - a^1)^2 + 32(b^2 - a^2)^2}$

5. Show that this distance cannot be given by the formula in the Remark (the top of page 287) whatever $g_{ij}$ we may choose.

Page 293

Exercise 10.2

2. $\sqrt{27/7}$

4. Line perpendicular to the triangle ABC through its center of gravity G.

6. $$\sqrt{\sum_{i=1}^{n-r} (a, a_i)^2}$$

Exercise 10.2 (cont'd)

8. Use Proposition 10.5 together with $\overrightarrow{f(P_0)\,f(P_i)} = A(\overrightarrow{P_0\,P_i})$, where A is the linear transformation associated to f.

Page 299

Exercise 10.3

3. The assumption implies that the linear mapping associated to f is orthogonal.

Page 307

Exercise 10.4

2. (a) rotation; fixed point $(a/2,\ b/2)$

(b) reflection; fixed line $(\cos\theta - 1)x + \sin\theta\, y = 0$

(c) rotation; fixed point $\left(\frac{3+\sqrt{2}}{2},\ \frac{2\sqrt{2}+1}{2}\right)$

4.
$$\begin{bmatrix} \frac{1}{3} & -\frac{2}{3} & \frac{2}{3} \\ -\frac{2}{3} & \frac{1}{3} & \frac{2}{3} \\ \frac{2}{3} & \frac{2}{3} & \frac{1}{3} \end{bmatrix},$$

$$\begin{bmatrix} 1 & & \\ & 1 & \\ & & -1 \end{bmatrix} \text{ or } \begin{bmatrix} 1 & & \\ & -1 & \\ & & 1 \end{bmatrix} \text{ or } \begin{bmatrix} -1 & & \\ & -1 & \\ & & -1 \end{bmatrix}$$